Dynamics of Hydrology

Dynamics of Hydrology

Editor: Allison Sergeant

R CALLISTO REFERENCE

www.callistoreference.com

Callisto Reference,
118-35 Queens Blvd., Suite 400,
Forest Hills, NY 11375, USA

Visit us on the World Wide Web at:
www.callistoreference.com

ISBN: 978-1-64116-049-0 (Hardback)

Cataloging-in-Publication Data

Dynamics of hydrology / edited by Allison Sergeant.
   p. cm.
Includes bibliographical references and index.
ISBN 978-1-64116-049-0
   1. Hydrology. 2. Hydrology--Environmental aspects. 3. Earth sciences. I. Sergeant, Allison.
GB665 .D96 2019
551.48--dc23

Table of Contents

Preface

The study of the quality, movement and distribution of water is known as hydrology. It helps in analyzing and solving issues related to water resources and their management. It branches out into sub-fields of hydrogeology, groundwater hydrology, surface hydrology, hydrometeorology, etc. Hydrology plays a crucial role in the calculation and analysis of rainfall, precipitation, water balance, surface run-off, floods, etc. This book attempts to understand the multiple branches that fall under this discipline and how such concepts have practical applications. Coherent flow of topics, student-friendly language and extensive use of examples make this book an invaluable source of knowledge. Those in search of information to further their knowledge will be greatly assisted by this book.

All of the data presented henceforth, was collaborated in the wake of recent advancements in the field. The aim of this book is to present the diversified developments from across the globe in a comprehensible manner. The opinions expressed in each chapter belong solely to the contributing authors. Their interpretations of the topics are the integral part of this book, which I have carefully compiled for a better understanding of the readers.

At the end, I would like to thank all those who dedicated their time and efforts for the successful completion of this book. I also wish to convey my gratitude towards my friends and family who supported me at every step.

Editor

A New Approach to Evaluate the Ecological Status of a River by Visual Assessment

Mary Ann Pandan¹* and Florencio Ballesteros²

¹Department of Chemical Engineering, University of St. La Salle, La Salle Avenue, Bacolod City 6100, Negros Occidental, Philippines
²Department of Chemical Engineering, University of the Philippines, Diliman, Quezon City, 1100, Metro Manila, Philippines

Abstract

Most water management methodologies require comprehensive studies and thus, entail voluminous data, time, and scientific expertise. Sensorial evaluation techniques were thus, considered as these represent methods with minimal cost and can involve the local communities. This study applied the Sarno River Visual Assessment Protocol (SRVAP), a modified version of the Stream Visual Assessment Protocol developed by the United States Department of Agriculture, to Sarno River, Italy and tested its reliability as a river assessment tool. SRVAP scores has a statistically significant positive correlation with Chemical Oxygen Demand (COD) and shows that local knowledge is important and increases the viability of incorporating public participation in the evaluation. Correlation between SRVAP and organic content greatly increased barring seasonal variability and a significant positive relationship was found between SRVAP score and Biochemical Oxygen Demand (BOD) and COD during spring, as well as during summer. The resulting regression equations may be used as rapid estimates of COD and BOD levels in Sarno River for the seasons of spring and summer.

Keywords: Ecological indicators; Environmental monitoring; Public participation; River water quality

Introduction

The degradation in the quality of surface waters has become a major environmental concern due to continued industrialization and the impact of rising population. As a necessary step in the protection of such resources, particularly of rivers, various methodologies have been developed to assess their conditions. In Europe, the evaluation of water sources is based on the European Water Framework Directive (WFD) [1]. This directive unifies the water management approaches in 32 countries in order to achieve the common goal of "good" water quality in all surfaces, underground and coastal waters [2-6]. In Italy, the surface water quality regulation based on WFD is defined by Legislative Decree No. 152/2006 [7]. For river systems, the law requires the determination of the environmental status of every homogenous reach based on the physical, biological, and chemical characteristics of the study area [8].

Nevertheless, the assessment methodologies available suffer from critical challenges primarily stemming from the use of numerous parameters that involve extensive data and require time and scientific expertise. This has led to the development of new strategies that allow the rapid assessment of water bodies at minimal cost owing to the use of fewer analytical procedures and the involvement of the local communities. One such strategy is the use of sensorial evaluation such as the Stream Visual Assessment Protocol (SVAP) developed by the United States Department of Agriculture [9]. This method uses visual indicators easily identifiable by the local population which can then provide preliminary assessment and a warning mechanism for intervention, should the need arise. However, due to the specific and unique conditions pertaining to the Sarno River in Italy, it becomes necessary to modify the SVAP. The end product is the Sarno River Visual Assessment Protocol (SRVAP). The assessment involves both technical and non-technical evaluations and the overall assessment score is used as basis for river quality classification. This study applied SRVAP in determining the ecological condition of the Sarno River in Italy using both technical and local knowledge in the assessment. The applicability of the protocol was tested by correlation of SRVAP scores and actual river organic content, a required parameter stipulated by Legislative Decree No. 152/2006 [7].

Materials and Methods

Study area and sampling stations

The area for the study is the Sarno River, considered as one of the most polluted rivers in Italy. With a length of 24 km, the river traverses three provinces in the Campania Region and affects between 750,000 and one million inhabitants. Degradation of the river results from the combination of high population and presence of highly-polluting industries in the area [10,11]. Water sampling was conducted at five monitoring stations labeled A, B, C, D, and E located at Ponte San Michele, Scafati, Cavalcavia Del Sarno, Castellamare di Stabia and Torre Annunziata, respectively (Figure 1). The stations form part of the monitoring network set up by The Italian National Environmental Protection Agency for Campania Region (Agenzia Regionale Protezione Ambient Campania or ARPAC).

Water Sampling Protocol and Analytical Methods

Water sampling and visual assessment of the river were done simultaneously in May and August, 2013 to obtain data representing the spring and summer seasons, respectively. At each site, 5-L samples were collected midstream and analyzed within 24 h. BOD5 measurement was done using Oxitop® Manometric BOD Measuring Device while

---

*Corresponding author: Mary Ann Pandan, Department of Chemical Engineering, University of St. La Salle, La Salle Avenue, Bacolod City 6100, Negros Occidental, Philippines, E-mail: maryannpandan@usls.edu.ph

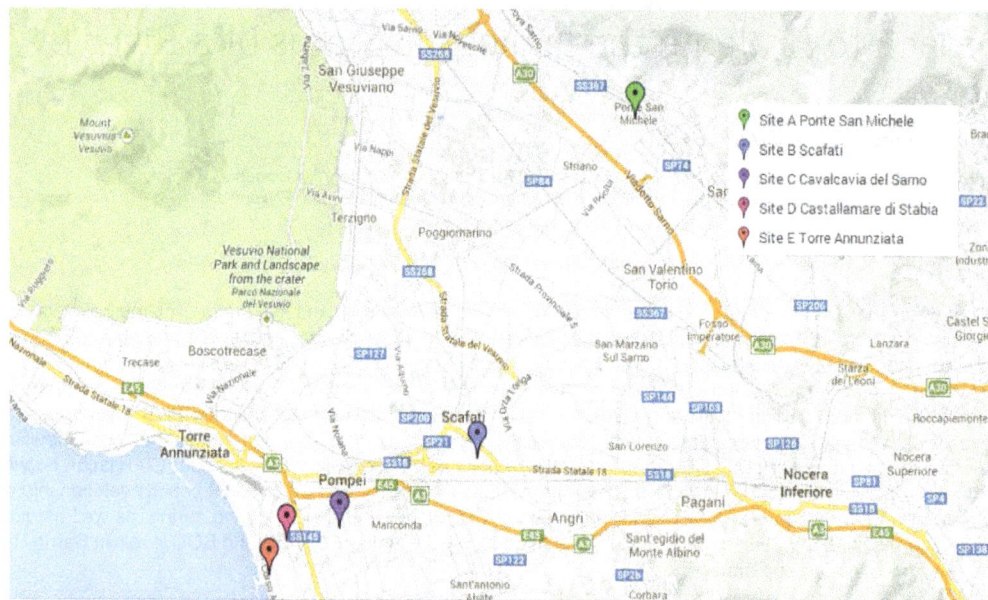

Figure 1: Geographic location of Sarno River and sampling sites1 75 1 Map Source: maps.google.com; geographic coordinates from ARPAC.

COD was determined using standard Open Reflux method specified by Standard Methods for the Examination of Water and Wastewater Section 5220B [12]. The focus on organic content was decided because Sarno River has consistently shown elevated levels ofthese parameters and also because BOD and COD allow analysis of pollutant load and identification of sources.

The sarno river visual assessment protocol (SRVAP)

The SRVAP involves both technical and non-technical evaluation. The technical evaluation is based on scientific knowledge of the environmental conditions and processes through instantaneous observations. On the other hand, the non-technical evaluation is based on temporal and experiential knowledge of the conditions of the river. The technical evaluation is composed of eleven (11) visual elements, namely, channel condition (ChC), hydrologic alteration (HA), riparian zone (RZ), bank stability (BS), water appearance (WA), nutrient enrichment (NE), barriers to fish movement (FB), in stream fish cover (IFC), invertebrate habitat (IH), canopy cover (CaC) and manure presence (MP). In consideration of the knowledge and availability of the local people, the non-technical evaluation was limited to six (6) elements, namely, ChC, HA, BS, W A, FB, and MP. The instrument used was a questionnaire composed of multiple choice questions, translated into the local knowledge. Pictures were also incorporated in order to facilitate ease and consistency of the answers. The scores for the visual elements were also used to determine the aspects of ecological condition, namely, energy sources (ES), chemical variables (CV), flow regime (FR) and habitat structure (HS). These are based on the factors that influence the integrity of streams defined by Karr [13] as cited by USDA [9] and USDA (2009).

Field evaluation of Sarno river visual assessment protocol

The two seasonal technical evaluations were undertaken by a panel of four experts. Meanwhile, the non-technical evaluation was done by respondents from the local population. For each site, 10 respondents were interviewed on three separate days (Tuesday, Wednesday, Thursday) for three weeks (1st,3rd,4th) in July, 2013. Scores for the technical evaluation were assigned by the panel of experts. For the non-

technical evaluation, the scores were based on the median scores of the respondents. A scale from 1 (Worst condition) to 10 (Best condition) was adopted in the evaluation of each measured element and the total assessment score was computed using the weighted-average scores for each element. Weights were pre-determined from Pair-comparison Analysis (PCA) Method [14-16]. The total assessment scores obtained is then used to classify the sites into 5 classes, specifically, bad, poor, fair, good and excellent condition.

Data analysis

Correlation among the parameters was analyzed by pair-wise linear regression analysis using JMP 10® [17].

Results and Discussion

Water quality of sarno river

The organic content of the river based on BOD and COD measurements is shown in (Figure 2). From the figure it can be seen that the organic loading of the river is lower in the upstream stations (Site A and Site B) than in the downstream stations (Site C, Site D and Site E). BOD values ranged from 0 mg/L to 42 mg/L while COD values ranged from 0 mg/L to 108.9 mg/L. The highest BOD and COD values were measured during the summer season at Site C and Site D, respectively. Both values indicate bad river quality based on Italian regulation for water quality DL 152/2006. A significant increase in organic content was seen in all stations during the summer season. This can be attributed to the increased agricultural activities as well as increased temperature during this season. Additionally, the highest COD value detected in Site D during summer may be due to sewage and other discharges from a highly-populated community near the sampling site. The highest BOD level in Site C is also attributed to urban discharges in the area.

Visual assessment of sarno river

The results of the SRVAP evaluation are shown in (Table 1). The total assessment scores ranged from 4.31 to 8.18, indicating poor to good river quality. Several physical infrastructures are present along

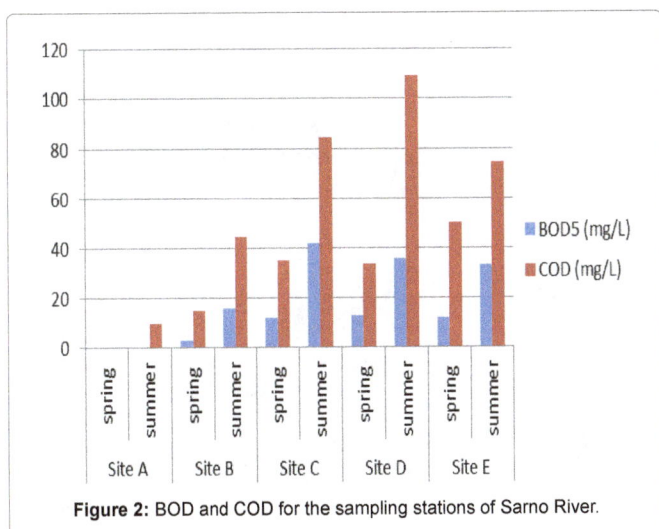

Figure 2: BOD and COD for the sampling stations of Sarno River.

Sarno River and these directly impacted the condition of the stream. In Sites A, C and E, the presence of dikes and pathways has affected the natural flow of the water while the drop structures in Site A has functioned as fish barriers and hindered biotic migration. Hydrologic alteration was observed during the summer season due to lowering of water level especially in Site A and Site C. As for the quality of the Riparian Zone in the river, it was observed that the highest quality was present in Site A and gradually deteriorated downstream. Bank Stability was high in Sites A, B and E and low in Sites C and D. However, bank stability improved in Site C during summer as more vegetation lined the banks of the stream. Canopy Cover was sufficient in the upstream stations (Sites A, B and C) and least in Site E where no shading was observed. Site D had poor cover although vegetation improved its score during summer. These observations showed that the conditions of the river were affected by seasonal variations.

Among the elements tested, Water Appearance and Nutrient Enrichment were worst as evidenced by the dark green coloration and turbidity along the stretch of the river, with the sole exception of the uppermost portion (Site A). Animal and human wastes were also observed throughout the river stretch. These factors affected the quality of the water and are reflected in the upward trend in the river's organic content.

The availability of space for Habitat for the biotic community in Sarno River decreased as the flow went downstream. Still, a decrease in availability was observed in Site A due to the lower water level (HA). At the same time, an increase in Site C was observed due to the increased vegetation in the riparian zone. These elements, together with Canopy Cover, define the energy sources and habitat structure in the river. They showed moderate variation with seasonal changes.

As observed, the major elements that contributed to the poor ecological condition of the river were water appearance, nutrient enrichment, manure presence, bank stability and riparian zones. While the first three elements are affected by land-use activities in the area such as agricultural run-off and municipal waste discharges, the last two elements can be related to management practices in the river. Therefore, anthropogenic activities near the river are the major causes of the deterioration in the ecological condition of the Sarno River. Based on the scores of the different aspects, Site A consistently exhibited good conditions. Site B had good condition as far as Flow Regime (FR) and Habitat Structure (HS) were concerned but had an overall fair evaluation due to low scores in Energy Sources (ES) and Chemical Variables (CV). The downstream stations (Sites C, D, and E) all had poor classification owing to low scores in ES and CV and fair conditions for FR and HS. The primary contributing factor for low water quality was identified as water appearance, indicating that people tended to judge the quality of the river in terms of visual cue. From the foregoing observations, it can be seen that the determination and analysis of the various visual elements and aspects are useful in obtaining more accurate information on the processes and interactions in the river. The consideration of factors such as the different pollution acceptor sources (water and soil) and energy flow in the evaluation are advantageous as this gives a holistic view of the condition of the river [18,19].

	Spring (May 8, 2013)					Summer (August 8, 2013)				
	Site A	Site B	Site C	Site D	Site E	Site A	Site B	Site C	Site D	Site E
Elements										
Channel Condition	4.75	10	4	8.5	4.75	4.75	10	4	8.5	4.75
Hydrologic Alteration	8.25	8	8.63	8	8.5	3	8	6.38	8	8.5
Fish Barriers	5.75	9.5	9.25	8.75	9.25	5.75	9.5	9.25	8.75	9.25
Bank Stability	8.5	8.5	4	3	6.5	8.5	8.5	4	7.5	6.5
Water Appearance	10	2.5	2.5	1	1	10	2.5	2.5	1	1
Manure Presence	6.13	5.25	4.75	5.25	4.75	6.13	5.25	4.75	5.25	4.75
Nutrient Enrichment	10	1	1	1	1	10	1	1	1	1
Canopy Cover	10	10	10	10	1	10	10	10	10	1
Riparian Zone	10	6.5	3	1	2	10	6.5	3	1	2
Instream Fish Cover	5	5	5	5	3	5	5	5	5	3
Invertebrate Habitat	10	7	7	7	7	10	7	7	7	7
Aspects										
Energy Sources score	8.71	3.92	3.75	3.92	2.25	8.71	3.92	3.75	3.92	2.25
Chemical Variable score	8.06	3.38	3.13	3.13	2.88	8.06	3.38	3.13	3.13	2.88
Flow Regime Score	7.58	8.35	5.25	6.25	5.88	5.48	8.35	4.75	6.7	5.88
Habitat Structure Score	8.49	7.21	5.45	5.33	5.44	7.7	7.21	6.21	6.3	5.44
Total score										
Weighted SRVAP Score	8.18	5.77	4.43	4.45	4.31	7.48	5.77	4.59	5.03	4.31
Classification	Good	Fair	Poor	Poor	Poor	Good	Fair	Poor	Fair	Poor

Table 1: Results of SRVAP Evaluation.

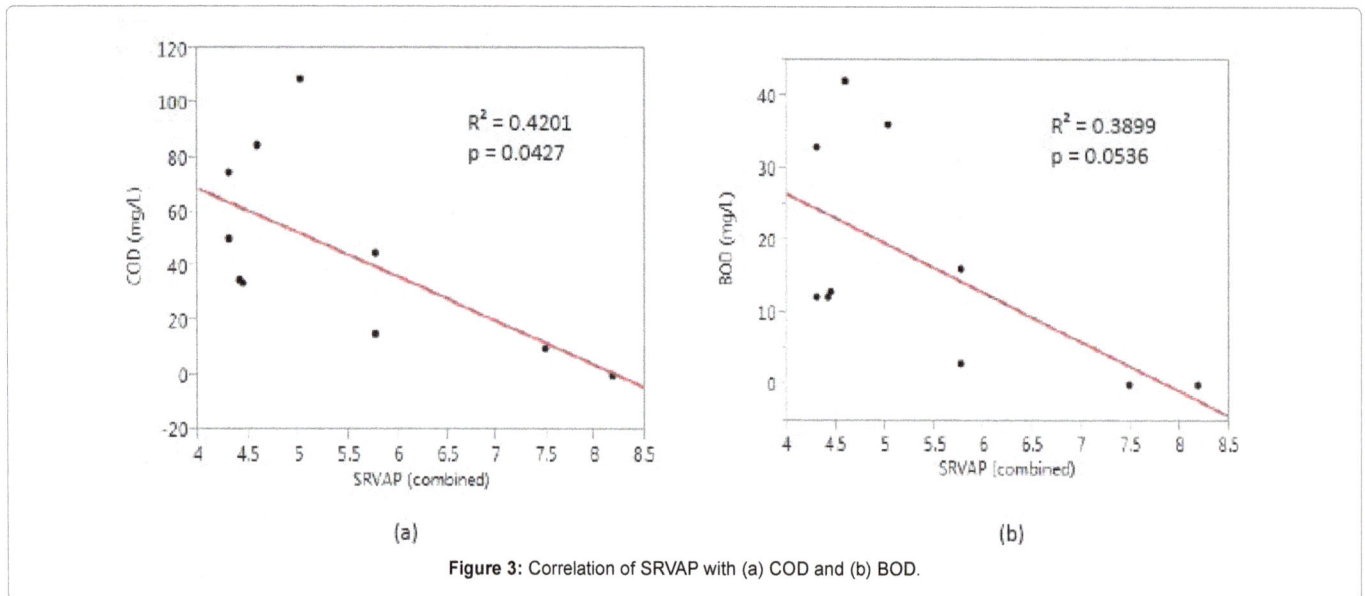

Figure 3: Correlation of SRVAP with (a) COD and (b) BOD.

Spring evaluation	SRVAP Score
BOD (mg/L)	0.8642 (p=0.0222)
	BOD=26.48 - 3.40*SRVAP
COD (mg/L)	0.8542 (p=0.0247)
	COD=85.74 - 10.88*SRVAP
Summer evaluation	
BOD (mg/L)	0.8843 (p=0.0173)
	BOD=94.56 - 12.72*SRVAP
COD (mg/L)	0.7156 (p=0.0709)
	COD=202.71-25.45*SRVAP

Table 2: Regression coefficients and equations of water quality parameters and SRVAP score.

Data analysis

Correlation of river water quality with SRVAP: Linear regression was used to determine possible relationships between SRVAP scores and BOD and COD values. Plots for the linear regression are shown in (Figure 3). SRVAP scores has a statistically significant positive correlation with COD (R^2=0.4201; p=0.0427) while no significant correlation was observed with BOD. In terms of monitoring, visual assessment can act as a rapid estimation for the organic content of the Sarno River. Based on the results of regression, an equation that can be used is COD=133.76 -16.24*SRVAPscore. This suggests that incorporating non-technical evaluation can be identified as a plausible predictor of COD in the river. This shows that local knowledge is important and increases the viability of the evaluation. This also shows the feasibility of adding public participation as a tool for evaluation of the river which is consistent with the recommendation of Silvano [20]. Public Participation is a key element in the implementation of the WFD [3] and this assessment method can be an innovative way of involving local people in watershed management.

In order to identify possible factors for the low correlation, analysis was also done to study seasonal trends. The results are shown in (Table 2). It was observed that correlation greatly increased between SRVAP scores and organic content (in terms of BOD and COD) during spring and SRVAP scores with BOD during summer. This shows that the seasonal variability affected negatively the regression and thus, the derived linear regression equations were deemed more suitable

for seasonal evaluation. Regression for COD during summer is not significant because COD values exhibited a very high increase during this time which indicates the contribution of agricultural activities such as tomato canning to the river.

Correlation of river water quality with SRVAP aspects/elements: Pair-wise linear regression analysis were also done for all SRVAP aspects and elements. Riparian zone was also found to be significantly and positively correlated with BOD (R^2=0.4051; p=0.0479) and COD (R^2=0.4833; p=0.0256). This suggests that riparian zone is the most influential element in determining water quality in Sarno River. This is highly relevant in terms of prioritization of river management plans because this signifies that changes in the riparian zone of Sarno River will bring a great impact on its ecological condition. A statistically significant relationship was also seen between chemical variables and COD, having an R^2 value of 0.4067 (p=0.0473) and this element may also be considered as one driving force in determining river condition. Furthermore, water appearance was related to COD (R^2=0.4383; p=0.0370) and this can be a possible indicator that local people can utilize as an early warning mechanism suggesting the need for management interventions in the river.

Conclusions

The modified SRVAP procedure using weighted-average technical evaluation proved to be representative of the general condition of Sarno River, with classification of good to poor for the five (5) sampling sites. The condition of the river was found to be deteriorating from the upstream to downstream stations. The results show that public knowledge should be incorporated in the evaluation. A statistically significant relationship was found for SRVAP and COD and this can be used as a basis for using SRVAP as alternative indicator of COD content in the river. Important elements to consider are riparian zone and water appearance which should be prioritized in management programs. High correlation between SRVAP and organic content was ascertained when seasonal variation is considered and linear regression equations which were generated may be used as rapid estimates of COD and BOD levels in Sarno River. However, it is recommended that additional validation studies are done to check the accuracy of the equations.

Acknowledgements

This study was made possible through the support of the Engineering Research and Development for Technology (ERDT) Program under the Department of Science and Technology, Philippines and the Sanitary Environmental Engineering Division (SEED) Laboratory, University of Salerno, Italy. The authors would like to thank the ARPAC (Campania Environmental Protection Regional Agency) for the use of their monitoring dataset. We would also like to express our gratitude to to P. Napodano, G. Carpentieri, N. Flammia and S. Giuliani of the Department of Civil Engineering in the University of Salerno who provided valuable assistance in the conduct of this research.

References

1. European Commission. (2000). The EU Water Framework Directive - integrated river basin management for Europe (No. 2000/60/EC).

2. Barth F, Fawell J (2001) The Water Framework Directive and European water policy. See comment in PubMed Commons below Ecotoxicol Environ Saf 50: 103-105.

3. Mostert E (2003) The European Water Framework Directive and water management research. Physics and Chemistry of the Earth 28: 523-527.

4. Naddeo V, Scannapieco D, Zarra T, Belgiorno V (2013) River water quality assessment: Implementation of non-parametric tests for sampling frequency optimization. Land Use Policy 30: 197-205.

5. Haro-Martinez M, Beiras R, Bellas J, Capela R, Coelho JP, Lopes I, Marques JC (2015). A review on the ecological quality status assessment in aquatic systems using community based indicators and ecotoxicological tools: what might be the added value of their combination? Ecological Indicators 48: 8-16.

6. Schneider SC, Lawniczak AE, Picinska-Faltynowicz J, Szoszkiewicz K (2012) Do macrophytes, diatoms and non-diatom benthic algae give redundant information? Results from a case study in Poland. Limnologica 42: 204-211.

7. Italian Ministry of the Environment (2006) DECRETO LEGISLATIVO 3 aprile 2006, n. 152 Norme in materia ambientale (No. D.L. 152/2006).

8. Naddeo V, Zarra T, Belgiorno V (2007) Optimization of sampling frequency for river water quality assessment according to Italian implementation of the EU Water Framework Directive. Environmental Secience and Policy 10: 243-249.

9. United States Department of Agriculture (USDA) (1998) Stream Visual Assessment Protocol (No. NWCC-TN-99-1). Portland, Oregon: National Water and Climate Center.

10. Agenzia Regionale Protezione Ambiente Campania ARPAC. (2000) Fiume Sarno.

11. Montuori P, Triassi M (2012) Polycyclic aromatic hydrocarbons loads into the Mediterranean Sea: estimate of Sarno River inputs. See comment in PubMed Commons below Mar Pollut Bull 64: 512-520.

12. Clesceri LS, Greenberg AE, Eaton AD (1998) Standard Methods for the Examination of Water and Wastewater (20thedn.). Washington D.C.: American Public Health Association.

13. Karr JR, Fausch PL, Angermier PR, Yant PR, Schlosser L (1986) Assessing biological integrity in running waters: a method and its rationale. Illinois Natural History Survey Special Publication 5.

14. Brown TC, Peterson GL (2009) An Enquiry Into the Method of Paired Comparison: Reliability, Scaling and Thurstone's Law of Comparative Judgment (No. RMRS- GTR-216WWW) (p. 98). Fort Collins, CO: United States Department of Agriculture, Forest Service, Rocky Mountain Research Station.

15. Ngo D (2014) Paired Comparison analysis. HumanResources.

16. Pavey S (2014) Paired Comparison Analysis: Working Out relative Importances.

17. SAS Institute Inc. (2012) JMP 10 Basic Analysis and Graphing Second Edition. Cary,NC: SAS Institute Inc.

18. Valipour M, Mousavi SM, Valipour R, Rezaei E (2012) Air, Water and Soil Pollution Study in Industrial Units Using Environmental Flow Diagram. Journal of Basic and Applied Scientific Research, 2: 12365-12372.

19. Valipour M, Mousavi SM, Valipour R, Rezaei E (2013) Deal with Environmental Challenges in Civil and Energy Engineering Projects Using a New Technology. Journal of Civil and Environmental Engineering 3.

20. Silvano RAM, Udvardy S, Ceroni M, Farley J (2005) An ecological integrity assessment of a Brazilian Atlantic Forest watershed based on surveys of stream health and local farmers' perceptions: implications for management. Ecological Economics 53:369-385.

Performance Evaluation of SWAT Model for Land Use and Land Cover Changes in Semi-Arid Climatic Conditions

Gebremedhin Kiros[1,2]*, Amba Shetty[1] and Lakshman Nandagiri[1]

[1]*Department of Applied Mechanics and Hydraulics, National Institute of Technology Karnataka, Surathkal, Mangalore - 575 025, India*
[2]*Department of Soil Resources and Watershed Management, Aksum University, Shire campus: Shire Endaselassie P. O. Box -314, Ethiopia*

Abstract

Evaluation of land use land cover changes on the hydrological regime of river basins is one of the concerns in the global climate change. With plethora of tools available in the literature choosing of an appropriate tool that can quantify and analyze the impact of land use land cover changes on the hydrological regime in a systematic and planned manner is important. Soil and Water Assessment Tool (SWAT) integrated with Geographic Information System (GIS) based interfaces and its easy linkage to sensitivity, calibration and uncertainty analysis tools made its applicability more simple and has great potential in simulation of the past, present and future scenarios. A number of standards were used to appraise the model set-up, model performances, physical representation of the model parameters, and the accuracy of the hydrological model balance to assess the models that are defined in journal papers. On the basis of performance indicators, the mainstream of the SWAT models were categorized as providing satisfactory to very good. This review debates on the application of SWAT in analyzing land use land cover changes in semi-arid environment. Application of SWAT and land use land cover simulation models for impact assessment in semi-arid region improves accuracy, reduces costs, and allows the simulation of a wide variety of conservation practices at watershed scale. It is also observed that different researchers and/or model versions bring about in different outcomes while a comparison of SWAT model applications on similar case study was applied. This review determines the interactive role of SWAT and GIS technologies in improving integrated watershed management in semi-arid environments.

Keywords: Impact assessment; Land use land cover changes; Semi-arid environment

Abbrevations: SWAT: Soil and Water Assessment Tool; GIS: Geographic Information System

Introduction

To study sustainable water resources and land use planning and development understanding the consequences of changes in land use and land cover scenarios is required. Human activities can affect the integrity of natural resources and the output of goods and services in the ecosystem. The development of new patterns of land use and land cover conditions can be enhanced by careful planning for the well-being of people [1]. The scientific framework for the analysis of land use systems have changed by the modelling tools which can addresses both spatial and temporal dynamics. It is a universal concern the changes in land use and land cover in river basins resulted in flooding events that has increased sediment loads [2-6]. There are some proportional alterations in the basin condition and hydrological response as a result of changes in land cover and land use scenarios. This is appropriately becoming one of the main existing land management issues [7].

The response of hydrological processes of river basins influenced by human activities and climate changes have been widely studied [8-12] [2,3]. In recent years, understanding the occurrences of natural processes at the watershed scale by the application of the model became an essential tool [13]. Geographic Information System (GIS) based spatial modeling has grown into an important tool to assess the effect of land use land cover changes on runoff and soil erosion studies and, consequently in advancement of suitable soil and water conservation strategies. Among several models SWAT linked with GIS has been extensively used in earlier studies.

Gassman et al. [14] investigated that the historical development, application and future research directions using SWAT model for a wide range of scales and environmental conditions across the globe and over a long period of time. The Soil and Water Assessment Tool (SWAT) model [15] has been proved to be an effective tool for assessing land use land cover changes, water resources and nonpoint-source pollution problems. This paper aims to review performance evaluation of SWAT model for land use and land cover changes in Semi-arid environments. An overview on the efficiency analysis of SWAT and its integration with land use and land cover simulation models are also presented.

SWAT Model

SWAT is readily applicable through the development of geographic information system (GIS) based interfaces and is attributed to the fact that the tool is freely available and easy linkage to sensitivity, calibration and uncertainty analysis tools makes it a very popular model. In data-scarce areas the online and free availability of basic GIS data made SWAT model applicability more straightforward [16]. Conservation practices such as riparian buffers and vegetative filter strips can be adequately simulated whilst SWAT is being altered to account for landscape spatial positioning [15]. One of the main advantages of SWAT is that it can be used to model watersheds with less monitoring data. For simulation, SWAT needs digital elevation model (DEM), land use and land cover

*Corresponding author: Gebremedhin Kiros, Department of Applied Mechanics and Hydraulics, National Institute of Technology Karnataka, Surathkal, Mangalore-575 025, India, E-mail: k.gebremedhin@yahoo.com

map, soil data and climate data of a specific study area. These data are used as an input for the analysis of hydrological simulation of surface runoff and groundwater recharge.

The Simulation of the hydrology of a watershed is done in two separate divisions. One is the land phase of the hydrological cycle that controls the amount of water, sediment, nutrient and pesticide loadings to the main channel in each sub-basin [17]. Hydrological components simulated in land phase of the hydrological cycle are canopy storage, infiltration, redistribution, evapotranspiration, lateral subsurface flow, surface runoff, ponds, tributary channels and return flow. The second division is routing phase of the hydrologic cycle that can be defined as the movement of water, sediments, nutrients and phase of hydrological cycle. SWAT simulates the hydrological cycle based on the water balance equation.

$$SW_t = SW_o + \sum_{i=1}^{t} \left(R_{day} + Q_{surf} - E_a - W_{seep} - Q_{qw} \right)_i \qquad (1)$$

Where SW_t is the final soil water content (mm), SW_o is the initial soil water content on day i (mm H_2O),t is the time (days), R_{day} is the amount of precipitation on day i (mm H_2O), Q_{surf} is the amount of surface runoff on day i (mm), E_a is the amount of evapotranspiration on day i (mm H_2O), W_{seep} is the amount of water entering the vadose zone from the soil profile on day i (mm H_2O) and Q_{qw} is the amount of return flow on day i (mm H_2O).

Surface runoff occurs whenever the rate of precipitation goes beyond the rate of infiltration. SWAT suggests two methods for estimating surface runoff: the SCS curve number procedure [18] and the Green & Ampt infiltration method [19]. Using daily or sub daily rainfall, SWAT simulates surface runoff volumes and peak runoff rates for each HRU. In most cases, the SCS curve number method was used to estimate surface runoff because of the unavailability of sub daily data for Green & Ampt method.

The SCS curve number equation is:

$$Q_{surf} = \frac{\left(R_{day} - 0.2S \right)^2}{\left(R_{day} + 0.8S \right)}, \text{ if } R_{day} > 0.2S; Q_{surf} = 0 \text{ otherwise} \qquad (2)$$

Where: Q_{surf} is the accumulated runoff or rainfall excess (mm), R_{day} is the rainfall depth for the day (mm); S is the retention parameter (mm). The retention parameter is defined by equation 3.

$$S = 25.4 \left(\frac{100}{CN} - 10 \right) \qquad (3)$$

Where; CN is the curve number for the day.

Erosion caused by rainfall and runoff is computed using the Modified Universal Soil Loss Equation (MUSLE) (Williams, 1975). MUSLE is a modified version of the Universal Soil Loss Equation (USLE) developed by [20]. It calculates the average annual gross erosion as a function of rainfall energy. In MUSLE, the rainfall energy factor is substituted with a runoff factor which advances the sediment yield prediction and permits the equation to be applied to discrete storm events. This advances sediment yield prediction because runoff is a function of antecedent moisture condition as well as rainfall energy [21].

$$Sed = 11.8 * \left(Q_{surf} * q_{peak} * area_{hru} \right)^{0.56} * K_{USLE} * C_{USLE} * P_{USLE} * LS_{USLE} * CFRG \qquad (4)$$

Where Sed is the sediment yield on a given day (metric tons), Q_{surf} is the surface runoff volume (mm H_2O/ha), q_{peak} is the peak runoff rate (m^3/s), $area_{hru}$ is the area of the HRU (ha), K_{USLE} is the soil erodibility factor (0.013 metric ton m^2hr/(m^3-metric ton cm)), C_{USLE} is the cover

and management factor, P_{USLE} is the support practice factor, LS_{USLE} is the topographic factor and CFRG is the coarse fragment factor. The details of the USLE factors and the descriptions of the different model components can be found in [15].

Impact Analysis of LUCC by SWAT Model

Hydrologic response is an integrated indicator of watershed condition, and significant changes in land cover may affect the overall health and function of a watershed. [7] used SWAT model for evaluating the effects of land cover change and rainfall spatial variability on watershed response of the Walnut Gulch experimental watershed encompassing approximately 150 km^2 located in southeastern Arizona, USA. The authors evaluated the impact of land cover change on the runoff depth of different land cover classes. The simulation results showed that the runoff responses of the watershed due to changes of land use and land cover.

Mango et al. [22] analyzed the sensitivity of model outputs to land use change for a sub-basin (700 km^2) on the Nyangores tributary of the Mara River basin, Kenya, used three hypothetical scenarios: partial deforestation, complete deforestation to grassland, and complete deforestation to agriculture. Simulations under all land use change scenarios indicated that various land use pattern should have various impacts on rainfall-streamflow interactions. For example, the conversion of forest land to agricultural land indicated an increased overland flow and a decreased subsurface flow and average flow over the period of simulation, while evapotranspiration shows a small positive increase. These outcomes are disagreed to the results achieved by [23], where a reduction in forest cover directed to a decrease in evapotranspiration, an increase in both surface and sub subsurface flow and a large increase in water yield.

Pikounis et al. [24] investigated that the hydrological effects of specific land use changes in a catchment of the river Pinios in Thessaly (Ali Efenti catchment, 2976 km^2), Greece, through the application of the SWAT model on a monthly time step. It should be noted that although the model was run for 23-years (1970 to 1993), the first 5 years of simulated output were disregarded in the calibration process, since they are required by the model as a warm-up period. This period was essential for the stabilization of parameters, as the results sometimes vary significantly from the observed values. The authors investigated the effect of land use change by using three land use scenarios which are: expansion of agricultural land, complete deforestation and expansion of urban area in the Trikala sub-basin. All the three scenarios resulted in an increased in streamflow during wet season and decreased during the dry season. Thus, the final calibration period was from April 1975 to December 1993. The result can be quite satisfactory.

SWAT application for assessing LUCC impacts on sustainable development and watershed hydrological status is gaining momentum worldwide due to enormous anthropogenic activities on the natural systems of river basins [25-32]. In a case study at the Little Miami Watershed, USA, it was recognized that there is significant reduction in flow, sediments, and nutrients were detected as the land use in the watershed shifts from predominantly agricultural to mixed rural and residential lands [33]. To simulate the main components of the hydrological cycle, SWAT model was also used in order to study the effects of land use changes in 1967, 1994 and 2007, in the Zanjanrood Basin, Iran [11]. The results indicated that the hydrological response was nonlinear and exhibited a threshold effect to overgrazing and changing of rain-fed agriculture and bare ground from rangelands where more than 60% of the rangeland was removed, the runoff increased considerably.

Singh and Gosain [34] used SWAT model for the valuation of the total amount of available water, as well as prediction of the impact of changes in the land management practices on the availability of water in Cauvery basin, India. Nine reservoirs in the basin where data was available were also modeled as impoundments structures in this study. When monthly streamflow values were considered, the values of NSE and R2 was 0.934 and 0.936, respectively, which reveals that the model had captured the system and there was no need to calibrate the model. These authors produced a series of scenarios for the participants to analyze the above-mentioned categories of changes in land use and land cover. This paper reported that as the percentage of forest decreases, the water yield increases in the basin. Similar scenario generation approach was made in Kaneri basin, Maharashtra by [35]. Four scenarios were considered in model simulations. The first one was the base scenario; second with sensitive parameters flow was calibrated to advance the outcomes. Each sub-basin was provided with pond in the third scenario and the impact was studied. For the fourth condition, the Best management practices (BMP's) like farm terracing, contouring, residue management and generic conservation practices were included and the impact was studied. The BMP's provided results in the reduction of surface runoff in the range of 62% to 75%, decrease of water yield in the range of 33% to 53% and reduction of sediment yield to nearly 98%. In four time periods (1973, 1986, 1992, and 1997) in the upper San Pedro watershed, USA another application of SWAT, hydrological modeling was conducted for each of the land use map [36]. Results verified that major environmental stressors affecting local water resources were urbanization and mesquite invasion. In the Chi River basin, Thailand in another place land use change were evaluated in five scenarios [37]. These scenarios have incorporated conversion of farmland to rice and sugarcane plantation and three scenarios involving a conversion of forested area, expansion of farmland, switching of rice paddy fields to energy crops. Results have shown that not worth mentioning changes shown on water flows and evapotranspiration (ET) due to the conversion of forested area and farmland. In the dry season there is reduced water flows and increased ET when paddy fields are substituted by sugarcane plantation. Predominantly, small changes occur on annual flow and ET but more significant effects occur on seasonal flows in case of expansion of rice paddy fields to farmland. In the dry season period the results showed there is an increase of water yield as a result of decreasing ET leading to a significant effect on seasonal ET showed as the conversion of farmland to sugarcane plantation for bio-fuel production, but small changes on water yields.

Using land use maps in the upper Huaihe River, China over three phases the 1980s, 1990s and 2000s the effect of land use change on the sediment yield characteristics were explored [38]. The results have shown that there is increasing rate for sediment yield and the sensitivity of rainfall–sediment yield relationship to rainfall changes move down by woodland, paddy field and farmland under the same condition of soil texture and terrain slope in the area. [39] assessed impact of two small scale Slovenian watersheds by using historical land use maps from 1787, 1827, 1940, 1984 and 2009 land use map depicting present situation for LUCC. Results showed statistically insignificant for both watersheds the influence of land use change on total and green water quantity, but would have considerable effects on the seasonal flow. Shao Y, Lunetta R, et al. [40] investigated in the Laurentian Great Lakes Basin, USA including the conversion of all "other" row crop types to corn and hay/pasture to corn they considered two future agricultural scenarios compared with the current baseline condition. Significant increases in average annual sediment yields were noticed when compared with the baseline condition.

Several researchers have investigated several studies on the combined impact of land use change with climate change. In the Loess Plateau of China, [10] among others, quantified the influences of the land use change and climate variability by comparing the SWAT outputs of the four scenarios, i.e., S1 (1985 land use and 1981–1990 climate), S2 (2000 land use and 1981–1990 climate), S3 (1985 land use and 1991–2000 climate), and S4 (2000 land use and 1991–2000 climate). Results revealed that the surface hydrology is influenced more significantly by the climate variability than the land use change within the watershed during the period 1981–2000. In the Be River Watershed, Vietnam through climate change scenarios three land use scenarios were considered in examining the impact of land use change on streamflow and sediment yield [41]. All current shrub lands were converted into perennial cropland, and the remaining land use types were kept constant as shown in the first scenario. The second scenario assumed that shrub land substituted all productive forest lands. All shrub land and productive forest land were replaced by perennial cropland while using the third scenario. Generally, Streamflow, sediment load, and water balance components response to the separate impacts of climate and land use changes were offset by one another. On the other hand, surface runoff and few components of subsurface flow were less sensitive to climate change than to land use change. However, surface runoff and few components of subsurface flow were more sensitive to land use change than to climate change. In addition, the results underlined increased soil erosion during the wet season and water scarcity during the dry season [42] showed that for runoff variations the climate conditions, especially precipitation, played an important role while land use change during the period 1970-2000 was secondary across the Taoerhe River basin, China. Furthermore, monthly runoff was larger in the wet season due to the effects of changes in land use land cover conditions. In the Biliu River basin, China [43] generalized the characteristics of the human activities to forecast future runoff using land use land cover change conditions. The Results indicated that under normal human activities and future land use land cover change scenarios; there will be approximately 10% future increase in annual flow from 2011 to 2030, as suggested by land use land cover change scenarios with a particularly wet year in the next 20 years.

SWAT results showing improved tillage practices could result in reduced sediment yields of almost 20% within the Rock River in Wisconsin, USA, [44]. In the Walnut Creek watershed in central Iowa, USA, [45] found that adoption of no tillage, changes in nitrogen application rates, and land use changes could greatly impact nitrogen losses. Large sediment reductions could be obtained, depending on the choice of Best Management Practice as indicated by [46] on their analysis of Best Management Practices for the Walnut Creek and Buck Creek watersheds in Iowa. The impacts of Best Management Practices of three 25-year SWAT scenario simulations for two small watersheds in Indiana, USA, were studied [28], and indicated that for streamflow, sediment, and total phosphate Best Management Practices in varying conditions, and best management practices in good conditions are reported. [47] reported that within the 3000 km² Delaware River basin in northeast Kansas, USA in response to simulated shifts of cropland into switch grass production large nutrient and sediment loss reductions are occurred.

In east Africa watersheds, [48] investigated that the net influence of land cover conversion was as estimated an overall slight increase in water yield, articulated as the total streamflow from the outlet of the river resulting from both overland flow and subsurface flow that happened when the soil is comparatively well inundated. Although the overall impact on water yield was relatively small, the amount of water yield resulting from overland flow increased considerably at the expense of soil water flow. Overland flow increased while lateral flow was reduced significantly. The increase in surface water was offset by an appropriate decline in groundwater recharge. In addition, these changes are due to two reasons: (1) declines in evapotranspiration due to the reduction in vegetation cover, and (2) greater fraction of rainfall actuality transformed into overland flow instead of going down into the soil and drifting to the aquifer and concluded that hydrologic changes were highly inconstant both spatially and temporally, and the streams in the uppermost of the forested highlands were most considerably affected and these variations have negative consequences for the environmental health of the river system.

Application of SWAT Model in Semi-arid Environments

Numerous studies have been conducted in the past two decades that pointed out the application of the SWAT model and has been used widely. Examples of studies carried out include those of [49-53] who studied and predict the potential impacts of climate change on water resources and yields. To predict various impacts of land management on water quantity [54,55]; assess the watershed response impact to land use/cove changes on the annual water balance and temporal runoff dynamics [32,24,56,57]; to predict streamflow which were compared favorably with measured data for a variety of watershed scales [58-61]. All these studies have shown varied results due to the different regions considered, and also have employed different methodologies to construct land use/cover change and scenarios on the impacts on the hydrological responses. However, most of these studies concluded that SWAT is suitable for long-term simulations (monthly, seasonal and yearly) have been preferred for use in impact assessment and that daily flows are simulated with lower efficiencies.

Van Griensen et al. [16] stated that researchers in the Nile countries are adopting SWAT for several integrated water resources studies such as erosion modeling, land use and climate change impact modeling and water resources management. The majority of the studies were focused on locations in the tropical highlands of Ethiopia and around Lake Victoria. The majority of the SWAT models were categorized with results satisfactory to very good on the basis of performance indicators. On the other hand, the hydrological mass balances as reported in a number of articles controlled losses that might not be acceptable.

Mengistu et al. [62] investigated the sensitivity of SWAT simulated streamflow to climatic changes within the three major sub-basins of Abbay (Blue Nile), Baro- Akobo and Tekeze in Eastern Nile River basin. Those sensitivity parameters ranking were CN2, SOL_AWC, Sol_K and ESCO. Calibration and validation periods used for model simulations were 1990-1996 and 1997-2004 respectively. However the curve number (CN2) was the main sensitive parameter for all the outlets. This is due to the fact that the curve number depends on several factors including soil types, soil textures, soil permeability and land use properties etc., Good agreements between simulated and observed flows in both of daily and monthly time scale were also noticed.

Easton et al. [63] used a modified version of SWAT model (SWAT Water Balance) tool to quantify the hydrologic and sediment fluxes in the Blue Nile Basin, Ethiopia. They modeled SWAT to simulate runoff and erosion in the Blue Nile basin with source of runoff from Ethiopia. The model was initialized for eight sub basins ranging in size from 1.3 km^2 to 174,000 km^2. This new version of SWAT, SWAT-WB, calculates runoff volumes based on the available storage capacity of a soil and distributes storages across the watershed using soil topographic wetness index [64]. In place of CN for each HRU to predict runoff losses, SWAT-WB model used water balance. To obtain good hydrologic predictions the model requires very little direct calibration. The authors selected

the most sensitive parameters controlling erosion in the watershed were those used for calculating the maximum amount of sediment that can be entrained during channel routing. The channel properties, channel erodibility factor (CH_EROD), channel cover factor (CH_COV), channel manning's n (CH_N) and channel saturated hydraulic conductivity (CH_K). The model prediction showed reasonable accuracy of NSE 0.53-0.9.

Bertie et al. [65] set up the SWAT model to simulate spatial distribution of soil erosion/sedimentation processes at daily time step and to assess the impact of three Best Management Practice (BMPs) scenarios on sediment reductions in the upper Blue Nile River basin in an area of 184,560 km^2. They found the most sensitive parameters for surface flow prediction were the surface flow parameters CN2, ESCO, SOL_AWC, SOL_K, SULAG, SLSUBBSN; baseflow parameters were ALPHA_BF, GW_DELAY, GWQMN, GW_REVAP, REVAPMIN, RCHRG_DP; channel routing parameters were CH_K2 and CH_N2. Taking different scenarios for best management practices at sub-basins scale and revealed that a wide-ranging spatial variability on sediment decrease. The sediment reduction was varied from 29% to 68% by buffer strip (Scenario-1), 9% to 69% by stone-bund (Scenario-2) and 46% to 77% by replantation of trees (Scenario-3) applying the Modified Universal Soil Loss Equation (MUSLE) which is embedded in the model SWAT. However, their results did not show the effects of gully erosion.

Integration of SWAT with LULC Simulation Models

Based on different coherent scenarios land use models have a common objective of simulating landscape dynamics in the future at multiple scales [66]. Within land use patterns they improve understanding and sensitivity of key processes [67]. Consideration of socio-ecological dynamics and performance has been facilitated by scenarios building based on land use models [68]. Therefore, it is new and thoughtful way for hydrological assessment of future and hypothetical land use and land cover scenarios by integrating of SWAT with land use simulation models. In a case study, to estimate the impact of land cover change on runoff in a tropical watershed in Kenya, [23] integrated the SWAT with Conversion of Land Use and its Effects at Small-regional-extent (CLUE-S) model. Sensitivity of the basin's hydrological system attributable to alterations in land cover and this study offers a practical insight. In the upstream watershed of Miyun Reservoir in Beijing, China, SWAT coupled with CLUE- S to simulate pollution loads under different land use scenarios [69].

A dynamic combination of land use changes with a hydrologic model offers a more truthful representation of the progressive development of land use changes, is probable to advance the temporal predictive ability of the model [70], and permits for a temporally categorical analysis of hydrologic impacts emphasized that a close-fitting temporal assimilation of the dynamics of land use change and hydrology is needed to accurately represent the interfaces between land use, climate, and hydrology. A measureable investigation on the multi-scale land use changes in space specifically predicting probable changes under land use conditions in the forthcoming, and taking into account vicinity factors, driving forces and land suitability associated to land use condition design by the dynamic land use simulation model of CLUE-S [71]. The consolidation of CLUE-S and SWAT can provide complete play to the benefits of model coupling, which both improves the rationality and accurateness of the model for land use scenario simulation and successfully appraises non-point source pollutions under different conditions.

Even though more sophisticated methodologies to describe land use change conditions by using land use change models are available, these are rarely dynamically incorporated with hydrologic impact assessments. Land use change scenarios may be derived as a result of simple assumptions [72-74]. Therefore, they provide a basis to predict land use change in a more complex technique. Several land use change models have been developed and are used for several purposes, comprising empirical-statistical, stochastic, optimization, process-based, and integrated modeling approaches [67]. A thorough review of land use change models and their specific characteristics is provided by [72]. However, the significance of a dynamic representation of land use changes has been acknowledged [57] a dynamic combination of spatially categorical models of land use change and hydrologic models is seldom found in the literature.

Conclusion

This paper emphasizes that SWAT is a very flexible and strong tool that can be used to simulate a variety of land management problems in different catchments with various climatic and land cover conditions. SWAT model is a potential and powerful model once calibrated and validated effectively for wide range of applications. The development of GIS-based interfaces, which provide a simple means of translating digital land use, topographic, and soil data into model inputs, has greatly facilitated the process of configuring SWAT for a given catchment. Furthermore, advancement of a new era in SWAT application for LUCC simulation with the highest possible accuracy as a result of the new facilities for SWAT auto-calibration and uncertainty analysis was presented. Simulation of hypothetical, real and future scenarios in SWAT has proven to be an effective method of evaluating alternative land use effects on runoff and sediment losses which made the SWAT robust and flexible framework that allows the simulation of a wide variety of conservation practices. This capability via the integration of SWAT with LULC simulation models has been strengthened to the best of possible. Therefore, the successful evaluation of SWAT model in semi-arid environments as demonstrated in this review provides the opportunity for expanding the model application to other similar climatic locations where there is limited number of gauge stations.

References

1. Millennium Ecosystem Assessment (2005) Millennium Ecosystem Assessment, Ecosystems and Human Well-Being: General Synthesis, Millennium Ecosystem Assessment Series, Island Press.

2. Memarian H, Balasundram S, Talib J, Sood A, Abbaspour K (2012a) Trend analysis of water discharge and sediment load during the past three decades of development in the Langat basin, Malaysia. Hydrological Sciences Journal 57: 1207-1222.

3. Memarian H, Balasundram S, Talib J, Teh Boon Sung C, Mohd Sood A, et al. (2012b) KINEROS2 application for land use/cover change impact analysis at the Hulu Langat Basin, Malaysia. Water and Environment Journal 27: 549-560.

4. Zhang X, Cao W, Guo Q, Wu S (2010) Effects of land use change on surface runoff and sediment yield at different watershed scales on the Loess Plateau. International Journal of Sediment Research 25: 283-293.

5. Zhang S, Lu X, Higgitt D, Chen C, Han J, et al. (2008) Recent changes of water discharge and sediment load in the Zhujiang (Pearl River) Basin, China. Global and Planetary Change 60: 365-380.

6. Garcia-Ruiz J, Regues D, Alvera B, Lana-Renault N, Serrano-Muela P, et al. (2008) Flood generation and sediment transport in experimental catchments affected by land use changes in the central Pyrenees. Journal of Hydrology 356: 245-260.

7. Hernandez M, Miller S, Goodrich D, Goff B, Kepner W, et al. (2000) Modeling runoff response to land cover and rainfall spatial variability in semi-arid watersheds. Environmental Monitoring and Assessment 64: 285-298.

8. Nearing M, Jetten V, Baffaut C, Cerdan O, Couturier A, et al. (2005) Modeling response of soil erosion and runoff to changes in precipitation and cover. Catena 61: 131-154.

9. He H, Zhou J, Zhang W (2008) Modeling the impacts of environmental changes on hydrological regimes in the Hei River Watershed, China. Global and Planetary Change 61: 175-193.

10. Li Z, Liu W, Zhang X, Zheng F (2009) Impacts of land use change and climate variability on hydrology in an agricultural catchment on the Loess Plateau of China. Journal of hydrology 377: 35-42.

11. Ghaffari G, Keesstra S, Ghodousi J, Ahmadi H (2010) SWAT-simulated hydrological impact of land-use change in the Zanjanrood basin, Northwest Iran. Hydrological Processes 24: 892-903.

12. Ouyang W, Hao F, Skidmore A, Toxopeus A (2010) Soil erosion and sediment yield and their relationships with vegetation cover in upper stream of the Yellow River. Science of the Total Environment 409: 396-403.

13. Setegn SG, Srinivasan R, Dargahi B (2008) Hydrological Modelling in the Lake Tana Basin, Ethiopia Using SWAT Model. The Open Hydrology Journal 2: 49-62.

14. Gassman P, Reyes M, Green C, Arnold J (2007) The soil and water assessment tool: historical development, applications, and future research directions. Center for Agricultural and Rural Development, Iowa State University 50: 1211-1250.

15. Arnold J, Srinivasan R, Muttiah R, Williams J (1998) Large area hydrologic modeling and assessment part I: Model development1. JAWRA Journal of the American Water Resources Association 34: 73-89.

16. Van Griensven A, Ndomba P, Yalew S, Kilonzo F (2012) Critical review of SWAT applications in the upper Nile basin countries. Hydrology and Earth System Sciences 16: 3371-3381.

17. Neitsch S, Arnold J, Kiniry J, Williams J (2011) Soil and water assessment tool theoretical documentation, version 2009. Texas, USA.

18. US. Department of Agriculture-Soil Conservation Service (1972) Hydrology: Soil Conservation Service National Engineering Handbook, Section 4. Washington, D.C.

19. Green W, Ampt G (1911) Studies on soil physics: The flow of air and water through soils. Journal of Agricultural Sciences 4: 11-24.

20. Wischmeier W, Smith D (1965) Predicting Rainfall-Erosion Losses from Cropland East of the Rocky Mountains - Guide for Selection of Practices for Soil and Water Conservation. USDA Handbook No. 282.

21. Abbaspour K, Yang J, Maximov I, Siber R, Bogner K, et al. (2007) Modeling hydrology and water quality in the pre-alpine/alpine Thur watershed using SWAT. Journal of hydrology 333: 413-430.

22. Mango L, Melesse A, McClain M, Gann D, Setegn S (2011) Land use and climate change impacts on the hydrology of the upper Mara River Basin, Kenya: results of a modeling study to support better resource management. Hydrology and Earth System Sciences 15: 2245.

23. Githui F, Mutua F, Bauwens W (2009b) Estimating the impacts of land-cover change on runoff using the soil and water assessment tool (SWAT): case study of Nzoia catchment, Kenya. Hydrological sciences journal 54: 899-908.

24. Pikounis M, Varanou E, Baltas E, Dassaklis A, Mimikou M (2003) Application of the SWAT model in the Pinios River Basin under different Land use Scenarios. Global Nest: the international journal 5: 71-79.

25. Yevenes M, Mannaerts C (2011) Seasonal and land use impacts on the nitrate budget and export of a meso-scale catchment in Southern Portugal. Agricultural Water Management 102: 54-65.

26. Lam Q, Schmalz B, Fohrer N (2011) The impact of agricultural Best Management Practices on water quality in a North German lowland catchment. Environmental Monitoring and Assessment 183: 351-379.

27. Mishra A, Kar S, Singh V (2007) Prioritizing structural management by quantifying the effect of land use and land cover on watershed runoff and sediment yield. Water Resources Management 21: 1899-1913.

28. Bracmort K, Arabi M, Frankenberger, Engel B, Arnold J (2006) Modeling long-term water quality impact of structural BMPs. Transactions of the ASABE 49: 367-374.

29. Heuvelmans G, Garcio-Qujano J, Muys B, Feyen J, Coppin P (2005) Modeling

the water balance with SWAT as part of the land use impact evaluation in a life cycle study of CO_2 emission reduction scenarios. Hydrological Processes 19: 729-748.

30. Haverkamp S, Fohrer N, Frede H (2005) Assessment of the effect of land use patterns on hydrologic landscape functions: a comprehensive GIS-based tool to minimize model uncertainty resulting from spatial aggregation. Hydrological Processes 19: 715-727.

31. Eckhardt K, Breuer L, Frede H (2003) Parameter uncertainty and the significance of simulated land use change effects. Journal of Hydrology 273: 164-176.

32. Fohrer, N, Haverkamp S, Eckhardt K, Frede (2001) Hydrologic response to land use changes on the catchment scale. Physics and Chemistry of the Earth 26: 577-582.

33. Tong S, Liu A, Goodrich J (2009) Assessing the water quality impacts of future land use changes in an urbanizing watershed. Civil Engineering and Environmental Systems 26: 3-18.

34. Singh A, Gosain AK (2011) Scenario generation using geographical information system (GIS) base hydrological modelling for a multi-jurisdictional Indian River basin. Journal of Oceanography and Marine Science 2: 140-147.

35. Swami VA, Kulkarni SS (2014) Evaluation of the Best management practices using SWAT model for the Kaneri micro-watershed, southern Maharashtra, India. International Journal of Advanced Technology in Engineering and Science 2: 9-16.

36. Nie W, Yuan Y, Kepner W, Nash M, Jackson M, et al. (2011) Assessing impacts of Land use and Land cover changes on hydrology for the upper San Pedro watershed. Journal of Hydrology 407: 105-114.

37. Homdee T, Pongput K, Kanae S (2011) Impacts of land cover changes on hydrologic responses: A case study of Chi river basin, Thailand. Journal of Japan Society of Civil Engineers 67: 31.

38. Cai T, Li Q, Yu M, Lu G., Cheng L, et al. (2012) Investigation into the impacts of land-use change on sediment yield characteristics in the upper Huaihe River basin, China. Physics and Chemistry of the Earth (A/B/C) 53-54: 1-9.

39. Glavan M, Pintar M, Volk M (2012) Land use change in a 200-year period and its effect on blue and green water flow in two Slovenian Mediterranean catchments lessons for the future. Hydrological Processes 27: 3964-3980.

40. Shao Y, Lunetta R, Macpherson A, Luo J, Chen G (2013) Assessing Sediment Yield for Selected Watersheds in the Laurentian Great Lakes Basin Under Future Agricultural Scenarios. Environmental Management 51: 59-69.

41. Khoi D, Suetsugi T (2012) The responses of hydrological processes and sediment yield to land-use and climate change in the Be River Catchment, Vietnam. Hydrological Processes 28: 640-652.

42. Li L, Jiang D, Hou X, Li J (2012) Simulated runoff responses to land use in the middle and upstream reaches of Taoerhe River basin, Northeast China, in wet, average and dry years. Hydrological Processes 27: 3484-3494.

43. Zhang C, Shoemaker C, Woodbury J, Cao M, Zhu X (2012) Impact of human activities on stream flow in the Biliu River basin, China. Hydrological Processes 27: 2509-2523.

44. Kirsch K, Kirsch A, Arnold J (2002) Predicting sediment and phosphorus loads in the Rock River basin using SWAT. Trans ASAE 45: 1757-1769.

45. Chaplot V, Saleh A, Jaynes D, Arnold J (2004) Predicting water, sediment and NO_3-N loads under scenarios of land-use and management practices in a flat watershed. Water, Air, and Soil Pollution 154: 271-293.

46. Vache K, Eilers J, Santelmann M (2007) Water quality modeling of alternative agricultural scenarios in the US Corn Belt. Journal of the American Water Resources Association 38: 773-787.

47. Nelson R, Ascough J, Langemeier M (2006) Environmental and economic analysis of switch grass production for water quality improvement in northeast Kansas. Journal of environmental management 79: 336-347.

48. Baker TJ, Miller SN (2013) Using the Soil and Water Assessment Tool (SWAT) to assess land use impact on water resources in an East African watershed. Journal of Hydrology 486: 100-111.

49. Tadele K, Forch G (2007) Impact of Land Use and Cover Change on Streamflow: The Case of Hare River Watershed, Ethiopia. Catchment and Lake Research, LARS.

50. Van Liew M, Garbrecht J (2001) Sensitivity of hydrologic response of an experimental watershed to changes in annual precipitation amounts. ASAE Annual International Meeting, Sacramento, CA.

51. Varanou E, Gkouvatsou E, Baltas E, Mimikou M (2002) Quantity and quality integrated catchment modeling under climate change with use of soil and water assessment tool model. Journal of Hydrologic Engineering 7: 228-244.

52. Jha M, Pan Z, Takle E, Gu R (2004) Impacts of climate change on stream flow in the Upper Mississippi River Basin: A regional climate model perspective. Journal of Geophysical Research 109: 10-29.

53. Gosain A, Rao,S, Basuray D. (2006) Climate change impact assessment on hydrology of Indian River basins. Current Science 90: 346-353.

54. Srinivasan R, Arnold J (1994) Integration of a basin-scale water quality model with GIS. Water Resources Bulletin 30: 453-462.

55. Muttiah R, Wurbs R (2002) Scale-dependent soil and climate variability effects on watershed water balance of the SWAT model. Journal of Hydrology 256: 264-285.

56. Kepner W, Semmens D, Bassett S, Mouat D, Goodrich D (2004) Scenario Analysis for the San Pedro River, Analyzing Hydrological Consequences of a Future Environment. Environmental Monitoring and Assessment 94: 115–127.

57. Fohrer N, Haverkamp S, Frede H (2005) Assessment of the effects of land use patterns on hydrologic landscape functions: development of sustainable land use concepts for low mountain range areas. Hydrological Processes 19: 659–672.

58. Saleh A, Arnold J, Gassman P, Hauck L, Rosenthal W, et al. (2000) Application of SWAT for the Upper North Bosque River Watershed. Trans ASAE 43: 1077-1087.

59. Santhi C, Arnold J, Williams J, Dugas W, Srinivasan R, et al. (2001) Validation of the SWAT model on a large river basin with point and nonpoint sources. Journal of the American Water Resources Association 37: 1169-1188.

60. Govender M, Everson C (2005) Modelling streamflow from two small South African experimental catchments using the SWAT model. Hydrological Process 19: 683-692.

61. Mao D, Cherkauer K (2009) Impacts of land-use change on hydrologic responses in the Great Lakes region. Journal of Hydrology 374: 71-82.

62. Mengistu D, Sorteberg A (2012) Sensitivity of SWAT simulated streamflow to climatic changes within the Eastern Nile River basin. Hydrology and Earth System Sciences 16: 391-407.

63. Easton Z, Fuka D, White M, Collick E, Biruk Ashagre A, et al. (2010) A multiple basin SWAT model analysis of runoff and sedimentation in the Blue Nile, Ethiopia. Hydrology and Earth System Sciences. 14: 1827-1841.

64. Easton Z, Fuka D, Walter M, Cowan D, Schneiderman E, et al. (2008) Re-conceptualizing the SWAT model to predict runoff from variable source areas. Journal of Hydrology 348: 279-291.

65. Betrie G, Mohamed Y, van Griensven A, Srinivasan R (2011) Sediment management modeling in the Blue Nile Basin. Hydrology and Earth System Sciences 15: 807-818.

66. Kok K, Verburg P, Veldkamp T (2007) Integrated Assessment of the land system: The future of land use. Land use Policy 24: 517-520.

67. Lambin E, Rounsevell M, Geist H (2000) Are agricultural land use models able to predict changes in land use intensity? Agriculture, Ecosystem and Environment 82: 321-331.

68. Veldkamp A, Lambin E (2001) Editorial: predicting land use change. Agriculture, Ecosystem, and Environment 85: 1-6.

69. Zhang P, Liu Y, Pan Y, Yu Z (2011) Land use pattern optimization based on CLUE-S and SWAT model for agricultural non-point source pollution control. Mathematical and Computer Modeling 58: 588-595.

70. Pai N, Saraswat D (2011) SWAT2009_LUC: a tool to activate land use change module in SWAT 2009. Trans ASABE 54: 1649-1658.

71. Birhanu BZ (2009) Hydrological Modelling of the Kihansi River Catchment in South Central Tanzania using SWAT Model. International Journal of Water Resources and Environmental Engineering 1: 1-10.

72. Kim J, Choi J, Choi C, Park S (2013) Impacts of changes in climate and land

use/land cover under IPCC RCP scenarios on streamflow in the Hoeya River Basin, Korea. Science of the Total Environ 452: 181-195.

73. Li Z, Deng X,Wu F, Hasan SS (2015) Scenario analysis for water resources in response to land use change in the middle and upper reaches of the heihe river basin. Sustainability 7: 3086-3108.

74. Wagner PD, Bhallamudi SM, Narasimhan B, Kantakumar LN, Sudheer KP, et al. (2015) Dynamic integration of land use changes in a hydrologic assessment of a rapidly developing Indian catchment. Science of the Total Environment 539: 153-164.

3

Characterization of a Typical Mediterranean Watershed Using Remote Sensing Techniques and GIS Tools

Mohamed Elhag

Department of Hydrology and Water Resources Management, Faculty of Meteorology, Environment and Arid Land Agriculture, King Abdulaziz University, Jeddah, 21589, Kingdom of Saudi Arabia

Abstract

Geographic Information Systems (GIS) were used to establish an information data base to characterize a watershed in Northern Greece, analyze the distribution of the drainage network according to the different characteristics of the watershed using a drainage density index based on GIS. The drainage network was delineated from ASTER GDEM and Landsat-8 OLI data. Digital image processing was based on enhancement techniques. GIS characterized the watershed easily and efficiently. The drainage density index, based on the number of pixels, was appropriate for analyzing the distribution of the drainage network in relation to other characteristics of the watershed. The possibility of using GIS to generate buffer zones around linear and area features helped to quantify sensitive areas close to streams. The problem of cell resolution was overcome by reference to the mapping scale and other factors. Landsat-8 OLI data gave promising results, closely accurate to those from a 1:50,000 topographic maps. The number of streams and total stream lengths of all orders from ASTER GDEM data were higher than from the other sources. Geometric characteristics of the watershed derived from ASTER GDEM data were almost the same as from the 1:50,000 topographic maps. Best results were from a new band index based on Bands 2 and 5. Both techniques, GIS and Remote Sensing, are suitable for application to watershed management in the Mediterranean region.

Keywords: ASTER GDEM; Digital image processing; GIS; Mediterranean; Remote sensing; Watershed management

Introduction

A watershed is a land unit which drains into a stream system and includes a major part of the natural resources. From these resources, water is of vital importance; the development of a nation is intimately connected with its water resources. In the Mediterranean area, all natural resources are in strong relationship to each other's. However, their state is man-induced, not only because of the long history of this area but also because of the vital link between man and land in a sociodynamic context, which in turn has led to the degradation and erosion of soils and watersheds, as shown in studies in Egypt, Greece, Italy and Spain [1-5].

In particular, accelerated land or soil erosion, in contrast to physical erosion, especially in semiarid areas of the Mediterranean, is mainly due to the action of man on his natural environment [5]. In Greece, this problem has increased through the pressure of man on the natural resources that are located in the watershed. Over four million hectares have been either completely or partially eroded, and it is estimated that around 1,000 cubic meters per square kilometer of material is eroded by streams each year [6].

Sound management of degraded watersheds requires not only the study of their characteristics, but also the investigation of their interrelationships. The study of this balances, whether influenced or not, requires such dealing, because in non-induced conditions a watershed forms an ecologically balanced system. Watershed management deals with all land resources, such as forest lands, range lands, areas destroyed by erosion, or others that can serve as protection areas [7]. Like other kinds of management, watershed management needs the sophisticated tools which have been developed in recent decades, Remote Sensing and Geographic Information Systems (GIS) have an important contribution to make.

A water shed is suitable for applications of GIS and Remote Sensing not only because it represents a natural hydrologic and topographic unit [8,9], but also because watershed management is a broad field that includes a multitude of subjects in the biophysical domain [10]. These two techniques are the most suitable for natural resources management in general [11] and for watershed management specifically.

Superficial deposits and the solum overlying the solid rock are often described in terms of the percentage of several soil types extending over the area of the watershed [12]. The soil is important in watershed management, as it consists of mineral particles, organic material and moisture. It is distinguished from the parent material or rock below, in that it supports the roots of vegetation and the un decomposed organic material above. The depth, texture, structure and porosity are hydrologically important.

A watershed has not been regarded as an integrated system: characterization, analysis of the characteristics and their interrelationships, and emphasis on the problems threatening it. In a GIS, answers about how, why and what to characterize, analyze and emphasize in a watershed are needed. In particular, the drainage network, as a skeleton of the watershed, has to be studied and analyzed. Erosion is a main problem in the watershed, but, most studies have concentrated on the use of empirical formulae [13], such as the

*Corresponding author: Mohamed Elhag, Department of Hydrology and Water Resources Management, Faculty of Meteorology, Environment and Arid Land Agriculture, King Abdulaziz University, Jeddah, 21589, Kingdom of Saudi Arabia E-mail: melhag@kau.edu.sa

universal loss equation (USLE), which is not always the appropriate tool for erosion studies. Further, new developments in data collection systems such as the Landsat-8 OLI system may answer the need for drainage network delineation and update. As in the case of soil types, vegetation character needs to be specified precisely and this can often be achieved by distinguishing major land use types such as broad leaf, conifer, agricultural and urban. Vegetation is an essential component as through plants, land becomes productive. Vegetation protects soil against erosion, is one of the factors by which a watershed can be manipulated and regulates runoff and infiltration. In some cases, one particular form of land use may be of paramount significance. In this case, the percentage of this single type within a basin may be used as an index of land use character. The objective of this study was to investigate the watershed as an integrated system using Remote Sensing and GIS, taking the drainage network as the skeleton of the watershed.

Materials and Methods

Study area

The study area was the Olynthos watershed, located in the Halkidhiki peninsula, near Polygyros south east of Thessaloniki in Northern Macedonia, Greece (Figure 1). It lies approximately between 23°17' to 23°E and 40°15' to 40°30'N. The main river is called the Olynthos and the watershed area is approximately 244 km², with an altitude ranging from sea level to 1100 m. The area is of particular interest because of its typical Mediterranean bioclimate, characterised by hot and dry summers and cold humid winters, and because Halkidhiki is of particular interest to many natural resource managers because of manmade problems [4,14-19]. The area is also of interest for dam construction on the upper part of Olynthos River which carries large load of debris after heavy rains, the result of soil erosion, a serious problem in the area. Water supply can be also, at least in part, solved. The climate ranges from semiarid to temperate and humid Mediterranean

[20]. Data from the Taxiarchis meteorological station, located in the northern part of the watershed at an altitude of approximately 1000 m (outside the watershed) shows that the mean annual rainfall is 759.5 mm and the mean annual air temperature is 10.7°C. South of the area, the rainfall decreases and the temperature increases. The major part of the area is characterized by moderate humidity and temperature, a mild and wet winter and a warm and dry summer. The geology of the area varies from south to north. The southern part of the area is dominated by tertiary calcareous deposits; the northern part is mainly composed of phyllites, quartzites and epigneisses. Close to the sea, there are mainly alluvial deposits [21]. The geological structure and the numerous lithological types, combined with the effects of human activities on the land, are the primary factors in determining soil variability and productivity. Climatic and edaphic factors and human activities are important factors affecting vegetation in most parts of the study area. In particular, during the dry summer, only plants adapted to the unfavorable conditions are able to survive. Fire, wood cutting and heavy stock grazing are the factors contributing to the degradation of the vegetation stands, accounting for the dominant maquis vegetation which is mainly composed of sclerophyllous evergreen shrubs adapted to the Mediterranean ecosystem.

Data collection

Two field visits to the area were necessary to acquire general information on the Land Use Land Cover (LULC) characteristics of the area, and to become familiar with the aerial photographs in the field. Development of LULC classification categories and accuracies were based on the field trips. The determination of the various classification categories was based on the hydrological importance and the frequency of occurrence. The classification system included twelve categories; the methodology followed was that of Karteris and Pyrovesti [22] with some modifications (Figure 2). Delineation of the watershed boundary on the orthophotomaps, after enlargement of the topographic maps

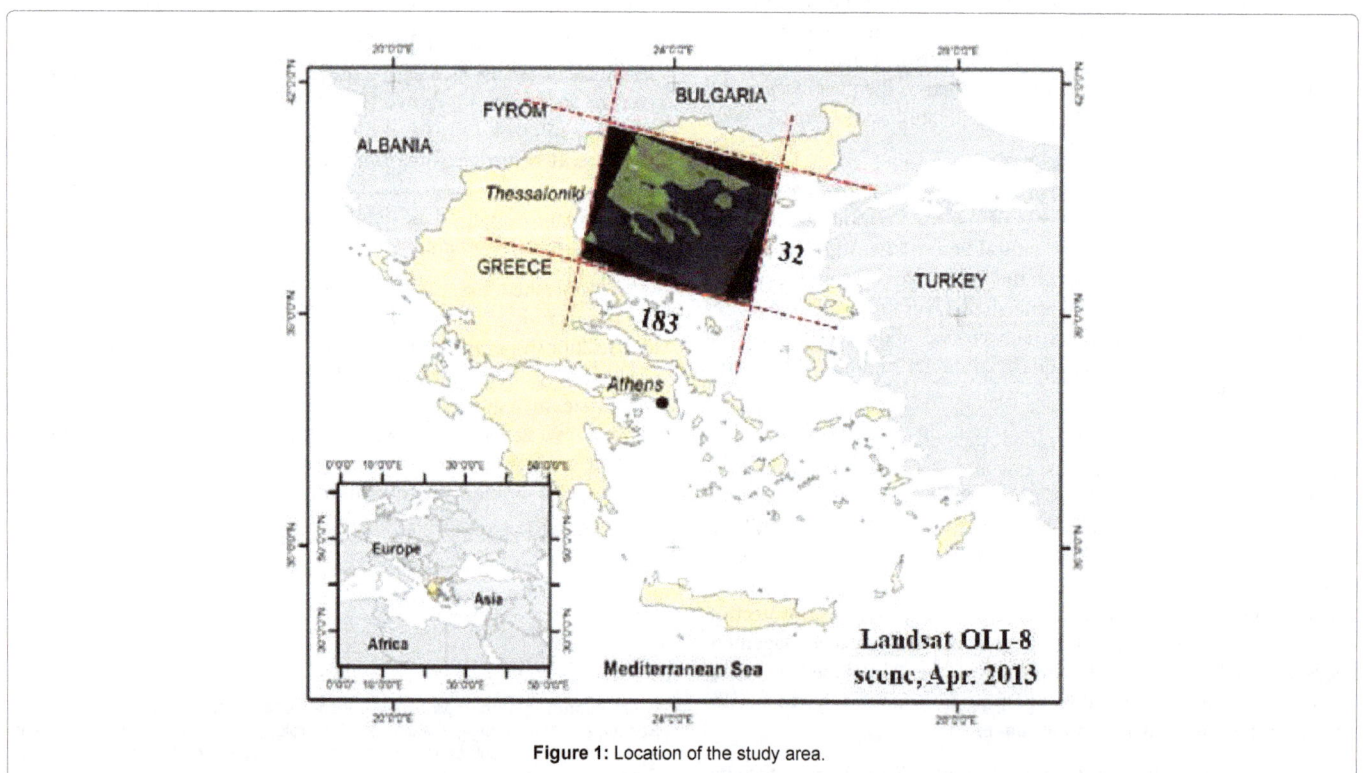

Figure 1: Location of the study area.

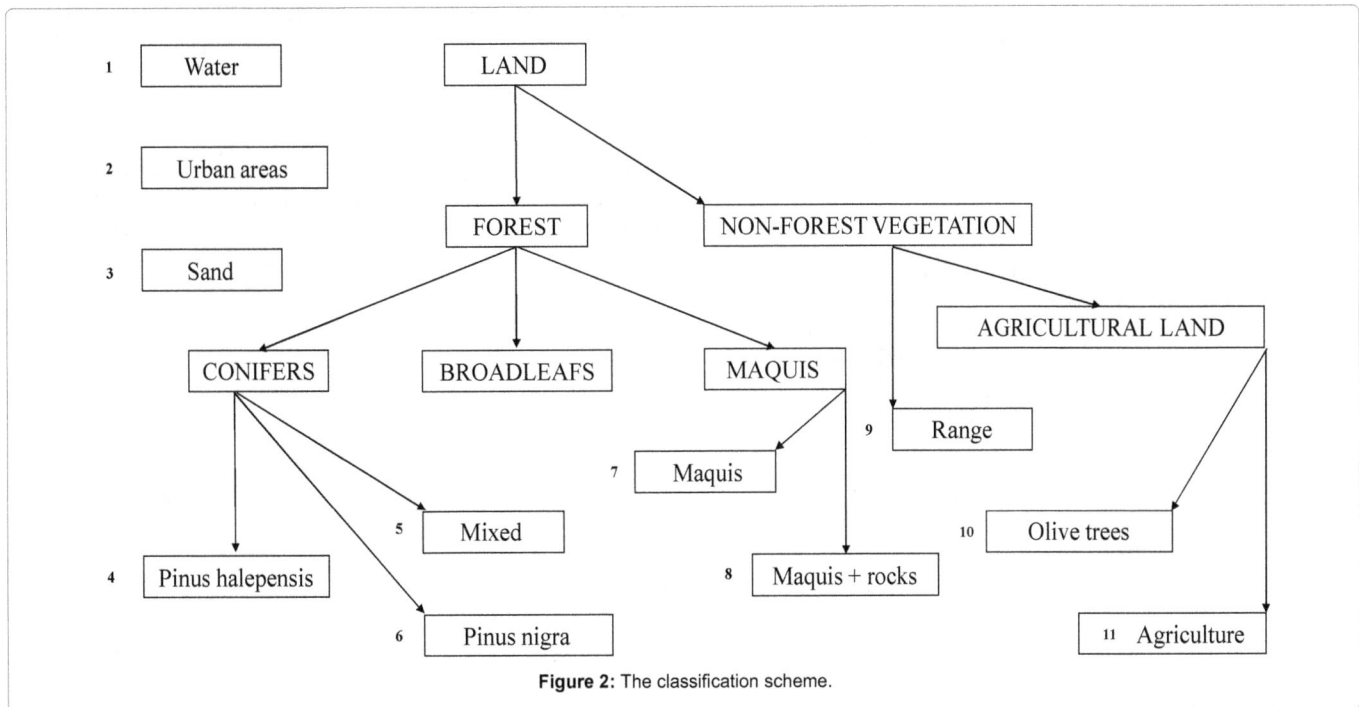

Figure 2: The classification scheme.

at 1:20,000 was carried out to cross check the outcomes of remote sensing data analysis. The orthophotomaps gave a preliminary LULC distribution. The preliminary map from the orthophotomaps was modified by the use of aerial photographs and adjusted to the new classification system.

Watershed characterization

This was based on the independent analysis of ASTER GDEM data. Under ArcGIS environment, Arc Hydro tools were specifically used. The Arc Hydro tools were used to derive several data sets that collectively describe the drainage patterns of a catchment. Raster analysis is performed to generate data on flow direction, flow accumulation, stream definition, stream segmentation, and watershed delineation [23]. These data are then used to develop a vector representation of catchments and drainage lines. Using this information, a geometric network is constructed.

Topographic characteristics

Area (A): the total watershed surface, generally expressed in square kilometers; Order (u): in a drainage network, there is a need to apply ordering procedures for further statistical analysis; many ordering systems have been developed. Strahler's [24] method was adopted for this study as it is the most objective and straight forward system [12,25]. Assuming that the channel-network map includes all intermittent and permanent flow lines located in clearly defined valleys, the smallest finger-tip tributaries are designed Order

1. Where two Order 1 channels join together, a channel segment of Order 2 is formed; where two of Order 2 joins together, a segment of Order 3 is formed and so forth [26]. Number of streams of order u (Nu) is the total number of streams of any order u. Stream length (Lu) is the length (in kilometers) of any stream of order u. Density of the stream network (Dd) is the sum of length of all streams per unit of drainage area: Dd=LLu/A. Drainage density is independent of order.

Digital image processing

Multiband enhancement band combination, principal component analysis (PCA) and band rationing were the main techniques applied to different bands. PCA was used to reduce image dimensionality and improve the appearance of the image on many bands together. Rationing consisted of dividing the digital number (DN) of each pixel in one band by the corresponding DN in another band. Normalized Difference Vegetation Index (NDVI) was created to enhance the drainage network, using the Algebra algorithm available in the ENVIS system (Band 4-Band 3/Band 4+Band 3). The ratio and index images were subjected to spectral and spatial enhancement to delineate the drainage network.

Results and Discussion

Supervised classification using different classification algorithms were performed. Both statistical and graphical analyses of feature selection were conducted. All visible and infrared bands (except for the thermal infrared band) were included in the analysis. SVM showed better classification results than the other classification algorithms [27]. Figure 3 shows that only 6 classification categories were performed following the adopted classification scheme. The absence of the other different categories is explained by limited surface heterogeneity found in the study area [27].

The overall accuracy assessments ranged from 78.25% to 98.51% and were considered reasonable for forestry purposes by Congalton [28] and Congalton and Green [29]. When Mahanaholbis distance was used the overall accuracy was the lowest (78.25%). The category of conifers was classified with an accuracy of 83.54% (*Pinus halepensis*). Also there were misclassification errors between the categories of *Pinus halepensis* and mixed conifers, Olive cultivation and agriculture according to the statistical separability index.

Support Vector Machine gave the highest overall accuracy (98.51%). The categories of *Pinus halepensis*, mixed conifers, olive trees,

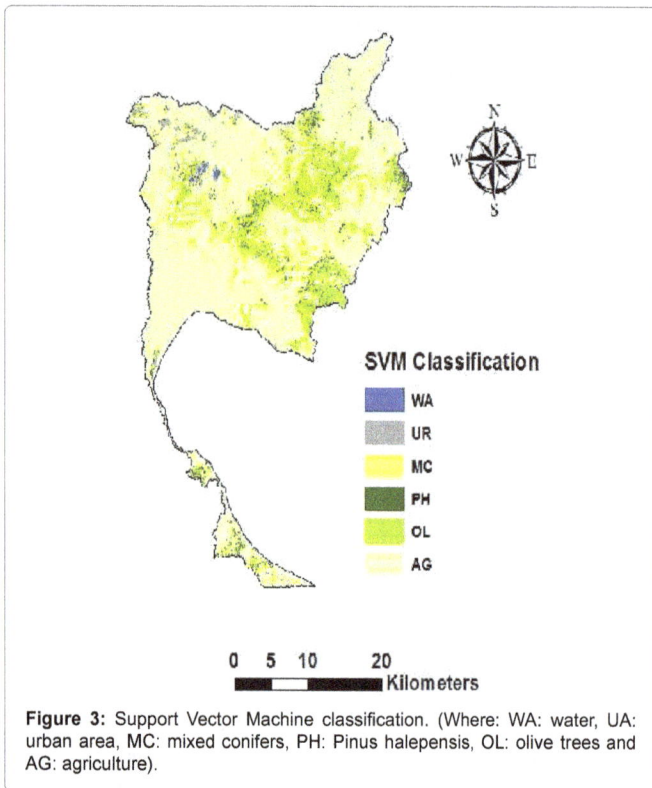

Figure 3: Support Vector Machine classification. (Where: WA: water, UA: urban area, MC: mixed conifers, PH: Pinus halepensis, OL: olive trees and AG: agriculture).

slope category report and tabulated in degrees (Table 1) then converted to percentage and reclassified (Table 2). The slope categories scheme deviates from that of Nakos [33], in which: 0 to 40%: gentle, 41 to 75%: middle, 76% and higher: severe.

Total number of pixels in each altitude category is given in the last column of Table 3, and the drainage density (Ddc) calculated for each altitude category (Table 4), computed as the total number of pixels in all drainage network tributaries divided by the area of each altitude category. The variation of the drainage density index in relation to altitude is explained as following: at the highest elevation, this index is low because of the presence of dense forest cover in these parts of

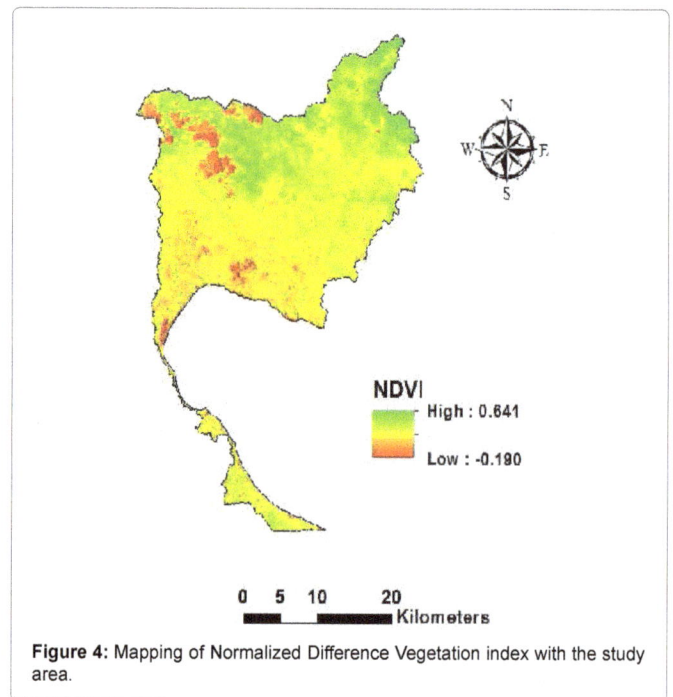

Figure 4: Mapping of Normalized Difference Vegetation index with the study area.

agriculture and Urban gave the highest producer's accuracy amongst the various channels combinations.

The NDVI composes a measurement for the photosynthetic activity and is strongly in correlation with density and vitality of the vegetation. The normalizing reduces topographically and atmospheric effects and enables the simultaneous examination of a wide area. Red and Near Infra-Red reflectance measurements are used to compute a vegetation index such as NDVI [30]. Figure 4 demonstrated that the NDVI values decreased gradually from the upstream toward downstream.

Fills sinks in a surface raster to remove small imperfections in the data. A sink is a cell with an undefined drainage direction; no pixels surrounding it are lower. The pour point is the boundary cell with the lowest elevation for the contributing area of a sink. If the sink were filled with water, this is the point where water would pour out. Digital Elevation Model shown in Figure 5, was undergone fill sinks for better watershed characterization results [31,32].

The concept is to create a raster of flow direction from each cell to its steepest downslope neighbor as it illustrated in Figure 6. If a cell has the same change in z-value in multiple directions and is not part of a sink, the flow direction is assigned according to the most likely direction.

The tool is to create a raster of accumulated flow into each cell. A weight factor can optionally be applied. Pixels of undefined flow direction will only receive flow; they will not contribute to any downstream flow. A cell is considered to have an undefined flow direction if its value in the flow direction raster is anything other than E, SE, S, SW, W, NW, N, or NE. Figure 7 shows that there are two main accumulated flows located within the study area. From the digital elevation model (DEM), both slope and aspect were automatically obtained from Arc Hydro Tool. Slope categories were obtained from

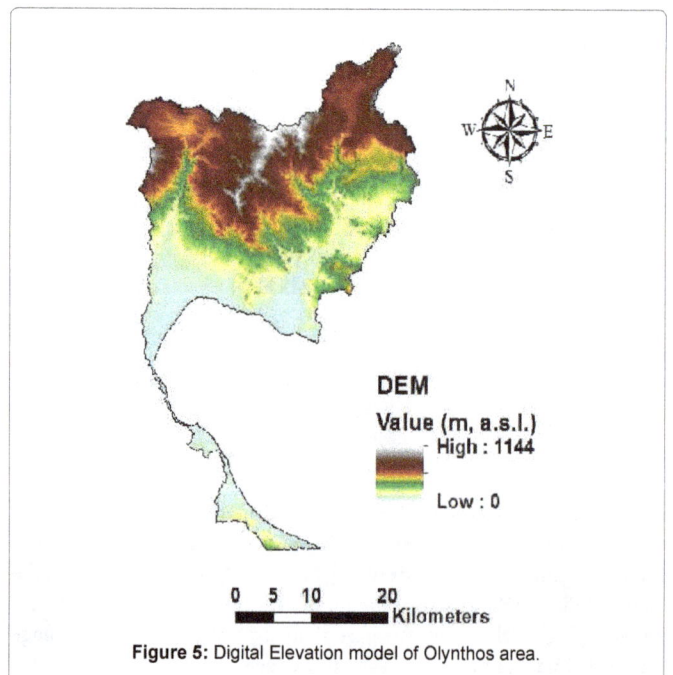

Figure 5: Digital Elevation model of Olynthos area.

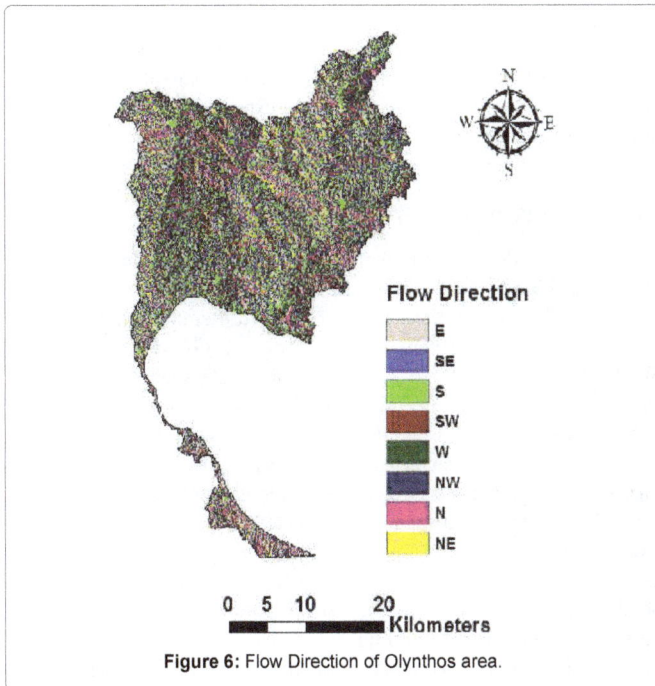

Figure 6: Flow Direction of Olynthos area.

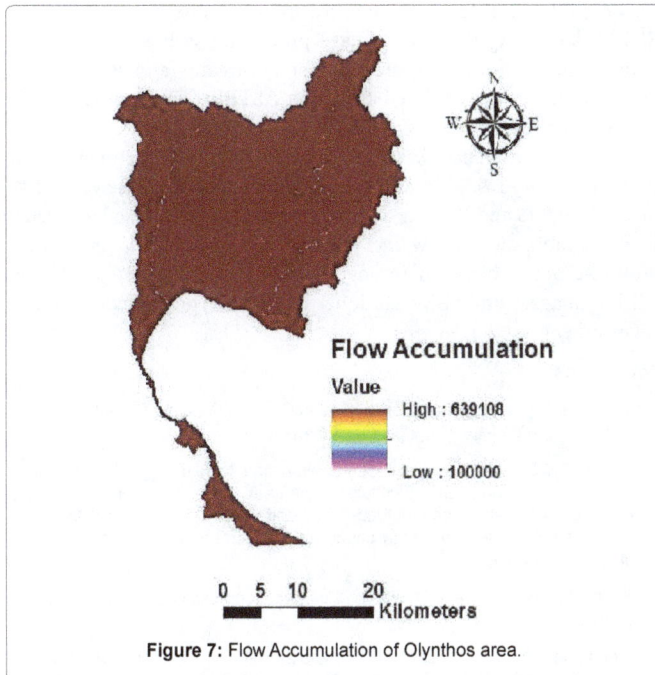

Figure 7: Flow Accumulation of Olynthos area.

the watershed. At middle elevations, the value reaches a maximum and then decreases, due the increase of the area of altitude categories.

The tool is to calculate the upstream or downstream distance, or weighted distance, along the flow path for each cell. A primary use of the Flow Length tool is to calculate the length of the longest flow path within a given basin. This measure is often used to calculate the time of concentration of a basin. In Figure 8, downstream distance in (m) of the designated study area of Oylnthos is demonstrated.

The output of Stream Order will be of higher quality if the input stream raster and input flow direction raster are derived from the same surface. If the streamraster is derived from a rasterized streams

dataset, the output may not be usable because, on a cell-by-cell basis, the direction will not correspond with the location of stream pixels. The results of the Flow Accumulation tool can be used to create a raster stream network by applying a threshold value to select pixels

Slope category	Percentage	km²	Number of 30 × 30 m pixels
0-15%	50.20	122.39	48957
16-40%	2.47	6.01	2406
41-65%	1.13	2.76	1103
66-85%	2.38	5.81	2324
86% and higher	43.82	106.82	42729
Total	100.00	243.80	97519

Table 1: Slope categories of the watershed: percentages of total, km² and number of 30 × 30 m pixels.

Aspect category	Percentage	Km²	Number of 30 × 30 m pixels
No aspect	50.20	122.39	48957
Southeast facing	1.55	3.77	1508
East facing	9.95	24.26	9704
Northeast facing	6.83	16.65	6660
North facing	6.37	15.54	6215
Northwest facing	5.52	13.47	5387
West facing	6.05	14.75	5900
Southwest facing	7.00	17.08	6831
South facing	6.52	15.89	6357
Total	100.00	243.80	97519

Table 2: Aspect categories of the watershed: percentages of total, km² and number of 30 × 30 m pixels.

Altitude category	1st	2nd	3rd	4th	5th	Total
1	4	0	0	0	0	4
2	52	0	0	0	0	52
3	269	40	0	0	0	309
4	642	213	45	0	0	900
5	1216	575	178	66	0	2035
6	1659	981	320	189	11	3193
7	803	534	369	222	16	1944
8	391	171	130	92	87	871
9	409	140	119	74	86	828
10	632	337	140	89	282	1480
11	274	299	311	30	154	1068
Total	6351	3290	1612	762	471	12739

Table 3: Distribution of the drainage network tributaries according to altitude.

Category	Ratio values	Ddc value
1	4/224	0.018
2	52/692	0.075
3	309/4242	0.073
4	900/8816	0.102
5	2035/16442	0.124
6	3160/22644	0.139
7	2082/13438	0.155
8	871/5861	0.149
9	846/5051	0.167
10	1284/8348	0.154
11	1196/1176	0.102

Table 4: Drainage density index computed for the altitude categories of the watershed.

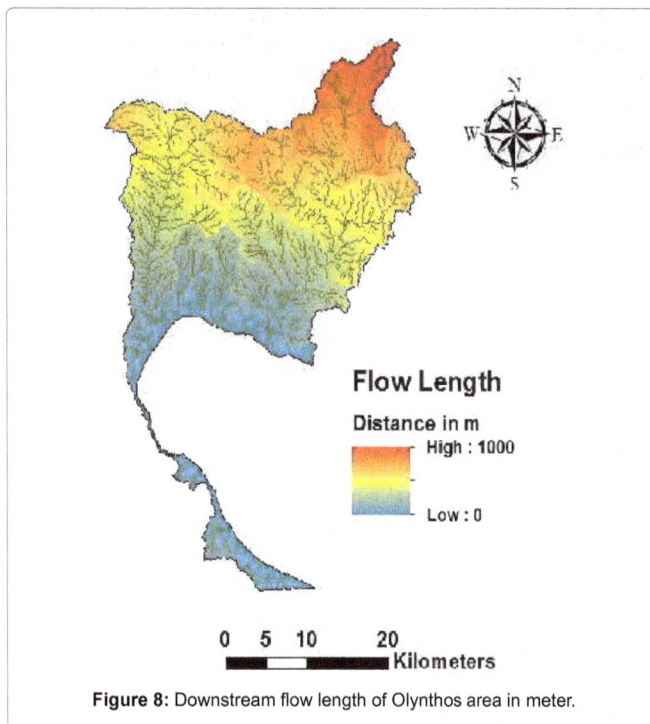

Figure 8: Downstream flow length of Olynthos area in meter.

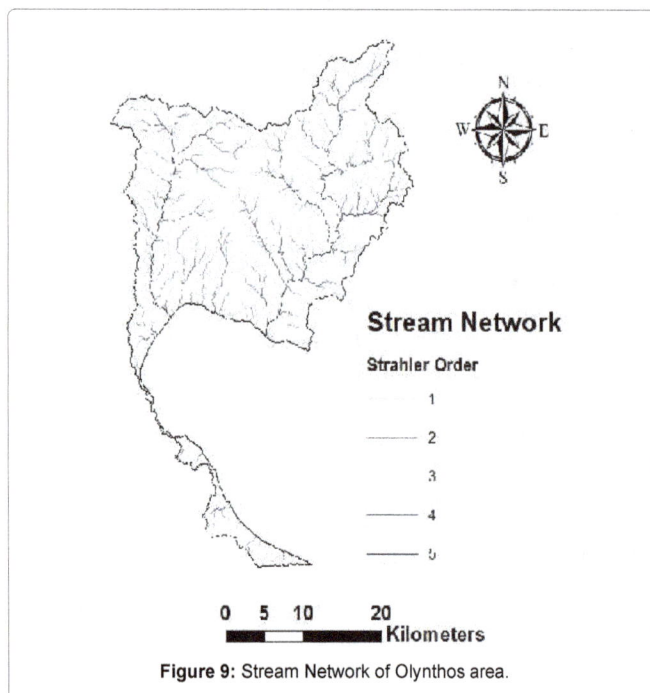

Figure 9: Stream Network of Olynthos area.

three sources. The lowest value was obtained from the Landsat OLI-8 data, which may have been due to the few streams all orders. The highest bifurcation ratio (Rb=6), which was obtained in the fourth order of the Landsat OLI-8 image, can be explained by the low number of fourth order streams. Bifurcation ratios (Rb) from the topographic maps (1:50,000 and 1:100,000) and Landsat OLI-8 image.

Conclusions and Prospects

Watershed degradation; which lead to reduced productivity, even on marginal lands, needs information which should be supplied in an accurate and timely effective manner. GIS and Remote Sensing are the best answer to this problem. The application of GIS using ArcMap in this study shows that characterization of a watershed can be simple and efficient. The drainage index, based on the number of pixels, was a good indicator of the distribution of the drainage network tributaries in relation to the characteristics of the watershed. It substituted the former drainage index based on measurement of total stream length which is a time consuming method. The ability of ArcMap to generate buffer zones around streams permitted the quantification and the assessment of the impact of land use activities. Better results may be obtained if a complex model for erosion potential assessment is used. This model integrated several outputs to locate precisely water stream networks and sink areas, most of which were located on agricultural land and Olive cultivation, indicating their sensitivity and the need for erosion protection. The drainage network was delineated from ASTER GDEM data using the ENVI image processing system. Solid results were obtained; moreover, the number of streams and total stream length in all orders obtained from the ASTER GDEM were closely equaled to those from the 1:50,000 topographic maps. A common Normalized Difference Vegetation Index for image enhancement gave satisfactory results, mainly at high altitudes with dense vegetative cover. Both GIS and Remote Sensing techniques can be used separated or in combination in watershed analysis. Future research must study the possibility of mapping and monitoring soil erosion from satellite imagery and GIS techniques in such a typical Mediterranean environment.

References

1. Vouzaras G (1988) Soil erosion and watershed restoration in Greece. National Report to the Ministry of Agriculture, Athens: p. 8.

2. Chisci G (1980) Physical soil degradation due to hydrological phenomena in relation to change in agricultural systems in Italy. Paper presented at a meeting on Research on indigenous soil and water conservation (SWC) and water harvesting in developing countries. Mediterranean Agronomic Institute of Chania, Greece: p. 13.

3. Rubio JL, Sala M (1991) Erosion and degradation of soil as a consequence of forest fire.

4. Stephanidis P, Kotoulas D (1992) Accelerated erosion after the forest fires in Greece. Internationales Symposion Interpravent Bern Tagungspublikation 1: 365-376.

5. Elmahdy SI, Huat BB, Mahmod AR (2012) Structural geologic control with the limestone bedrock associated with piling problems using remote sensing and GIS: a modified geomorphological method. Environmental Earth Sciences 66: 2185-2195.

6. Ministry of Agriculture of Greece (1988) Greece.

7. FAO (1976) Conservation in arid and semi-arid zones. FAO Cons. Guide 3. FAO, Rome, p. 125.

8. Salamonson VV, Rango A (1980) Water resources. In: Salamonson VV (ed), Remote Sensing In Geology. John Wiley and Sons, New York: 607-633.

9. Meijerink AMJ (1990) Summary report on ILWIS development. ITC Journal, 3: 205-214.

with a high accumulated flow. An analytical method for determining an appropriate threshold value for stream network delineation is presented in Tarboton [34]. Stream order only increases when streams of the same order intersect. Therefore, the intersection of a first-order and second-order link will remain a second-order link, rather than creating a third- order link as it shown in Figure 9.

The bifurcation ratio, is the total number of streams in any order (u) divided by the total number of streams of the next order (u+1) (Rb=Nu/Nu+1), showed that there was no great difference between the

10. Meijerink AMJ (1988) Remote sensing applications to watershed management. In: Sarma A (ed), Remote sensing applications to water resources. FAO, Rome: 229-280.

11. Karteris MA, Gatzoyannis S, Galanos F (1988) Use of second generation earth observation satellites in the implementation of forestry management models in Greece. Final Report to National Forest and Nature Agency, Danish Ministry of Environment p. 93.

12. Jenson SK, Domingue J (1988) Extracting Topographic Structure from Digital Elevation Model Data for Geographic Information System Analysis. Photogrammetric Engineering and Remote Sensing 54: 1593-1600.

13. Bocco G, Valenzuela CR (1988) Integration of GIS and image processing in soil erosion studies using ILWIS. ITC Journal 4: 309-319.

14. Papamichos N, Alifragis D (1980) Drought damage to plantations, in Polygyros, Halkidhiki. Topos and Parartima 4: 77-106.

15. Vardavakis E, Pavlides G, Lavrentlades G (1987) On the vegetation of a typical xerothent soil of Polygyros area (SE Thessaloniki). Feddes Repertorium 98: 253-264.

16. Papanastasis V (1988) Rehabilitation and management of vegetation after wild fires in maquis- type brushlands. Anatupo apo to tefhos 2: 77-90.

17. Pavlides G, Vardavakis E, Lavrentiades G (1988) Notes on the vegetation and soil profiles near Polygyros (Halkidiki Peninsula, Northern Greece). Journal of Botany 37: 19-47.

18. Kritikos G, Karteris MA (1991) Assessment of forest fire damages in Greece using remote sensing techniques. Symp. On Bilateral Cooperation in Environmental Resarch, Germany: p 6.

19. Karteris MA, Meliadis IM, Kritikos GS, Nikolaidis A (1992) Forest classification and mapping of Mediterranean Forests and Maquis with satellite data. Report to the Mediterranean Agronomic Institute of Chania, Greece: p. 153.

20. Carranza EJ (2009) Objective selection of suitable unit cell size in data-driven modeling of mineral prospectivity. Computers & Geosciences 35: 2032-2046.

21. Rao SY, Jugran KD (2003) Delineation of groundwater potential zones and zones of groundwater quality suitable for domestic purposes using remote sensing and GIS. Hydrogeolgical Science Journal 48: 821-833.

22. Karteris MA, Pyrovesti M (1986) Land cover/use analysis of Prespa National Park, Greece. Environmental Conservation 13: 319-330.

23. Fujisada H, Bailey GB, Kelly GG, Hara S, Abrams MJ (2005) ASTER DEM Performance. IEEE transactions on Geoscience and Remote Sensing 43: 2707-2714.

24. Strahler AN (1957) Quantitative analysis of watershed geomorphology. Transactions - American Geophysical Union 38: 913-920.

25. Dahal RK, Hasegawa S, Nonomura A, Yamanaka M, Masuda T, et al. (2008) GIS-based weights-of-evidence modelling of rainfall-induced landslides in small catchments for landslide susceptibility mapping. Environmental Geology 54: 311-324.

26. Jha MK, Peiffer S (2006) Applications of Remote Sensing and GIS Technologies in Groundwater Hydrology: Past, Present and Future. Bayreuth University Press, Bayreuth, Germany.

27. Elhag M, Psilovikos A, Sakellariou M (2013) Detection of Land Cover Changes for Water Recourses Management Using Remote Sensing Data over the Nile Delta Region. Environment, Development and Sustainability 15: 1189-1204.

28. Congalton R (1991) A review of assessing the accuracy of classification of remote sensed data. Remote Sensing of Environment 37: 35-46.

29. Congalton R, Green K (2008) Assessing the Accuracy of Remotely Sensed Data: Principles and Practices. 2nd Edition. CRC/Taylor & Francis, Boca Raton, FL pp. 183.

30. Elmahdy SI, Mostafa MM (2013) Remote sensing and GIS applications of surface and near-surface hydromorphological features in Darfur region, Sudan. International Journal of Remote Sensing 34: 4715-4735.

31. Elhag M, Bahrawi J (2014a) Cloud Coverage Disruption for Groundwater Recharge Improvement Using Remote Sensing Techniques in Asir Region, Saudi Arabia. Life Science Journal 11: 192-200.

32. Elhag M, Bahrawi J (2014b) Conservational use of remote sensing techniques for novel rainwater harvesting in arid environment, Environmental Earth Sciences 72: 4995-5005.

33. Nakos G (1983) The land resource survey of Greece. Journal of Environmental Management 17: 153-169.

34. Tarboton DG, Bras RL, Rodriguez–Iturbe I (1991) "On the Extraction of Channel Networks from Digital Elevation Data." Hydrological Processes 5: 81-100.

4

Geoelectrical Exploration in South Qantara Shark Area for Supplementary Irrigation Purpose-Sinai-Egypt

Mostafa Said Barseem*, Talaat Ali Abd El Lateef, Hosny Mahomud Ezz El Deen and Abd Allah Al Abaseiry Abdel Rahman

Geophysical exploration department, Desert Research Center, 1 Matahaf El Matariya, Cairo, Egypt

Abstract

This research paper is dealing with Geoelectrical Exploration as a Geophysical method used, Vertical Electrical Sounding (VES) and 2D profile imaging to find a solution of the problems affecting the research station in South of Qantara Skark. This research station is one of the desert research center stations used to develop the desert for agriculture. The area of study is suffering from the shortage of irrigation water whereas, it depends on the water flow of the tributary of Salam Canal which being not available all the time. The appropriate solutions of these problems have been delineated by the results of 1D and 2D geoelectrical measurements. It exhibits the subsurface sedimentary sequences and extension of subsurface layers in horizontal and vertical directions especially in the groundwater aquifer. Moreover, the most suitable locations of drilling water wells could be detected. The surface and subsurface layers of the quaternary deposits consists of sand, sandy clay and clay facies.

Nineteen Vertical Electrical Sounding (VES) are arranged as a grid to cover the study area and two 2D geoelectrical imaging profiles are acquired. The results are represented through different contour maps and cross sections that exhibit the horizontal distribution of successive layers which reflect the lithology and changes in all directions. The water bearing layers consisted of two zones. The upper one was less salty than the lower one. The thickness of the upper zone ranges from 5 to 7 meters, but the lower zone ranges between 15 and 30 meters. The last detected layer is clay that decreases in depth towards the Southwest of the study area, causing the phenomenon of water logging. The thickness of the upper zone of the water bearing layer is inadequate for irrigation. Recommended basins to be constructed and filled through nearby drilled wells to overcome this problem. The most suitable location to dig a channel for water drainage is in the Southwest, where there is a less depth to the clay layer and all the layers are dipping toward this side.

Keywords: Supplementary irrigation; Vertical electrical soundings (VES); Electrical resistivity tomography (ERT); Qantara shark

Introduction

Research stations of desert research center are considered as productive stations to solve most agriculture problems. One of these stations is South Qantara Shark that constructed in West Sinai. The present study area of this station has length reach 1600 m. and width 850 m nearly eight hundred Feddan in West of the Sinai region at East Qantara. It lies east Suez Canal between latitudes 30° 47' and 30° 49' N and longitudes 32° 27' and 31° 24' E and act as a model for neighboring areas (Figure 1). This station suffers from shortage water supply for agriculture in some seasons especially summer whereas it depends on one of tributary El Salam canal that not full of water all time. There is one drilled well for human activity just South of the study area. Because of not existence good drainage system, some patches of water logging appear in low land at Southwest of area. Geoelectrical resistivity techniques are used in the present study to deal with pervious mention conditions.

The geoelectrical resistivity survey technique is used to solve many problems related to groundwater assessment, investigation, exploration and salinity. Some uses of this method in groundwater are; determination of the thickness, boundary and depth of different layers of the aquifer [1], determination of boundary line between saline water and fresh water [2,3], exploration of ground water quality [4,5] and detecting the impact of geologic setting on the groundwater occurrence [6]. Khaled, et al. [7] studied the impact of salt water intrusion on the groundwater occurrence.

The target of this study is solving the problems suppressed developments by carrying out geoelectrical techniques. It comprises a grid of Vertical Electrical Sounding (VES) covering the study area. It can be used to detect form results the successive subsurface layers horizontally and vertically, also detect the water bearing layers with depth to water and its flow direction. For delineation, the detail lateral changes in lithological content, two dimensional imaging profiles are carried out to detect the different zones of water quality according to resistivity values. Finally, it can be detect the best side for drilled productive wells, do suitable safe discharge for drilled wells and suppose a suitable drainage system to reduce or prevent the extension of marshes.

Northwestern Sinai is located within the semi-arid belt of Egypt and is locally affected by the Mediterranean climate. This aridity is manifested by the occurrence of sand dunes and sand sheets, saltmarshes and ponds as well as lack of vegetation.

Geomorphologic setting

Northwestern Sinai comprises five distinctive geomorphologic units according to Al Hussein [8]. It includes coastal area, El-Bardawil Lagoon, aeolian sand, mobile sand dunes and salt marshes and sabkhas.

*Corresponding author: Mostafa Said Barseem, Geophysical exploration department, Desert Research Center, 1 Matahaf El Matariya, Cairo, Egypt
E-mail: Barseem2002@hotmail.com

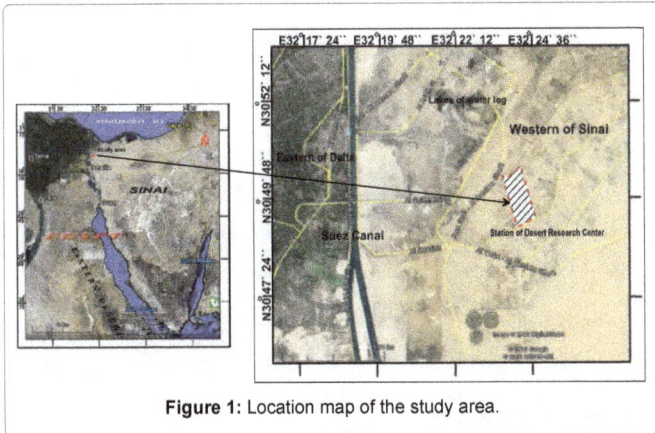

Figure 1: Location map of the study area.

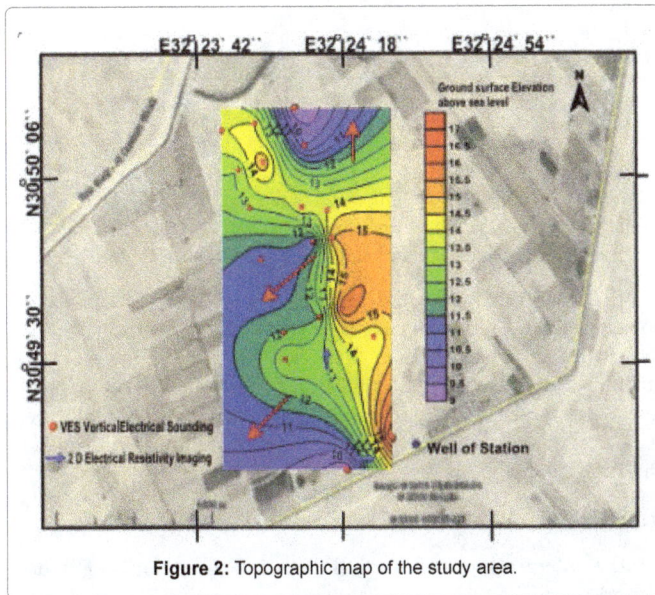

Figure 2: Topographic map of the study area.

The study area is entirely covered by quaternary sediments of littoral, alluvial and Aeolian origin which show a variation in their texture and composition ranging from unconsolidated sands to sand and clay. Ball and Said [9,10] studied the Northwestern part of Sinai that covered by aeolian sand and gravels with occasional clay interbeds of the Holocene and Pleistocene deposits. The sand dune deposits are deflected and diverted from Northwest to Southeast direction, most likely due to local winds. Sandy inland sabkhas are situated in low areas between hummocky surface and sand dunes. It formed as a result of high evaporation in low relief areas characterized by shallow groundwater and occasional rain fall water.

The study area is a flat to slightly undulated surface of aeolian sand with low ground elevation which ranges from 9 m to 17 m. a. s. l. (Figure 2). In low relief area at the Southwest direction, a Salt marshes and sabkhas are composed mainly from medium to coarse sands that are sometimes covered by salt crust.

Geologic setting

Northwestern Sinai is covered by quaternary deposits of littoral, alluvial and aeolian origin which show a variation in their texture and composition ranging from unconsolidated sands to sand and clay (Figure 3). The Pleistocene deposits include Sahl El-Tineh

formation (a mixture of black and white sands with silt), Al-Qantara formation (sand and grits with minor clay interbeds, coquina deposits, conglomerates) and alluvial hamadah deposits [11]. According to this map, the Holocene deposits are classified into coastal sand dunes which extend parallel to the Mediterranean Sea coast, inland sand dunes and sheets that cover large areas of Northwestern Sinai (the main water bearing formation for groundwater), coastal and inland sabkhas, and interdunal playa deposits; consist of fine sand and silt associated with evaporates [12]. The sand dune deposits change direction from northwest to Southeast, most likely due to local winds. To the West, near the Suez Canal, Northeast trending linear dunes grade progressively into cresentic (transverse and barchans dunes) and complex cresentic dunes that are homogenous and continuous.

Hydrogeologic setting

According to different authors such as Said, Shata A, El Shamy and Khaled [7,9,13,14] the Northwest of Sinai area is covered by quaternary deposits which are composed of sand, gravel, clay and sand dunes. Either clay or sand is saturated with saline water underlies the aquifer. The groundwater resource in the study area and its vicinity is represented by the unconfined aquifer of the quaternary deposits. There is a drilled well in the study area with total depth 21 m, depth to water 11 m and its salinity reached 2528 ppm.

Geoelectrical Studies

Geoelectrical field work in the area of study is represented by Vertical Electrical Soundings (VES) and Electrical Resistivity Tomography (ERT) profiles.

Methodology

The process of vertical electric sounding takes sequential measurements of the resistance by increasing the virtual distance between the poles of the current deployment, while the center of array and the trend remains constant. The ratio between the depth of

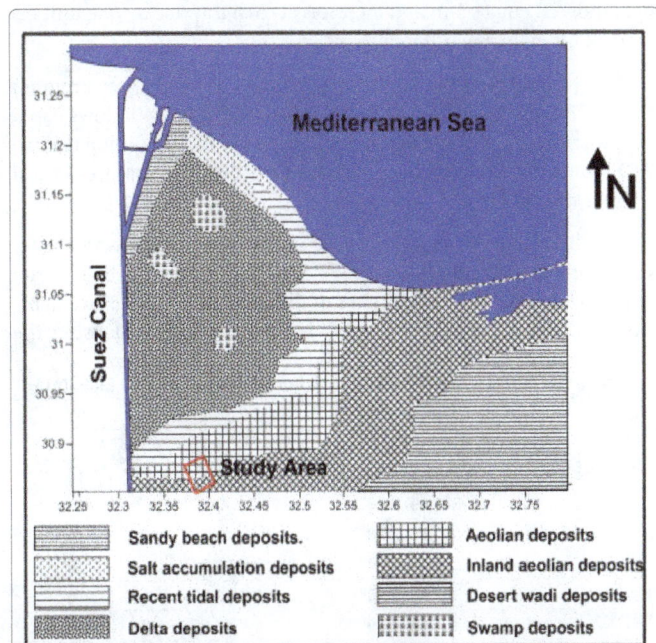

Figure 3: Geological map of the study area (After GSE 1992 and AL Hussein, 2012).

current penetration and the distance between the electrodes is called penetration factor. The depth of current penetration is of about (1/4 to 1/3) the distance between the poles of power. Rock resistance values are ranging from one to a few tens (ohm-m) in the mud and marl, and (10-1000) ohm-m in sand and sandstones.

Electrical Resistivity Tomography (ERT) is a useful tool to determine variations with depth in soil resistivity. The resistivity changes along the vertical and horizontal directions can be more accurate using the 2D model. The survey technique involves measuring a series of constant separation traverses.

Vertical electrical soundings (VES): The study area is covered by 19 Vertical Electrical Soundings (VES) (Figure 4). The Schlumberger configuration is applied in the present investigation. The current electrode separation (AB) start from 1 m and extended in successive manner to reach maximum distance 600 m. This electrode separation is found to be sufficient to reach reasonable depth range that fulfills the aim of the study. One of these soundings is conducted beside a drilled well in order to parameterize and verify the geoelectrical interpretation.

The RESIST computer program [15] is applied for the quantitative interpretation of the geoelectrical sounding curves. It is an interactive, graphically oriented, forward and inverse modeling program for interpreting the resistivity curves in terms of a layered earth model. An arbitrary initial model has been constructed in view of the overall shape of the sounding curves and refers to data of drilled well.

Electrical resistivity tomography (ERT): Two imaging profiles are conducted (Figure 4) to verify the results of Vertical Electrical Sounding (VES) especially the boundary between geoelectrical layers. One of these imaging profiles measures from West to East and other from South to North. The resistivity changes along the vertical and horizontal directions can be more accurate using the 2D model. The survey technique involves measuring a series of constant separation traverses. In the present study the Wenner electrode array is applied where the measurements start at the first traverse with a unit electrode separation "a" equals 5 m and increases at each traverse by one unit i.e. 10, 15, 20,. ., n. to reach 105 m.

For the interpretation of the imaging data, the computer program RES2DINV, ver 3.4 written by Loke [16] is used. It is a Windows based computer program that automatically determines a two-dimensional (2-D) subsurface resistivity model for data obtained from electrical imaging surveys [17].

The direct current resistivity meter "Tetrameter" model SAS 1000 C is used for measuring the resistance "R" with high accuracy. The accurate locations (coordinates) of the sites of the geoelectrical measurements and their elevations relative to sea level are determined using the

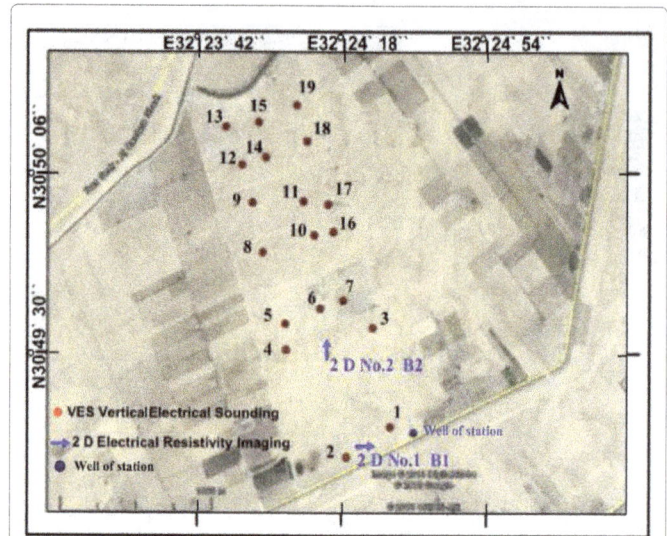

Figure 4: Location map of the Vertical Electrical Sounding (VES) and 2-D electrical resistivity imaging in the study area.

geographic positioning system (GPS) GPS apparatus (Trimble type) contact with nine satellites and topographic map scale 1:50,000. The obtained locations and ground elevations are listed in Table 1.

The interpretation of the geoelectrical resistivity data depends on determining and following up the geoelectrical parameters i.e. resistivities and thicknesses of a series of layers. The interpretation includes also correlation of similar layers where a layer or some layers may be absent because of lithology variations. The results are a geological model that can be reflected in terms of lithological variation and stratigraphy.

Results and Interpretation

The interpretation of Vertical Electrical Sounding (VES) data comprises qualitative and quantitative processes. The qualitative interpretation includes comparison of the relative changes in the apparent resistivity and thickness of the different layers. It gives information about the number of layers, their continuity throughout the area or in a certain direction and reflects the degree of homogeneity or heterogeneity of the individual layer. The quantitative interpretation, on the other hand, involves the determination of the number of the geoelectrical layers as well as the true depth, thickness and resistivity of each layer.

The computer program automatically determines a two-dimensional (2-D) resistivity model for the subsurface using the

VES No	Latitude (N.)	Longitude (E.)	Elev. (m.)	VES No.	Latitude (N.)	Longitude (E.)	Elev (m.)
1	30° 49' 15.3"	32° 24' 30.1"	17	11	30° 50' 0.1'	32° 24' 8.3"	14
2	30° 49' 9.3"	32° 24' 19.2"	9	12	30° 50' 7.4'	32° 23' 52.8"	13
3	30° 49' 35"	32° 24' 25.7"	14	13	30° 50' 15'	32° 23' 48.8"	14
4	30° 49' 30.6"	32° 24' 4.1"	13	14	30° 50' 8.9'	32° 23' 58.8"	15
5	30° 49' 35.8"	32° 24 3.8"	12	15	30° 50' 15.9'	32° 23' 57"	14
6	30° 49' 40.4"	32° 24' 18.3"	12	16	30° 50' 54.1'	32° 24' 15.8"	15
7	30° 49' 38.8"	32° 24' 12.5"	16	17	30° 50' 59.5'	32° 24' 14.1"	14
8	30° 49' 49.9"	32° 24' 58.2"	11	18	30° 50' 15.3'	32° 24' 9.1"	10
9	30° 49' 59.9"	32° 24' 55.5"	13	19	30° 50' 19.3'	32° 24' 6.5"	9
10	30° 49' 53.4"	32° 24' 11"	11				

Table 1: Coordinates of the measured sounding stations.

data obtained from the imaging survey. The results of interpreting geoelectrical field data indicate the following:

Interpretation of the vertical electrical sounding (VES) Data

The delineation of the subsurface sequence of the geoelectrical layers according to the qualitative and quantitative interpretation is as follow:

Qualitative interpretation: A preliminary qualitative interpretation of the sounding curves using partial curve matching [18] provides the initial estimates of the resistivities and thickness (layer parameters) of the various geoelectrical layers. The qualitative interpretation of the field curves (Figure 5) indicate generally QQ types of the vertical electrical sounding curves exhibiting homogeneity in resistivity values. It shows decrease in resistivity values with depth due to increases clay content with depth and high resistivity values reflect sand deposits. Generally, it can be detected from curves homogeneity in thickness but there is a variation at last cycle curves of VESes No 17 and 19 at North West direction

Quantitative interpretation: The quantitative interpretation involves the determination of the number of the geoelectrical layers as well as the true depth, thickness and resistivity of each layer. The geologic setting and relevant information are visualized and described in view of a number of generated geoelectrical cross sections crossing the concerned sites in different directions and contour maps. Figure 6 shows the interpretation of the modeled resistivity sounding VES No. 1 beside drilled well.

The results of the geoelectrical interpretation (Table 2) are correlated with available lithological information obtained from well that found in the study area. Three geoelectrical layers can be detected and their parameters (resistivity and thickness) are listed in Table 3.

The detailed interpretation results of the geoelectrical resistivity sounding measurements in the study area are discussed as follows:

The subsurface geoelectrical succession

The geoelectrical succession is formed of a number of layers being grouped together in three main layers In order to make the above mentioned description more illustrative the geoelectrical parameters of the interpreted layers. The first layer is surface layer "A". The second is

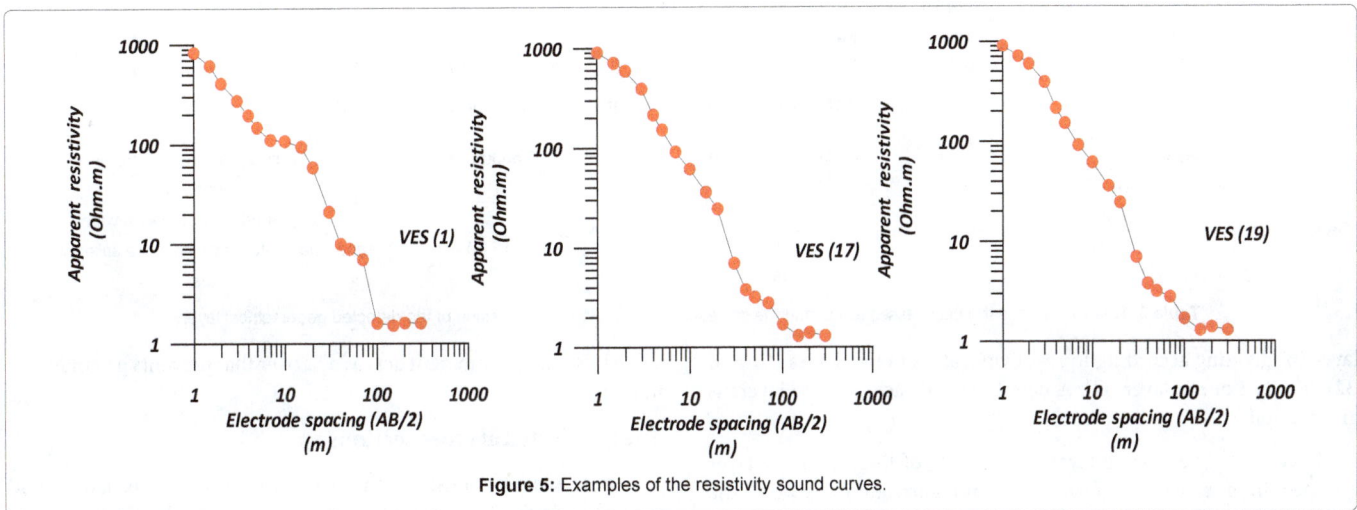

Figure 5: Examples of the resistivity sound curves.

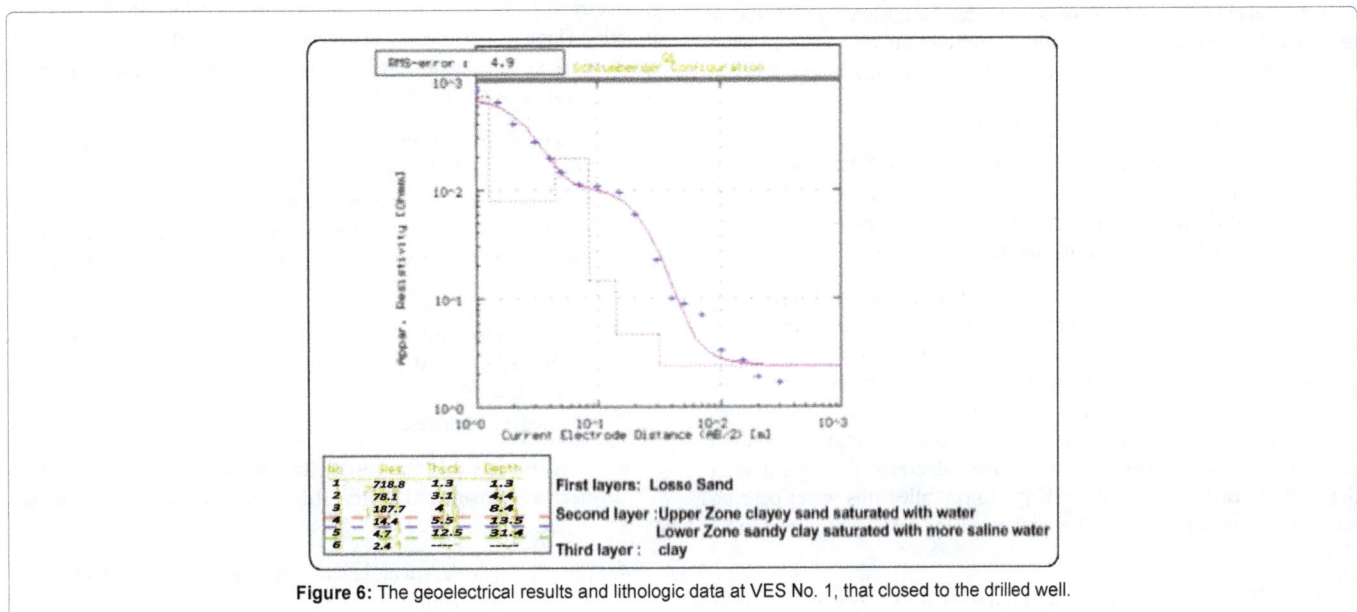

Figure 6: The geoelectrical results and lithologic data at VES No. 1, that closed to the drilled well.

VES No	Geoelectrical layer (A) (Dry loose sand)						Geoelectrical later (B) (Clayey sand saturated with water)				Geoelectrical layer (C) (clay)	
							Upper zone (B1)		Lower zone (B2)			
	$\rho1$	h1	$\rho2$	h2	$\rho3$	h3	$\rho4$	h4	$\rho5$	h5	$\rho6$	h6
1	718.8	1.3	79.1	3.1	197.7	4	14.4	5.5	4.7	17.5	2.4	-
2	600.2	1	57.5	1.7	168	0.7	13	5.5	4.1	16.3	0.7	-
3	736.7	0.4	252.8	2.3	137.8	5.9	37.7	6.2	4.1	16.6	1.2	-
4	156.6	3.2	11.2	1.5	92.8	3.4	17.6	5.5	8.6	18.8	1.4	-
5	595.9	0.6	217.2	2.1	118.5	2.4	18.2	5.2	7.6	19.6	1.6	-
6	656	0.9	367.5	1.1	213.1	5.4	23.9	5.6	6	14.1	1.3	-
7	1200.3	0.4	431.9	1.7	108.7	6.2	20.6	7.2	9.3	16.4	1.6	-
8	1999.6	0.4	146.9	1.6	80.5	3	19	5.3	6.8	18.6	1.4	-
9	1864.5	0.5	317.1	0.9	89.6	4.8	17.3	5	3.8	18.8	1.3	-
10	876.5	0.5	382.6	1.5	133.3	3	32	4.8	5.1	14.3	1.7	-
11	2607.7	0.4	215.9	1.4	84.3	5.5	13.7	5.8	7.9	22	2.3	-
12	1169.9	0.4	223.6	1.2	141.2	3.4	31.8	4.8	8.4	18.6	2.1	-
13	803.5	0.6	91.4	1.1	118.8	6.4	37	5.4	9.8	29	1.8	-
14	817.6	0.7	116.6	1	162.1	5.9	37.5	5.5	7.6	18.1	2	-
15	1407.8	0.3	408	1.9	159.2	6.2	33.7	5.3	6.2	16.5	1.4	-
16	1300.9	0.6	449.3	1	133	6	31	5.2	8.2	21	1.5	-
17	2015.6	0.4	246	0.4	90.4	7.5	30	4.9	7.7	27	0.8	-
18	3648.3	0.3	188.1	1.4	87.7	4.8	28.7	5.1	7.7	22	0.9	-
19	1022.2	0.7	670.6	0.7	103.1	4	34.5	4.7	6.7	16.5	1.3	-

Table 2: Resistivities and corresponding thicknesses at vertical electrical sounding stations.

Layer No.		Resistivity (Ohm.m)	Thickness	Corresponding lithology
Geoelectrical layer (A)		>50	1-5	Dry loose sand
Geoelectrical layer (B)	Upper zone (B1)	10-50	5	Clay Sand saturated with water
	Lower zone (B2)	5-10	15-30	sandy clay saturated with more saline water
Geoelectrical layer (C)		<5	---	Clay

Table 3: Resistivities and the thicknesses range and the corresponding lithologic composition of the detected geoelectrical layers.

layer "B" dividing according to resistivity values in two zones (B1 and B2). The last one is layer "C". A description of each of these layers is given as follow:

Layer "A": The surface layer (A) consists of three thin dry layer grouped in one layer to reach optimum correlation between the geoelectrical layers and the predominant geologic units. The resistivity of such a layer is plausibly expressed in terms of the average transverse resistivity (ρt) (Equation 1) [19]. This parameter can be calculated from the resistivities and thicknesses of the group of thin layers as follows

$$pt = \sum(\rho i.hi)/\sum hi....i=1 \text{ to } n \quad (1)$$

Where; ρi is the resistivity of the ith layer, hi is its thickness and n is the number of layers. This layer formed from sand and clay content as the data of drilled well and the exhibited lithological surface with resistivity larger than 50 Ohm.m. and their thickness not exceed 12 m.

Layer "B": The second layer (B) consists of two zones (B1 and B2) according to resistivity values and formed of saturated sand. The upper zone (B1) exhibits resistivity less than 50 Ohm.m and thickness reached 5m. But the lower zone (B2) represented resistivity ranges from 5 to 10 Ohm.m and thickness varies 15 - 30m. The changes in resistivity values reflect the groundwater quality so the lower zone (B2) more saline than the upper one (B1). The resistivity values decrease downward with depth then reach to the last layer. It can be called this water bearing layer (B) as a perched aquifer due to the presence of lower impermeable clay layer.

Layer "C": The last detected layer (C) consists of clay with resistivity

value less than 5 Ohm.m. It acts as a barrier that prevents groundwater more passage.

The geoelectrical cross sections

These sections illustrate the geoelectrical sequence, lateral and vertical variation for different layers along the profile direction. Two geoelectrical cross sections are constructed, section South - North direction (Figure 7) while, the other section has the direction of West - East (Figure 8). The detailed description of the geoelectrical layers from top to bottom can be described as follows:

1. Generally, the geoelectrical cross sections (A-A') and (B-B') (Figures 7 and 8) consist of three geoelectrical layers "A", "B", and "C". The surface layer (A) is formed of sand deposits, the second layer (B) is divided into zone (B1 and B2) acting as water bearing. It is formed of sand and the last layer (C) consists of clay.

2. These layers exhibited a regular thickness although changed in the relief of last clay layer at VESes No. 17 and 19 in the Northwest direction of the study area due to lateral in lithological changes.

3. It is noticed that the resistivity values of geoelectrical layers decrease downward in the study area due to increasing clay intercalation.

4. The first geoelectrical layers decrease in their thicknesses Southward direction.

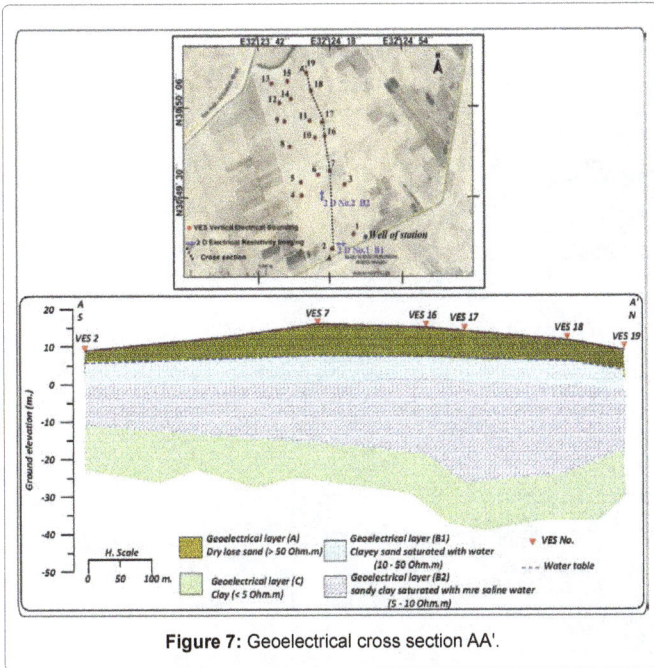

Figure 7: Geoelectrical cross section AA'.

5. The thickness of saturated water bearing layer (B) is generally increases toward the Southeast direction.

6. The groundwater is shallower toward the South due to low ground elevation but it becomes deeper at VESes No. 6, 16 and 17 in center of the study area.

Interpretation of the electrical resistivity tomography (ERT) data:

Examination of the imaging profiles at the selected two sites (B1 and B2) in the study area (Figures 9 and 10) indicates deposits of high resistivity corresponds to the dry coarse grain sand dominate in the upper parts. These images shows obvious downward resistivity decrease that represents saturated sand deposits. It can be detected two geoelectrical layers. The upper layer acts as surface layer of the study area with high resistivity values (larger than 70 Ohm.m) reflecting homogeneity content of dry sand. The lower layer is saturated with water and divided into two zones according to resistivity differentiation. The first zone exhibited resistivity values ranges from 10 to 50 Ohm.m and relatively high than the second zone which exhibits resistivity values less than 10 Ohm.m. This means that the lower zone is more saline than the upper one.

The depth to water reaches 8 m from the ground surface. The thickness of the upper zone is 5 m and composed of sand and clayey sand. The thickness of the lower zone is not detected and composed of sandy clay and clay. These results are compatible with the interpretation of Vertical Electrical Sounding (VES).

Groundwater Occurrences

According to the limited hydrogeological information in the study area, the applied geoelectrical methods are integrated to collect the common features that may suggest groundwater occurrence. The quaternary aquifer dominates the area of study consisting of sand gravel (unconfined aquifer). There is one drilled well in area of study with total depth 21 m. and depth to water reach 11 m. Its salinity has record 2528 ppm. As mention above, the study area belong research station for agriculture and depends on the tributary of El Salam Canal during irrigation. The problem in this area that the water for irrigation not sufficient. So, the results of geoelectrical data deal with this problem to suggest a suitable solution.

The interpretation of geoelectrical data of both Vertical Electrical

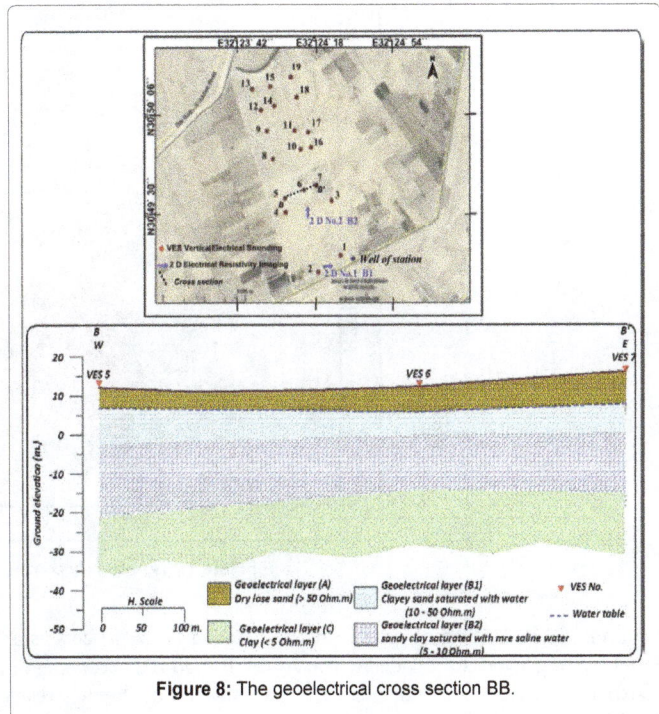

Figure 8: The geoelectrical cross section BB.

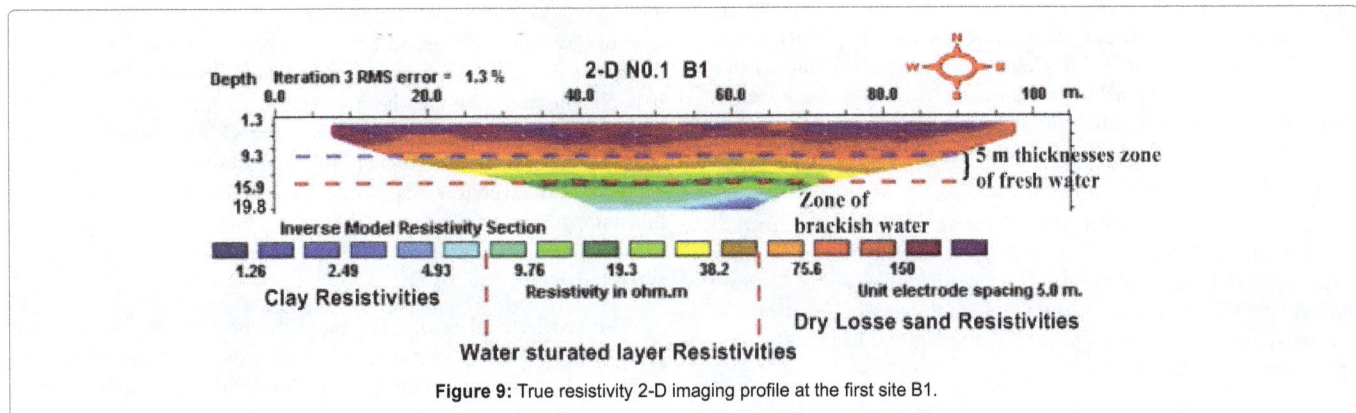

Figure 9: True resistivity 2-D imaging profile at the first site B1.

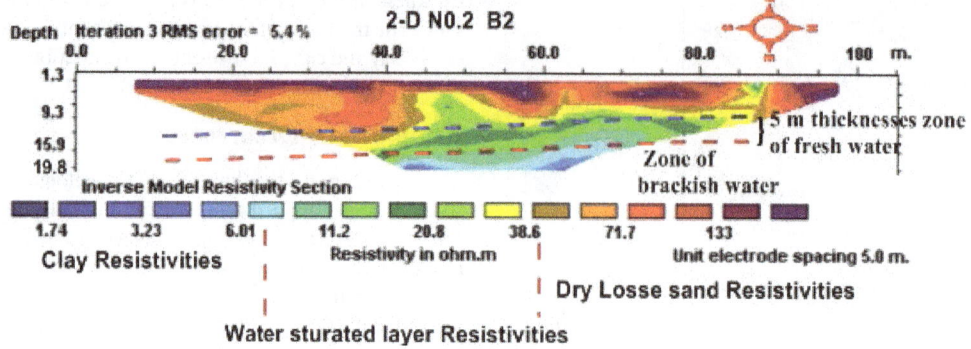

Figure 10: True resistivity 2-D imaging profile at the first site B2.

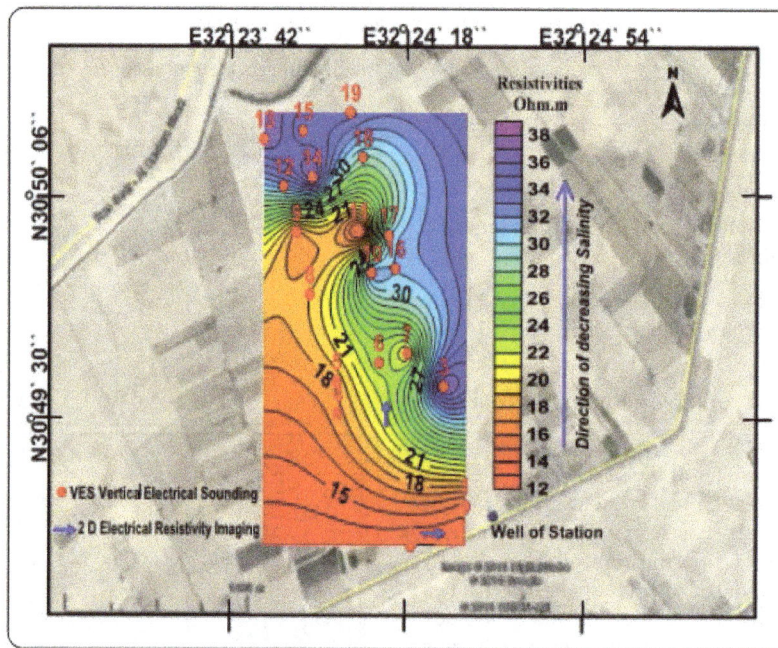

Figure 11: Isoresistivity contour map of upper saturated zone (B1).

Sounding and 2D imaging reveals that the second layer (B) acts as a water bearing layer. It is divided in two zones (B1 and B2) according to resistivity values reflecting the degree of groundwater quality whereas, the lower one more saline than the upper. The isoresistivity contour maps (Figures 11 and 12) of the two saturated zones (B1 and B2) are constructed for more details. They exhibited high resistivity values in the Northeastern and South Western directions. This means that the priority sites of the drilled wells can be chosen in these trends. Generally, the upper saturated zone is better groundwater quality than the lower saturated zone whereas, the resistivity values of the upper zone is higher than the lower

The thickness of saturated layer (B) is an important geoelectrical parameter judging suitable sites for drilled well. The isopach contour maps (Figures 13 and 14) are constructed. They represented increase in thickness values in South-eastern and South-western direction and generally, the lower saturated zone (B1) has larger thickness than the upper saturated zone (B2).

The level of the water bearing layer related to sea level from

VES results is used to construct contour map (Figure 15). It shows groundwater flow direction. The level contour map clarifies the groundwater flow in two directions. The first flow direction is toward South Western trend of the study area. This is caused accumulation of ground water in this site that appears as a pond in the area according to raise of clay layer. The second flow direction is towards North Eastern trend whereas; the depth to clay layer has large record in this location. It is obvious that, the clay layer is played an important role in the ground water condition so; the contour map (Figure 16) of the depth of clay layer will be constructed. This map shows generally increase depth of clay layer towards North Eastern direction. According to this information, the best suitable place for a drainage system is constructing in South Western direction whereas, the dipping layer and less depth of clay.

The geoelectrical results represent the best priority sites of the productive drilled wells for supplementary irrigation in the study area in North Eastern and South Eastern direction. These wells must be drilled with total depth not exceed 30 m. The suitable technique for

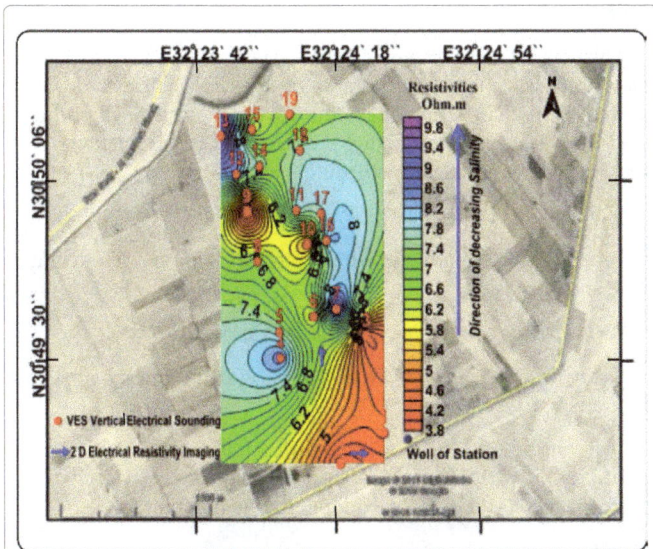

Figure 12: Isoresistivity contour map of lower saturated zone (B2).

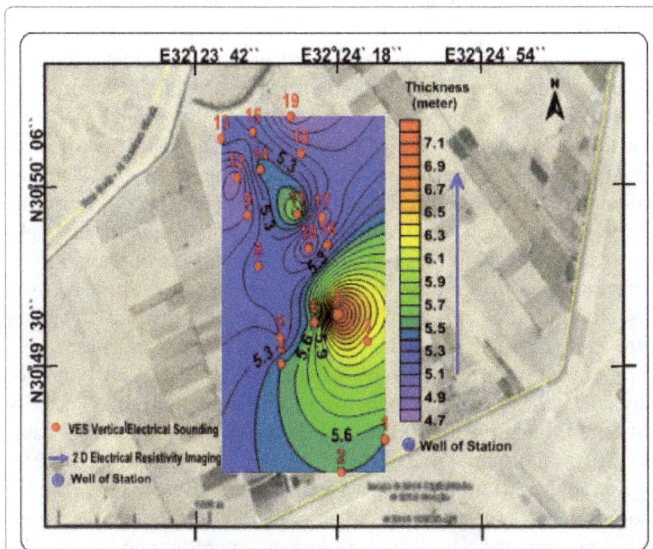

Figure 13: Isopach contour map of upper saturated zone (B1).

quality groundwater and low salinity degree than the lower saturated zone (B2). So it must be noticed that the safe yield for preventing a mixing of the groundwater of both saturated zones (B1 and B2) during discharge from wells and consequence harmful effect on cultivation. When an aquifer contains an underlying layer of saline water such as the study area and is pumped by a well penetrating only the brackish water of the upper part of the aquifer, a local rise of the interface between the saline and brackish water below the well occurs. This phenomenon is known as up-coning, by pumping. This generally necessitates that the well has to be shut down because of the influence of the saline water. Up-coning is a compiler phenomenon and only in recent years has significant headway been made in research to enable criteria to be formulated for the design and operation of wells for skimming brackish water from the saline water.

There is an analytical solution for safe yield of drilled wells. Calculation of safe yield needs to know the type of the well, total

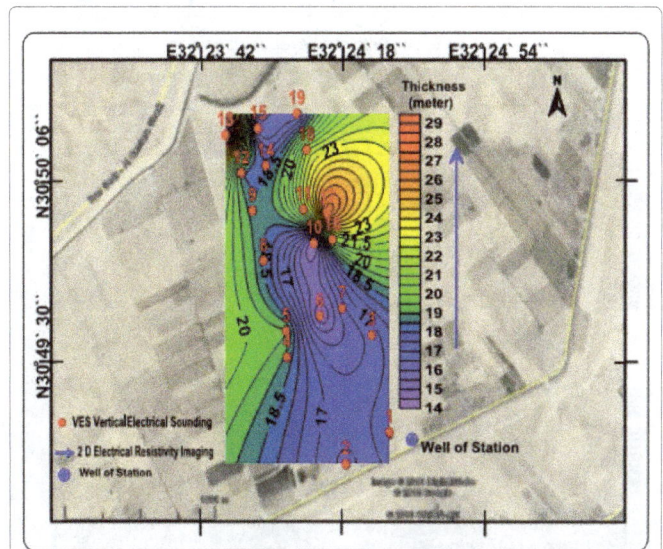

Figure 14: Isopach contour map of lower saturated zone (B2).

drilling is hand dug but possible rotary drill is to detect depth. It must be observing the mission of drilled wells to construct a suitable casing. Also, pumping test is important for safe yield and serves good quality of groundwater recharge.

According to the small thickness value of the upper saturated zone (B1) which is approximately 5 m. of good quality groundwater, suggestion is considered to construct three cement treasurers with distance 3 × 20 × 20 m. These treasurers permit collecting water approximately 1200 m³. This collecting water can be used during shortage of water in tributary of El Salam Canal. The distribution of these treasurers may be one constructed in South Western direction of the study area and using the existing water in the ponds for recharge. The other two treasurers are constructing in the center and the end of the study area as showing in Figure 17.

Geoelectrical measurements indicate two saturated zones (B1 and B2) with different salinities. The upper saturated zone (B1) is in good

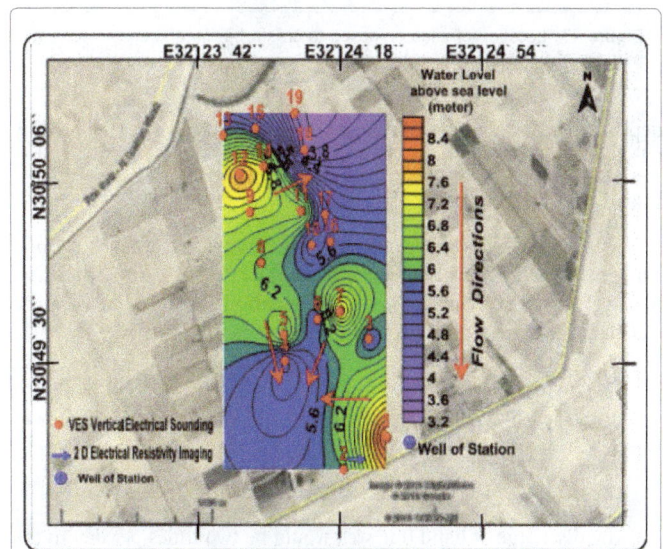

Figure 15: Water table contour map.

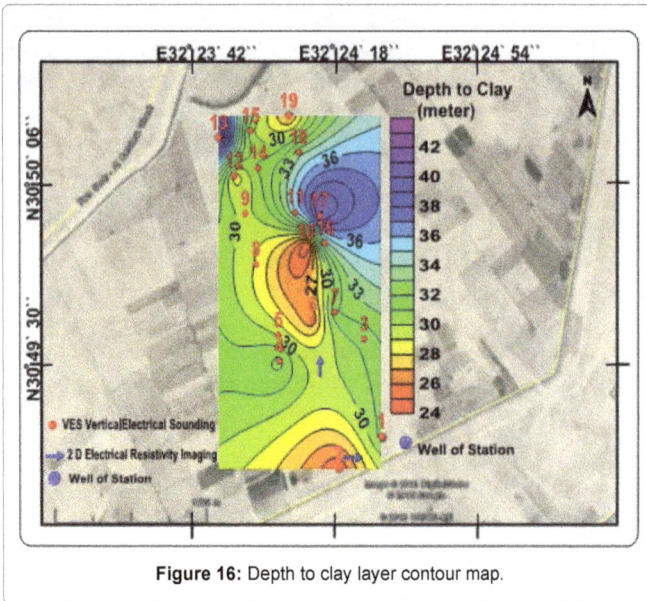

Figure 16: Depth to clay layer contour map.

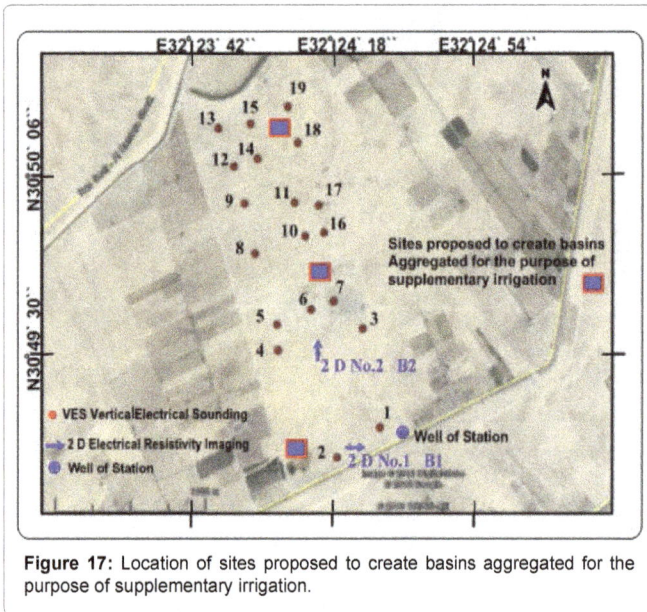

Figure 17: Location of sites proposed to create basins aggregated for the purpose of supplementary irrigation.

depth and well design. Well design and safe yield can be calculated by Ghyben-Herzberg relation [20] and discussed as follows:

Hand dug well

This type of wells is recommended at an area lying near to shore. These wells have 3 m in diameter (2r) and thickness of water (Z) reaches to 5 m (Figure 18a). The safe yield (Q) (Equation 2) of these wells can be calculated by the following equation:

$$Q = \pi r^2 Z m^3 \qquad (2)$$

Where: r is the half diameter of the well. And Z is the thickness of water

Then, $Q = 3.14 \times (1.5)2 \times 5 = 35.33 m^3$

The discharge of hand dug well would be two times per day in the morning and in the evening. The total safe yield of every hand dug well is (Q) 35.33 m³/day.

Drilled well

This type of wells (Figure 18b) is recommended and can be dug where thickness of the brackish water exceeds 25 m and depth to interface more than 35 m. Due to the upconing of the brackish-saline water interface during pumping, the safe yield (Q) (Equation 2) for every well can be calculated according the following:

$$Z = Q/[2\pi d2k(\Delta\rho/\rho b)] \qquad (3)$$

Where $\Delta\rho = \rho s - \rho b$; ρs (1.025), ρb (1.0) is the specific weight of saline and brackish water.

K=Hydraulic conductivity. Z=the critical rise.

d=the distance between the end of the well and interface between saline and brackish water.

If the upcoming exceeds a certain critical rise (z), it accelerates upwards in the well (Figure 18b). Critical rise (z) has been estimated to approximate S=0.3 to 0.5 Thus, adopting an upper limit of Z / d =0.5.

Conclusions and Recommendation

The Sinai peninsula has enormous development evidence especially in agricultural activities to face growing settlements and communities. This development process increased the demand for water. The continuous research services and scientific guidance becomes important for development. Many research stations are constructed by desert research center covering desert land in Egypt. One of these stations is South Qattara Shark that constructed in West Sinai for solving problems related to agriculture and also considered as productive station. The present study concentrates on the area of this station having length reach to 1600 m. and width 850 m., nearly eight hundred Feddan in West of the Sinai region at East of Qattara. The surface and subsurface layers belonging to quaternary deposits consists of sand, sandy clay and clay facies of Pleistocene. This station suffers from shortage water supply for agriculture in some seasons especially summer whereas it depends on a tributary of El Salam Canal that not fully of water all the time. There is one drilled well for human activity just South the study area with total depth 21 m and depth to water reach 11 m. Its salinity has record 2528 ppm. Some patches of water logging appear in low land at South west of area because of not existence good drainage system. Geoelectrical resistivity techniques were used in the present study to deal with pervious mention conditions. The appropriate solutions to these problems have been delineated by the results of 1D and 2D geoelectrical measurements. It exhibits the

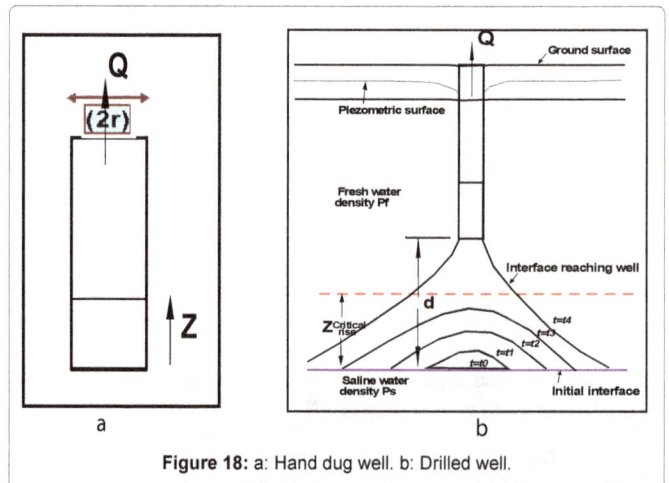

Figure 18: a: Hand dug well. b: Drilled well.

subsurface sedimentary sequences and extension of subsurface layers in horizontal and vertical directions. As well as the choice of the most suitable places to drill production wells with good possibilities and quality.

Nineteen Vertical Electrical Sounding (VES) arranged in a grid to cover the study area and two 2D geoelectrical imaging profiles are carried out. The results of geoelectrical successions are formed of a number of layers being grouped together in three main layers. The first layer is surface layer "A" and the second is layer "B" which divided according to resistivity values in two zones (B1 and B2) that acts as water bearing layer, while the last one is layer "C". The results are represented through different contour maps and cross sections that exhibit the horizontal distribution of successive layers which reflect the lithology and extent of change in all directions. The results declare that the thickness of the water bearing layers consists of two zones. The upper one was less salty than the lower zone. The thickness of the upper zone ranged from 5 to 7 meters but the lower zone ranged from 15 to 30 meters. The last detected layer is clay that decreases in depth towards the Southwest of the study area causing the phenomenon of water logging.

As the results of geoelectrical measurements, the two saturated zones (B1 and B2) exhibited different salinities. The upper saturated zone (B1) is in good quality groundwater and low salinity if compared with the lower saturated zone (B2) according to resistivities values. It is recommended that the safe yield prevent to mixing of the groundwater of both saturated zones (B1 and B2) during discharge from wells and consequence harmful effect on cultivation. This phenomenon is known as up-coning, by over pumping. Generally, necessitates that the well has to be shut down because of the influence of the saline water by a local rise of the interface between the saline and brackish water.

The best priority of the sites of productive wells for supplementary irrigation in the study area is at North eastern and South Eastern directions. These wells must be drilled with total depth not exceed 30 m. The suitable technique for drilling is hand dug but possible rotary drill to detect depth. It must be observing the mission of drilled wells to construct a suitable casing. Also, pumping test is important for safe yield and serves good quality of groundwater recharge. The discharge from hand dug well would be two times per day in the morning and in the evening. The total safe yield of every hand dug well is (Q) 35.33 m³/day when these wells have 3 m in diameter and thickness of water reaches to 5 m.

It can be recommended that a construction of three cement treasurer with distance 3 × 20 × 20 m according to the small thickness of the upper saturated zone (B1) approximately 5 m of good quality groundwater. These treasurers permit collecting water approximately 1200 m³. The collecting of water can be used during shortage of water in branch of El Salam Canal. The distribution of these treasurers may be one which is constructing in South-western direction of the study area and using the existing water in the ponds for recharge. The other two treasurers are constructed in the center and the end of the study area at North trend.

The best suitable place for a drainage system is constructing in South-western direction whereas, the dipping layer and less depth of clay.

References

1. Zohdy AAR (1989) A new method for automatic interpretation of Schlumberger and Wenner sounding curves. Geophysics 54: 244-253.

2. El-Waheidi MM, Merlanti F, Paven M (1992) Geoelectrical Resistivity Survey of the Central Part of Azraq Basin (Jordan) for Identifying Saltwater/Freshwater Interface. J Applied Geophysics 29: 125-133.

3. Choudhury K, Saha DK, Chakraborty P (2001) Geophysical study for saline water intrusion in a coastal alluvial terrain. J Appl Geophysics 46: 189-200.

4. Barseem MS (2011) Delineating the conditions of groundwater occurrences in the area south Baloza- Romana road -North West Sinai- Egypt, Egyptian Geophysical Society EGS Journal 9: 135-143.

5. El Austa MM (2000) Hydrogeological study for Evaluation on the area between El Qantaraand Ber El-Abd, North Sinai-Egypt. Geol Dep Fac of Sci Minufiya Univ. p. 162.

6. Barseem MS, Ayman M, Tamamy El, Milad HZ, Masoud (2013) Hydrogeophyical Evaluation of water occurrences in El Negila area, Northwestern coastal zone- Egypt. Journal of Applied Sciences Research 9: 3244-3262.

7. Khaled MA, Galal GH (2012) Study of groundwater occurrence and the impact of salt water intrusion in East Bitter lakes area, Northwest Sinai, Egypt, by using the geophysical techniques, Egyptian Geophysical Society EGS Journal 10: 1-12.

8. Basheer AA, Salah SO, Ayman IT (2012) Assessment of the Saline-Water Intrusion through the Fresh Groundwater Aquifer by Using ER and TEM Methods at the Qantara Shark Area, Sinai Peninsula, Egypt. International journal of innovative research & development 3: 398-406.

9. Ball J (1939) Contribution to the Geography of Egypt. Survey Dep Egypt Cairo 77: 334-335.

10. Said R (1962) The Geology of Egypt. Elsevier Publishing Company. Amsterdam. p. 377.

11. Geological Survey of Egypt (GSE) (1992) Geological map of Sinai, A.R.E. Sheet No. 5, Scale 1: 250,000.

12. Deiab AF (1998) Geology, pedology and hydrogeology of the Quaternary deposits in Sahl El Tinah area and its vicinities for future development of North Sinai, Egypt.Geol Dept Fac Sci Mansoura Univ Egypt. p. 242.

13. Shata A (1956) Structural development of the Sinai peninsula, Egypt. Bull Inst desert Egypt 6: 117-157.

14. El Shamy IZ (1983) On the Hydrology of West Central Sinai. Egypt. J Geol 27: 1-2.

15. Verlpen BPA (1988) RESIST, version 1.0, a package for the processing of the resistivity sounding data. M.Sc. Research project. ITC, Delft, the Netherlands.

16. Loke MH (1998) RES2DINV V.3.4, "Rapid 2-D resistivity inversion using the least-square method. ABEM instruments AB, Bromma, Sweden.

17. Griffiths DH, Barker RD (1993) Two-dimensional resistivity imaging and modeling in areas of complex geology. Jour of Applied Geoph 29: 211-226.

18. Orellana E and Mooney HM (1966) Master Tables and Curves for Vertical Electrical Sounding over Layered Structures. Inteciencis Madrib. p 34.

19. Barseem MS (2006) Geophysical contribution to groundwater exploration in carbonate rocks, West Sidi Barani area, northwestern coast, Al Azhar University. p.102.

20. Todd DK (1980) Groundwater Hydrology (2ndedn). Johnswiley and Sons, New York. p. 277-296.

5

The Upper Orange River Water Resources Affected by Human Interventions and Climate Change

Mahasa Pululu S[1]*, Palamuleni Lobina G[2] and Ruhiiga Tabukeli M[2]

[1]*Department of Geography, Faculty of Natural and Agricultural Sciences, Qwaqwa Campus, University of the Free State, Phuthaditjhaba, South Africa*
[2]*Department of Geography & Environmental Sciences, School of Environmental and Health Sciences, Mafikeng Campus, North West University, Mmabatho, South Africa*

Abstract

The major problem in the study area is the unlawful water abstractions for irrigation use. In South Africa, indications show that about 240 million m³/a of illegal water use is due to unauthorised withdrawals or violations of water use licenses. The status of water use for irrigation in the Orange-Senqu Basin also shows that insufficient information exists such that work needs to be done to understand the potential for increased efficiency of water use, taking into account issues pertaining to crop type, soil type and technological options. Studies like this one could also shed light on the potential impact of climate change on water use in the basin as this area may well experience significant impacts from rising temperatures and changing rainfall patterns. The processes of validation and verification will determine the extent of existing lawful water use. The use of remote sensing techniques (satellite, aerial photographs, etc.) could be employed to determine if the volume of water use registered by irrigators is accurate, i.e. valid and that the volume of water use registered is lawful (verification). Currently, ecological requirements for the river mouth are met through releases from Vanderkloof Dam and amount to just 290 million m³/a. However several recent studies including the Gesellschaft für International Zusammenarbeit - Integrated Water Resources Management (GIZ – IWRM) study highlight that this is based on a fairly outdated methodology. The more recent Lower Orange Management study found a high level estimate of ecological requirements to be in order of 1 062 million m³/a.

Keywords: Water demand; Water allocation; Water user association; Irrigation board; Orange river; Lesotho; South Africa

Introduction

This paper focuses on the different water users in the study area. According to Meyer [1], South Africa uses 97% total water withdrawal from the Orange River thus making it the largest water user. It is further mentioned that although Lesotho contributes to over 40% of the stream flow, it only uses 1% of the water resources, further downriver outside the study area Botswana accounts for less than 1%, and Namibia uses about 2%. Agriculture is the main activity in the basin and this account for 61% of water demand in the area such that agriculture-inclined employment accounts for more than 50% of the basin's population. Most agricultural activities are along the fertile strips next to the river, but most of the commercial agriculture is artificially irrigated, using water both from the river and from groundwater due to the region's aridity [2-6]. Large parts of the basin are used for commercial rain-fed agriculture (e.g. for maize and wheat production)[7]. Other users in the area are large urban residential centres and industries in the Greater Bloemfontein area and to a limited smaller extent the rural centres scattered across the study area. This Caledon-Modder System supplies water to the Mangaung-Bloemfontein urban cluster (largest urban centre in the study area)[8].

Geographical overview

The Upper Orange Water Management Agency (WMA) covers 103 671 km² and is part of the Orange River watercourse. Lesotho has been included in the study area and covers 30 492 km². The total area is 134 163 km² as shown in Figure 1. This area lies between Latitudes (28⁰ 0' 0" and 32⁰ 0' 0" S) and Longitudes (24⁰ 0' 0" and 30⁰ 0' 0" E). The Orange River, (called the Senqu River in Lesotho), originates in Lesotho *Maluti Mountains,* close to the Lesotho's highest peak, *Thabana Ntlenyana* at 3.482 m above sea level. If there were no developments of any nature in the river basin, the average natural run-off would be more than 12000 million m³/a, representing the average river flow that would be evidenced. It now happens that less than half of the natural run-off reaches the river mouth at Alexander Bay due to high levels of developments in the basin[9,10].

Lesotho

Table 1 shows Lesotho farmers' categories and crops produced.

South Africa

On a national level in South Africa, estimated volume of water per sector is summarised below and this information from Water Authorization and Registration Management System (WARMS) [11].

Proportion of water use per main economic sector	
Agriculture / Irrigation	60%
Municipal /Domestic	27% (i.e. 24% Urban, 3% Rural)
Industrial	± 3% (If not part of Urban Domestic)
Hydro-electricity generation	2%
Mining and associated activities	± 2%
Livestock and Nature conservation	2.5%
Afforestation	3%

---

*Corresponding author: Mahasa Pululu Sexton, Department of Geography, Faculty of Natural and Agricultural Sciences, Qwaqwa Campus, University of the Free State, Private Bag X13, PHUTHADITJHABA 9866, South Africa
E-mail: mahasapululusexton@gmail.com

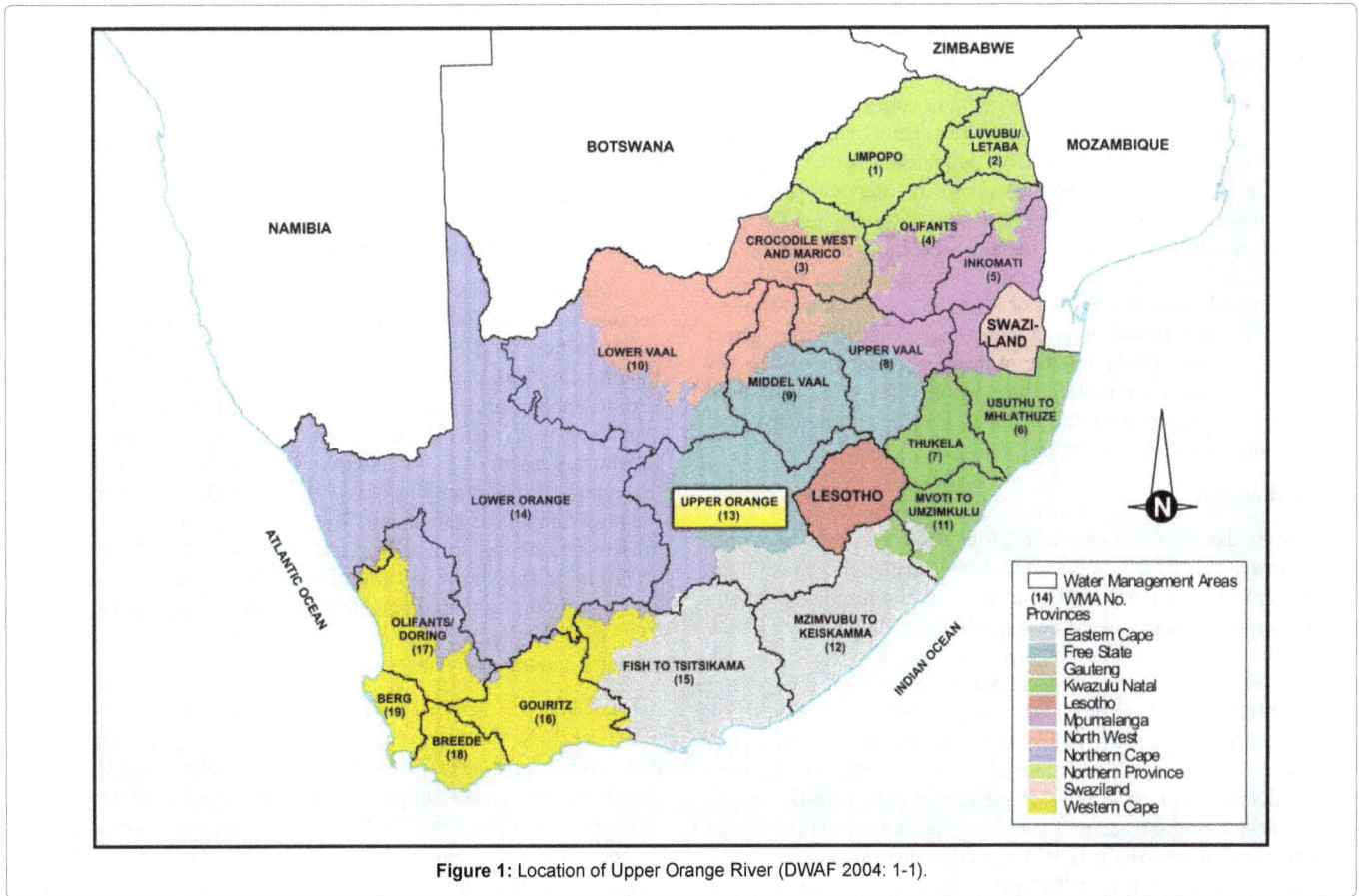

Figure 1: Location of Upper Orange River (DWAF 2004: 1-1).

Type of Farmer	Irrigation System	Crops	Size of Farm (ha)
Subsistence Farmer	None	Maize, sorghum, beans	0.1 – 0.2
	None	Cereals, maize, sorghum, legumes, potatoes	>0.2 - 1
Micro-irrigating Farmer	Watering can, hose pipe/low pressure sprinklers	Vegetables, fruit trees	0.025 (home garden)
	Gravity-fed irrigation	Vegetables, fruit trees	0.1–0.5
Small-scale semi-commercial Farmer	High-pressure irrigation system	Vegetables, fodder	1-4
Medium-scale commercial Farmer	High pressure system with travelling guns	Vegetables	10-20

Table 1: Lesotho farmers' categories and crops produced (ORASECOM 2011: 29).

Three things are worth-noting about the various user sectors above:

- Ecological Reserve (environmental-in-stream flow requirements). This water is not consumed.

- Alien vegetation has not been included above though provision of water for its existence should be allowed for.

- Hydropower. This water is not consumed apart from losses. It should be noted that hydropower is a secondary determinant of the weekly releases from Gariep and Vanderkloof Dams, irrigation requirements are the primary concern.

Irrigation within the study area

In South Africa the irrigation schemes are categorised as Government Water Schemes (GWS), Irrigation Boards (IBs) or Water User Associations (WUAs) [12]. From the Lesotho border to the Orange River delta, the Orange River is divided into 22 River Reaches to cater for irrigation. Irrigation is only facilitated from canal infrastructure that runs asymptotic the river or as direct abstraction from the river. Irrigation mostly occurs on commercial farms with freehold tenure on farming units that are 50 ha on average. However, numerous farms have acquired larger irrigated areas when irrigation units were consolidated under one owner so as to improve or maintain financial viability of irrigated farming. There is also another initiative from Government aimed at introducing and developing resource poor farmers on smallholder irrigation schemes [12]. Calibrated sluice gates facilitate the distribution to irrigation water to the farmers within schemes that are in close proximity to the river, while in some schemes located further afield or more centrally in the basin inline flow meters with telemetry are used. The irrigation infrastructure that distributes water and especially the lined open canals are old warranting widespread rehabilitation as an immediate requirement. Approximately 85000 ha of irrigation is done in these schemes. Specific challenges pertinent to this area warranted the need to identify water users in the study area; and to identify actual allocation of water needs by Department of Water Affairs (DWA)-Republic of South Africa (RSA), Lesotho Highlands Development Authority (LHDA) and Department of Water Affairs (DWA)-Lesotho.

Methodology

Sampling procedure

One expert from each water department or entity was interviewed. The interviews were conducted on a four tier setting namely national, water management agency (WMA) or provincial, district and local municipal levels. In total on the South African side, interviews conducted were two at national, three at WMA level, four at district and twelve at local municipality levels. On the Lesotho side, interviews conducted were two at national, one (i.e. Water and Sewerage Authority-WASA) at department level and three at local levels. The sampling strategy and network was designed to cover wide range of determining factors (i.e. water allocation needs, number of licenses issued, etc.) at the key sites, which reasonably represented the whole study area. It involved retrieving information from the national, provincial, municipal and various users' inventories in the area.

Data collection

Primary data was collected through field surveys. Two secondary data, namely, archival data was made available in digital form from the Department of Water Affairs, Department of Environmental Affairs (DEA) and various stakeholders dealing with water management issues. Similar secondary data was obtained from the Lesotho Highlands Development Authority (LHDA) of the Lesotho Highlands Water Project (LHWP), Ministry of Water, Energy and Mining (WEMMIN), (Water and Sewerage Authority-WASA) and the Maseru Municipal Council. In order to achieve these objectives, structured interview was conducted amongst selected transnational departments, namely DWAF–RSA, Lesotho Highlands Development Authority (LHDA) and Department of Water Affairs (DWA – Lesotho) in order to identify the various users in the study area. The water users were determined based on data availability and their relative role in local economy (i.e. relative to agriculture, industrialisation, residential consumption and tourism). Experts from three WMAs (namely Bloemwater and Sedibeng Water – Republic of South Africa (RSA) side) and WASA (-on the Lesotho side) was used for obtaining this information.

Results

In total the area has seven water user associations and/or IBs. The last two are outside the study area but are supplied from the area through Inter-basin Transfers (IBTs) in the Orange River upper reaches of the area and "which is notably much interconnected" [13]. According to ORASECOM [12], "it is a tremendously integrated water resource system that is highly complex with numerous large intra-and inter-basin water transfers. In the world the Orange River basin is the most complicated and integrated river basins and is operated using highly sophisticated system models which have been developed over a period of more than 25 years."

- Kalkfontein water user association
- Orange Riet water user association
- Lower Modder River water user association
- Orange Vaal water user association
- Christiana water user association (Bloemhof Dam)
- Boegoeberg water user association
- Vaalharts water user association
- Sand Vet water user association

Characteristics of individual water user association

Individual characteristics of water user associations are discussed in this section. The last two water user associations are not in the study area but due to the complex inter-transfers between basin and sub-divisions it could be arguably correct that they are supported from the Orange River due to an upstream transfer into the Ash River and finally into the Liebenbergsvlei River.

Kalkfontein water user association

Originating as a GWS the Kalkfontein scheme, in 1994 it became an IB. In 1998 in terms of the new Water Act, it happens to be the first IB to be transformed to a water user association. The Kalkfontein Dam receives its water from three basins, The Riet River, Kromellenboogspruit and Tierpoort Dam catchments. The Kalkfontein Dam receives its water directly from the Riet River catchment which generates a runoff of 83.96 million m^3/a, provides 49.97 km^2 of water in support of irrigation purposes with 31.34 million m^3 remaining in the reservoirs of the catchment. This dam is also receives water from the Kromellenboogspruit catchment (i.e. runoff of 101.89 million m^3/a, 20.71 km^2 in support of irrigation and 26.58 million m^3 remain in the reservoirs of the catchment) and the Tierpoort catchment (i.e. runoff of 23.23 million m^3/a, 23.35 km^2 in support of irrigation and 17.25 million m^3 remain in the reservoirs of the catchment) [14-16].

Irrigators are supplied with water by lined canals on either sides of the Riet River downstream. Other irrigators pump directly from the Riet River. According to ORASECOM [12], initially 3526 ha was under irrigation, with canals supplying water for 3046 ha while pumping direct from the river supplies 480 ha. The total serviced area of 3502.9 ha now remain after pumping from the river was reduced to 456.9 ha. Originally the intention was that canals should provide water for flood irrigation but now about 10% is flood irrigated while irrigation via centre pivots is now 90%. A total of 120 irrigators are supplied from the Kalkfontein water user association.

Measurement of all water for irrigation use is carried out including those that use telemetry to determine water demand. Measurement of water consumed by canal users is done using calibrated sluices. Line water meters are used for irrigators that extract directly from the river. On average maize occupies 60% of the area is planted on a two year rotational basis with wheat and three other crops. Lucerne is planted on 20% of the land when a variety of other crops occupy the remaining 20% [12].

Together with Kalkfontein water user association which supplies irrigators, other users in the area include urban areas of Koffiefontein, Jacobsdal, Jagersfontein and Fauresmith. Water is also supplied to two De Beers mines at Jagersfontein and Koffiefontein [14].

Water allocated is 11 000 m^3/ha/yr. However billing on users is done on the actual use on a volume basis.

It is a common phenomenon for water to be short supply in this WUA. On the basis of the Kalkfontein Dam, farmers may receive as little as 15% of their allocation in some years. Again there is a noticeable deterioration of water quality because the Riet River flows rarely. Farmers in this part of the study area are highly conscious of Water Conservation and Water Demand Management (WC/WDM) because of regular shortages of water [12].

Orange riet water user association

Of all of the schemes, formed in 2000 this happens to be the most complicated. It is the result of the transformation of the Riet River GWS, formerly consisting of the Riet River Settlement, the Ritchie Irrigation Area and Scholzberg Irrigation Area, and the Lower Riet IB. Numerous lined canals serve the scheme while other irrigators pump directly from the Riet River. Because it was common for water to be in short supply in the Riet River, this led to the construction of the Orange Riet Transfer Scheme so as to convey water from the Vanderkloof Dam, on the Orange River, to service these areas. Water from the Orange River is pumped to a point high enough so that it could move under the influence of gravity via a canal to the serviced area. Irrigators along this transfer canal draw water directly from the canal. Initially the 16903 ha were under irrigation but the area has been expanded to 17050 ha after allocation purchases from the Eastern Cape. The canals were initially meant for flood irrigation purposes. At present about 90% of the area is under irrigation through centre pivots while 9% still maintains flood irrigation and the remaining 1% using other systems. In total 190 irrigators are supplied from the Orange Riet water user association.

All users are measured by either in-line water meters or calibrated sluice. Telemetry is connected to a central 24 hr control station and has to monitor all measuring stations thus making water to be available only on demand. The level of expertise shown by the Orange Riet water user association in South Africa for water measurement and use is the best so far [12].

Wheat (37%) and lucerne (26%) are the main crops. The remaining 37% caters for like maize, potato, barley, oats, groundnut, grapes and other crops. Allocations are done on a volume basis such that 11 000 m³/ha/yr is allocated. However billing of actual use on users is on volume basis. Similar to the Kalkfontein water user association's case, the water in the Riet River does not always flow and there is drastic deterioration of water quality further down the river.

A virtual water bank is operated by the Orange Riet water user association. Subject to decreased use irrigators may trade their water allocations or hand the water back to the water user association who may sell the water to willing buyers at a premium [12].

Lower modder river water user association

In 2010 the Lower Modder River water user association was established and it incorporated the former Modder River GWS. Because it is next to the well-managed Orange Riet water user association, the Lower Modder River Board decided to "piggyback" on the expertise of the Orange Riet water user association and which also manages the Lower Modder River water user association. These two water user associations utilise the services of a common CEO [12].

The water user association obtains its water from the Krugersdrift Dam on the Modder River and irrigators abstract directly from the Modder River to provide for irrigation of 3 526 ha. According to ORASECOM [12]. Measurement of water uses in-line water meters fitted with telemetry and about 90% of the irrigators are supplied. In 2011 plans were put in place to measure the remaining 10%. The annual allocations and crop mix are similar to those of the Orange Riet water user association. The water user association is managed in a similar manner as the Orange Riet water user association by using the same operating procedures and personnel.

Orange vaal water user association

In 2007 when the Orange Vaal IB's were converted to a water user association, the Orange Vaal water user association was formed. According to ORASECOM [12] initially the total irrigated area was 8113 ha and was increased to 11058 ha when purchases of water allocations from outside the scheme were made.

The scheme is situated at the confluence of the Orange and Vaal Rivers. Initially its water was extracted from the Douglas Weir situated on the Vaal River. It was deemed necessary to construct a transfer scheme from the Orange River to the Douglas Weir due to an increase in water use upstream in the Vaal River basin. This transfer scheme, known as the Orange Vaal Transfer Scheme, has an installed pumping system to covey water into the Bosman Canal flowing into the Douglas Weir in another catchment area. The Vanderkloof Dam supplies the entire scheme's water allocation via the Orange Vaal Transfer Scheme. Under is irrigation about 90% of the area, which pumps directly from the Douglas Weir, the Bosman Canal or the Orange River downstream of the Douglas Weir. About 10% of the area is serviced by water from two canals, the Atherton and the Buckland Canals. Originally these canals were intended for flood irrigation supply. However 90% irrigation is done through centre pivots, 7% by other systems and only 3% still maintains flood irrigation [12]. It is further mentioned that 180 irrigators are supplied by the Orange Vaal water user association.

Measurement using calibrated sluices is used to determine allocations made to canal users. Pertaining to water use by irrigators, measurement is not done physically but it is calculated on a pre-season basis. About 90% is under irrigation through centre pivots that are known for their application rates for each area known. Since operational plans of each irrigator relating to planting have to be communicated to the water user association, then the water user association determines the allocation. It is calculated as a product of average weekly evaporation rates for the past five years multiplied by a crop factor to determine the amount of water required for the crop and the area. In consultation with the irrigator the areas are then adjusted until the annual requirement is equal to the irrigator's allocation and both parties are in agreement. Planting is only limited to areas that have been agreed upon by the irrigator and the water user association usually validation of this information follows afterwards.

Maize is planted to about 5000 ha (45%) and this in rotation with wheat and three other crops that are grown every two years. Lucerne occupies 2000 ha (18%) while cotton occupies 2000 ha (18%). The remaining area (19%) may be used for a wide variety of other crops [12].

Users are charged on a volume per area per year (i.e. m³/ha/annum) basis. Originally allocations were about 9140 m³/ha/yr but have increased due to purchased allocations to 10000 m³/ha/yr [12]. A result of poor quality water from the Vaal River is a major water quality problem to the Orange Vaal water user association. Industrial pollution and acid mine drainage from the Vaal River catchment, and nutrient rich sub-surface drainage water from the upstream Vaalharts scheme results in as high as 861 of total dissolved solids (TDS), while an alarmingly high value of 1500 TDS is produced by the drainage water from the Orange Riet and Kalkfontein schemes upstream on the Riet River. In contrast the water from the Orange River has a TDS of only 145. Water that emanates from all three sources is deposited in the Douglas Weir [12]. It is further alluded that research has shown that as a result of different densities the good and poor quality water do not mix but remain in envelops which cause problems for irrigators when the envelop of poor quality water pass the extraction points [17].

Christiana water user association (Bloemhof Dam)

Submission of a proposal for establishing a water user association has been delayed until so far. This is because a consensus could not be reached on registration of water use over such area and also because questions arose as how registered volumes of irrigation water could be utilised without exceeding the permit abstraction flow-rate [12].

Boegoeberg water user association

Established in 2003 the Boegoeberg water user association came into being when the Boegoeberg GWS, the Gariep IB, the Northern Orange IB, a portion of the Middle Orange Irrigation Area, and the Karos Geelkoppan Water Board amalgamated. Under irrigation in the scheme now the total area is 9198 ha. In 1931 the Boegoeberg GWS was formed [12]. Field observations revealed similarities with other schemes in the area in that lined canal on one or both sides of the Orange River and do so downriver. Most farmers' irrigation activities are fed from canals while some farmers pump directly from the rivers. From the Boegoeberg water user association is supplied 306 irrigators also nine livestock farmers for domestic and animal purposes with water. Livestock farming occupies 60000 ha of the area in this scheme [12]. The initial design of the scheme was to be flood irrigated such that even now flood irrigation accounts for 90% of the area while irrigation with micro and drip irrigation is done on the remaining 10%. Laser levelling has been done on 30% of the flood irrigated area to improve on irrigation efficiency. There are 306 irrigators in the scheme. Calibrated sluices are used to measure consumption by canal users (i.e. 297 users) and only 9 river users are not measured. Grapes account for 80%, maize and lucerne (10%), with other crops such as peas, wheat, cotton and pecan accounting for 10% of the area [12].

While irrigation water is rarely in short supply, users are billed on a volume/area/annum basis and presently the allocation is 15000 m³/ha/yr for users on the Boegoeberg, Gariep and Northern Orange portions while the Middle Orange portion is allocated 10000 m³/ha/yr. The infrastructure is very old and the entire scheme needs immediate rehabilitation. Consequently high losses of water are experienced.

The operating philosophy is that the water user association has preserved the natural setting in terms of flow-rate by constructing a divergence of discharge and retaining losses as if the water is still in the river-course.

Vaalharts water user association

In 2001 the Vaalharts water user association was established when the Vaalharts GWS and the Harts GWS were amalgamated. This WUA receives irrigation water from the KB canals and the Taung Irrigation Scheme. The Bloemhof Dam on the Vaal River is the main source of supply and with some water supplied from the Spitskop Dam on the Harts River. The Vaalharts Weir collects water released from the Bloemhof Dam and then distributes it via lined canals to the Vaalharts, Taung and KB canal areas. Water from the Spitskop Dam is distributed via a canal to users downstream of the dam. Certain users also pump directly from the Vaal and Harts rivers. The Vaalharts water user association serves 900 users on a total area of 35700 ha. According to ORASECOM [12], initially it was designed as flood irrigation scheme with 40% still remains under flood irrigation, 40% is under irrigation by centre pivot and the remainder (20%) uses micro and drip irrigation systems. Water released to irrigators is measured using calibrated sluices except for Taung which is not measured at all. Maize, wheat, vegetables, pecans, lucerne, groundnuts, citrus, cotton, olives and grapes are crops grown in the water user association. Billing on irrigators is done on a cubic meter per hectare basis. ORASECOM [12] mentioned that the Vaalharts and KB canals areas jointly obtain

9140 m³/ha/a, Taung 8470 m³/ha/a and the Harts area 7700 m³/ha/a. Infrastructure in the water user association is very old (i.e. >80 years) requiring major investment for rehabilitation.

Acid drainage from the Gauteng area causes the water quality from the Vaal River to deteriorate rapidly. High nutrients in the Vaal River lead to excess algae and water hyacinth growth which often blocks canals and structures used for measuring water use. Salinity in the Vaalharts scheme is also a problem. Consequently installations of sub-surface drainage had to be done on large portions of the scheme. Nutrient rich sub-surface drainage water from Vaalharts is collected in the Spitskop Dam resulting in water quality problems for users from this dam [18].

In 2014, there was a massive project aimed at the development of a joint business case by DWA, Vaalharts Water User Association, Department of Agriculture and Rural Development, 7 municipalities served by the scheme, National Planning Commission and Northern Cape and province and interested private sector.

Sand vet water user association

In 2007 the Sand Vet water user association came into being when two GWS amalgamated. This occurred between the Sand Vet GWS (i.e. supplied from the Allemanskraal Dam on the Sand River) and the Vet River GWS (i.e. supplied from the Erfenis Dam on the Vet River). Raw water for irrigation purposes is supplied to more than 700 privately owned and government owned settlement properties by means of the Allemanskraal Dam and the Erfenis Dam with a 651.84 km long system of channels and drains[19]. An area of 12317 ha is under irrigation consisting of 7162 ha on the Vet River and 5155 ha on the Sand River. Both schemes comprise of a lined canals on one or both sides of the river tracking the river course. The majority are supplied from these canals while some farmers pump directly from the rivers. In the Vet scheme 301 irrigators are supplied and 234 in the Sand scheme giving 535 irrigators in all.

Initially the design of the scheme was to be flood irrigated and only about 1% maintains the status quo. Irrigation by centre pivots accounts for 90% and the remaining 9% is under micro and drip irrigation [18]. Canal users are supplied with irrigation water by means of calibrated sluices while river users are supplied by means of in-line water meters.

The prominent crops grown are maize and wheat with some potatoes and groundnuts and also sunflower, oats, lucerne, onions and vegetables are cultivated under irrigation to some extent. DWA [19] also mention that the Sand-Vet also supplies raw water to be purified for commercial and household use to Theunissen, Bultfontein, Brandfort and Virginia as well as Harmony Mine, Correctional Services at Virginia and the Agricultural experimental farm. The annual operating cost for the Sand-Vet Water Scheme is approximately R 11 million. The scheme suffers from regular water shortages and users seldom receive their full annual allocation [12]. Users are charged on a cubic meter per hectare basis using calibrated sluice gates.

Water Requirements

At present most land use in the area is under natural vegetation with livestock farming (sheep, cattle, goats and some game) and conservation areas occupy large parts. Dry land cultivation to produce grains covers extensive areas and this lie in the north-eastern parts of the water management areas. Large irrigated areas for grain production, fodder crops, a wide variety grapes etc., have been developed along the main rivers, mostly downstream of dams.

Irrigation accounts for representing 88% of the total gross water use of 1996 million m³/ and happens to be the dominant water use sector in the Orange River WMAs based on estimates for the year 2012. This amount excludes the transfers out of the WMAs. The urban, industrial, mining and rural sectors account only for 12% of the water use. Mainly from Lesotho and the Upper Orange WMA, transfers from the Orange River account for 2159 million m3/annum of water use. Expected future growth will mainly be as result of 12000 ha allocated to resource poor farmers and limited growth in urban/ industrial and mining sectors which will mainly be as result of developments in the Bloemfontein, Thaba 'Nchu area. The projected water use required for 2025 is 2134 million m³/a and this does not include the transfers. Until 2025 no new transfer schemes may be expected out of this area.

Understanding growth in water requirements

Usually the prediction of water requirements for purposes planning is based on the primary drivers of water demand, which are population growth and local economic growth. These two factors are intertwined to some extent, as economic growth may stimulate population growth as a result of migration from the rural areas or other urban area with a poor economy. There are also numerous other contributing factors that can impact future water requirements, and specifically for the Greater Bloemfontein area, these may include:

- Change in the level of service, as improvements in the water services, sanitation, and health awareness will most likely impact on future requirement scenarios. Typical initiatives in the study area include the eradication of water and sanitation backlogs linked to the UN Millennium Goals, as well as the delivery of houses to the poor to meet SA National target with regard to housing.

- The impact of HIV/AIDS is a significant factor, with the highest occurrence in the rural areas of South Africa.

- Improvement in water management in terms of water meter coverage, the extent and accuracy of meter reading and billing, and the effectiveness of credit control policies.

The historic growth in water requirement has not been consistent, and has fluctuated quite significantly. The water growth has included periods of negative or relatively flat growth possibly as a result of above average rainfall being experienced in these specific years.

Water use, when expressed on a per capita basis, is in the region of 200 litres per person per day. There are uncertainties, however, associated with the future population growth rate figures as described below.

Population growth rates

Population growth rates are based on the birth rate, mortality rate, and migration. The following sources and references were found which described the historic and possible future population growth rates.

- Information taken from the IDP report 2007/2008 for Mangaung Metropolitan Municipality indicated that the future population growth rate for Bloemfontein was 3.1% per annum. The growth in population between 1996 and 2001 based on 2001 Census figures for the Bloemfontein areas was estimated to be 3.1% per annum.

- A report entitled "Identification of Bulk Engineering Infrastructure in Support of Housing Development in Mangaung, Master-plan prepared for Mangaung Metropolitan Municipality determined that the anticipated population growth figures for Bloemfontein up to 2030 would be 1% per annum.

- Population projection scenarios were also developed for the All Towns study for Central Region (June 2009). This study proposed two alternative population growth scenarios, a High Population Growth Scenario and a Low Population Growth Scenario. The high population growth scenario translates to an aggregate population growth rate for Bloemfontein, Botshabelo and Thaba Nchu of 1% per annum, whilst the low population growth scenario translates to an aggregate population growth rate of 0% per annum.

- Migration is proportional to economic growth rate, implying a strong economic growth will result in "immigration" whereas a decline in economic growth will result in "emigration". Migration figures that could be relevant to the study area were sourced from Provincial trends as abstracted from the "2009 StatsSA Mid-Year Projections for the Orange Free State Province (2006 to 2011 Projection)". Migration affects the rural and smaller towns more significantly, as a result of people seeking economic and employment opportunities in the larger urban centres. Migration is assumed to vary between 0.00% and 0.25% for Bloemfontein and Botshabelo, assuming more people migrating to, and residing in these towns. For the smaller towns with less economic opportunity, the migration rates vary from between -0.4% and 0.0%. The assumption is that current residents could be leaving the smaller towns to reside and seek opportunities in the larger centres.

- The impact of HIV/AIDS is a significant factor when estimating population projections, and more specifically, its influence on the mortality rate. The impact of HIV/Aids relevant to the study area has been based on National statistics, where the highest occurrence is in the rural areas of South Africa. The mortality rate as a result of HIV/Aids has been assumed to be as high as 0.4% for the urban towns, and as high as 0.75% for the rural towns and villages.

Economic growth rates

The largest urban centre in the area is Bloemfontein, Botshabelo follows next and finally Thaba Nchu and most private and public investment occurred in these areas. The latest Integrated Development Plan (IDP) suggests that Bloemfontein shall still be the focus for future development as it is anticipated that Bloemfontein will be home to about 65% of the total population by 2016.

The economy of the MMM plays a significant role in the Motheo District economy (92.5%) as well as the Free State economy (25.5%), but it is relatively small when compared to the national economy (1,6%). Of importance is the relatively small share of the local agriculture, mining and manufacturing sectors compared to the province and the country. Mining's small share is understandable as the Mangaung area competes with the Goldfields area, which is very strong in mining; however the share of agriculture and manufacturing is disturbingly low. On the other hand, the tertiary sector of the local economy is very significant within the context of the province.

Approximately 87% of economic production in the MMM area occurs in Bloemfontein while only 7% and 6% respectively occur in Botshabelo and Thaba Nchu.

The overall annual economic growth rate for the Mangaung area was 3.59% between 2001 and 2004 and a significantly higher growth of 9.5% occurred between 2004 and 2007. In Bloemfontein an economic growth rate between 2004 and 2007 of 9.86% was recorded compared with 8.55% in Botshabelo, while that of Thaba Nchu was considerably less at 5.08% per annum. This confirms the fact that the Bloemfontein economy is and will be increasing its proportional share of the economy.

While community services contribute to over a third of Mangaung's economy, other prominent sectors include finance, retail and trade, transport, and manufacturing. The remaining sectors such as agriculture and mining are very small and make a minor contribution to the local economy. Community services contributes 35% to the city's economy, finance 18%, trade 16%, transport 13%, manufacturing 8%, agriculture 4%, construction 3% and utilities 3%.

Growth in the transport sector, given the strategic central location of Bloemfontein, is likely to be stimulated by increasing economic activity elsewhere in the country.

Future water requirement scenarios

The following assumptions were made for the development of the future water requirement scenarios from the Greater Bloemfontein Water Supply System.

- High Growth Water Requirement Scenario will take place on account of high population growth rate and high economic growth rates. Given the relatively low population projection growth rates and the contrasting relatively high historic growth in water requirements (the authorised billed and unbilled water consumption figures for the last 3 years have grown at a rate of 3% per annum) it was decided to use long term historical growth rate of 3% per annum as the basis for the high growth scenario.

- Low Growth Water Requirement Scenario will take place on account of low population growth and low economic growth. It was decided to base the low growth scenario on a growth in water requirement of 1% per annum.

The high and low water requirement projections have been projected from the 2009 base for the following reasons:

- There were significant summer rains in the 2011 and this may have resulted in a depressed demand.

- It is still too early to ascertain whether or not the drop in 2010 can be ascribed to structural reasons (e.g. improved metering, WC/WDM) or is as a result of climatic influences.

- It is conservative to plan from a higher base. As future years actual water requirements become known, the base from which the projections are made can always be changed.

Important qualification

It is important to note that the water requirement scenarios presented above were developed during a global economic crisis. The global recession and a slow recovery from this recession are likely to have significant implications for water requirement growth projections for the Greater Bloemfontein Water Supply System.

The implications of the recession for the strategy to meet future water requirements are as follows:

- The economic uncertainty increases uncertainty concerning the growth in water requirements.

- Water use must be continuously and carefully monitored;

- Future scenarios/projections need to be revised frequently, based on updated information;

- Planning to increase water availability needs to be as flexible as possible; and

- Interventions that are more flexible in terms of timing should be favoured, all other considerations being equal.

Agricultural water requirements

The only expected growth in irrigation requirements is the allocation of 12000 ha to resource poor farmers. The effect of the 12000 ha (4000 ha for the Upper Orange WMA, 4000 ha for the Lower Orange WMA, and 4000 ha for the Fish-Tsitsikama WMA) is estimated to be in the region of 114 m³/a. The Implementation Strategy for the development of 3000 ha irrigation in the Free State Province indicates that there is ± 200 ha available near Ficksburg (Caledon River) and ± 2000 ha available next to the Orange- Riet Canal, which starts at the Vanderkloof Dam. The agricultural water requirement for the 200 ha near Ficksburg was taken into account in the determination of the available yield [20].

Water balance reconciliation

The Upper Orange WMA is a component of the extended Orange and Vaal River System. This has been the subject of various water balance and reconciliation studies. The latest water balance from the Orange River System indicated a surplus of 274 million m³/a for the year 2008. Subsequent planning to supply water to emerging farmers and for the growth in water requirements in the Upper Orange, Lower Orange and the Fish to Tsitsikamma WMAs would reduce this surplus to only 40 million m³ by 2025. Furthermore the proposed developments under Phase 2 of the Lesotho Highlands Water Project would reduce the yield of the Orange River downstream by approximately 283 million m³/a (proposed Polihali Dam and transfer). Based on a conceptual estimate of the mass balance across the Orange River system, it can be inferred that a system deficit of about 243 million m³/a could be expected by 2037. It can be concluded that there is currently surplus water available in the Orange River system (including the Caledon River) which can be allocated to the Greater Bloemfontein Area. Other water resource development options on the Orange River will only become feasible after the water requirements from the Vaal WMA have increased to such an extent that they reduce the availability of water in the Orange River, and a new supply intervention is implemented to augment the loss in yield [21,22].

The future Polihali Dam site is situated on the Senqu River approximately 1.5 km downstream of the confluence of the Senqu and Khubelu Rivers. Polihali Dam would increase the water delivered from Lesotho Highlands Water Project to the high value industries in the Vaal catchment, but would, in the long term, result in a reduction in the water available at downstream Gariep and Vanderkloof dams. It is envisaged that the Polihali Dam would reduce the yield of the Orange River downstream by approximately 283 million m³. This is based on the assumption that overall yield of the system increases by 182 million m³/a but an additional 465 million m³/a might be transferred to Gauteng, causing a shortfall of 283 million m³/a (46-182 = 283) [22]. Table 2 shows a mass water balance of the Upper Orange WMA.

The Upper Orange WMA has a large commitment to support the local water requirements and transfers to the Upper Vaal WMA, the Fish to Tsitsikama WMA, as well as release obligations to the Lower Orange WMA. A number of augmentation interventions have been identified to provide additional yield to the Orange River System to make up the envisaged shortfall caused by transfers from Polihali Dam to the Gauteng area. Some of the interventions identified include: using the lower level storage in Vanderkloof Dam; the construction of Bosberg/Boskraai Dams; and the raising of Gariep Dam. It is the

	Surplus Yield (million m³)-2004	Surplus Yield (million m³)-2012
Year Surplus Yield	333 (2000)	274 (2008)
Less Transfer to Gauteng from the Mohale Dam (impact on Orange River)	-175	-175
Net Available Yield	158	158
Less Allocation for Resource Poor Farmers	-114	-144
Net current available yield for growth in urban water requirements	44	44
Less growth in urban , industrial and mining sectors in the Upper Orange WMA, the Lower Orange and allocation for resource poor farmers of the Fish-Tsitsikama area (NWRS 2025)	-90	-90
Net deficit in yield in 2025	-46	-40
Less Transfer to Gauteng from the Polihali Dam (impact on Orange River in 2053)	-283	-283
Anticipated net yield (will be higher with additional growth in urban water requirements)	-329 (2053)	-243 (2037)

Table 2: Orange River Water Balance (DWA 2012e:19).

intention of the DWA to initiate a separate reconciliation strategy study on the Orange River System, which will draw on the information from the Greater Bloemfontein Reconciliation Strategy Study.

The Greater bloemfontein area

The anticipated surplus yield in the Orange River System (including the Caledon River) is approximately 44 million m³/a. According to the Internal Strategic Perspective for the Upper Orange River WMA, this surplus is reserved for the growth in demands in the urban, industrial, and mining sectors in the Upper Orange WMA, the Lower Orange, and the Fish to Tsitsikama WMAs. It is not anticipated that there will be any further growth in agricultural water requirements in the Greater Bloemfontein Area (with the exception of the allocation made to the resource poor farmers) [23]. As the agricultural sector and urban sector in the Greater Bloemfontein Area and surrounds do not share any yield from a common surface water resource, it is possible to undertake a reconciliation of supply and requirement based on the current urban water requirements and available yield of the surface water schemes serving the Greater Bloemfontein area and surrounds.

Figure 2 illustrates the comparison of available surface water supply and current water requirements for the High and Low water requirement scenarios in the Greater Bloemfontein Area. The current water requirement (based on 2009 data) is approximately 83 million m³/a while the available supply is 84 million m³/a (Historical Firm Yield).

It appeared that the 2009 water requirement was in balance with available supply (historical firm yield) and any increase in use (as predicted by the high and low water requirement scenarios) would put the system at risk. The higher the growth in water requirements, the higher the risk would be. It is clear that measures to increase the surety of supply need to be implemented as soon as possible. This includes measures to increase the supply of water as well as WC/WDM measures to reduce the demand.

Issues which could impact on the reconciliation of supply and requirement

There are a number of issues which could impact on the reconciliation of supply and requirement in the longer term. These issues are listed below:

- Effectiveness of WC/WDM;

- Existing bulk water supply infrastructure capacity i.e. bulk water pipelines and water treatment works.

- Illegal use of water;

- Impact of HIV/Aids;

- Migration of people from the rural areas to the urban centres, particularly Bloemfontein;

- Sedimentation; and

- Surface and groundwater quality.

Discussion

There are tremendous differences in determining the users in the study area and equally so are methods of determining allocation schedules across the whole study area. This is because no pro-active water demand management is employed. One issue which came up repeatedly was the importance of measuring water usage at the farm and distributer level. This step is fundamental to successful implementation of a whole range of practices which can lead to improved water conservation and water demand. Directly linked to the issue of measurement is how water is paid for. Once paid for volumetrically, a whole range of incentives can be put in place to encourage farmers to use their allocations of water more efficiently. Another important factor was that although identification and description of best management practices can be done individually, concerted efforts through working together in a holistic approach to scheme and farm management could bring about the full realisation of benefits of each best management practice. Finally, and perhaps most importantly, within the South African context, there was generally agreement that the South African Water Act and its provisions for water user associations and water management plan provided an excellent framework for water conservation and water demand management to take place. The establishment of the irrigation database has shown that significant changes in irrigation practices are taking place and that additional areas are being put under irrigation.

Conclusions

A reduction in water consumption largely due to the presence of a virtual water bank was evident in the area resulting in greater accountability resulting from volumetric metered payment system. It was noted that increased level of awareness and expertise amongst irrigators tied to demonstration of best practices led to other water user associations improving their water use efficiency.

Recommendations

There should be accelerated and supported implementation of the South African Water Act and catchment management agencies and its provisions for water user associations. This framework provides valuable lessons which can be used even from the other basin states. It is a necessity to improve irrigation water measurement at both the distributor and irrigator level through encouragement and be mandatory as well. It is a further recommendation that investigating

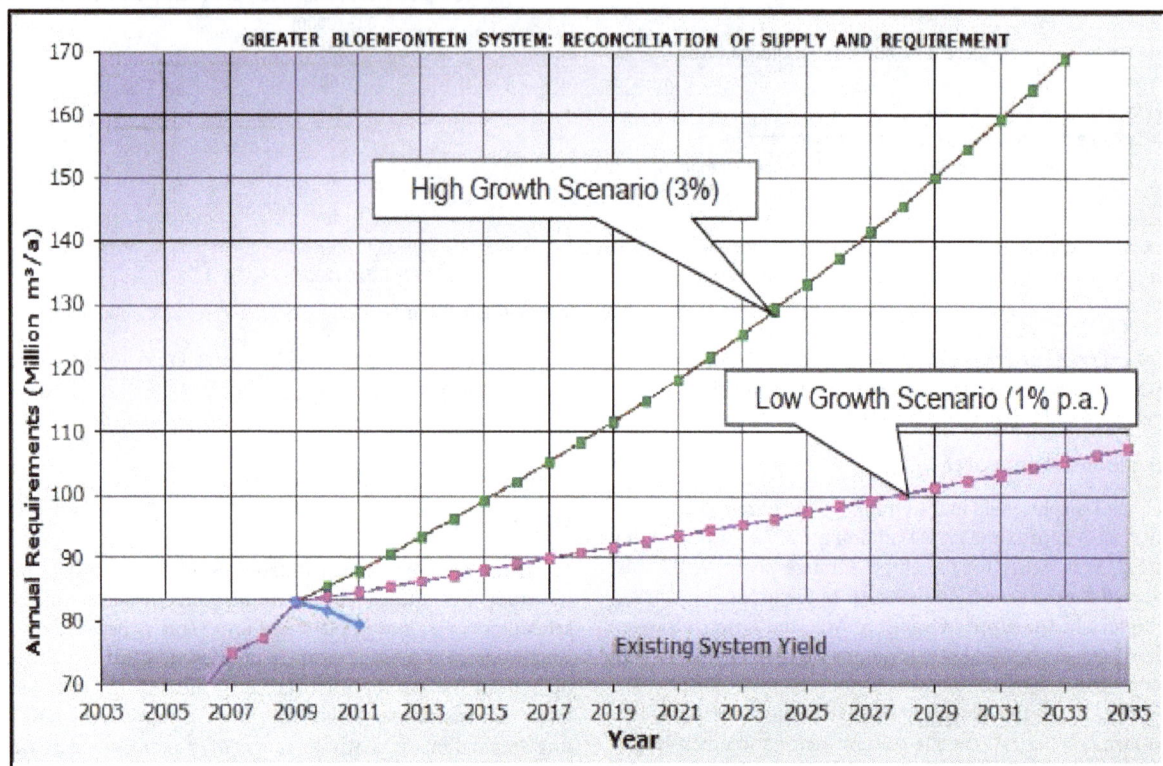

Figure 2: Surface Water Balance for Study Area (DWA 2012: 20).

the cost and extent of this requirement may be expedited and funding options determined.

Attempts have to be made to advocate for investigation of the possibility of offering assistance to Water User Associations so that they may establish appropriate GIS-type data-base of irrigated lands, irrigation, cropping patterns, requirements etc. (including the training of personnel) for the improvement of water management and water use. Data acquisition could be useful at the Water User Association level and lead to the improvement on estimating parameters related to irrigation in the catchment area.

Demonstrations of the best practices at site-locations (for both irrigators and suppliers of irrigation water) could enhance the dissemination of information on efficiency of these irrigation practices. It is another recommendation that the database of areas under irrigation is updated regularly on a yearly basis and that classification using the remote sensing tools is developed, complimented by GIS ground truth database of Water user associations such that wasteful water demands on the overall can curbed, or keeping growth rates at low levels could prove beneficial. During this process continued capacity-building should be advocated for.

References

1. Meyer C (2013) Integrated Water Resources Management-The Orange Senqu River Basin in South Africa. Master's Thesis. University of Hamburg, Germany.

2. Makurira H, Mapani B, Mazvimavi D, Mul M, Magole L, et al. (2014) Transboundary water cooperation building partnerships (Part 2). Physics and Chemistry of the Earth, Parts A/B/C, 76: 1-2.

3. Ramollo P (2014) Lower Orange River forum-maintaining South Africa's largest river. Water Wheel 13: 40-41.

4. Malherbe W, Mahlangu S, Ferreira M, Wepener V (2015) Fish and macroinvertebrate community composition of a floodplain wetland associated with the Harts River, South Africa, in relation to water quality and habitat parameters. African Journal of Aquatic Science: 1-7.

5. Matthews S (2015) Orange River mouth-saving the integrity of one of SA's most important estuaries: water resource management. Water Wheel 14: 30-34.

6. Munyika S, Kongo V and Kimwaga R (2015) River health assessment using macroinvertebrates and water quality parameters: A case of the Orange River in Namibia. Physics and Chemistry of the Earth, Parts A/B/C. 76-78: 140-148.

7. ORASECOM-Orange-Senqu River Commission (2008) Orange-Senqu River Basin. Preliminary transboundary Diagnostic Analysis.

8. DWA-Department of Water Affairs, South Africa (2012) Development of Reconciliation Strategies for Large Bulk Water Supply Systems Orange River: Surface Water Hydrology and System Analysis Report. WRP Consulting Engineers Aurecon, Golder Associates Africa, and Zitholele Consulting. Report No: PRSA D000/00/18312/7.

9. DWA-Department of Water Affairs, South Africa (2012b) Inception Report for the Large Bulk Water Supply Systems of the Greater Bloemfontein Area. Prepared by Aurecon in association with GHT Consulting Scientists and ILISO Consulting as part of the Water Reconciliation Strategy Study for the Large Bulk Water Supply Systems: Greater Bloemfontein Area. DWA Report No: P WMA 14/C520/00/0910/01.

10. DWAF-Department of Water Affairs and Forestry, South Africa (2004) Internal Strategic Perspective: Orange River System Overarching. Prepared by PDNA, WRP Consulting Engineers (Pty) Ltd, WMB and Kwezi-V3 on behalf of the Directorate: National Water Resource Planning. DWAF Report No: P RSA D000/00/0104.

11. DWA-Department of Water Affairs. South Africa (2013c) National water resource strategy second edition: Water for an equitable and sustainable future. Pretoria: Department of Water Affairs, Republic of South Africa.

12. ORASECOM-Orange-Senqu River Commission (2011) Overall Project Executive Summary. Support to Phase 2 of the ORASECOM Basin-wide Integrated Water resources Management Plan. WRP Consulting Engineers in

association with Golder Associates, DMM, PIK, RAMBOLL and WCE. Report number: ORASECOM 013/2011.

13. DEA-Department of Environmental Affairs, South Africa (2013) Long-Term Adaptation Scenarios Flagship Research Programme (LTAS) for South Africa. Climate Change Implications for Water Sector in South Africa. Pretoria. South Africa.

14. DWA-Department of Water Affairs, South Africa (2013b) Kalkfontein Scheme Operating Rule Establishment of Drought Operating Rules for Stand Alone Dams typical of Rural/Urban Municipal Water Supply Scheme (Central Region) Pretoria. WRP Consulting Engineers (Pty) Ltd. Report P RSA 000/00/14311/ Central/Kalkfontein.

15. DWA-Department of Water Affairs, South Africa (2013a) Development of Reconciliation Strategies for Large Bulk Water Supply Systems: Irrigation Demands and Water Conservation/Water Demand Management. Pretoria. WRP Consulting Engineers (Pty) Ltd., Aurecon, Golder Associates Africa, and Zitholele Consulting. Report P RSA D 000/00/18312/6.

16. DWA-Department of Water Affairs South Africa (2013d) Water allocation reform. Portfolio committee on water and environmental affairs.

17. Viljoen MF, Armour RJ, Oberholzer JL, Grosskopf M, van der Merwe B, et al. (2006) Multi-dimensional Models for the Sustainable Management of Water Quantity and Quality with reference to the Orange-Vaal-Riet Convergence system. WRC Report No: 1352/1/06. Pretoria. Water Research Commission.

18. ORASECOM-Orange-Senqu River Commission (2009) Feasibility Study for the Development of a Mechanism to Mobilise Funds for Catchment Conservation. Conservation Fund Assessment Report - Identification of key mitigation measures. Report number: ORASECOM 003/2009.

19. DWA-Department of Water Affairs, South Africa (2011) Development of Reconciliation Strategies for all Towns in the Central Region. Lejweleputswa District Municipality in the free State Province: Reconciliation Strategy for Brandfort Town Area consisting of Brandfort, Page Park, Majwemasweu, Mountain View and Somerset settlements in Masilonyana Local Municipality, in the Middle Vaal WMA. Pretoria. WRP Consulting Engineers (Pty) Ltd in association with DMM, Golder, KV3, Zitholele and Sub consultants. Contract WP 9713.

20. DWA-Department of Water Affairs, South Africa (2012c) Interim Reconciliation Strategy Report for the Large Bulk Water Supply Systems of the Greater Bloemfontein Area. Prepared by Aurecon in Association with GHT Consulting Scientists and ILISO Consulting as part of the Water Reconciliation Strategy Study for the Large Bulk Water Supply Systems: Greater Bloemfontein Area. DWA Report No: P WMA 14/C520/00/0910/02

21. DWA-Department of Water Affairs, South Africa (2012d) Interventions Report for the Large Bulk Water Supply Systems of the Greater Bloemfontein Area. Prepared by Aurecon in association with GHT Consulting Scientists and ILISO Consulting as part of the Water Reconciliation Strategy Study for the Large Bulk Water Supply Systems: Greater Bloemfontein Area. DWA Report No. P WMA 14/C520/00/0910/03.

22. DWA-Department of Water Affairs, South Africa (2012e) Reconciliation Strategy Report for the Large Bulk Water Supply Systems of the Greater Bloemfontein Area. Prepared by Aurecon in association with GHT Consulting Scientists and ILISO Consulting as part of the Water Reconciliation Strategy Study for the Large Bulk Water Supply Systems: Greater Bloemfontein Area.

23. DWA-Department of Water Affairs, South Africa (2012f) Water Quality Assessment Study for the Large Bulk Water Supply Systems of the Greater Bloemfontein Area. Prepared by Aurecon in association with GHT Consulting Scientists and ILISO Consulting as part of the Water Reconciliation Strategy Study for the Large Bulk Water Supply Systems: Greater Bloemfontein Area.

6

Water Quality Assessment and Apportionment of Pollution Sources of Selected Pollutants in the Min Jiang, a Headwater Tributary of The Yangtze River

Jian Zhao*, Guo Fu, and Kun Lei

Chinese Research Academy of Environmental Sciences, Beijing 100012, China

Abstract

This paper analyzed the spatial-temporal variations of surface water quality along the middle and lower reaches of the Min Jiang between 2003 and 2012 and investigated its pollution sources by analyzing the data from 4 water quality monitoring stations. The results showed that surface water quality was higher polluted in the middle reaches of the Min Jiang than that in the lower reaches and its tributary. Seasonal and spatial differences were found for DO, CODmn and NH3-N, whereas for TP the differences were mainly due to the water quality station. The level of organics (CODmn) was higher in summer (high flow period), and the level of NH3-N was higher in winter (low flow period). In the middle reaches of the Min Jiang, point sources (from wastewater treatment plants and industrial effluents) were found to be the dominant inputs of organics (CODmn) and nutrients (NH3-N and TP) to river. In the lower reaches of the Min Jiang, diffuse sources (from agricultural fertilizer, soil erosion, etc.) were the dominant contributor of organics and TP to river, while point sources were the dominant input of NH3-N. In tributary, diffuse sources were the dominant organics and TP input, both point and diffuse sources were dominant NH3-N inputs. Overall, these results reinforced the notion that pollution control by periods and regions was important for effective water quality management, and it is necessary to enhance the treatment of industrial effluent, to strictly carry out the discharge standard for water pollutants and the total amount control system, to incorporate NH3-N in the total amount control system in the Min Jiang.

Keywords: Water quality; Spatial variations; Seasonal variations; Pollution control; Water management; Min Jian

Introduction

Nowadays, the surface water pollution is of great environmental concern worldwide because of it has hastened the scarce of water resources, and affected sustainable development and human healthy. Rivers are highly vulnerable water bodies to pollution due to their roles in assimilating or carrying off the municipal and industrial wastewater and run-off from agricultural land in their vast drainage basins. Surface water quality is controlled by complex anthropogenic activities and natural factors. The discharge of municipal and industrial wastewater is kind of a constant polluting source. However, the surface run-off is a seasonal phenomenon and highly affected by weather condition. Seasonal variations in precipitation, surface run-off, interflow, groundwater flow and pumped inflow and outflow have strong impacts on river discharge and subsequently on the concentration of pollutants in rivers [1]. Due to the complexity of water environments, water quality specialists and decision-makers are confronted with significant challenges in their efforts to manage surface water resources [2]. Therefore, identifying temporal and spatial changes in water quality in river basins is an imperative task, so as to provide an improved understanding of the environmental conditions and help researchers establish priorities for sustainable water management [3-6].

The temporal pattern of total pollutant load inputs to a river from point and diffuse sources is fundamentally different. Loadings from point sources, such as STWs and industrial effluents, tend to be relatively constant throughout the year, and are generally independent of river flow. Diffuse sources (from agricultural fertilizer, soil erosion, septic tank soak-aways and atmospheric deposition) are principally flow dependent, and should occur intermittently, particularly during the periods of the year with high precipitation. This temporal difference in the mode of pollution delivery results in clear differences in the relationship between pollutant concentration and river discharge [5,7]. Furthermore, the role of pollution source from the point and diffuse inputs can be identified. In rivers that are point source dominated, the constant rate of input of contaminations means that pollutant concentrations will be highest at low flow, and this concentration will decrease reciprocally with increasing river flow rate, due to dilution. Conversely, rivers that receive pollutant primarily from diffuse sources will tend to show an increase in pollutant load and concentration with increasing river flow [7].

The Min Jiang River has recently been the focus of attention due to recognition of the increasing stress being placed on its water resources and of the resulting environmental degradation in the Yangtze River basin. Since the mid-1970s, the Sichuan Basin area has undergone rapid development. Locally, industry, agriculture, and domestic activities have posed great pressure on the ecological environment, especially the aquatic environment [8]. At the end of the 1990s, the percentage of water quality poorer than Grade III was 30.1% in the Min Jiang based on the Environmental Quality Standards for Surface Water GB3838-2002 in China, see (Table 2). In contrast, the percentage of water quality poorer than Grade III was 41.26% in 2006. In order to suspend the deterioration of water environments and improve the surface water quality, Sichuan province has been urged to take serious actions during the 11th Five Years program (2006-2010). Surface water quality of the Min Jiang has been improved in recent years, but according to

---

*Corresponding author: Jian Zhao, Chinese Research Academy of Environmental Sciences, Beijing 100012, China, E-mail:zj103823@163.com

the comprehensive evaluation of 2011, 29.4% of total length was still over grade III. In order to have an effective, long-term management and reduce the constituent concentrations in the Min Jiang we need to acquire understanding of the behavior and the variation of water quality parameters and major pollution sources within the catchments. However, only a few studies on water quality in the Min Jiang have focused on nitrogen contamination [9] and sediment yields [10]. Recent studies of surface water quality in the Min Jiang have reported on the chemical and physical weathering [8,11]. Further investigation of water contamination and pollutant sources is needed.

The objective of this study is to analyze the temporal and spatial variations of the water quality of the Min Jiang and to identify the pollution sources by exploring the concentration-flow relationships. Findings from this study will extend the available information for effective water management for the watershed.

Material and Methods

Study area

In this work, we studied the middle and lower reaches of the Min Jiang, from Dujiangyan to YiBin city, with a length of approximately 370 km (Figure 1). The middle and lower reaches of the Min Jiang has a subtropical climate, with average annual temperatures between 15~18°, and annual precipitation between 1200~1500 mm. Precipitation is concentrated from May to October, during which about 75% of the total annual precipitation occur.

The river serves as a major source of domestic and industrial water supply for nearby cities: Chengdu (population of 11,120,000), Meishan (population of 3,450,000), Leshan (population of 3,530,000) and Yinbin (population of 5,270,000), which are the major urban settlements on the banks of the Min Jiang. Subsequently, the river receives domestic and industrial wastewater from these cities and numerous minor settlements along the river.

Monitoring stations and data

In this study, eight water quality monitoring stations were selected under the river quality monitoring network of Min Jiang basin (Figure

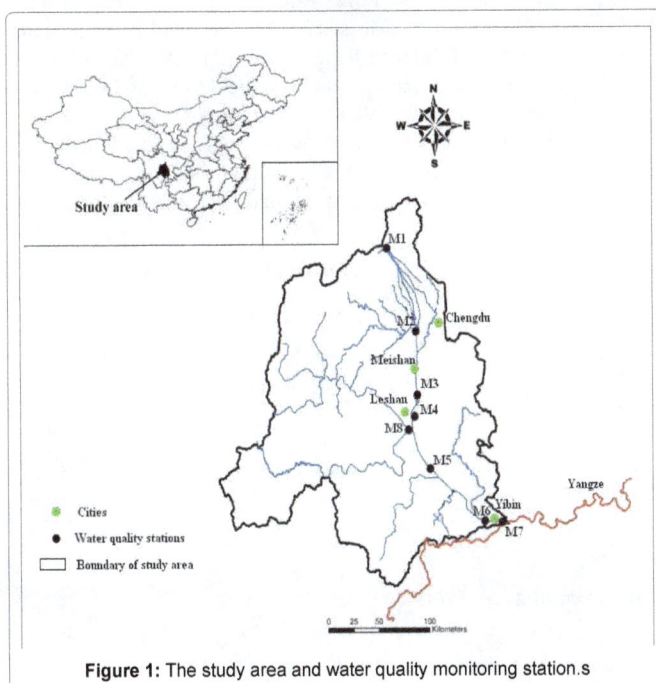

Figure 1: The study area and water quality monitoring station.s

1). Four stations, M1, M2, M3 and M4, on the middle reaches of the main stream of the Min Jiang, three stations M5, M6 and M7, on the lower reaches of the main stream of the Min Jiang, and one station (M8), on the Dadu River, which is the largest tributary of the Min Jiang. These water quality stations were selected because of the completeness of the hydrology data series. Table 1 gives a description of the water quality monitoring stations with their locations and types.

The monthly water quality data for eight water quality monitoring stations between 2003 and 2012 were obtained from Environmental Monitoring Center of Sichuan province. The analyzed water quality parameters include dissolved oxygen (DO), potassium permanganate index (COD_{mn}), ammonia nitrogen (NH_3-N) and total phosphorus (TP). The sampling, preservation, transportation and analysis of the water samples were carried out following standard methods [12]. Daily flow data at four water quality monitoring stations (M4, M5, M7 and M8) between 2006 and 2012 were obtained from Data-sharing network of China hydrology.

Load estimation method

The method used to estimate the pollution loads was based on averaging estimators, also called integration or interpolation methods, use the means of concentrations and flows over a time interval, and has been widely accepted. All the available flow data in the sampled period were used in this method. The equations used were:

$$L_w = \frac{\sum_{i=1}^{n} A_i C_i}{\sum_{i=1}^{n} A_i} \cdot \frac{\sum_{i=1}^{n} Q_i}{n} \cdot n = \overline{C} \cdot \mu_q \cdot n \qquad (1)$$

$$\text{With: } \mu_q = \frac{\sum_{i=1}^{n} Q_i}{n}$$

where A_i represents the indicator for availability of concentration data (1 if data is available, 0 if not), C_i the concentration on day i, Q_i the average flow on day i, n the total number of days for the period of load estimation. Over bars denote sample arithmetic means, and L_w the resulting load.

Furthermore, to study the point and diffuse source load, L_w were considered for each monthly period. With the average load of low flow months is multiplied by 12 to estimate the annual point source load. Annual load minus point source load is the load of diffuse.

Statistical procedures

Two-way analysis of variance (ANOVA) was performed to estimate the temporal and spatial differences of water quality in the River. Previously, normality and homogeneity of data were checked by means of the Kolmogorov-Smirnov and Levene tests, respectively. The non-parametric Kruskal-wallis'test (K-W) was used for DO, COD_{mn}, NH_3-N and TP, as they were either not homogeneous or not normal distributed [13].

In this study, the relationships between concentration (C) and river flow (Q) were studied by means of regression techniques for different models such as power ($C=aQ^b$), hyperbolic ($C=a+b/Q$), exponential ($C=ae^{bQ}$), linear ($C=a+bQ$) and logarithmic ($C=a+b\ln(Q)$). Different models were proposed to describe the relationship between concentration-flow [14-16] The best regression model was chosen according to maximum correlation coefficient [16].

Results and Discussion

Water quality status

The mean value and standard deviation of the eight water quality parameters from eight water quality monitoring stations in the middle and lower reaches of the Min Jiang during 2003-2012 were summarized in Table 2 Based on the "Environmental Quality Standards for Surface Water" of China, the surface water environment is divided into five grades, and each grade has its corresponding standard value (Table 2). In the area, the values of pH were within permissible range of surface water (6.0-9.0) at all stations. Both DO and BOD_5 reached grade Ⅰ at stations M1, M2, M5 M6, M7 and M8, grade Ⅲ at station M4, while DO reach grade Ⅳ at station M3. The COD_{mn} reached grade Ⅰ at station M1, grade Ⅱ at stations M2, M5, M6, and M7, grade Ⅲ at station M4, and grade Ⅳ at station M3. The NH_3-N reached grade Ⅰ at station M6, grade Ⅱ at four stations (M1, M5, M7 and M8), grade Ⅲ at stations M2 and M3, while over grade Ⅴ at station M4. The TP reached grade Ⅱ at station M8, grade Ⅲ at stations M1, M5, M6 and M7, and grade Ⅳ at station M2, M3 and M4.

It must be emphasized that average concentrations of some variables such as DO, COD_{mn}, NH_3-N and TP in the middle reaches of the Min Jiang are over grade Ⅲ or Ⅴ, whereas only water with a grade lower than grade Ⅲ was allowed in this region, and over grade Ⅴ is the worst score in the national standard for water quality in China. Therefore, the water resource of the middle reaches of the Min Jiang was not suitable for human consumption or industrial purposes.

Temporal trends of water quality

Two-way ANOVA and K-W test showed significant temporal differences ($P<0.05$) for all water quality parameters except TP among the 120 sampling times over the nine year period (2003-2012). Temporal variations of the eight variables were illustrated by box-whiskers plots (Figure 2).

The Water temperatures (Figure 2) demonstrated a seasonal pattern. The DO reflected the same temporal patterns showing higher

values in winter and lower values in summer, with a range of 6.42 to 8.70 mg/L and 0.2 to 11.9 mg/L, respectively. In contrast, the COD_{mn} reflected the same temporal patterns showing higher values in summer and lower values in winter, with a range of 0.2 to 13.2 mg/L and 0.4 to 38.6 mg/L, respectively. Regarding nutrients, the concentrations of NH_3-N varied from 0.05 to 9.50 mg/L, with lower values in summer (July and August), and higher values in March. TP was absence significant temporal various, with a range of 0.03-8.23 and 0.01-1.20 mg/L, respectively.

The higher values of COD_{mn} and lower values of DO in summer are influenced by various factors. Higher rainfall and river-flow cause the wash of organic matter into the surface water, which decreases the concentration of dissolved oxygen with biodegradation, whilst the increase in temperature causes a decrease in oxygen solubility, thus causing a further reduces the DO concentrations [17]. In addition, as the amount of available DO decreases, undergoes anaerobic fermentation processes leading to formation of organic acids. Hydrolysis of these acidic materials causes a decrease of water pH values [1]. Furthermore, NH_3-N in river mainly derived from point source, which provides a relatively constant input of constituent to the river throughout the year [14], NH_3-N concentrations will decrease reciprocally with increasing flow rate in summer, due to dilution effect[18].

Spatial distribution of water quality

Results from spatial two-way ANOVA and K-W test displayed significant spatial differences ($P<0.05$) for all water quality parameters at the eight water quality stations. Spatial variations of eight variables were illustrated by box-whiskers plots (Figure 3).

The lowest water temperature was found at station M1, which located in the mountain area, and other stations (M2-M8) showed less variation (Figure 3). The DO was lower in middle reaches (station M2 to M4) than in lower reaches (station M5 to M7) of the Min Jiang (Figure 3). On the contrary, the COD_{mn}, NH_3-N and TP concentrations were higher in the middle reaches (station M2 to M4) of the Min Jiang, and all of them were lower in tributary (station M8) than in the main stream of the Min Jiang, except NH_3-N.

High values of DO and lower values of COD_{mn} and nutrients (NH_3-N and TP) were found in station M1 indicated that the water from upper reaches of the Min Jiang, relatively unaffected by human activities, was good of quality. However, the values of COD_{mn} and nutrients (NH_3-N and TP) followed the sharply increasing trends from station M2 to M4 (Figure 3), where high population density and main industrial were located. Chengdu city is one of the major cities in this area. The population of Chengdu city account for 73% of

Stations	Locations	Section character	Water quality target
M1	Middle reaches	Aba prefecture boundary section	I
M2	Middle reaches	Chengdu boundary section	
M3	Middle reaches	Meishan boundary section	Ⅲ
M4	Middle reaches	central section of Leshan city	Ⅲ
M5	Lower reaches	Leshan boundary section	Ⅲ
M6	Lower reaches	Cuiping district of YinBin city	Ⅲ
M7	Lower reaches	Near entry to Yangze River at YinBin city	Ⅲ
M8	Tributary	central section of Leshan city	Ⅲ

Table 1: Water quality monitoring stations in the middle and lower reaches of the Min Jiang and its tributary.

Parameters	Middle reaches								Environmental guidelines [a]				
	M1		M2		M3		M4						
	Mean	Std.	Mean	Std.	Mean	Std.	Mean	Std.	I	Ⅱ	Ⅲ	Ⅳ	Ⅴ
DO	8.62	1.14	6.55	1.39	4.5	1.49	5	1.7	≥7.5	6	5	3	2
COD_{mn}	1.51	0.72	3.27	1.58	6.94	5.15	5.37	2.09	≤2	4	6	10	15
NH_3-N	0.26	0.19	0.75	0.48	0.98	1.03	2.17	1.93	≤0.15	0.5	1	1.5	2
TP	0.17	0.19	0.25	0.18	0.27	0.14	0.28	0.19	≤0.02	0.1	0.2	0.3	0.4
Parameters	Lower reaches						Tributary		Environmental guidelines [a]				
	M5		M6		M7		M8						
	Mean	Std.	Mean	Std.	Mean	Std.	Mean	Std.	I	Ⅱ	Ⅲ	Ⅳ	Ⅴ
DO	8.17	1.12	8.38	1.02	8.48	1.06	8.79	1.12	≥7.5	6	5	3	2
COD_{mn}	3.06	1.13	3.79	1.98	3.55	1.58	2.41	0.92	≤2	4	6	10	15
NH_3-N	0.41	0.2	0.11	0.09	0.16	0.12	0.3	0.16	≤0.15	0.5	1	1.5	2
TP	0.12	0.04	0.14	0.06	0.15	0.06	0.07	0.04	≤0.02	0.1	0.2	0.3	0.4

Table 2: Statistics descriptive of selected water quality parameters in the middle and lower reaches of the Min Jiang during 2003-2012.

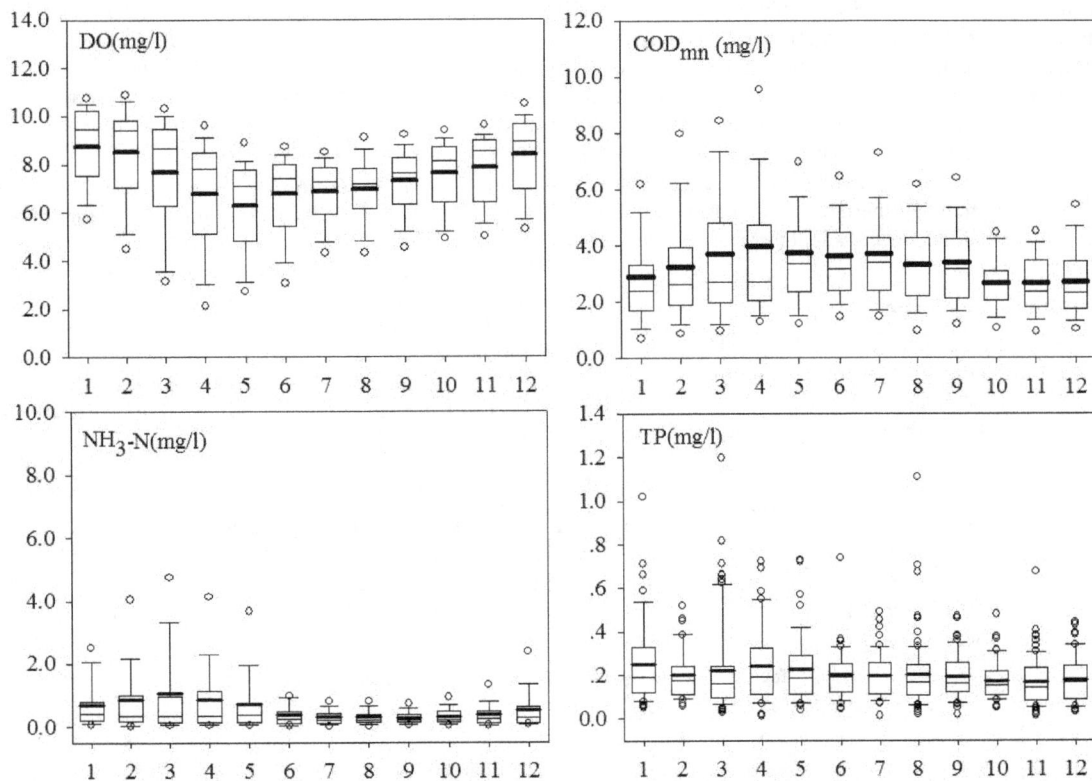

Figure 2: Seasonal variations of eight variables in the middle and lower reaches of the Min Jiang.

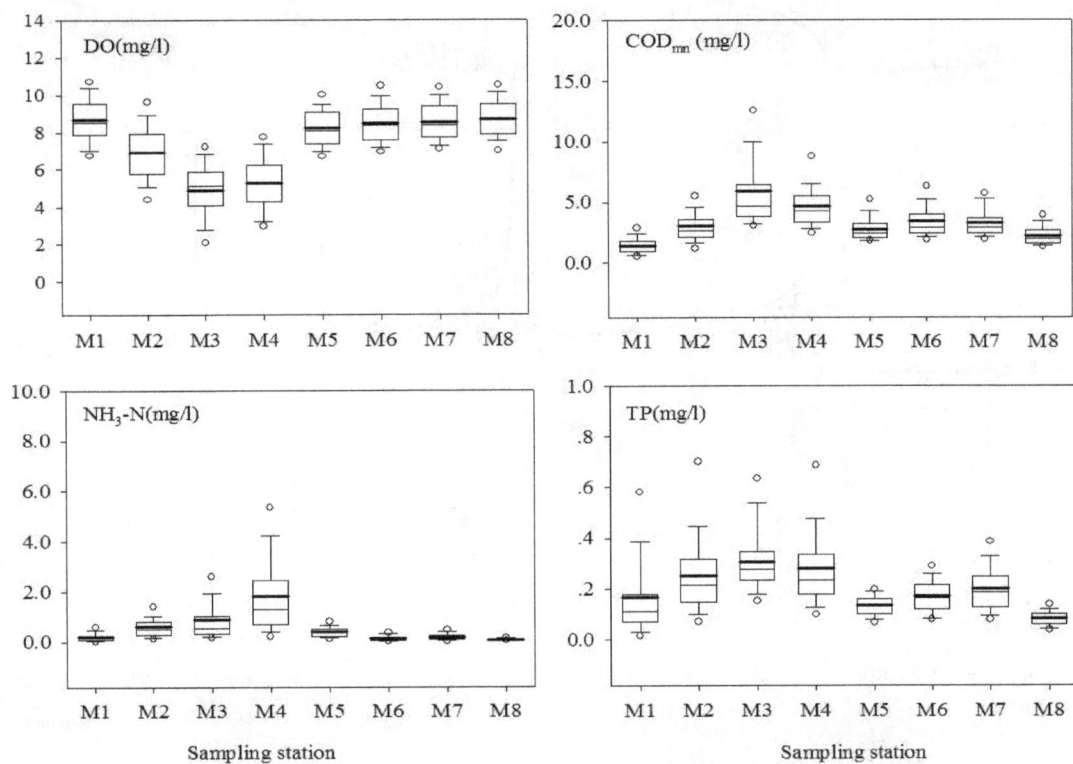

Figure 3: Spatial variations of eight variables in the middle and lower reaches of the Min Jiang.

the Min Jiang Basin, gross industrial output value account for 75%, and pollutant emission account for 73.8%. In addition, Meishan and Leshan city, where stations M3 and M4 were located respectively, have been experiencing a rapid economic development, but the lack of the treatment plants for the produced wastes and inefficient management results in direct contamination of local surface water systems. Less populated water was found in the lower reaches of the Min Jiang and its tributary due to lower anthropogenic activities and the better hydraulic conditions to dilute pollution. Furthermore, differences of surface water quality between the middle reaches and lower reaches due to the influence of tributary (Dadu River) .

Concentration-flow relationships

The relationships between pollutant (COD_{mn}, NH3-N and TP) concentration and river volumetric flow for the stations (M4, M5, M7 and M8), and their associated model solutions, were shown in Fig4. The parameter values produced from the modelling were given in Table 3.

For COD_{mn}, in the lower reaches of the Min Jiang (stations M5 and M7) and its tributary (station M8), COD_{mn} concentration showed an increasing relationship with river flow increasing (Figure 4). This feature implies diffuse source provides the major source controls on river COD_{mn} concentrations. The negative relationship between COD_{mn} and flow for station M4 implies that point source is the dominant input to the middle reaches of the Min Jiang.

The negative relationship between NH_3-N concentration and river flow was observed at all stations, except at station M8 (Figure 4). The power model describes better relationships between the NH_3-N concentration and river flow for stations M4, M7 and M5, with the correlation coefficient values are 0.67, 0.34 and 0.32, respectively (Table 3) No correlation was found between NH_3-N and river flow at station M8 (R^2=0.002). Regarding TP, the better correlation was found between TP and river flow at station M4 (0.461), following by stations M8 (R^2=0.44) and M7 (R^2=0.34), while no correlation was found at station M5 (R^2=0.02).

Station	COD_{mn}				NH_3-N				TP			
	Equation	a	b	R^2	Equation	a	b	R^2	Equation	a	b	R^2
M4	Power	17.52	-0.2497	0.34	Power	171.48	-0.8931	0.59	Power	0.7671	-0.178	0.08
M5	Power	0.7224	0.1911	0.10	Power	11.541	-0.4487	0.32	Linear	0.000003	0.1103	0.02
M7	Power	0.3098	0.3192	0.26	Power	14.337	-0.6304	0.34	Power	0.0115	0.3264	0.34
M8	Power	0.3818	0.2565	0.28	Power	0.3171	-0.0275	0.002	Linear	0.00002	0.0386	0.44

Table 3: Concentration-flow relationships for four stations.

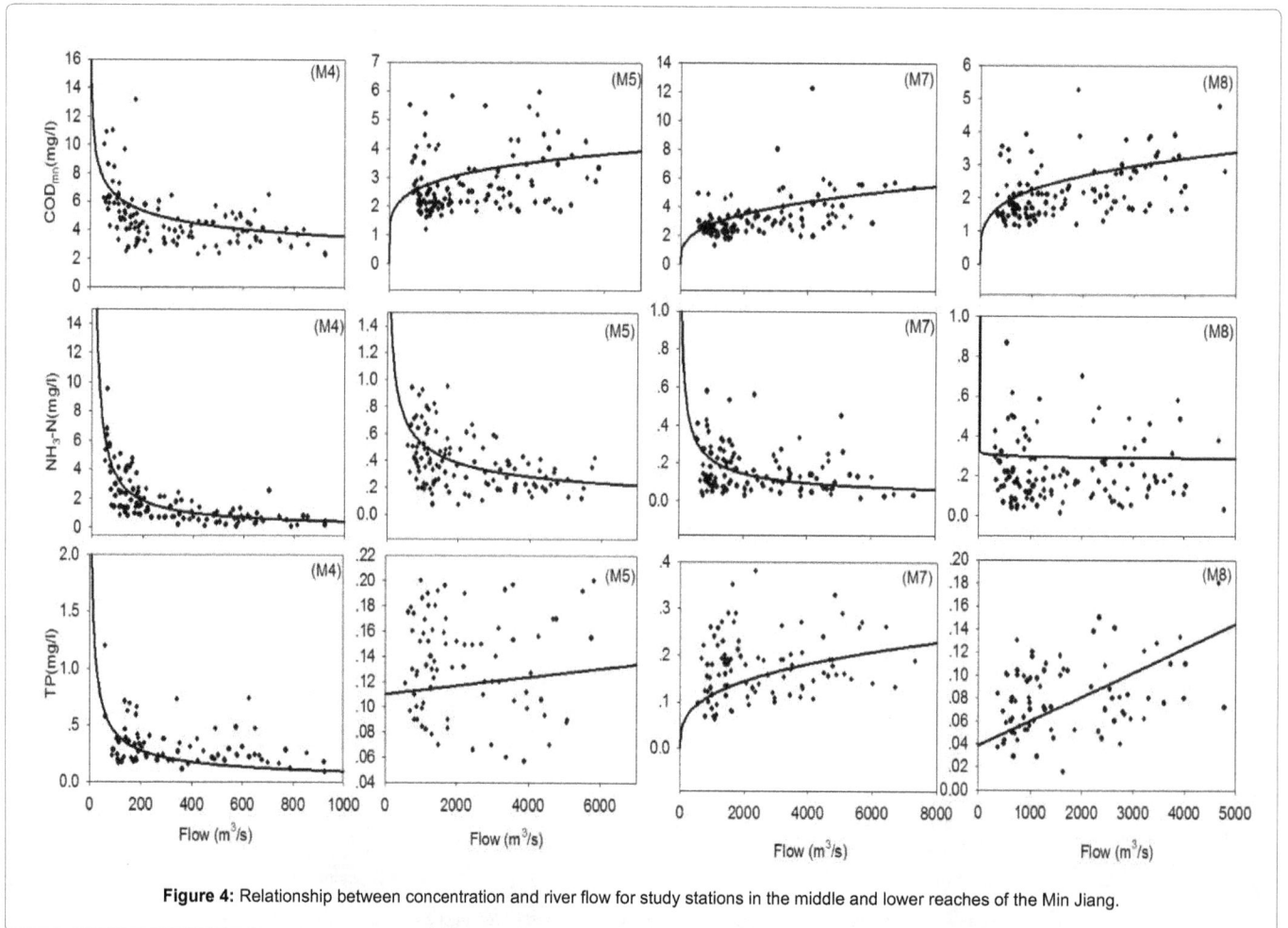

Figure 4: Relationship between concentration and river flow for study stations in the middle and lower reaches of the Min Jiang.

All water quality monitoring stations (M4, M5 and M7) in the main stream of the Min Jiang exhibited a decrease NH_3-N concentration with increasing river flow. This feature implies point source provides the major source controls on river NH_3-N concentrations. In addition, station M4 showed a dilution TP relationship with increasing river flow, indicating contributions from agricultural diffuse sources are minimal and point sources are the overwhelmingly dominant source of TP to rivers [5,19,20]. For stations M7 and M8, TP concentration showed an increasing with river flow increasing, indicating a lack of any significant point source contribution and a dominance of diffuse TP input.

Result from analyses of the concentration-flow relationship showed the organics (COD_{mn}) and nutrients (NH_3-N and TP) pollution in the middle reaches of Min Jiang dominated the point sources input, which attributed to these regions are predominantly domestic and industry activities, and have many wastewater treatment plants (WWTPs) input [21,22]. Furthermore, In the lower reaches of the Min Jiang (stations M5 and M7) and its tributary (station M8), a lack of any significant point source contribution or a dominance of diffuse pollutant input, which attributed to these regions are predominantly agriculture activities and. have no WWTPs input.

Source apportionment of pollution loads

Applying the load estimation model, the loads of COD_{mn}、NH_3-N and TP entering the Min Jiang, and the percentages of point source and diffuse sources loads to the total loads are shown in Table 4. Table 4 showed that the total diffuse source loads of both COD_{mn} and NH_3-N were less than point source loads in the study basin, and TP loads of point source were nearly equal to diffuse source loads.

The percentages of point source and diffuse sources loads to the total loads at each monitoring station are presented in Figures 5-7, respectively. Diffuse source loads were higher than point source loads at 50% monitoring stations. The point source loads of NH_3-N in M4 and M5 were greater than diffuse source loads, respectively. However, there were some difference between the main stream and tributaries. COD_{mn} loads of diffuse source were greater than those of point source loads M7 did not.

Management Options for Water Quality Improvement

Through the analyses of the spatial-temporal variations of water quality parameters and concentration-flow relationships, a number of control measures can be recommended to mitigate the pollution condition in the middle and lower reaches of the Min Jiang. In summer or high flow period, organics pollution should be as the main controlled target; in winter or low flow period, NH_3-N pollution should be as the main controlled target; TP should be as the main controlled target for the whole year. In the middle reaches of the Min Jiang, organic matters and nutrients pollution from point sources need to be reduced; Nevertheless, in the lower reaches, COD_{mn} and TP pollution control need to focus diffuse source input, while NH_3-N pollution control should target on reducing point source inputs. In tributary, controlling diffuse source inputs is the key to solve the pollution problem caused by COD_{mn} and TP, whilst NH_3-N pollution control needs to be carried out to reduce both point and diffuse source inputs.

Regarding to the quantification of point and diffuse source loads entering to the main stream and tributaries, point source, especially point source NH_3-N is the key pollution source in the middle reaches. It is necessary to enhance the treatment level of industrial effluent, to strictly carry out the discharge standard for water pollutants and the total amount control system, to incorporate NH_3-N in the total amount control system.

Together with the impact of some industrial effluents, agricultural activities and urban sewage caused a clear impact on water quality of the Min Jiang. The proportion of diffuse source pollution to total loads is generally rising, and will become the major source of the pollution loads to the river. Agricultural diffuse source pollution, such as erosion of cropland and the unreasonable application of agrochemicals to cropland, should be controlled and diminished firstly by land use planning and best management practices. Maintaining the natural geomorphologic features, especially the meandering pattern of the river is also compulsory for the good ecological condition of the river, and it is a key factor in preserving the self cleansing capacity of the river [23].

Source	COD$_{mn}$ (t/yr)	Percentage (%)	NH$_3$-N (t/yr)	Percentage (%)	TP(t/yr)	Percentage (%)
Diffuse	6119	70	749	78	317	49
Point	2640	30	211	22	334	51
Total	8759	100	961	100	651	100

Table 4: Loads of COD NH3-N and TP entering study area.

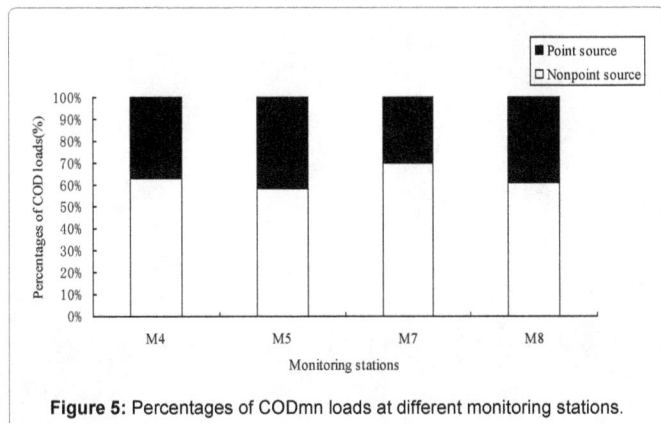

Figure 5: Percentages of CODmn loads at different monitoring stations.

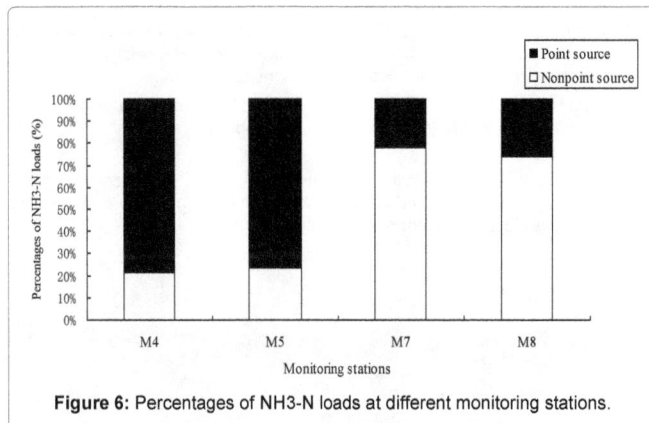

Figure 6: Percentages of NH3-N loads at different monitoring stations.

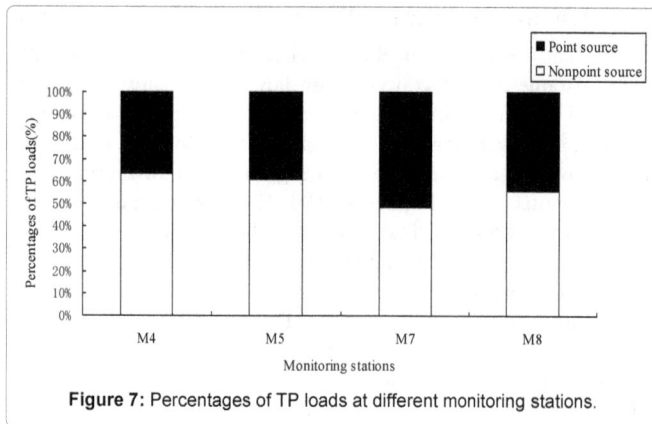

Figure 7: Percentages of TP loads at different monitoring stations.

Conclusions

This paper investigates the spatial-temporal variations and pollution source of surface water quality by analyzing the data from 8 water quality monitoring stations along the middle and lower reaches of the Min Jiang between 2003 and 2011. Seasonal and spatial differences were found for DO, COD_{mn} and NH_3-N, whereas for TP the difference was mainly due to the water quality station. The level of organics (COD_{mn}) was higher in summer (high flow period), and the level of nutrients (NH_3-N) was higher in winter (low flow period). Surface water was highly polluted in the middle reaches of the Min Jiang than that in the lower reaches and tributary. In the middle reaches of the Min Jiang, contributions from point sources (from WWTPs and industrial effluents) were the dominant inputs of organics (COD_{mn}) and nutrients (NH_3-N and TP) to river. In the lower reaches of the Min Jiang, diffuse sources were the dominant contributor of organics and TP to river; point sources were the dominant input of NH_3-N. In tributary, diffuse sources were the dominant organics and TP input, both point and diffuse sources were dominant NH_3-N input. Overall, pollution control by periods and regions is important for effective water management in the Min Jiang. It is necessary to enhance the treatment level of industrial effluent, to strictly carry out the discharge standard for pollutants and the total amount control system, to incorporate NH_3-N in the total amount control system.

Acknowledgements

This work was supported by the National Water Special Project (2012ZX07506-008). The authors sincerely thank Qian Jun and Tong Hong jin for their help in the data providing.

References

1. Vega M, Pardo R, Barrado E, Deban L (1998) Assessment of seasonal and pollutingeffects on the quality of river water by exploratory data analysis.Water Research 32: 3581-3592.

2. Elhatip H, Hinis MA, Gulgahar N (2007) Evaluation of the water quality at Tahtali dam watershed in Izmir-Turkey by means of statistical methodology. *Stochastic Environmental Research and* Risk Assessment 22: 391-400.

3. Bhangu I, Whitfield, Paul H (1997) Seasonal and long-term variations in water quality of the Skeena River at Usk, British Columbia. Water research 31: 2187-2194.

4. Antonopoulos VZ, Papamichail DM, Mitsiou KA (2001) Statistical and trend analysis of water quality and quantity data for the Strymon River in Greece. Hydrology and Earth System Sciences 5: 679-692.

5. Cooper DM, House WA, May L, Gannon B (2002) The phosphorus budget of the Thame catchment, Oxfordshire, UK: 1. Mass balance. The Science of the Total Environment 282: 233-251.

6. Ouyang Y, Nkedi-Kizza P, Wu QT, Shinde D, Huang CH (2006) Assessment of seasonal variations in surface water quality. Water Research 40: 3800-3810.

7. Bowes MJ, Smith JT, Jarvie HP, Neal C (2008) Modelling of phosphorus inputs to rivers from diffuse and point sources. The Science of the Total Environment 395: 125-138.

8. Li XD, Masuda H, Liu CQ (2006) Chemical and Isotopic Compositions of the Minjiang River, A Headwater Tributary of the Yangtze River. Journal of Environment Quality 37: 409-416.

9. Chen JS, Gao XM, He DW, Xia XH (2000) Nitrogen contamination in the Yangtze River system, China. Journal of Hazardous Materials 73: 107-113.

10. Higgitt DL, Lu XX(2001) Sediment delivery to the three gorges: I.Catchment controls. Geomorphology 41: 143-156.

11. Qin JH, Huh YS, Edmond JM, Du G, Ran J (2006) Chemical and physical weathering in the Min Jiang, a headwater tributary of the Yangtze River. Chemical geology 227: 53-69.

12. State Environmental Protection Administration (SEPA) (2002) Environmental Quality Standards for Surface Water of China.

13. Mendiguchía C, Moreno C, García-Vargas M (2007) Evaluation of natural and anthropogenic influences on the Guadalquivir River (Spain) by dissolved heavy metals and nutrients. Chemosphere 69: 1509-1517.

14. Edwards AMC (1973) The variation of dissolved constituents with discharge in some Norfolk rivers, Journal of Hydrology 18: 219-242.

15. Hirsch RM, Slack JR, Smith RA (1982) Techniques of trend analysis for monthly water quality data. Water Resource Research 18: 107-121.

16. Tsirkunov VV, Nikanorav AN, Laznik MM, Dongwel Z (1992) Analysis of long-term and seasonal river water quality changes in Latvia. Water Research 26: 1203-1216.

17. Kannel PR, Lee S, Lee YS (2008) Assessment of spatial-temporal patterns of surface and ground water qualities and factors influencing management strategy of groundwater system in an urban river corridor of Nepal. Journal of Environment Management 86: 595-604.

18. Edwards AC, Withers PJA (2008) Transport and delivery of suspended solids, nitrogen and phosphorus from various sources to freshwaters in the UK. Journal of Hydrology 350: 144-153.

19. Wood FL, Heathwaite AL, Haygarth PM (2005) Evaluating diffuse and point phosphorus contributions to river transfers at different scales in the Taw catchment, Devon, UK. Journal of Hydrology 304: 118-138.

20. Jarvie HP, Neal C, Withers PJA (2006) Sewage–effluent phosphorus: a greater risk to river eutrophication than agricultural phosphorus, The Science of the Total Environment 360: 246-253.

21. Neal C, Jarvie HP, Howarth SM, Whitehead PG, Williams RJ, et al. (2000) The water quality of the River Kennet: initial observations on a lowland chalk stream impacted by sewage inputs and phosphorus remediation. The Science of the Total Environment 251-252: 477-495.

22. Neal C, Jarvie HP, Love A, Neal M, Wickham H, et al. (2008) Water quality along a river continuum subject to point and diffuse sources. Journal of Hydrology 350: 154-165.

23. Kiedrzyńska E, Kiedrzyński M, Urbaniak M, Magnuszewski A, Skłodowski M, et al. (2014) Point sources of nutrient pollution in the lowland river catchment in the context of the Baltic Sea eutrophication. Ecological Engineering 70: 337-348.

7

Impact of Caspian Sea Drying on Indian Monsoon Precipitation and Temperature as Simulated by RegCM4 Model

Abhishek Lodh[1,2*]

[1]Indian Institute of Technology Delhi, Centre for Atmospheric Sciences, Hauz Khas, New Delhi 110016, India
[2]National Centre for Medium Range Weather Forecasting (NCMRWF) Earth System Science Organisation, Ministry of Earth Sciences, A-50, Sector-62, NOIDA- 201 309, India

Abstract

This study using a regional climate model, ICTP-RegCM4.0 simulations examines the impact of drying and shrinking of Caspian Sea on Indian summer and winter monsoon, particularly on precipitation over northern plains of India due to Western disturbances. Shrinking of Caspian Sea is a man-made catastrophe with serious environmental implications. To perform the sensitivity experiment the original landuse map in the model is altered where the "Caspian Sea" in Central Asia is changed to "semi-desert" in place of "inland water" type of vegetation. The model is forced with NNRP2 boundary conditions for year 2009, 2010. Analysis of sensitivity experiment output w.r.t baseline experiment says that rainfall over Northern India decreases (significant at 5% level), during the months of winter season (months of October to March) primarily from Western disturbances originating from Central Asia and Caspian Sea region. Also, it is found that minimum (maximum) temperature decreases (increases) particularly over Indian region during October to March and June to September. During June to September (for year 2009, 2010) from model simulations results it is found that over Central Asia (India) air temperature extending upto 700hPa increases (decreases).

Keywords: Regional climate model; Caspian sea; Precipitation; Western disturbances; Temperature

Introduction

Previous studies pertaining to Central Asia fails to draw a clear boundary between local and regional responses to global climate change and the trends caused by local land use changes, such as massive irrigation activities by constructing dams over the rivers feeding the Caspian Sea and the consequent desertification processes [1]. Though shrinking of the Aral Sea and Caspian Sea are considered as "Planet Earth's worst environmental disaster", hence in this research work it is planned to examine the impact of shrinking and drying of Caspian Sea on Indian monsoon circulation and precipitation. In the past few decades there is significant decrease in sea level of Caspian Sea attributed to human induced activities as threatened by pollution and climate change (http://www.naturalhistorymag.com/features/112161/fate-of-the-caspian-sea).

There are various factors, which determine the climate of India. India is divided by the Tropic of Cancer, north of it lies in sub-tropical and temperate zone and the part lying south lies in the tropical zone. Westerly jet streams or winds along the altitude of 9-13 km (~ 300hPa - 150hPa) flows from west to east through West and Central Asia. This westerly jet stream also blow across latitudes north of the Himalayas and reaches the Tibetan highlands. There are also shallow cyclonic depressions or western disturbances originating from Mediterranean Sea which travel eastwards across West Asia, Caspian Sea finally reaching Indian subcontinent, impacting weather of north India, northwestern India and Pakistan during the winter months (November, December, January, February and March). The westerly disturbances get moisture content from the Caspian Sea. Chaudhari et al. (2008) [2] while studying 2009 and 1994 Indian monsoons finds the cyclonic anomalous circulations develop over Caspian Sea near Central Asia during 2002 severe drought monsoon year whereas anticyclonic anomalous circulation develop during 1994 rain year. Rajeevan (1993) [3] found that anomalous cyclonic circulations with cold temperature over Caspian Sea develop during pre-monsoon season of Indian drought years, which persists during later monsoon season. Study by Rajeevan 1993 [3], strengthened the findings of Krishnamurthi et al. (1989) [4]. These cold cyclonic anomalies affect Indian summer monsoon by delaying the summer heating of the landmass [3]. Hence, the role of drying Caspian Sea in steering the westerly winds or disturbances towards India is addressed in this study in the context of climate change.

In this paper, the "methodology" of study is followed by "Result and Discussions" and then the "Conclusions" of the study.

Methodology

Dynamical sownscaling approach

The RegCM4.0 (regional climate model version 4.0) [5,6] coupled with BATS (Biosphere-Atmosphere Transfer Scheme) land surface scheme is used in this present study to test the objective. Parameterization schemes employed are same as reported in earlier studies [7-10] as Indian monsoon atmospheric circulations and regions of precipitation maxima are best simulated with these combination of parameterization schemes. The baseline landuse map file in the RegCM4.0 model is modified to represent the land use change map for the sensitivity experiment. The model is run at 90 km resolution for sensitivity study over the Cordex South-Asia domain (12°E to 138°E and 33°S to 55°N) with 18 vertical levels in the atmosphere. The model is run from 00GMT of 1st January 2009 to 24GMT of 31st December 2010. For the lateral and lower boundary conditions to force the model run, NNRP 2 (NCEP-DOE AMIP-II Reanalysis (R-2)) [11] 6-hourly data

*Corresponding author: Abhishek Lodh, Indian Institute of Technology Delhi, Centre for Atmospheric Sciences, Hauz Khas, New Delhi 110016, India
E-mail: abhishek.lodh@gmail.com

and Reynolds weekly sea surface temperature data [12] respectively, is used (Figure 1).

See Figure 1 for details of the changed land use map. Terrestrial variables like elevation, landuse, and sea-surface temperature are horizontally interpolated from a latitude-longitude world domain to a high-resolution (90 km) domain using Normer projection. BATS have 20 vegetation types in the model [13]. "Inland water" is the fourteenth vegetation (green color) and semi-desert is the "eleventh" vegetation (red color) in the landuse map. More details about the RegCM4.0 model and landuse map, its user guide can be found at the website: http://gforge.ictp.it/gf/project/regcm/frs/. In the design experiment *the Caspian Sea is converted from "inland water" type of vegetation in the original land use map to "semi-desert".* The lake model is not invoked here.

Results and Discussion

To study the impact of the drying Caspian Sea on Indian monsoons using a regional climate model, the differences in precipitation, wind magnitude and circulation at 850 hPa and 200 hPa, 2 metre-Temperature, air temperature (850 hPa and 700 hPa), maximum and minimum temperature, geopotential height are examined and chronologically the following findings are reported (Figure 2).

It is observed that during the January-February-March (JFM) season of 2009 and 2010 precipitation over north India, northwestern India and Pakistan decreases by-1 to-1.5 mm/day. But during March-April-May (MAM) season of 2009 and 2010 over Jammu and Kashmir precipitation decreases but over northwest India, northeast India precipitation increases. During the Indian summer monsoon (June-July-August-September, JJAS) season precipitation over Arabian Sea decreases whereas over Bay of Bengal increases. Over Indo-Gangetic plain, parts of east India (Orissa) precipitation decreases whereas over Nepal, Central India, northwest India, northeast India and Western Ghats precipitation increases. During the October-November-December (OND) season of 2009 and 2010 precipitation over parts of North India (Jammu Kashmir, Punjab) decreases. The above results reported are significant at 5% significance level (Figure 3).

Correspondingly during the JJAS season of 2009 it is observed that over Central India there is increase in wind magnitude, along with formation of anomalous cyclonic (anti-clockwise in northern hemisphere) circulations at 850hPa. As a result over Central India and Western Ghats precipitation is more. But over North India there is decrease in wind magnitude. During OND season of 2009 wind magnitude is seen to be decreasing over North, Central and peninsular

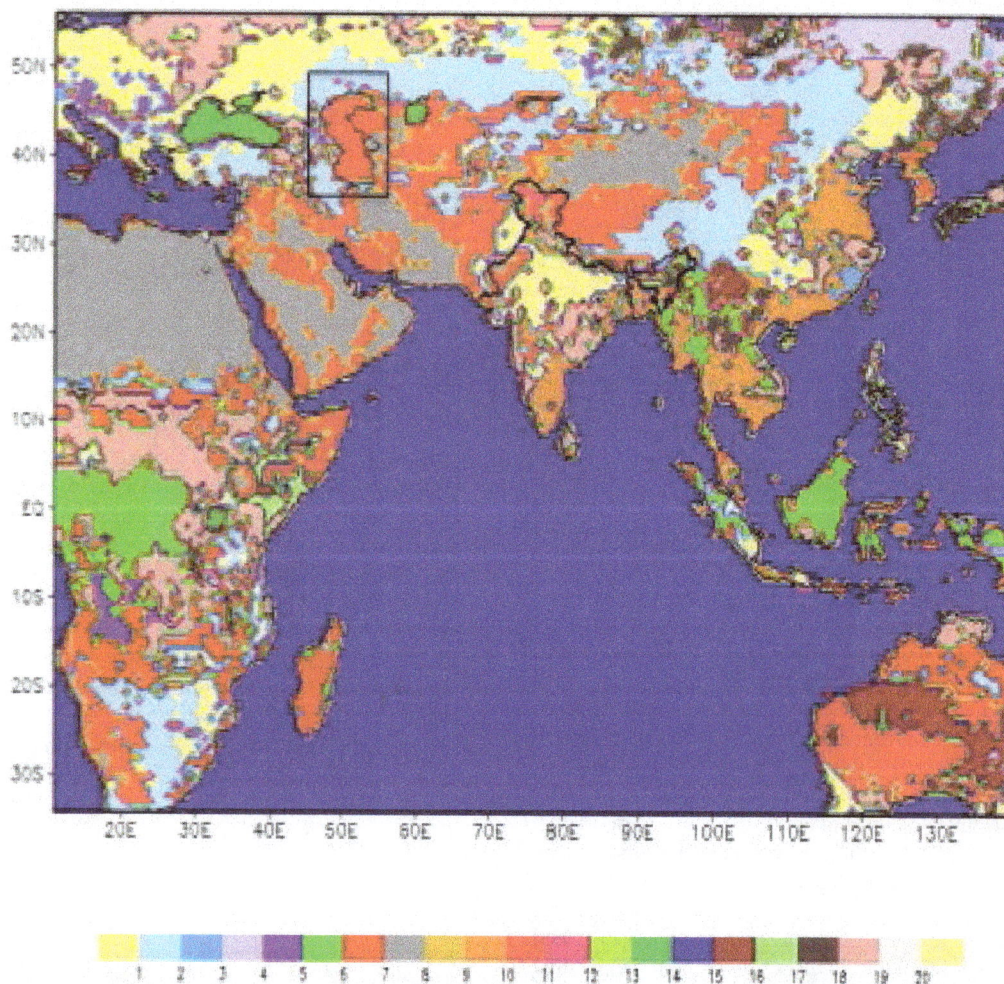

Figure 1: Designed Land use map for Caspian Sea sensitivity experiment, representing desert type of vegetation over Caspian Sea (marked and squared).

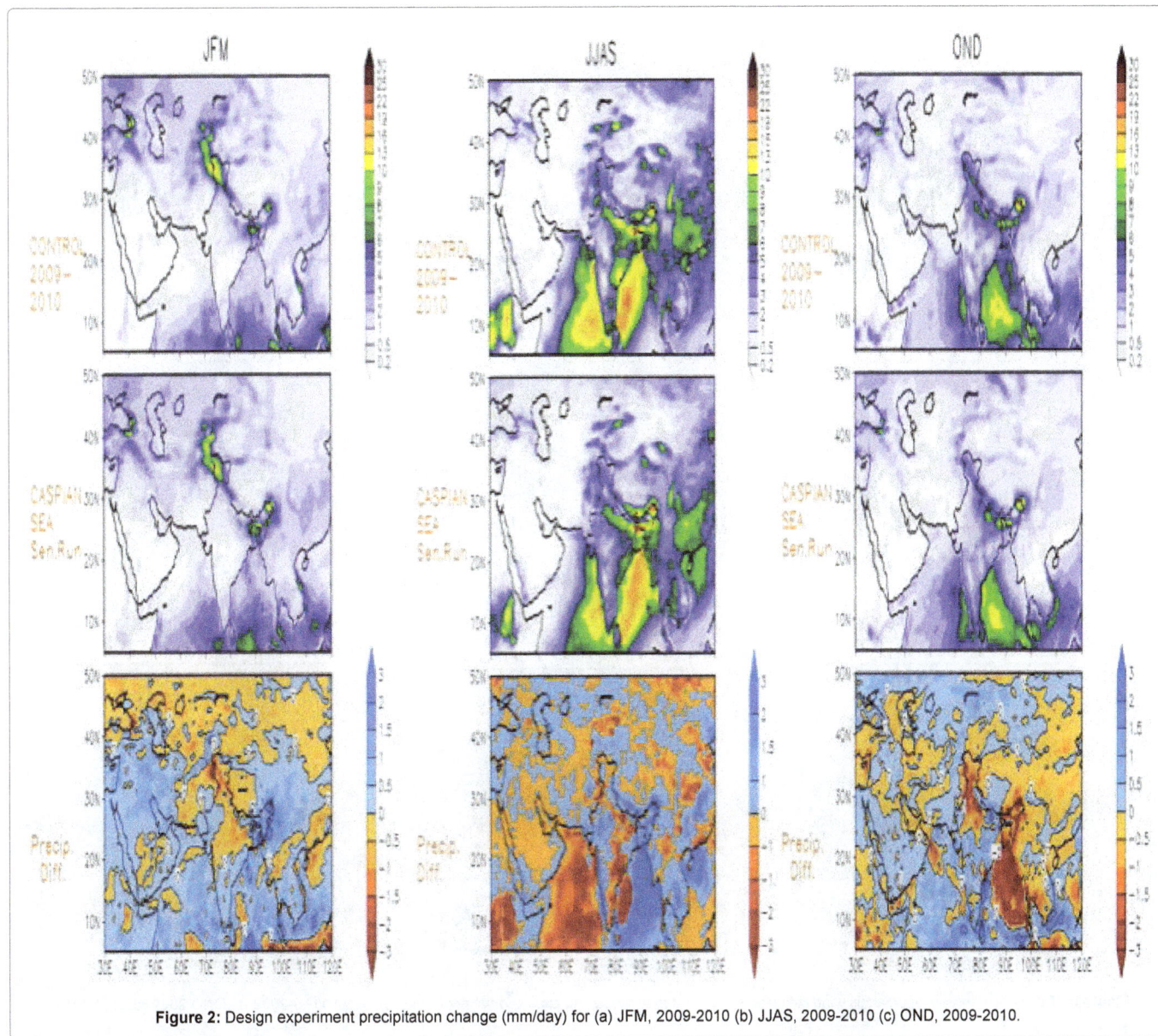

Figure 2: Design experiment precipitation change (mm/day) for (a) JFM, 2009-2010 (b) JJAS, 2009-2010 (c) OND, 2009-2010.

India. Also during JFM 2009 season, wind magnitude is seen to be decreasing over North India. Simultaneously over North India plains, particularly over Jammu and Kashmir, and Punjab it is reported that precipitation is decreasing by-1 to-1.5 mm/day. Similar results are arsing from the analysis of precipitation and wind pattern changes for pre-monsoon, monsoon and post-monsoon season of 2010. (Figures 4 and 5)

From Figure 6, change in 2-metre temperature is observed during winter (January–February–March, JFM), monsoon (JJAS) and post-monsoon (OND) season of 2009-10. During the winter season of January–February–March the 2 metre-temperature, ground temperature; air temperature from 925 hPa to 700 hPa is increasing by 1-2°C over North India, Jammu and Kashmir, and Central Asia (Figure 7). During monsoon (JJAS) season, temperature north of 30°N (over Central Asia) is seen to increase whereas south of it is seen to decrease over the Indian subcontinent and China (significant at 5% level). Over Caspian Sea temperature is seen to increase in all

the seasons. But during post monsoon (OND) season temperature is seen to decrease north of 30°N significantly. From the model output, it is observed that maximum temperature is increasing and minimum temperature is decreasing during the months, October to March and June to September. Also sea level pressure decreases north of 20°N and low-level geo-potential height (850hPa) increases over the Indian subcontinent during monsoon JJAS season (Figures 6-8).

Conclusions

Based upon the results of the study it is concluded that winter precipitation over Northern India will decrease due to drying and shrinking of Caspian Sea. The states of Punjab, Himachal Pradesh and Haryana usually get rainfall from Western disturbances during the post- monsoon and winter months starting from October to March. This is because moisture content of the westerly disturbances is augmented from the Caspian Sea, which is affected as strength of the low level winds decreases over regions north of 30°N, during

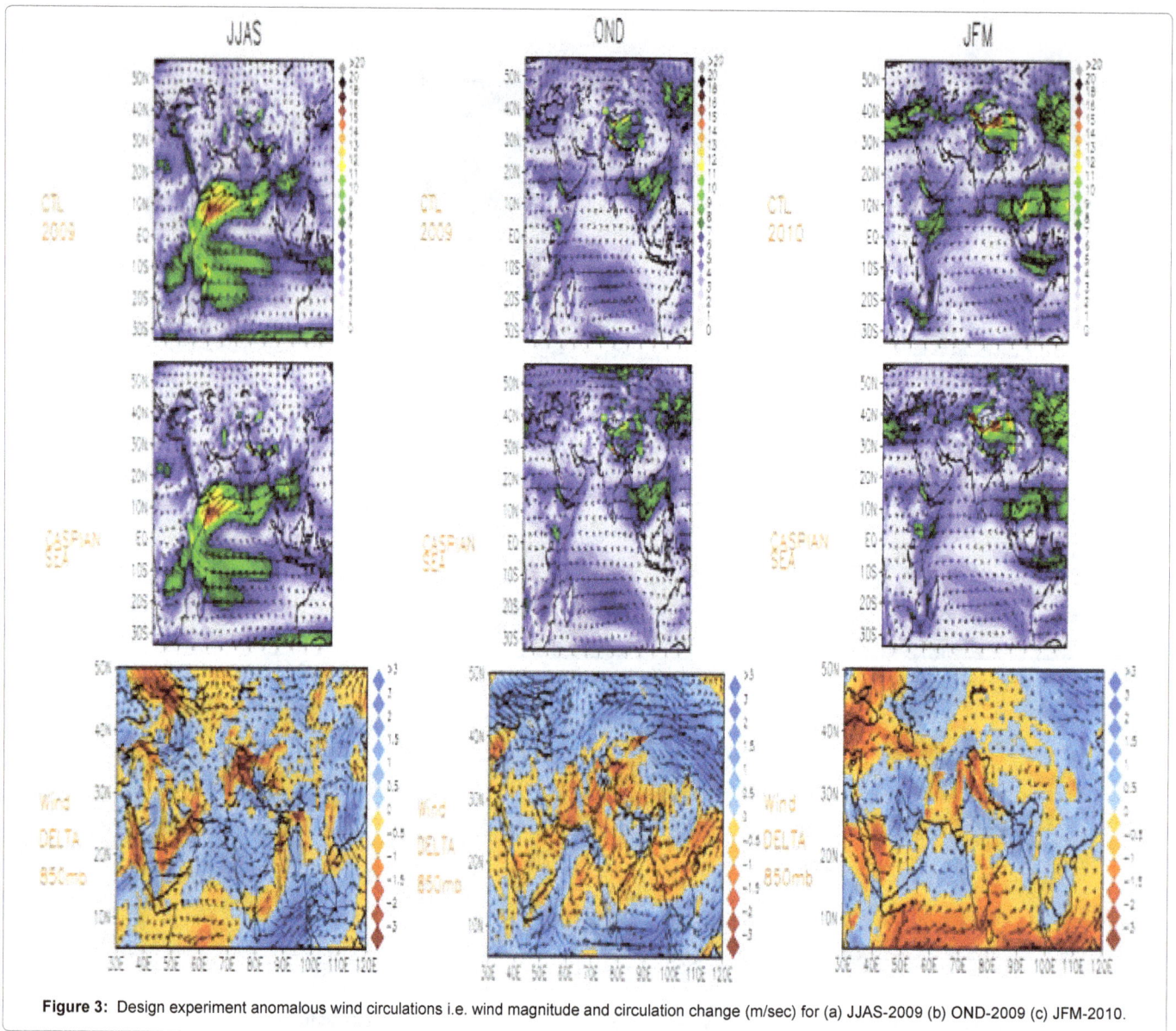

Figure 3: Design experiment anomalous wind circulations i.e. wind magnitude and circulation change (m/sec) for (a) JJAS-2009 (b) OND-2009 (c) JFM-2010.

the post-monsoon and winter months starting from October to March. It is also concluded that temperature (ground, 2metre and air temperature upto 500hPa) over Central Asia, Pakistan and northern regions of India will increase by 1-2°C, strengthened by previous climate model results as reported by Lioubimtsevaa et al. 2005 [1]. Decrease in ground temperature over Central India during monsoon (JJAS) season, increase in geopotential height coupled with dry air intrusions from Central Asia will impact the Indian summer monsoon. Hence, progress of Indian summer monsoon will be delayed as summer heating of Indian land mass is also delayed [3]. As it is observed from the sensitivity experiment results that maximum temperature is increasing and minimum temperature is decreasing, which mean days are getting hotter and nights are getting colder. For better understanding of the relationship between climate, ecosystem and hydrosphere, further modelling studies is required along with remote sensing studies. The experiments in future will be done for more number of years for better understanding.

Acknowledgment

The author is grateful to Institute student fellowship (GATE) provided by MHRD (Govt. of India). Also acknowledged the central cloud-computing facility available at Computer Services Centre, IIT Delhi for the regional climate model simulations. The author is equally thankful to ICTP, International Centre for Theoretical Physics, Trieste, Italy for making available the regional climate model codes of RegCM4.0 freely available for this research study and NCEP/NCAR for providing high-resolution meteorological datasets for setting the initial and boundary conditions to run the model. The Grid Analysis and Display System (GrADS) version 2.0 software is used for plotting. The author acknowledges Dr. H.C. Upadhyaya for his helpful suggestions to improve the write-up of the study and for permitting the submission of the paper to this journal. The author also acknowledges the research work done by researchers previously in the field of climate change, climate modelling and associated studies. Wealth of online resource available at www.google.com, scholar.google.com was also helpful.

References

1. Lioubimtsevaa E, Cole R, Adamsb JM, Kapustinc G (2005) Impacts of climate and land-cover changes in arid lands of Central Asia.Journal of Arid Environments 62: 285-308.

2. Chaudhari HS, Shinde MA, Oh JH (2008) Understanding of anomalous Indian

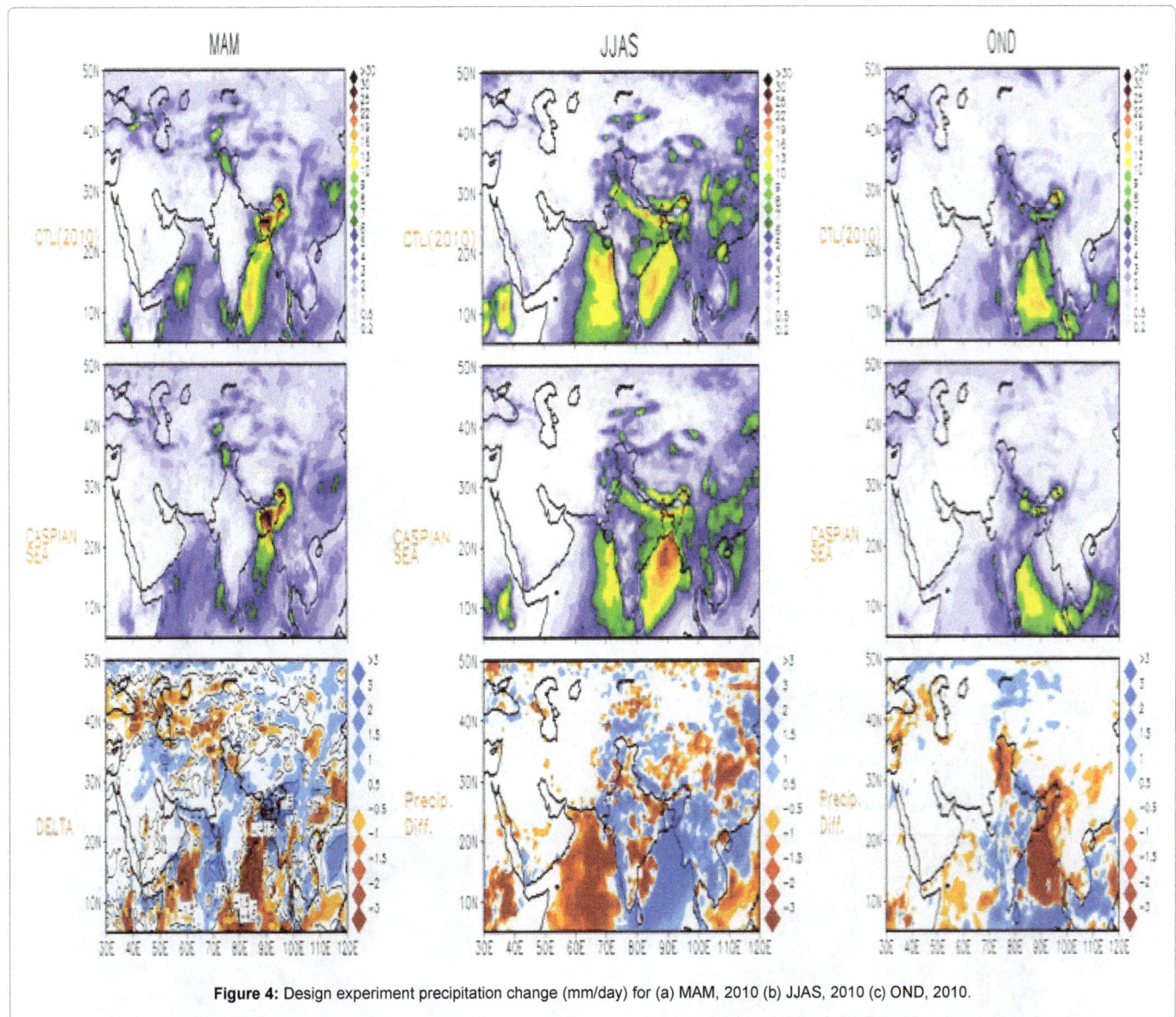

Figure 4: Design experiment precipitation change (mm/day) for (a) MAM, 2010 (b) JJAS, 2010 (c) OND, 2010.

Summer Monsoon rainfall of 2002 and Quaternary International, doi:10.1016/j.quaint.2008.05.009.

3. Rajeevan M (1993) Upper tropospheric circulation and thermal anomalies over central Asia associated with major droughts and floods in India, Research Communications 64: 244-247.

4. Krishnamurthi TN, Bedi HS, Subramaniom M (1989) The Summer Monsoon of 1987.Journal of Climatology 2: 321-340.

5. Giorgi F (2012) RegCM4: model description and preliminary tests over multiple CORDEX domains. Climate Research 52: 7-29.

6. Elguindi Nellie, Xunqiang Bi, Filippo Giorgi, Badrinath Nagarajan, Jeremy Pal, et al. (2010) RegCM Version 4.0 Core Description, The RegCM Team Trieste.

7. Lodh A, Jha S, Raghava R (2011) Impact of El Niño and La Niña on soil moisture - precipitation feedback of Indian monsoon over Central India. Int. Arch. Photogrammetry. Remote Sens 102-108.

8. Lodh Abhishek, Ramesh Raghava (2013) Soil moisture-precipitation feedback on Indian climatic zones in Indian summer monsoon regime, Vayumandal 38: 46-55.

9. Lodh A, Raghava R, Singh K and Kumar S (2014) Climatology of Atmospheric Flow and Land Surface Fields of Indian Monsoon Captured in High Resolution Global and Regional Climate Model, Earth Science and Climate Change S11.

10. Lodh A (2015) Studying Himalayan snow-Indian monsoon relationship by some LULCC change sensitivity experiments in RegCM4.0, International Journal of Geology and Earth Sciences.

11. Kanamitsu W, Ebisuzaki JW, Yang SK, Hnilo JJ, M. Fiorino, et al. (2002) NCEP-DEO AMIP-II Reanalysis (R-2) 21631-1643, Bulletin of the Atmos. Met. Soc.

12. Reynolds RW, Rayner NA, Smith TM, Stokes DC, Wang W (2002) An improved in situ and satellite SST analysis for climate. J Climate 15: 1609-1625.

13. Dickinson RE, Henderson-Sellers A, Kennedy PJ (1993) Biosphere-atmosphere transfer scheme (bats) version 1e as coupled to the NCAR community climate model, Tech. rep., National Center for Atmospheric Research.

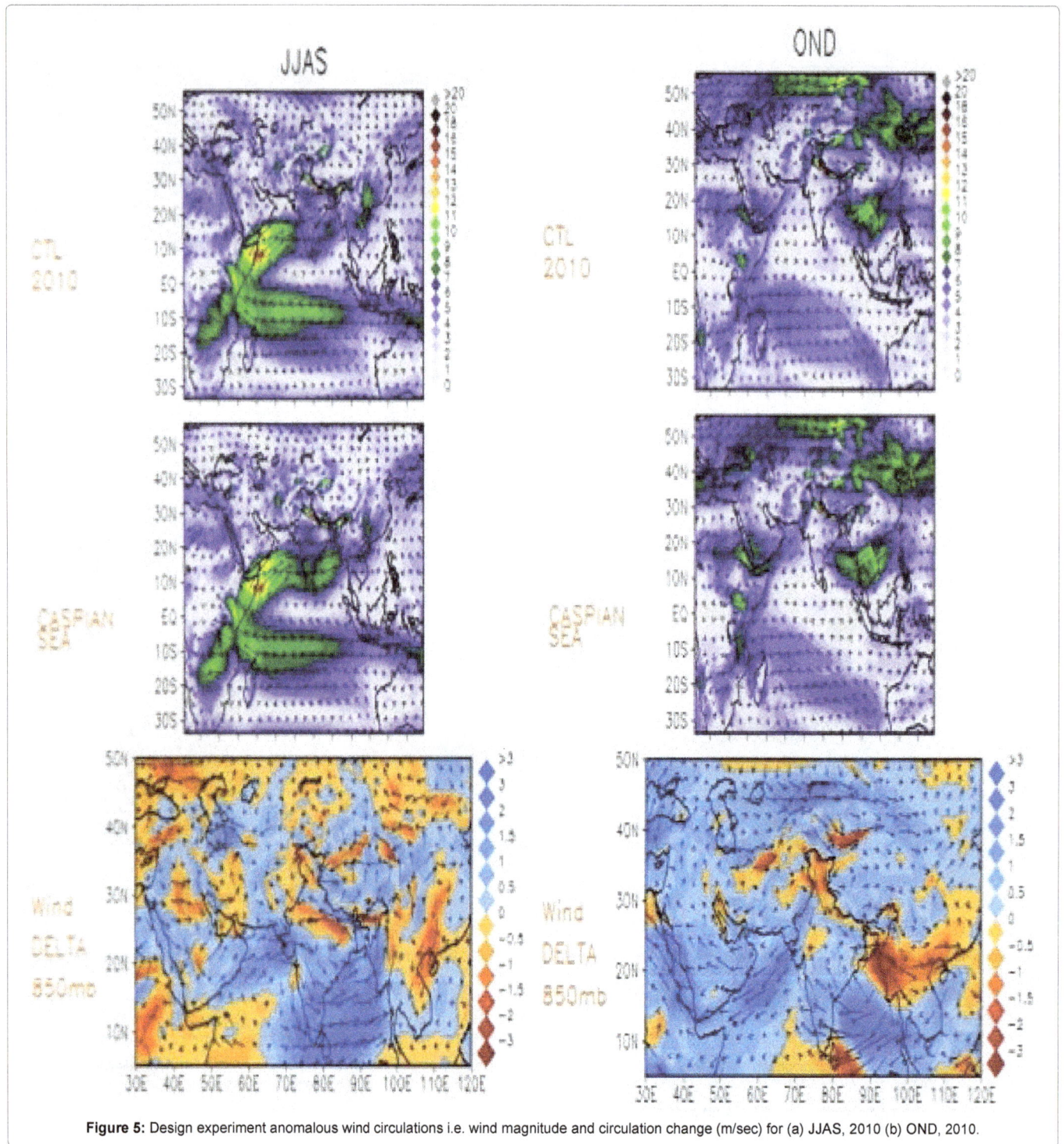

Figure 5: Design experiment anomalous wind circulations i.e. wind magnitude and circulation change (m/sec) for (a) JJAS, 2010 (b) OND, 2010.

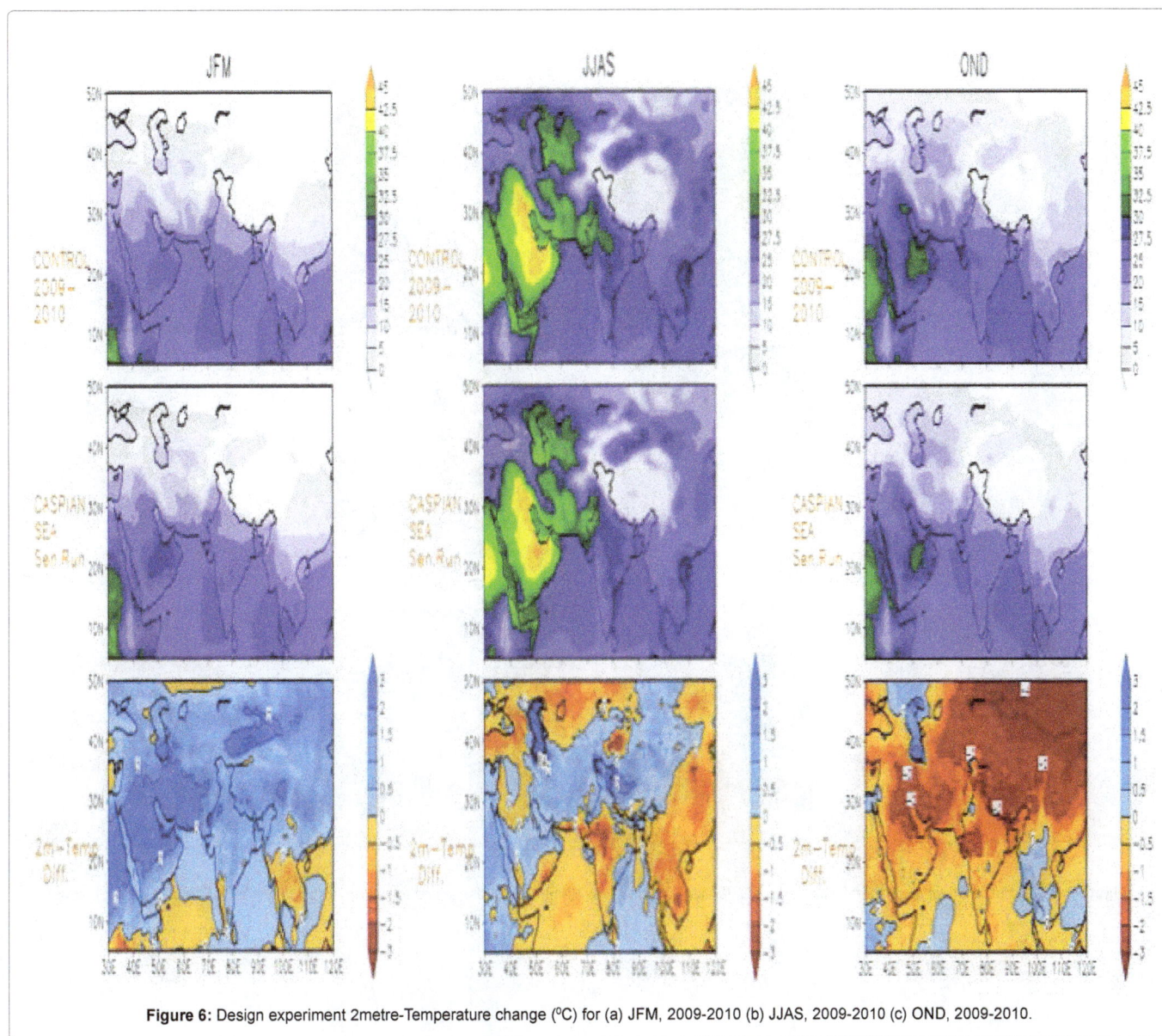

Figure 6: Design experiment 2metre-Temperature change (°C) for (a) JFM, 2009-2010 (b) JJAS, 2009-2010 (c) OND, 2009-2010.

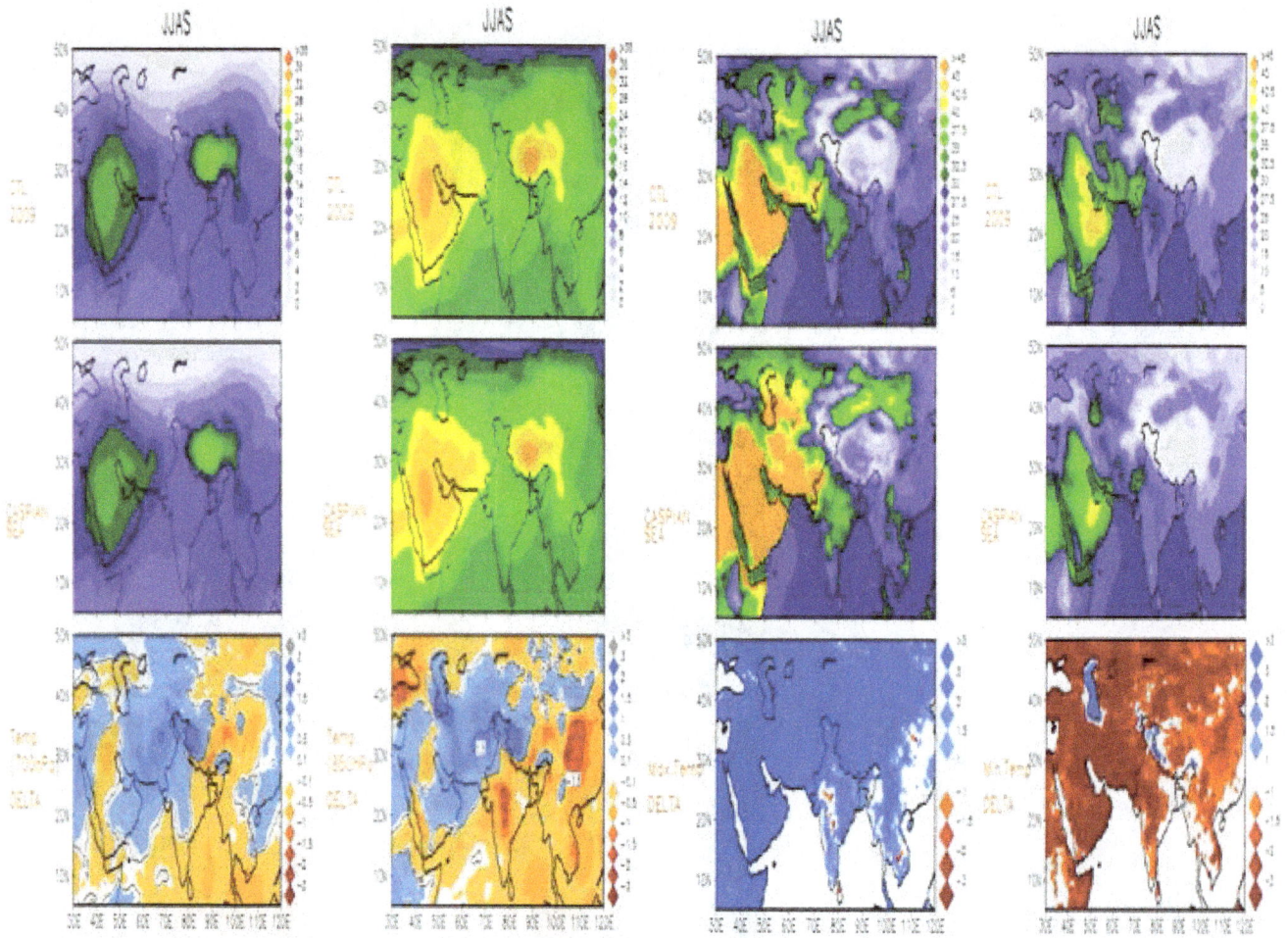

Figure 7: Design experiment air temperature change (°C) for JJAS, 2009 at (a) 700hPa (b) 850hPa, and (c) maximum and (d) minimum temperature change

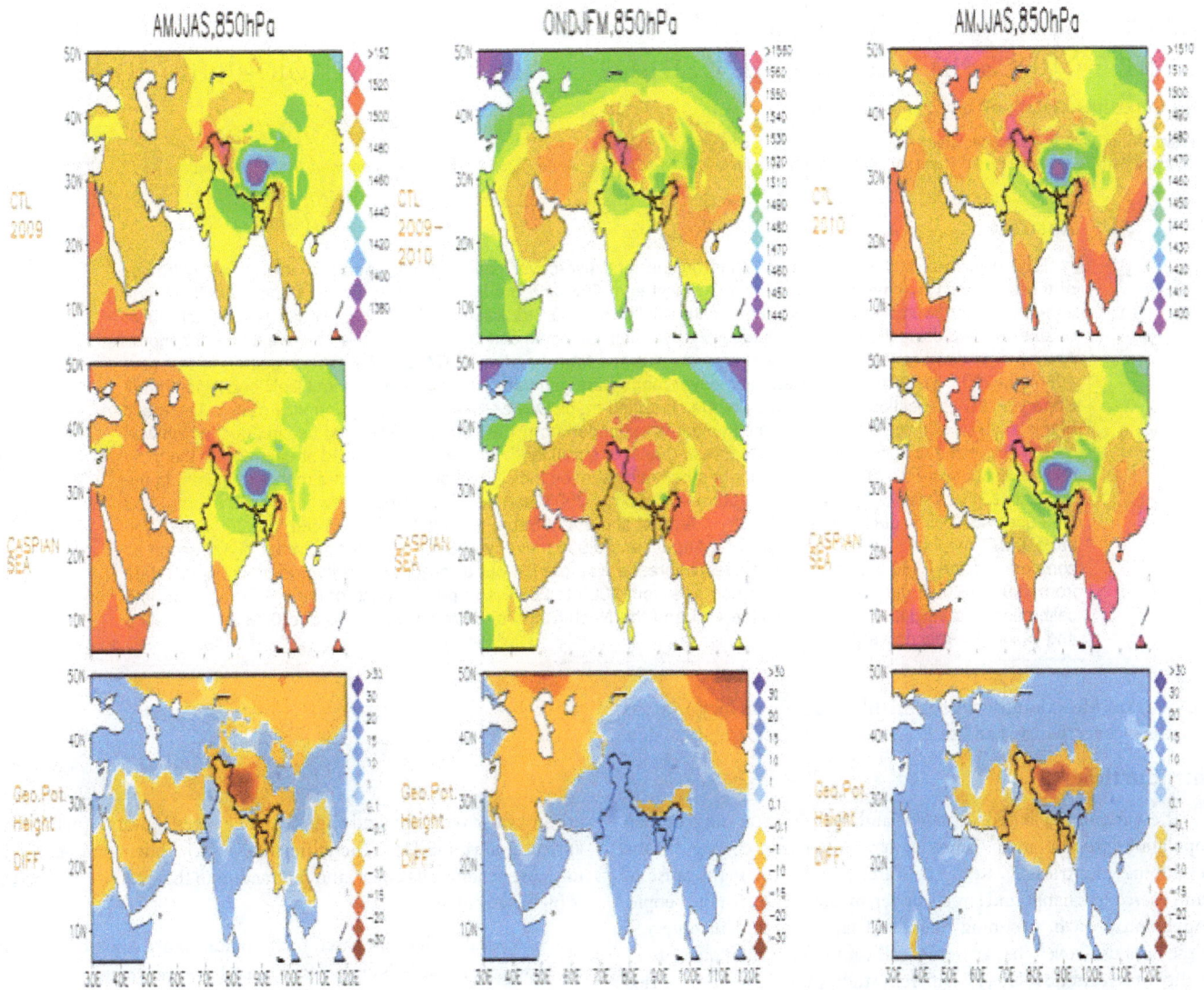

Figure 8: Design experiment Geopotential height change (m) at 850hPa for (a) AMJJAS, 2009 (b) ONDJFM, 2009-2010 (c) AMJJAS, 2010. (Here AMJJAS stands for April-May-June-July-August-September, and ONDJFM stands for October-November-December-January-February-March of 2009-2010)

8

Assessing the Impacts of Land Use-Cover Change on Hydrology of Melka Kuntrie Subbasin in Ethiopia, Using a Conceptual Hydrological Model

Yitea Seneshaw Getahun[1]* and Van Lanen HAJ[2]

[1]*Department of Natural Resources Management, Debre Berhan University, Ethiopia*
[2]*Hydrology and Quantitative Water Management Group, Centre for Water and Climate, Wageningen University, Wageningen, The Netherlands*

Abstract

The growth of population and its effect on the land use-cover change have been influencing the hydrology of the sub basin by changing the magnitude of stream flow and groundwater flow. In this paper, the likely land use-cover change impacts on hydrology of the Melka Kuntrie sub basin in the Upper Awash River Basin have been evaluated using the semi-distributed HBV hydrological model and Landsat imageries for two different periods. ArcGIS was used to generate the land use-cover maps from Landsat 5 TM and 7 ETM+ acquired, in the year 1986 and 2003, respectively. The land use-cover maps were generated using the Maximum Likelihood Algorithm of Supervised Classification. The accuracy of the classified maps was assessed using contingency matrix. The result of this analysis showed that the cultivated land has expanded from 1986 to 2003. The land use in 2003, which was mostly converted to agriculture land from forest, grass, or shrub land, showed an increased stream flow in the main rainy season, while the stream flow in dry or small rainy season indicted inconsistency from month to month. In the same time, there was a decrease in evapotranspiration in 2003 land use. The stream flow increased by the 2003 land use was 25% in June, 4% in July, 6% in August and 9% in September that corresponded to 0.065 mm/day in June, 0.077 mm/day in July, 0.07 mm/day in August and 0.039 mm/day in September for the main rainy season as compared to the 1986 land use. The model calibration was carried out using observed hydrometeorological data from 1991 to 2004 and the validation period was from 2005 to 2008. The performance of the HBV model for both calibration and validation was reasonable well and the Nash-Sutcliffe efficiency was 0.86 and 0.78 for calibration and validation, respectively.

Keywords: Melka kuntrie sub basin; HBV model; Hydrology; Impacts of land use-cover change; Landsat imageries; Hydrometeorological data

Introduction

The agricultural based economy and rapidly increasing human population are the main cause of land use-cover change in the developing countries [1]. Resource scarcity is also the main cause of Land use-cover change and largely driven by the decision of the people, population growth, declining household farm size and income [2]. These land use-cover change have significant influence on quantity or quality of stream flow [1,3,4]. Different studies that have been carried out in many parts of Ethiopia indicated that croplands have expanded at the expense of natural vegetation, forests and shrub lands [5-10]. The Land use-cover change has negative consequence in hydrological system of a sub basin [11].

High population growth, deforestation, traditional agriculture techniques, land use-cover change, and improper use of land have resulted in massive land degradation with water scarcity [12]. Population growth often results into an increase of water need and land for agriculture, while land use-cover change has an impact on the hydrology of a basin. Today, many rural families can barely make their living from agriculture because of the consequences of rapid deforestation, degradation of land resources, water scarcity, and loss of fertility in their agricultural land [12]. The major effects of land use-cover change is likely to alter the hydrologic response of sub basin and change in water availability [10,13,14]. The Land cover under little vegetation is subjected to high surface runoff and low water retention [1]. Whereas, the high vegetation covers increase, evapotranspiration and decrease the mean annual river flow. The Land use-cover plays a fundamental role in driving hydrological processes within a sub basin [15]. These include changes in water demands such as irrigation, changes in water supply from altered hydrological processes of infiltration,

groundwater recharge, and runoff, and changes in water quality from agricultural runoff [16]. Therefore, a far better understanding of land use-cover change, its effect, and interaction to the hydrology of a basin is highly essential.

Small-scale sub basin based hydrological information considering land use-cover change is crucial for stream flow assessment for irrigated agriculture or any use of water. It is very clear that water availability is becoming a critical factor in so many sectors, so that assessing the anticipated impacts of land use-cover change on hydrology is unquestionable [17,18]. Irrigation schemes in the Upper Awash River Basin have been very functional for many years, but reservoirs are becoming to be filled with sediment and therefore storage capacity is decreasing [19]. Water management in the basin is becoming very difficult that needs assessment on a regular basis because of the reduction in storage capacity, variability of rainfall, and high water demand. The basin also faced recurrent flood during the rainy season, which results in loss of life, and property damage, while at the end of the dry season there was insufficient water in the basin to meet the demand of irrigated agriculture or other purpose [20].

*Corresponding author: Yitea Seneshaw Getahun, Department of Natural Resources Management, Debre Berhan University, Ethiopia
E-mail: yiseneshaw@gmail.com

Assessing the Impacts of Land Use-Cover Change on Hydrology of Melka Kuntrie Subbasin in Ethiopia, Using...

57

The Landsat 5 TM and Landsat 7 ETM+ were Level 1T (terrain corrected), which means that those images have been corrected for geometric and radiometric correction. Band combination of 2, 3 and 4 from both imageries were used for the land use-cover analysis [21-23]. From the standard "false color" composite of band 2, 3 and 4 vegetation appears in shades of red, urban areas are cyan blue, and soils vary from dark to light browns. Ice, snow and clouds are white or light cyan [23]. For image, enhancement and classification the most common nonlinear Histogram Equalize Stretch and Supervised Maximum likelihood Classification were used, respectively [24-28]. Thematic image accuracy has been also evaluated how well the class name on the map correspond to what is really on the ground [29,30].

In summary, since the changing land use-cover and its impact on hydrological processes are a widespread concern and a great challenge, it is vital to understand the impact of land use-cover change on future hydrology in the Melka Kuntrie sub basin using a hydrological model that is fed with hydrometeorological data. The main objective of this study is to assess the expected changes in stream flow in the Melka Kuntrie sub basindue to the changing land use-cover using the conceptual rainfall-runoff hydrological HBV model and ArcGIS. Landsat imageries were analyzed to investigate land use-cover change using ArcGIS.

Description of study area

Location: The geographic location of the Awash River Basin is between 7°53'N and 12°N latitudes and 37°57'E and 43°25'E of longitudes [31]. The largest part of the Awash River Basin is located in the arid lowlands of the Afar Region in the northeastern part of Ethiopia. However, Melka Kuntrie sub basin located in the Upper Awash River Basin (Figure 1). The upper part of the Awash River Basin that is Melka Kuntrie sub basin is the study area of this research. It lies upstream of Koka dam. The Melka Kuntrie sub basin covers about 4456 km² with particular geographical location of 8:42: 0 N and 38:36: 0 E.

Climate: The movement of the inter-tropical convergence zone (ITCZ) and the influence of the Indian Monsoon throughout the year, mainly determine the climate pattern of the Melka Kuntrie sub basin [32]. There are three seasons in the Melka Kuntrie sub basin based on the movement of inter-tropical convergence zone (ITCZ), the amount of rainfall and the rainfall timing. The three seasons are Kiremt, which is the main rainy season (June-September), Bega, which is the dry season (October-January), and Belg, the small rainy season (February-May) [33]. The mean annual rainfall over the Meka kuntrie sub basin is 1216 mm.

As shown in Figure 1, the traditional climate classification based on a digital elevation model, indicated that there is a dominant Woinadega (Subtropical) climate in the southwestern, southeastern highlands and upper basin part of the river basin. The traditional climate classification zone of the region based on elevation, indicated that Kola (hot dry tropical) is between 1500-1800 m a.s.l, the WoinaDega (subtropical) is between 1800-2400 m a.s.l, the Dega (temperate) is between 2440-3500 m a.s.l, and the Wurch (alpine) is over 3500 m a.s.l [34].

Land use and soils: The common land use types in the Meka kuntrie sub basinare cultivated agricultural land, grassland, cropland with shrub land and forestland. The Meka kuntrie sub basinconsists of different soil types. The most common soil types are Cambisols and Vertisols. The Vertisolsare dominated by the montomorillonite clay mineral. This clay mineral expands when there is a wet condition and shrinks when there is a dry condition, causing cracks at the surface in the dry season [35].

Methodology and data: The conceptual semi-distributed rainfall–runoff HBV "Light" Model, ArcGIS 10.1, and otherstatistical tools were used to analyze Landsat imageries, GIS files, observed hydro meteorological data. The impact of land use-cover change on stream flow of the basin was assessed by statistical analysis of model output of hydrometeorological variables.

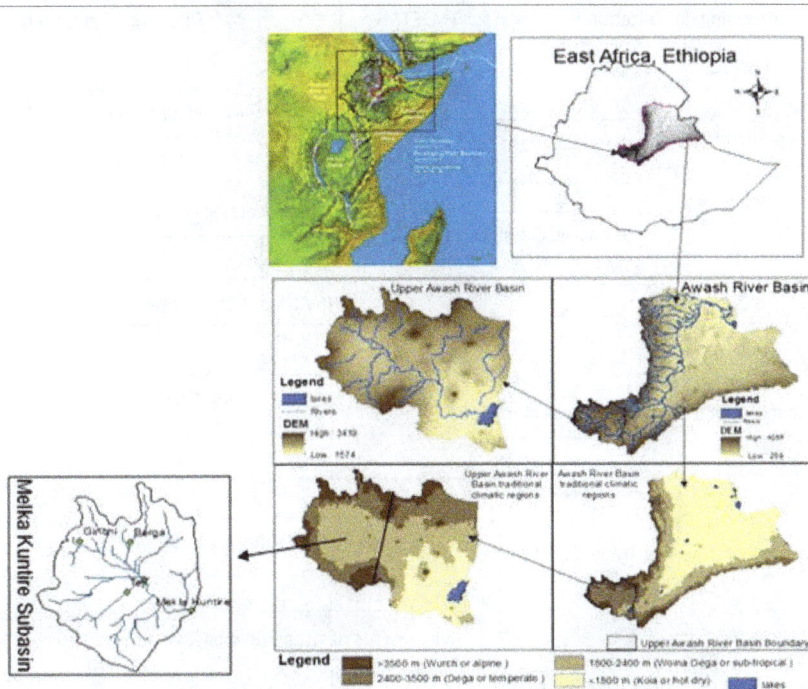

Figure 1: Study area with traditional climatic regions based on DEM (Upper left figure taken from James, 2014).

Approach: The Landsat TM and ETM+ imageries were selected for the year 1986 and 2003, respectively and processed using GIS for the land use-cover change analysis as shown in Figure 2. Accuracy of the classified land use-cover classes were verified with the set of referenced points. The observed hydro meteorological data were organized according to the requirements of the HBV model. Multiple elevation zones up to 18 with the interval of 100 m, and two vegetation zones were used for the HBV model calibration. After the HBV model calibration, validation and knowing that the model efficiency is good based on observed hydro meteorological datasets, the next step was the land use-cover change assessment using some reference period and changing the vegetation coverage based the change detection results. Finally, the land use-cover change impact on hydrology of the Melka Kuntrie sub basin were analyzed (Figure 2).

Data and data analysis

Landsat imageries: Landsat imageries of different bands were downloaded and analyzed using ArcGIS to identify the land use-cover change in the Melka Kuntrie sub basin. The two imageries of Landsat 5 TM and Landsat 7 ETM+ were downloaded from the United State Geological Survey (USGS) earth explorer in Geo TIFF format. The Digital Elevation Model (DEM) and Land use-cover were also

downloaded from the USGS and Corn Land Cover Facility (GLCF), respectively. Landsat 7 ETM+ image for the year 2003 and Landsat 5 TM image for the year 1986 were processed to detect the land use-cover change between those years.Landsat 5 TM was acquired on January 12th, 1986 and Landsat 7 ETM+ was acquired on February 23th, 2003, with the WRS-2 path/row for both imageries of 168 and 169/54.The timing of both images was as close as possible that is in the same annual season to circumvent a seasonal variation in vegetation pattern.

The spatial resolution is 30 m for all bands apart from band 6 that has a spatial resolution of 120 meters (TM 5) and 60 m (ETM+) (Table 1). Landsat 7 is equipped with an Enhanced Thematic Mapper Plus (ETM+), the successor of TM. The observation bands are essentially the same seven bands as TM, and the newly added panchromatic band 8, with a high resolution of 15 m was added.

The accuracy of a classified image refers to the extent to which it agrees with a set of reference data. Most quantitative methods to assess classification accuracy involve an error matrix built from the two datasets, which are remotely sensed map classification and the Google Earth reference data. The reference data was taken from Google Earth. Reference data were collected for each class type and compare against the classified image using a contingency matrix (Table 2). Overall map

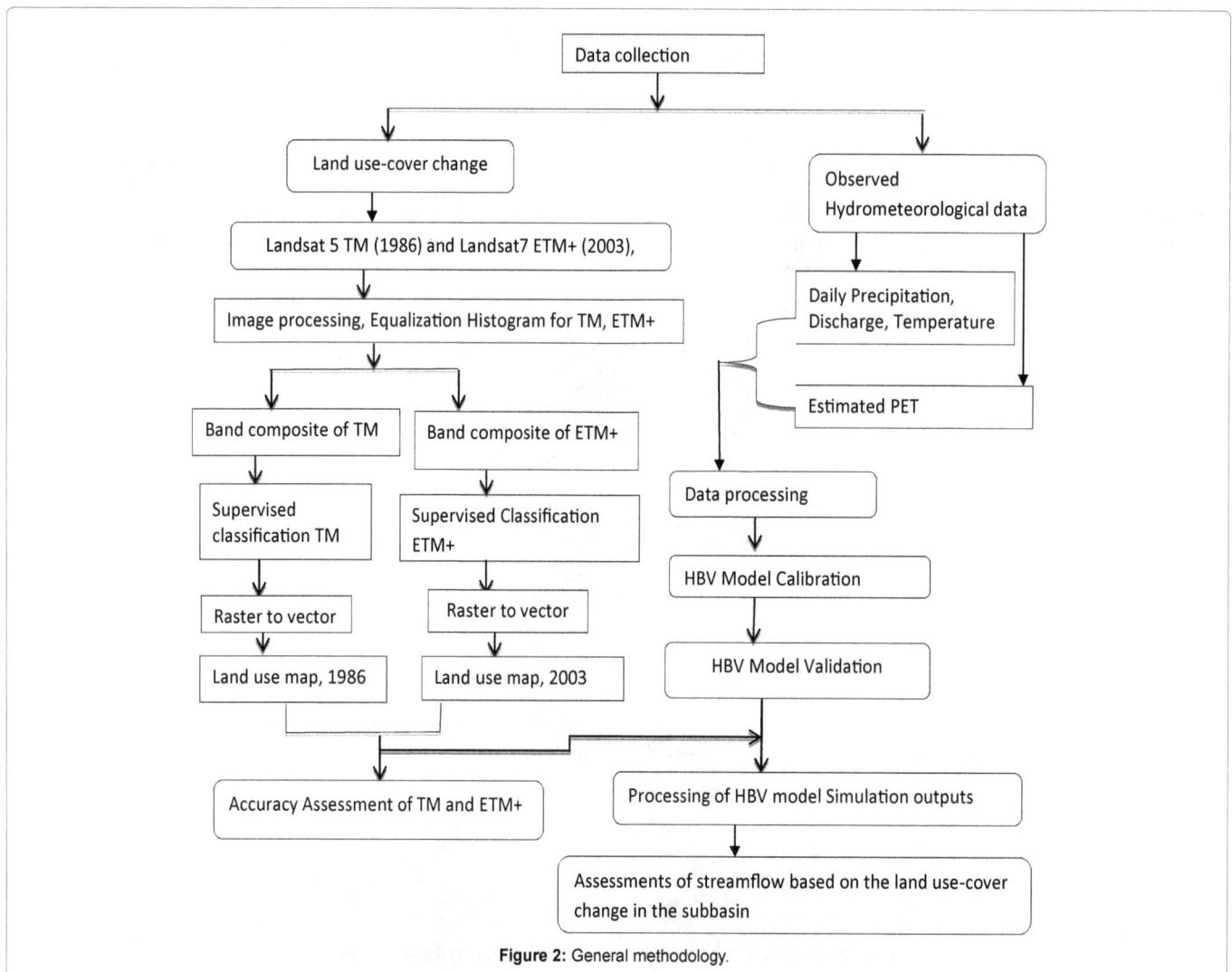

Figure 2: General methodology.

Table 1: Landsat 5 TM and ETM+ Sensor specification.

Bands	Description	Spatial resolution (m)	Spectral resolution (µm)	Temporal resolution (days)
1	Blue	30	0.45-0.515	16
2	Green	30	0.525-0.605	16
3	Red	30	0.63-0.69	16
4	Near Infrared	30	0.75-0.90	16
5	Short-wave Infrared	30	1.55-1.75	16
6	Thermal Infrared	120 (TM) 60 (ETM)	10.40-12.5	16
7	Short-wave Infrared	30	2.09-2.35	16
8	Panchromatic	15	0.52-0.90	16

Table 2: Accuracy assessment using a contingency matrix.

Classified image	Referenced data			
	Agriculture	Grass/shrub land	Forest	Total
Agriculture	24	3	0	27
Grass/shrub land	4	22	3	29
Forest	1	2	25	28
Total	29	27	28	84

accuracy was computed by dividing the total correct (obtained by summing the major diagonal of the error matrix) by the total number of pixels in the error matrix. Error of omission is the percentage of pixels that should have been put into a given class but were not. Error of commission indicates pixels that were placed in a given class when they actually belong to another (Figure 3).

Many ways are available to look at the thematic accuracy of a classified image. The overall, producer, and user accuracy criteria were investigated for the classified image.

Overall accuracy

Correctly classified is the diagonal values which is =24+22+25=71

Total number of reference =84

Overall accuracy = (71/84)*100%=84%

Errors of omission are the type on the ground is not that type on the classified image – the real type is omitted from the classified image.

Errors of omission for Agriculture land =1+4= (5/29)*100%=17%

For grass/shrub land =3+2= (5/27)*100%=18%

For Forest= 0+3= (3/28)*100%=10%

Error of omission for agriculture, grass/shrub land and forest were 17, 18 an 10 percent respectively, indicating that all 29, 27 and 28 reference pixels for agriculture, grass/shrub land and forest were categorized well which means above 80 % producers accuracy.

Producer's accuracy = 100%- error of omission (%)

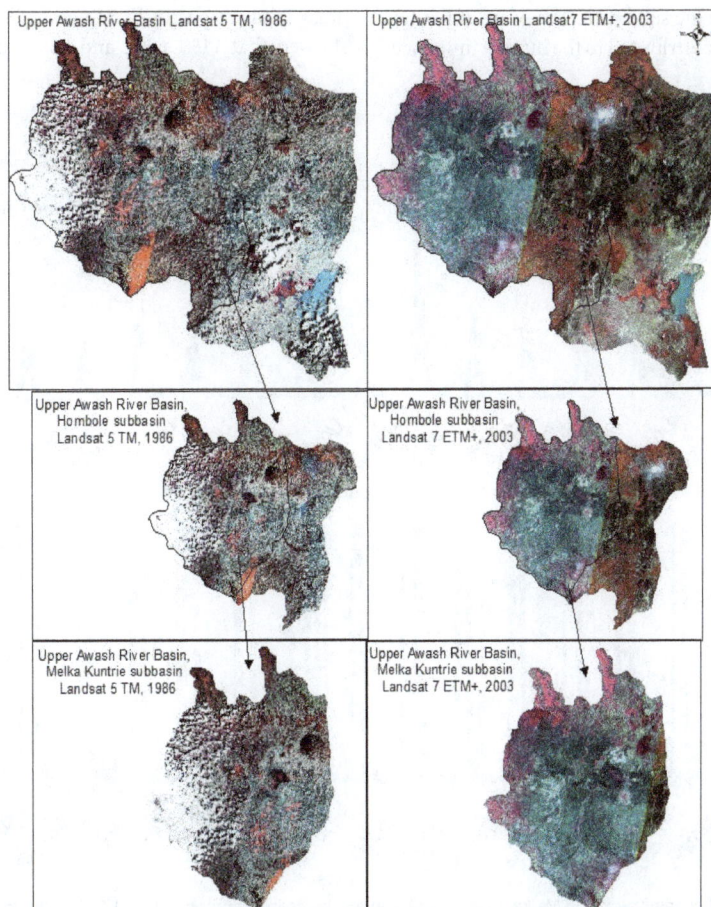

Figure 3: The Standard "False Color" composite image of the Hombole and Melka Kuntrie subbasin of the year 1986 and 2003.

Errors of commission represent pixels that belong to another class but are labeled as belonging to the class

Errors of commission for Agriculture =0+4= (4/30)*100%=13%

For grass/shrub land =3+3= (6/29)*100%=20%

For Forest= 1+2= (3/23)*100% =13%

User's accuracy = 100%- error of commission (%)

User's accuracy or reliability is indicative of the probability that a pixel classified on the map/image actually represents that category on the ground.

Model calibration (1991-2004) and validation (2005-2008)

The HBV model was calibrated using observed hydro meteorological data for the Melka Kuntrie sub basin from the period (1991-2004) and the Nash-Sutcliffe model efficiency (NS)was 0.87. Whereas, the validation Nash-Sutcliffe model efficiency was 0.78. The HBV model performance for the calibration and validation period was reasonably well for the Melka Kuntrie sub basin.

The observed versus simulated stream flow hydrograph for the calibrated period (Figure 4) indicate that the HBV model underestimated the high flow and overestimated the low flow in the Melka Kuntrie sub basin in most of the years, excluding the years 1995 and 2004 for high flow. The underestimation of high flow in many years may be attributed to data quality, the less ability of the HBV model to characterize the sub basin, orographic enhanced intense and high amount of precipitation, soil type in the sub basin. The overestimation that happened in the low flow may be attributed to the human influence

in the sub basin; there are a lot of small scale irrigation system and water extraction for different purposes. The Melka Kuntrie sub basin total mean stream flow was 199.56 mm/year and 196.34 mm/year for observed and simulated, respectively.

The model performance in the validation period were slightly poorer compared with the results of the calibration period. The observed versus simulated stream flow hydrograph for the validation period (Figure 5) indicate a similar pattern as for the calibration hydrograph of HBV model that underestimate the high flow and overestimate the low flow.

Results and Discussions

Forest, shrub/grass, and agricultural land were the major land use-cove types in the Melka Kuntrie sub basin. The classified forest, shrub/grass, and agricultural land use-cover types for the Melka Kuntrie sub basin indicated that most of the forest and shrub/grass lands that were in the 1986 converted into agricultural land in the year 2003 (Figures 6 and 7). Agricultural expansion was the major driver of land use change in the Melka Kuntrie sub basin. There was more shrub/grass land in 1986 than in 2003, which approached the area of agricultural land though the agricultural land was the largest in both years. Besides, the conversion of forest land into agricultural land there was also a largest area of shrub/grass land in 1986 that converted into agricultural land for the 2003 land use-cover (Figures 6 and 7).

The land use-cover map for the year 1986 in Figures 6 and 7 showed that there was about 19%, 45%, and 36% forest, agricultural and shrub/grass land, respectively. However, the land use- cover map for 2003 showed that 11%, 63%, and 23% was forest, agricultural and shrub/

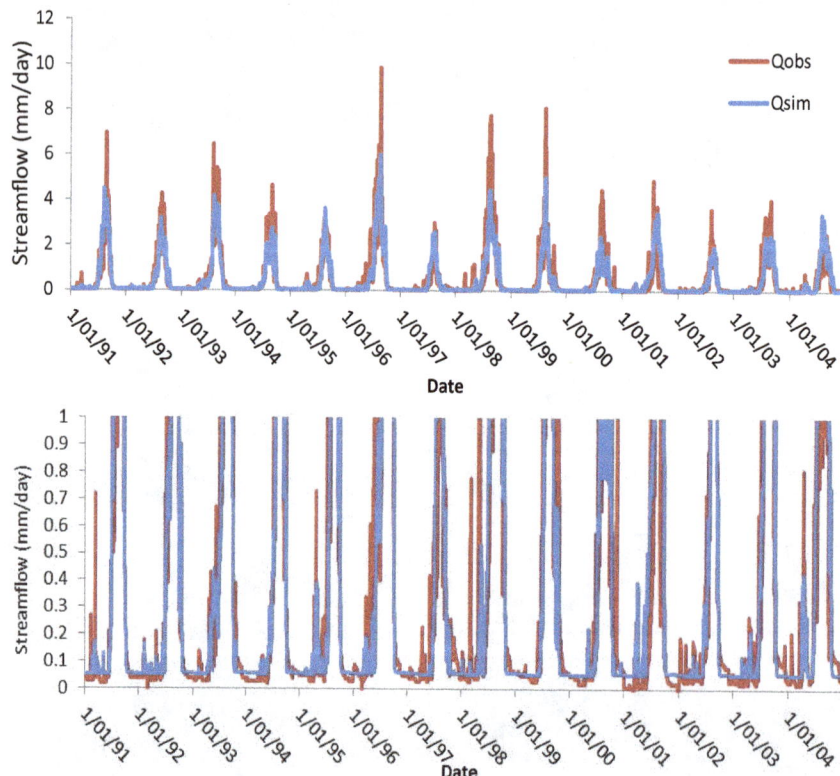

Figure 4: The observed and simulated streamflow in the Melka Kuntrie subbasin for the calibration period (1991-2004), the lower pannel indicates the zoomed in observed and simulated streamflow.

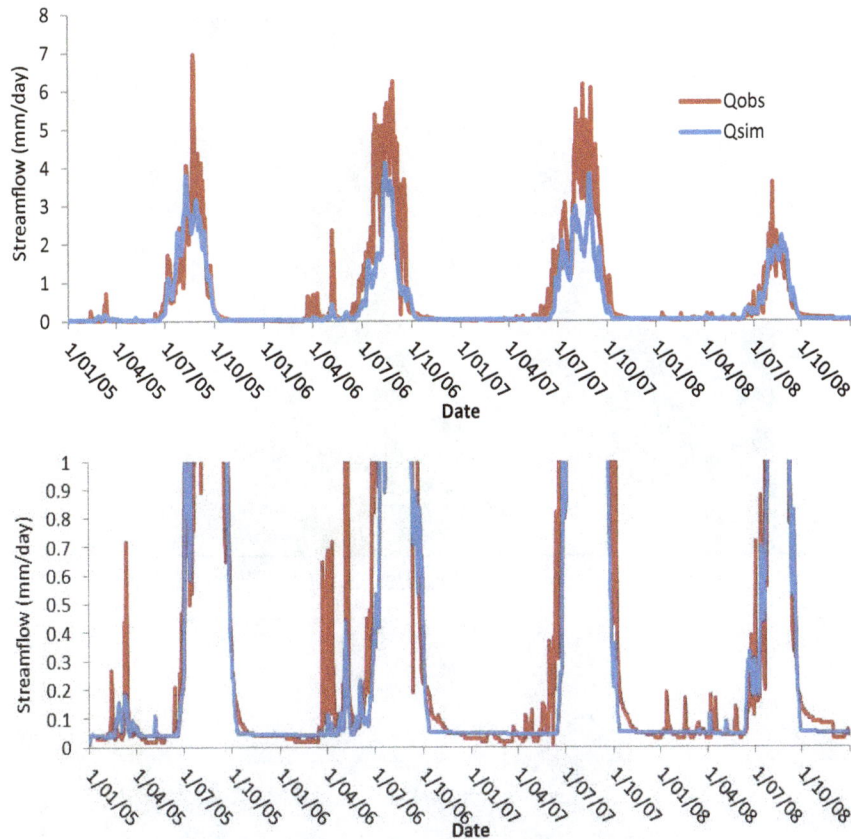

Figure 5: The observed and simulated streamflow in the Melka Kuntrie subbasin for validation periods (2005-2008),the lower pannel indicates the zoomed in observed and simulated streamflow.

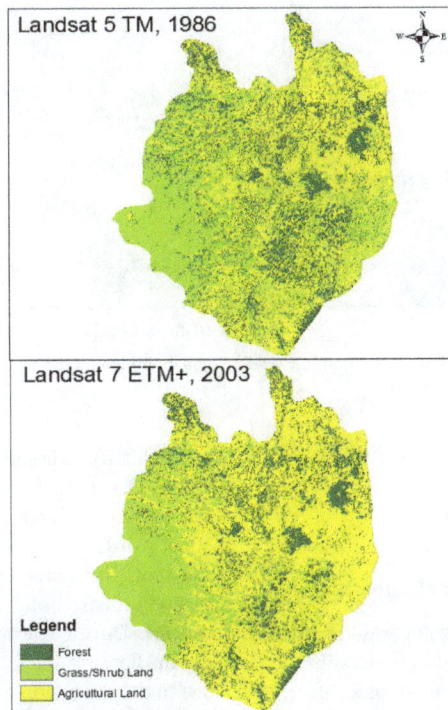

Figure 6: Land use cover map for the year 1986 (upper) and 2003 (lower) for the Melka Kuntrie subbasin derived from Landsat images.

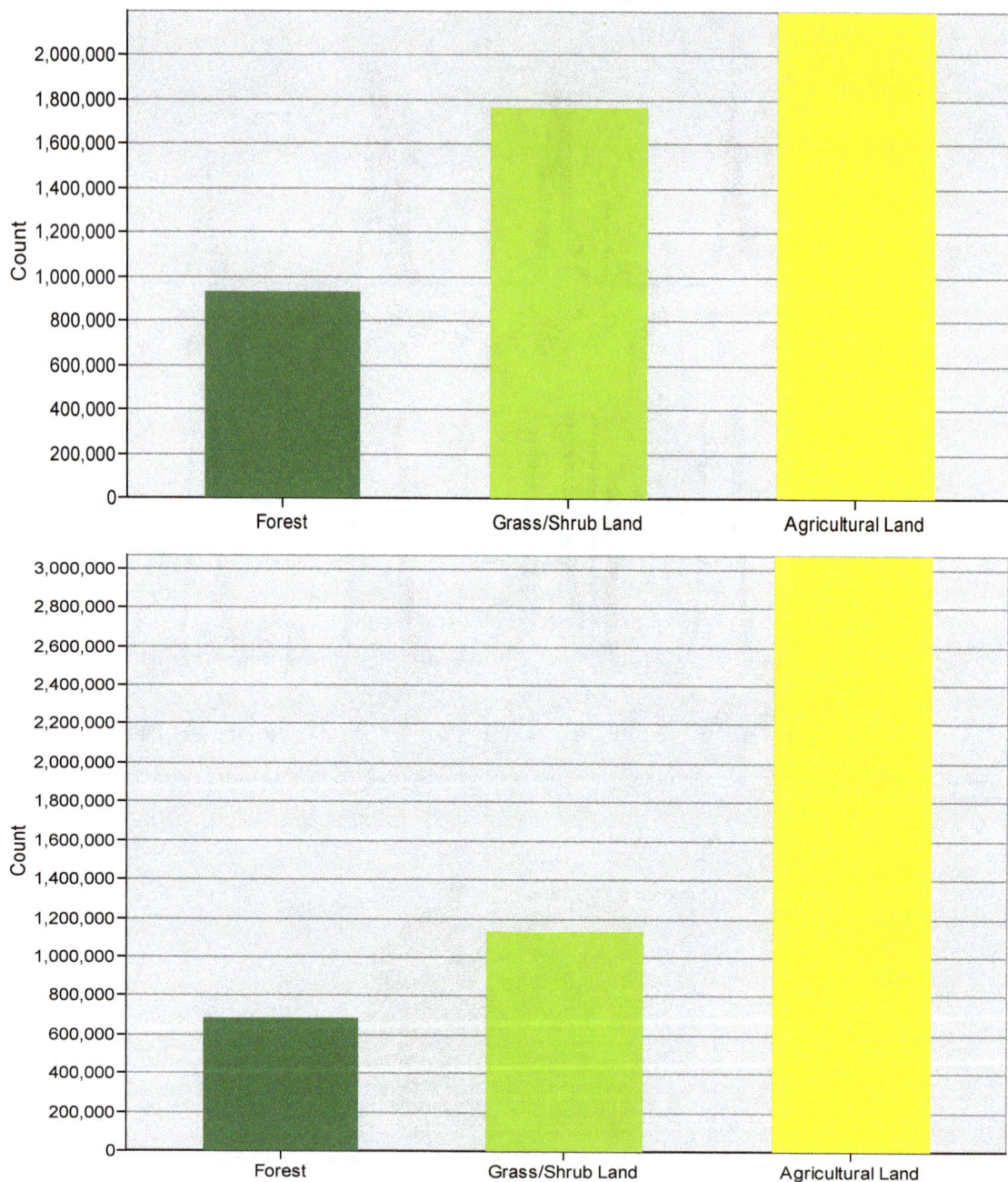

Figure 7: Land use-cover of the Melka Kuntrie subbasin obtained from Landsat images in the year 1986 for Landsat TM (upper) and in the year 2003 for Landsat ETM+ (lower).

grass land, respectively. The large area of forest and shrub/grassland area in the sub basin was partly changed into agricultural land. There was only small forest coverage left in the north and central part of the sub basin.

Hydrological response to the land use-cover change

The HBV model was re-calibrated for 1990-2008 period using the corn land use and the following runs were performed without calibration by only changing the land use-cover type. The meteorological forcing of this period was used for the HBV model to simulate stream flow and evapotranspiration using the 1986 and 2003 land use-cover types for

the Melka Kuntrie sub basin. Based on the land use-cover type in 1986 and 2003, the HBV model was forced by keeping the other parameters and meteorological variables as it was in the calibration period. In other words, the parameter and meteorological variables were kept constant. The simulated stream flow response due to the change in vegetation based on 1986 and 2003 data were processed for different seasons. During the main rainy season, there was slightly increased stream flow for the 2003 land use as compared to the 1986 land use due to the expansion of agriculture land (Figure 8). For the dry season, the simulated stream flow for the 1986 and 2003 land use showed some variability in the flow differences.

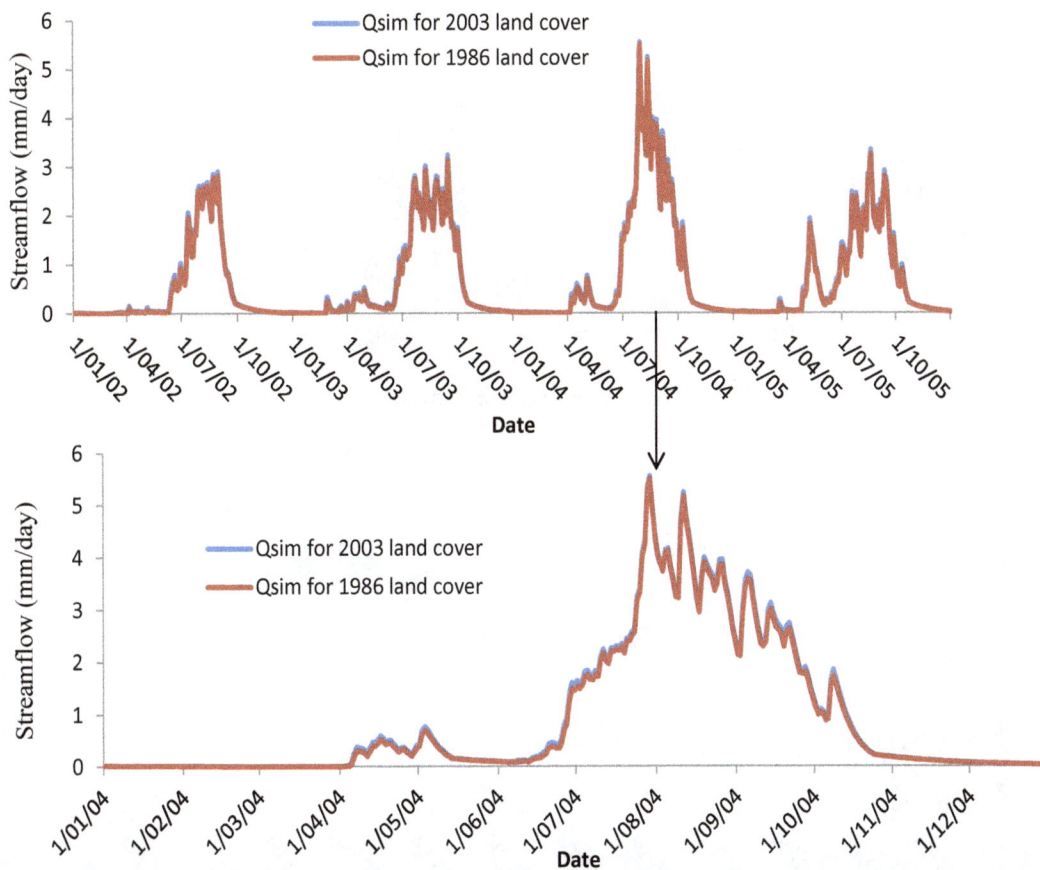

Figure 8: Simulated streamflows for the year 2004 (lower panel) and for 2002-2005 (upper panel), using the 1986 and 2003 land cover for the Melka Kuntrie subbasin.

Impacts of land use-cover change on stream flow and evapotranspiration

As indicated in Figures 8 and 9, there was a slightly higher stream flow for the 2003 land cover than for the 1986 land cover. Because of the decrease in forest or shrub/grass land cover the rainfall in 2003 land cover could easily increase the stream flow and there are slightly higher stream flow in 2003 land cover relative to 1986.In the case of the 1986 land cover type, the larger part of the rainfall was lost to evapotranspiration and that is why the stream flow was slightly lower than 2003 land cover especially in the main rainy season as shown in the (Figure 10).

For the main rainy season, the stream flow increase in the 2003 land use was 8% in June, 4% in July, 5% in August and 3% in September as compared to the 1986 land use, which corresponds to 0.3, 0.6, 0.11, and 0.5 mm/day, respectively. The evapotranspiration increase for the 1986 land use was 3% in June, 7% in July, 9% in August and 3% in Septembers compared to the 2003 land use, which corresponds to 0.06, 0.23, 0.37, and 0.11 mm/day, respectively. The small rainy season stream flow increase for the 2003 land use as compared to the 1986 land cover was 12% in April and 9% in May that corresponds to 0.01 and 0.02 mm/day daily mean stream flow. The rest small rainy season two months showed decreased stream flow for the 2003 land use as compared to the 1986 land use that was 9% in February and 11% in March corresponding to very slow stream flow 0.003 and 0.006 mm/day, respectively.

The dry season stream flow for the 2003 land use showed decreasing trends in most of the months as compared to the 1986 land cover for the period 1990-2008. The stream flow decrease for the 2003 land use as compared to the 1986 land use was 15% in January, 13% in November and 9% in December corresponding to very small flows 0.0007, 0.06 and 0.0007 mm/day, respectively. The month October showed an increasing stream flow for the 2003 land use cover as compared to the 1986 land use cover, which was 8% corresponding to the 0.005 mm/day.

Extreme Land use-cover change scenarios and their impact on simulated stream flow

Two extreme scenarios were defined that is a completely forested and a completely non-forested Melka Kuntrie sub basin. For the main rainy season, the completely non-forested scenario showed that the daily mean stream flow increased by 2% in June, 13% in July, 14% in August and 7% in September as compared to the completely forested sub basin. This implied that there was 0.01, 0.23, 0.35, and 0.12 mm/day increase of daily mean stream flow for June, July, August, and September, respectively due to the complete deforestation relative to the forested one as shown in the Figure 11. The change in stream flow was due to an increased evapotranspiration for the forested scenario as compared to the deforested scenario as shown in the Figure 12. The increased evapotranspiration loss for the completely forested scenario was 8% in June, 14% July 7% August and 9% in September that corresponds to 0.20, 0.52, 0.28 and 0.35 mm/day in stream flow loss, respectively for the forested scenario as compared to the completely deforested scenario.

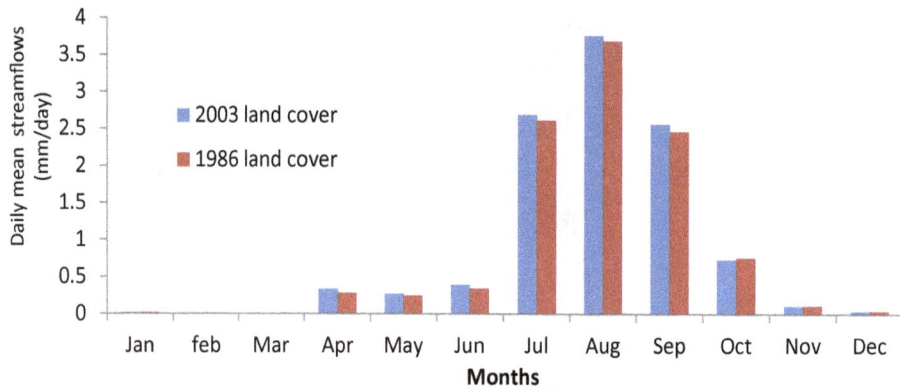

Figure 9: Simulated streamflow for the 2003 and 1986 land cover for the year 2004.

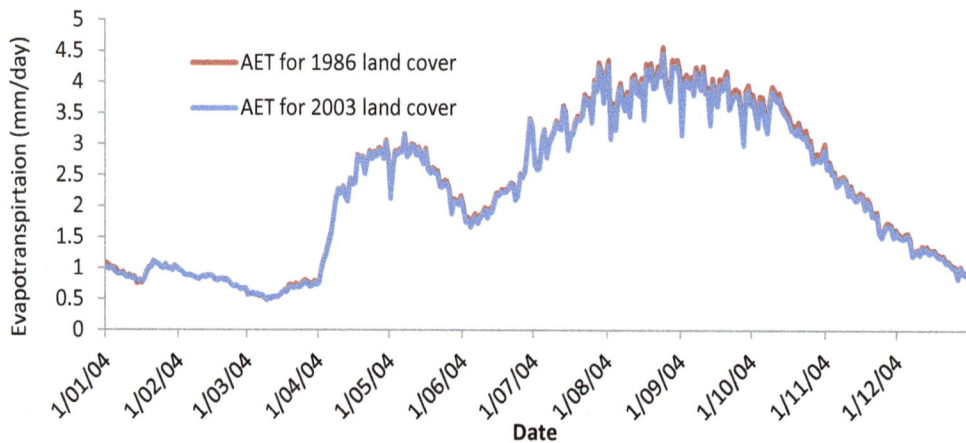

Figure 10: Evapotranspiration in the 2004 for 1986 and 2003 land cover.

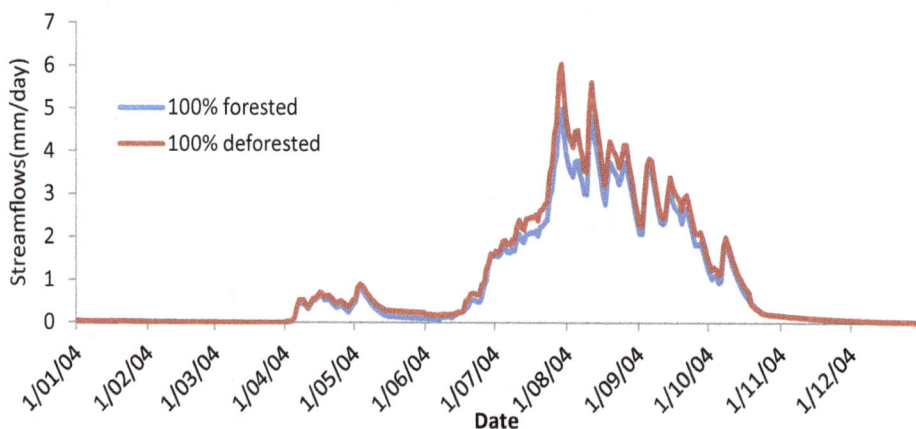

Figure 11: Simulated streamflow for the year 2004 for the completely forested and completely deforested land cover scenarios.

The daily mean stream flow for the small rainy season indicated that there was also increasing stream flow for the completely deforested scenario as compared to the completely forested scenario in all months, excluding February. In the dry season, the differences in flow between the forested and deforested scenarios were variable, but very small.

The 100% deforested extreme scenario indicated that there is higher stream flow in the main rainy season and decreased evapotranspiration, while the 100% forested scenarios showed the reverse. For the 100% forested extreme scenario the evapotranspiration change was more noticeable in the small and main rainy seasons. For the 100% deforested extreme scenario the stream flow increase was more noticeable in the main rainy season. Overall, for the main rainy season there was an increasing daily mean stream flow for the 2003 land use as compared

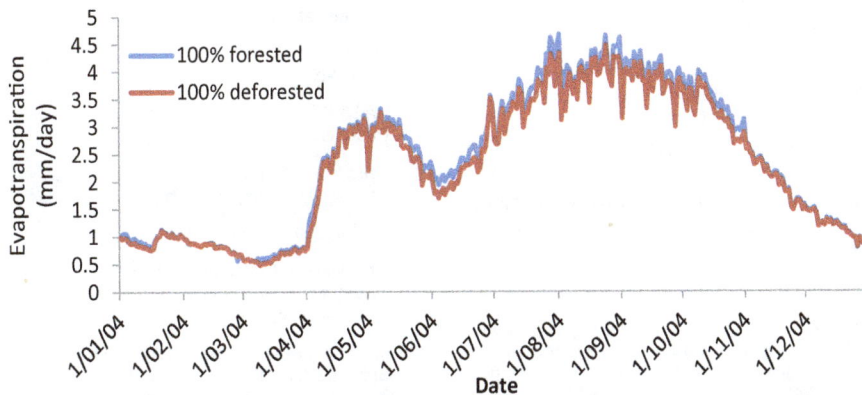

Figure 12: Simulated evapotranspiration for the year 2004 for the completely forested and completely deforested land cover scenarios.

to the 1986 land use, while the dry and small rainy season daily mean stream flow showed variability in the differences.

Discussions

If the land use-cover change in the entire basin is very small, clearly the stream flow at the basin outlet usually is insensitive to land use cover change [36]. Based on a study in the Rhine basin, the hydrological regime of the basin is expected to shift from a combined snowmelt-rainfall regime to a more rainfall-dominated regime and land use change may reinforce the effect of the shift along with all projected land use change scenarios indication of increased stream flow [36]. The decrease in stream flow in forestland can be attributed to the higher rate of water loss by evapotranspiration relative to the agricultural land [37]. Deep roots of trees can draw more moisture from the soil more than shallow rooted agriculture plants or bare soil. Besides, deciduous forest has a larger leaf area to transpire [37]. Based on a study in the Xinjiang River basin, China, an increase in stream floww as reported in the wet season and a decrease in stream flow in the dry season due to the land use change from forest into agricultural land [37]. Similarly, the studies that were carried out in the Angereb watershed (in the north) and in the Hare watershed (in the southern) parts of Ethiopia stated that the stream flow for the wet months had increased, while there was a decrease in the dry season due to the land use change [5,38].

The peak flow could also increase, for instance, due to land degradation, urbanization, which leads to a reduction of infiltration and storage. Studies carried out in different regions indicated that these processes lead to an increase in peak flow particularly in the rainy season and either a decrease or an increase of base flow particularly, in the dry season [4,39,40]. Similarly, based on a study carried out in Indonesia by Narulita [41] indicated that there was a constant baseflow decrease and a high flow increase after rapid land use change from natural vegetation cover to agricultural land. Overall, most of the studies conclude that there is a stream flow increase including peak flow, particularly in the main rainy season; however, in the dry season, there is inconsistence change in stream flow including baseflow, due to the land use change in a basin. According to Bewket and Strek [4] a study in Chemoga sub basin in north Ethiopia, the hydrology of a sub basin has been influencing negatively due to the clearance of natural vegetation cover, land degradation, and expanded agricultural land that reduce infiltration, decrease groundwater recharge, and increase runoff.

Conclusions and Recommendations

The analysis of impact of land use-cover change showed that there was a slightly higher stream flow for the 2003 land cover as compared to the 1986 land cover. In the case of the 1986 land cover type, a larger part of the rainfall was lost to evapotranspiration and that is why the stream flow was slightly lower than for the 2003 land cover, especially in the main rainy season. The dry season stream flow for the 2003 land use showed a decreasing magnitude in most of the months as compared to the 1986 land cover. For the main rainy season, the completely non-forested scenario showed that the daily mean stream flow increased by 2% in June, 13% in July, 14% in August and 7% in September as compared to the completely forested Melka Kuntrie sub basin. These implied that there were 0.01, 0.23, 0.35, and 0.12 mm/day increase of daily mean stream flow for June, July, August, and September, respectively due to the complete deforestation relative to the complete forested one. During the main rainy season, there was slightly increased stream flow for the 2003 land use as compared to the 1986 land use due to the expansion of agriculture land. For the dry season, the simulated stream flow for the 1986 and 2003 land use showed some variability in the flow differences.

The HBV model performance based on the Nash-Sutcliffe efficiency (NS) in the Melka Kuntrie sub basin was reasonably well. There was also a strong seasonal effect in the sub basin that sometimes increases the Nash-Sutcliffe efficiency or correlation coefficient.

Recommendations

Further research activities should be consider using different hydrological models in the region for the sake of further investigation of the impact of land use-cover change on the hydrology of sub basin.

There should be better land management programs or practices that encourage afforestation so that precipitation during the rainy season could easily infiltrate and recharge groundwater.

There should be strong encouragement and support in rainwater harvesting, afforestation, water and soil conservation practice in community level.

Land use-cover change hazard awareness at all levels (community, locall, regional and national levels) and appropriate response techniques.

It is recommendable to have more hydrometeorological data

measurement instruments in and around the sub basin that could provide adequate data with better quality.

Better data gathering techniques and dissemination process should be foreseen so that local and regional authorities can be involved in integrated and coordinated manner.

Acknowledgement

The authors gratefully appreciate to Ethiopian Ministry of Water and Energy (MoWE), and Ethiopian National Meteorology Service Agency (NMSA) for providing hydrometeorologicaland GIS data.

References

1. Tufa DF, Abbulu Y, Srinivasarao GVR (2014) Watershed Hydrological response to changes in land use/cover pattern of River Basin: A review. International Journal of Civil, Structural, Environmental and Infrastructure Engineering Research and Development 4: 157-170.

2. Hamza IA, Iyela A (2012) Land use pattern, climate change, and its implication for food security inEthiopia: Review. Ethiopia journal of environmental study and management 5: 1

3. Legesse D, Coulomb CV, Gasse F (2003) Hydrological response of a catchment to climate and land use changes in Tropical Africa: case study South Central Ethiopia. Journal of Hydrology 275: 67-85.

4. Bewket W, Sterk G (2005) Dynamics in land cover and its effect on stream flow in the Chemogawatershed, Blue Nile basin. Ethiopia, Hydrol Process 19: 445-458.

5. Kassa Tadele, GerdFörch (2007) Impact of Land Use/Cover Change on Streamflow: The Case of Hare River Watershed, Ethiopia. Catchment and Lake Research LARS

6. Diress Tsegaye, Moe Stein R, Paul Vedeld, Aynekulu E (2010) Land-use/cover dynamics in Northern Afar rangelands, Ethiopia. Agriculture, Ecosystems and Environment 139: 174-180.

7. Gerold D, DagnachewNegesse (2012) Land Use Land Cover Change in the Katar Catchment, Ziway Watershed, Etiopia.

8. Daniel Ayalew M (2008) Remote sensing and gis-based Land use and land cover change detection in the upper Dijoriver catchment, Silte zone, southern Ethiopia.

9. Daniel Shewangizaw, Yonas Michael (2010) Assessing the Effect of Land Use Change on the Hydraulic Regime of Lake Awassa. Nile Basin Water Science& Engineering Journal 3.

10. Getachew Haile E, MelesseAssefa M (2012) The impact of land use change on the hydrology of the Angereb Watershed, Ethiopia. International Journal Water Science.

11. Byragi RT, Mekonen Aregai (2011) Effect of Land Cover Dynamics and the Relative Response to Water Resources in Middle Highland Tigray, Ethiopia: the Case of Areas AroundLaelay-Koraro 2: 868-873.

12. Bishaw B (2003) Deforestation and land degradation on the Ethiopian highlands: A strategy for physical recovery 8: 7-25.

13. Shimelis G, Setegn, David Rayner, Melesse Assefa M, BijanDargahi, et al. (2011) Climate Change Impact on Agricultural Water Resources Variability in the Northern Highlands of Ethiopia. SpringerScience Business Media.

14. Kassa, TadeleMengistu (2009) Watershed Hydrological Responses to Changes in Land Use and Land Cover, and Management Practices at Hare Watershed, Ethiopia.

15. Gwate O, Woyessa Yali E, David Wiberg (2015) Dynamics of Land Cover and Impact on Stream flow in the Modder River Basin of South Africa: Case Study of a Quaternary Catchment. International Journal of Environmental Protection and Policy 3: 31-38.

16. DeFries R, Eshleman KN (2004) Land-use change and hydrologic processes: a major focus for the future. Wiley InterScience, Hydrol. Process 18: 2183-2186.

17. Tubiello (2007) Land and water use options for climate change adaptation and mitigation in agriculture. New York: FAO.

18. Agarwal A (2002) Integrated Water Resources Management. Stockholm: Global Water Partnership (GWP).

19. Kurkura (2011) Water balance of Upper Awash Basin based on satellite-derived data Addis Ababa.

20. Behailu, Shimelis (2004) Stream flow simualtion for the Upper Awash Basin Addis Ababa.

21. NASA (2006) How Landsat Images are made. NASA's Landsat, Education and Public Outreach team.

22. Federico Moran (2011) Land Cover Classification in a Complex Urban-Rural Landscape. PhotogrammEng Remote Sensing.

23. James (2001) Band combinations.

24. Paranjape RB, Morrow WM, Rancayan RM (1992) Adaptive-Neighborhood Histogram Equalization for Image Enhancement, Graphical models and image processing 54: 259-267.

25. Manpreet, KaurJasdeep Kaur, Jappreet Kaur (2011) Survey of Contrast Enhancement Techniques based on Histogram Equalization. International Journal of Advanced Computer Science and Applications.

26. Krishna Bahadur KC (2009) Improving Landsat and IRS Image Classification: Evaluation of Unsupervised and Supervised Classification through Band Ratios and DEM in a Mountainous Landscape in Nepal. Remote Sensing 1: 1257-1272.

27. Tammy, James (2013) Classification of a Landsat Image (Supervised), Remote Sensing Analysis in an ArcMap Environment.

28. Chao, Zhongfu (2005) Brightness Preserving Histogram Equalization with Maximum Entropy: A Variational Perspective. IEEE Transactions on Consumer Electronics 51: 1326-1334.

29. Congalton RG (1991) A review of assessing the accuracy of classifications of remotely sensed data. Remote Sensing of Environment 37: 35-46.

30. Yuan F, Bauer ME, Heinert NJ, Holden GR (2005) Multi-level land cover mapping of the Twin Cities (Minnesota) metropolitan area with multi-seasonal Landsat. Geocarto. Int 20: 5-14.

31. Taddese G, Sonder K, Peden D (2006) The water of the Awash River Basin a future challenge to Ethiopia. International Livestock Research Institute. Addis Abeba, Ethiopia.

32. Romilly, Gebremichael (2010) Evaluation of satellite rainfall estimates over Ethiopian river basins. Hydrol. Earth Syst Sci 15: 1505-1514.

33. Degefu W (1987) Some aspects of meteorological drought in Ethiopia, Cambridge University Press.

34. Alemayehu, Mengistu (2006) Country Pasture/Forage Resource Profiles.

35. Nederveen, Coenraads S (2010) Flood Recession Farming: An Overview and Case Study from the Upper Awash Catchment, Ethiopia. VU University.

36. Hurkmans RTWL, Terink W, Uijlenhoet R, Moors EJ, Torch PA, et al. (2009) Effects of land use changes on streamflow generation in the Rhine basin. Water resources research 45: 1-15.

37. Guo H, Qi Hu, Tong Jiang (2008) Annual and seasonal stream flow responses to climate and land-cover changes in the Poyang Lake basin, China. Journal of Hydrology 355: 106-122.

38. Getachew, Melesse (2013) The Impact of Land Use Change on the Hydrology of the Angereb Watershed, Ethiopia. International Journal of Water Sciences.

39. Rientjes THM, Perera JBU, Haile AT, Gieske ASM, Booij MJ, et al. (2011) Hydrological Balance of Lake Tana, Upper Blue Nile Basin, Ethiopia. Nile River Basin: Hydrology, Climate and Water Use 69.

40. Costa MH, Botta A, Cardille JA (2003) Effects of large-scale changes in land cover on the discharge of the Tocantins River, Southeastern Amazonia. Journal of Hydrology 283: 206-217.

41. Narulita I (2012) Streamflow fluctuation and land use changes in Bandung Basin. Retrieved 5 50, 2014, from Royal Netherlands Meteorogical instistiute, International workshop climate data, climate change anlysis DIDAH project.

9

The Water Isotopic Version of the Land-Surface Model ORCHIDEE: Implementation, Evaluation, Sensitivity to Hydrological Parameters

Camille Risi[1]*, Jerome Ogée[2], Sandrine Bony[1], Thierry Bariac[3], Naama Raz-Yaseef[4,5], Lisa Wingate[2], Jeffrey Welker[6], Alexander Knohl[7], Cathy Kurz-Besson[8], Monique Leclerc[9], Gengsheng Zhang[9], Nina Buchmann[10], Jiri Santrucek[11,12], Marie Hronkova[11,12], Teresa David[13], Philippe Peylin[14] and Francesca Guglielmo[14]

[1]*LMD/IPSL, CNRS, UPMC, Paris, France*
[2]*INRA Ephyse, Villenave d'Ornon, France*
[3]*UMR 7618 Bioemco, CNRS-UPMC-AgroParisTech-ENS Ulm-INRA-IRD-PXII Campus AgroParisTech, Bâtiment EGER, Thiverval-Grignon, 78850 France*
[4]*Earth Sciences Division, Lawrence Berkeley National Laboratory, Berkeley, USA*
[5]*Department of Environmental Sciences and Energy Research, Weizmann Institute of Science, PO Box 26, Rehovot 76100, Israel*
[6]*Biology Department and Environment and Natural Resources Institute, University of Alaska, Anchorage, AK 99510, USA*
[7]*Bioclimatology, Faculty of Forest Sciences and Forest Ecology, Georg-August University of Göttingen, 37077 Göttingen, Germany*
[8]*Instituto Dom Luiz, Centro de Geofísica IDL-FCUL, Lisboa, Portugal*
[9]*University of Georgia, Griffin, GA 30223, USA*
[10]*Institute of Agricultural Sciences, ETH Zurich, Zurich, Switzerland*
[11]*Biology Centre ASCR, Branisovska 31, Ceske Budejovice, Czech Republic*
[12]*University of South Bohemia, Faculty of Science, Branisovska 31, Ceske Budejovice, Czech Republic*
[13]*Instituto Nacional de Investigação Agrária e Veterinária, Quinta do Marquês, Portugal*
[14]*LSCE/IPSL, CNRS, UVSQ, Orme des Merisiers, Gif-sur-Yvette, France*

Abstract

Land-Surface Models (LSMs) exhibit large spread and uncertainties in the way they partition precipitation into surface runoff, drainage, transpiration and bare soil evaporation. To explore to what extent water isotope measurements could help evaluate the simulation of the soil water budget in LSMs, water stable isotopes have been implemented in the ORCHIDEE (ORganizing Carbon and Hydrology In Dynamic EcosystEms: the land-surface model) LSM. This article presents this implementation and the evaluation of simulations both in a stand-alone mode and coupled with an atmospheric general circulation model. ORCHIDEE simulates reasonably well the isotopic composition of soil, stem and leaf water compared to local observations at ten measurement sites. When coupled to LMDZ (Laboratoire de Météorologie Dynamique-Zoom: the atmospheric model), it simulates well the isotopic composition of precipitation and river water compared to global observations. Sensitivity tests to LSM (Land-Surface Model) parameters are performed to identify processes whose representation by LSMs could be better evaluated using water isotopic measurements. We find that measured vertical variations in soil water isotopes could help evaluate the representation of infiltration pathways by multi-layer soil models. Measured water isotopes in rivers could help calibrate the partitioning of total runoff into surface runoff and drainage and the residence time scales in underground reservoirs. Finally, co-located isotope measurements in precipitation, vapor and soil water could help estimate the partitioning of infiltrating precipitation into bare soil evaporation.

Keywords: Water isotopes; Land-surface model; Global models; Soil water budget; Rain infiltration; Runoff; Evapo-transpiration partitioning

Introduction

Land-surface models (LSMs) used in climate models exhibit a large spread in the way they partition radiative energy into sensible and latent heat [1,2] precipitation into evapo-transpiration and runoff [3-5], evapo-transpiration into transpiration and bare soil evaporation [6,7], and runoff into surface runoff and drainage [8-10]. This results in an large spread in the predicted response of surface temperature [11] and hydrological cycle [12,13] to climate change [11] or land use change [14,15]. Therefore, evaluating the accuracy of the partitioning of precipitation into surface runoff, drainage, transpiration and bare soil evaporation (hereafter called the soil water budget) in LSMs is crucial to improve our ability to predict future hydrological and climatic changes.

The evaluation of LSMs is hampered by the difficulty to measure over large areas the different terms of the soil water budget, notably the evapo-transpiration terms and the soil moisture storage [16,17]. Single point measurements of evapo-transpiration fluxes [18] and soil moisture [19] are routinely performed within international networks, but those measurements remain difficult to upscale to a climate model grid box due to the strong horizontal heterogeneity of the land surface [20,21]. Spatially-integrated data such as river runoff observations are very valuable to evaluate soil water budgets at the regional scale [22,23],

but are insufficient to constrain the different terms of the water budget. Additional observations are therefore needed.

In this context, water isotope measurements have been suggested to help constrain the soil water budget [24,25], its variations with climate or land use change [26], and its representation by large-scale models [27,28]. For example, water stable isotope measurements in the different water pools of the soil-vegetation-atmosphere continuum have been used to quantify the relative contributions of transpiration and bare soil evaporation to evapo-transpiration [29-32], to infer plant source water depth [33], to assess the mass balance of lakes [34-36] or to investigate pathways from precipitation to river discharge [37-40]. These isotope-based techniques generally require high frequency isotope measurements and are best suitable for intensive field campaigns at the local scale. At larger spatial and temporal scales, some

*Corresponding author: Camille Risi, LMD/IPSL, CNRS, UPMC, Paris, France
E-mail: crlmd@lmd.jussieu.fr

attempts have been made to use regional gradients in precipitation water isotopes for partitioning evapo-transpiration into bare soil-evaporation and transpiration [41-43].

To explore to what extent water isotope measurements could be used to evaluate and improve land surface parameterizations, water isotopes were implemented in the LSM ORCHIDEE (ORganizing Carbon and Hydrology In Dynamic EcosystEms [44,45]. This isotopic version of ORCHIDEE has already been used to explore how tree-ring cellulose records past climate variations [46] and to investigate the continental recycling and its isotopic signature in Western Africa [47] and at the global scale [48].

The first goal of this article is to evaluate the isotopic version of the ORCHIDEE model against recently-made-available new datasets combining water isotopes in precipitation, vapor, soil water and rivers. The second goal is to evaluate the isotopic version of the ORCHIDEE model when coupled to the atmospheric General Circulation Model (GCM) LMDZ (Laboratoire de Météorologie Dynamique Zoom [49]). The third goal is to perform sensitivity tests to LSM parameters to identify processes whose representation by LSMs could be better evaluated using water isotopic measurements.

After introducing notations and models in section 4, we present ORCHIDEE simulations in a stand-alone mode at measurement sites and global ORCHIDEE-LMDZ coupled simulations.

Notation and Models

Notations

Isotopic ratios ($HDO/H_2^{16}O$ or $H_2^{18}O/H_2^{16}O$) in the different water pools are expressed in‰ relative to a standard: $\delta = \left(\frac{R_{sample}}{R_{smow}}-1\right) \cdot 1000$, where R_{sample} and R_{SMOW} are the isotopic ratios of the sample and of the Vienna Standard Mean Ocean Water (V-SMOW) respectively [50,51]. To first order, variations in δD are similar to those in $\delta^{18}O$ but are 8 times larger. Deviation from this behavior can be associated with kinetic fractionation and is quantified by deuterium excess ($d=\delta D-8.\delta^{18}O$ [50,52]). Hereafter, we note $\delta^{18}O_p$, $\delta^{18}O_v$, $\delta^{18}O_s$, $\delta^{18}O_{stem}$ and $\delta^{18}O_{river}$ the $\delta^{18}O$ of the precipitation, atmospheric vapor, soil, stem, river water respectively. The same subscripts apply for d.

The LMDZ model

LMDZ is the atmospheric GCM (General Circulation Model) of the IPSL (Institut Pierre Simon Laplace) climate model [53,54]. We use the LMDZ-version 4 model [49] which was used in the International Panel on CLimate Change's Fourth Assessment Report simulations [55,56]. The resolution is 2.5 ° in latitude, 3.75 ° in longitude and 19 vertical levels. Each grid cell is divided into four sub-surfaces: ocean, land ice, sea ice and land (treated by ORCHIDEE) (Figure 1a). All parameterizations, including ORCHIDEE, are called every 30 min. The implementation of water stable isotopes is similar to that in other GCMs [57,58] and has been described in [59,60]. LMDZ captures reasonably well the spatial and seasonal variations of the isotopic composition in precipitation [60] and water vapor [61].

The ORCHIDEE (ORganizing Carbon and Hydrology In Dynamic EcosystEms: the land-surface model) model

The ORCHIDEE model is the LSM component of the IPSL climate model. It merges three separate modules: (1) SECHIBA (Schématisation des EChanges Hydriques a l'Interface entre la Biosphère et l'Atmosphère [44,62]) that simulates land-atmosphere water and energy exchanges, (2) STOMATE (Saclay-Toulouse-Orsay

Model for the Analysis of Terrestrial Ecosystems [45]) that simulates vegetation phenology and biochemical transfers ; and (3) LPJ (Lund-Postdam-Jena [63]) that simulates the vegetation dynamics. Water stable isotopes were implemented in SECHIBA, and we use prescribed land cover maps so that the two other modules could be de-activated.

Each grid box is divided into up to 13 land cover types: bare soil, tropical broad-leaved ever-green, tropical broad-leaved rain-green, temperate needle-leaf ever-green, temperate broad-leaved ever-green, temperate broad-leaved summer-green, boreal needle-leaf ever-green, boreal broad-leaved summer-green, boreal needle-leaf summer-green, C3 grass, C4 grass, C3 agriculture and C4 agriculture. Water and energy budgets are computed for each land cover type.

Figure 1b illustrates how ORCHIDEE (ORganizing Carbon and Hydrology In Dynamic EcosystEms: the land-surface model) represents the surface water budget. Rainfall is partitioned into interception by the canopy and through-fall rain. Through-fall rain, snow melt, dew and frost fill the soil. The soil is represented by two water reservoirs: a superficial and a bottom one [64,65]. Taken together, the two reservoirs have a water holding capacity of 300 mm and a depth of 2 m.

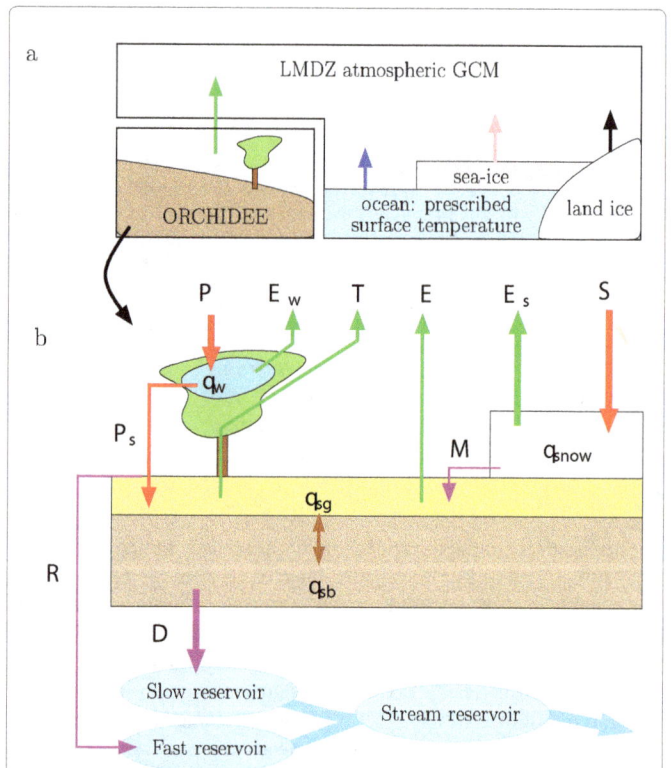

Figure 1: a) The four sub-surfaces in the LMDZ GCM: land, ocean, sea ice and land ice. Their relative fraction in each grid box is prescribed. The sea surface temperature of the ocean is prescribed, and interactively calculated for sea-ice and land-ice. Over land, the land-Surface model (LSM) ORCHIDEE calculates interactively the surface temperature and outgoing water fluxes. b) Water fluxes and pools represented in the ORCHIDEE LSM. Water pools are the soil water in the superficial (q_{sg}) and bottom (q_{sb}) layers, the water intercepted by the canopy (q_w) and the snow pack (q_{snow}). Fluxes onto the land surface are the total rain (P) and snow (S), and possibly dew or frost. As some rain is intercepted by the canopy, only throughfall rain (P_s) arrives at the soil surface. Evaporation fluxes are the evaporation of intercepted water (E_w), transpiration by the vegetation (T), bare soil evaporation (E) and snow sublimation (E_s). Snow melt may be transferred from the snow pack to the soil (M). Water from rainfall, melt (and possibly dew) exceeding the soil capacity is converted to surface runoff (R) and drainage (D). The routing model then transfers surface runoff and drainage to streams.

Soil water undergoes transpiration by vegetation, bare soil evaporation or runoff. Transpiration and evaporation rates depend on soil moisture to represent water stress in dry conditions. Runoff occurs when the soil water content exceeds the soil holding capacity and is partitioned into 95% drainage and 5% surface runoff [66]. Snowfall fills a single-layer snow reservoir, where snow undergoes sublimation or melt. By comparison, when not coupled to ORCHIDEE, the simple bucket-like LSM in LMDZ makes no distinction neither between bare soil evaporation and transpiration nor between surface runoff and drainage [67].

Surface runoff and drainage are routed to the coastlines by a water routing model [68]. Surface runoff is stored in a fast ground water reservoir which feeds the stream reservoir with residence time of 3 days. Drainage is stored in a slow ground water reservoir which feeds the stream reservoir with residence time of 25 days. The water in the stream reservoir is routed to the coastlines with a residence time of 0.24 days.

Implementation of water stable isotopes in ORCHIDEE

We represent isotopic processes in a similar fashion as other isotope-enabled LSMs [69-73]. Some details of the isotopic implementation are described in Risi [74]. In absence of fractionation, water stable isotopes ($H_2^{16}O$, $H_2^{18}O$, HDO, $H_2^{17}O$) are passively transferred between the different water reservoirs. We assume that surface runoff has the isotopic composition of the rainfall and snow melt that reach the soil surface. Drainage has the isotopic composition of soil water [24]. We calculate the isotopic composition of bare soil evaporation or of evaporation of water intercepted by the canopy using the Craig and Gordon equation [75] (Equation 3). We neglect isotopic fractionation during snow sublimation (Equation 2). We consider isotopic fractionation at the leaf surface (Equation 5) but we assume that transpiration has the isotopic composition of the soil water extracted by the roots (Equation 2).

In the control coupled simulation, we assume that the isotopic composition of soil water is homogeneous vertically and equals the weighted average of the two soil layers. However, transpiration, bare soil evaporation, surface runoff and drainage draw water from different soil water reservoirs whose isotopic composition is distinct [76-78]. Therefore, we also implemented a representation of the vertical profile of the soil water isotopic composition.

Stand-alone ORCHIDEE Simulations at MIBA (Moisture in Biosphere and Atmosphere: Network for Water Isotopes in Soil, Stem and Leaf Water) and Carbo-Europe Measurement Sites

First, we performed simulations using ORCHIDEE as a stand-alone model at ten sites. Using isotopic measurements in soil, stem and leaf water, simulations are evaluated at each site at the monthly scale. Sensitivity tests to evapo-transpiration partitioning and soil infiltration processes are performed.

Measurements used for evaluation

To first order the composition of all land surface water pools is driven by that in the precipitation [79]. Therefore, a rigorous evaluation of an isotope-enabled LSM requires to evaluate the difference between the composition in each water pool and that in the precipitation. Besides, to better isolate isotopic biases, we need a realistic atmospheric forcing. We tried to select sites where (1) isotope were measured in different water pools of the soil-plant-atmosphere continuum, during at least a full seasonal cycle and (2) meteorological variables were monitored at a frequency high enough (30 minutes) to ensure robust forcing for our model and (3) water vapor and precipitation were monitored to provide isotopic forcing for the LSM. Only two sites satisfy these conditions: Le Bray and Yatir. Relaxing some of these conditions, we got a more a representative set of ten sites representing diverse climate conditions (Table 1 and Figure 2).

Description of the ten sites: The ten sites belong to two kinds of observational networks: MIBA (Moisture Isotopes in the Biosphere and Atmosphere [80-82] or Carbo-Europe [83,84].

Le Bray site, in South-Eastern France, joined the MIBA and GNIP (Global Network for Isotopes in Precipitation) network in 2007. It is an even-aged Maritime pine forest with C3 grass understory that has been the subject of many eco-physiological studies since 1994, notably as part of the Carbo-Europe flux network [85]. In 2007 and 2008, samples in precipitation, soil surface, needles, twigs and atmospheric vapor were collected every month and analyzed for $\delta^{18}O$ following the MIBA protocol [82,86]. This site was also the subject of intensive campaigns where soil water isotope profiles were collected between 1993 and 1997, and in 2007 [87].

The Yatir site, in Israel, is a semi-arid Aleppo pine forest. It is an afforestation growing on the edge of the desert, with mean-annual precipitation of 280 mm [88,89]. It has also been the subject of many eco-physiological studies as part of the Carbo-Europe flux network [89] and joined the MIBA network in 2004. It. In 2004-2005, samples of soil water at different depth, stems and needles were collected following the MIBA protocol. The water vapor isotopic composition has been monitored daily at the nearby Rehovot site (31.9 ° N, 34.65 ° E, [90]) and is used to construct the water vapor isotopic composition forcing. We must keep in mind however that although only 66 km from Yatir, Rehovot is much closer to the sea and is more humid than Yatir. The precipitation isotopic composition has been monitored monthly at

Site name	Country	Location	Network	Years	Reference
Le Bray	France	44.70° N, 0.77° W	MIBA, Carbo- Euroe	2007-2008	[87]
Yatir	Israel	31.33° N, 35.0° E	MIBA, Carbo-Euroe	2004-2005	[89,104]
Morgan-Monroe	United States	39.32° N, 86.42° W	MIBA-US	2005-2006	[167,172]
Donaldson Forest	United States	29.8° N, 82.163° W	MIBA-US	2005-2006	[91,169]
Anchorage	United States	61.2° N, 149.82° W	MIBA-US	2005-2006	-
Mitra	Portugal	38.5° N, 8.00° W	Carbo-Euroe	2001-2002	[171]
Bily Kriz	Czech Republic	49.5° N, 18.53° E	MIBA, Carbo-Euroe	2005	[92]
Brloh	Czech Republic	49.80° N, 14.66° E	MIBA	2004-2010	[93]
Hainich	Germany	50.97° N, 13.57° E	Carbo-Euroe	2001-2002	[170]
Tharandt	Germany	51.08° N, 10.47° E	Carbo-Euroe	2001-2002	-

Table 1: Information on the 10 sites used in this study: geographical location, network the sites are part of, years during which the istopic measurements were made and are used in this study, reference.

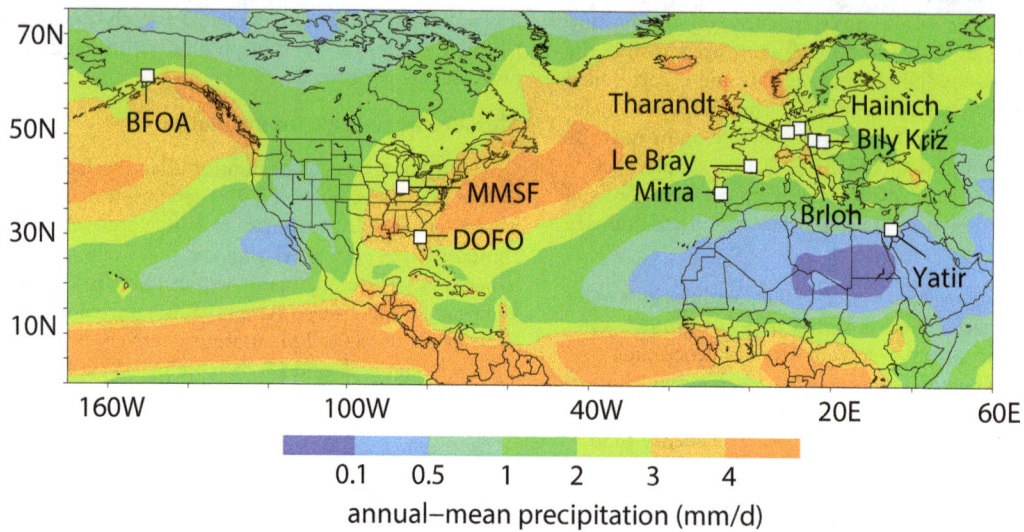

Figure 2: Location of the ten stations used in this study for single-point model-data comparison. The background represents the annual-mean precipitation from GPCP (Global Precipitation Climatology Project) to illustrate the diversity of climate regimes covered by the ten stations. Each station is described in more detail in Table 1.

the nearby GNIP station Beit Dagan (32 ° N, 34.82 ° E) and is used to construct the precipitation isotopic composition forcing.

The Morgan-Monroe State Forest, Donaldson Forest and Anchorage sites are part of the MIBA-US (MIBA-United States) network and are located in Indiana, in Florida and in Alaska respectively (Table 1). Sampling took place in 2005 and 2006 according to the MIBA protocols. The Donaldson Forest site, which jointed the MIBA-US network in 2005, is located at the AmeriFlux Donaldson site near Gainesville, Florida, USA. The site is flat with an elevation of about 50 m. It was covered by a forest of managed slash pine plantation, with an uneven understory composed mainly of saw palmetto, wax myrtle and Carolina jasmine [91]. The leaf area index was measured during a campaign in 2003 and estimated at 2.85. We use this value in our simulations.

The Mitra, Bily Kriz, Brloh, Hainich and Tharandt sites are part of the Carbo-Europe project. Hainich and Tharandt are located in Germany. The experimental site of Herdade da Mitra (230 m altitude, nearby Évora in southern Portugal) is characterized by a Mediterranean mesothermic humid climate with hot and dry summers. It is a managed agroforestry system characterized by an open evergreen woodland sparsely covered with *Quercus suber L.* and *Q. ilex* rotundifolia trees (30 trees/ha), with an understorey mainly composed of *Cistus* shrubs, and winter-spring C3 annuals. The isotopic samplings of leaves, twigs, soil, precipitation and groundwater were performed on a seasonal to monthly basis. All samples where extracted and analyzed at the Paul Scherrer Institute (Switzerland).

Bily Kriz and Brloh are both located on the Czech Republic. Bily Kriz is an experimental site in Moravian–Silesian Beskydy Mountains (936 m a.s.l.) with detailed records of environmental conditions [92]. It is dominated by Norway spruce forest. It joined the MIBA project in the season 2005. Brloh is a South Bohemian site in the Protected Landscape Area Blanskýles (630 m a.s.l.). It is dominated by deciduous beech forest and was used as MIBA sampling site from 2004 to 2010 [93].

Isotopic measurements: Samples of soil water, stems and leaves were collected at the monthly scale. The MIBA and MIBA-US protocols recommend sampling the first 5-10 cm excluding litter and the Carbo-

Europe protocol recommends sampling the first 5 cm [84], but in practice the soil water sampling depth varies from site to site. At some sites, soil water was sampled down to 1 m. For evaluating the seasonal evolution of soil water $\delta^{18}O$, we focus on soil samples collected in the first 15 cm only. Observed full soil water $\delta^{18}O$ profiles were used only at Le Bray and Yatir for evaluating the shape of simulated soil water $\delta^{18}O$ profiles.

Carbo-Europe samples were extracted and analyzed at the Department of Environmental Sciences and Energy Research, Weizmann Institute of Science, Israel. MIBA-US samples were extracted and analyzed at the Center for Stable Isotope Biogeochemistry of the University of California, Berkeley. Analytical errors for $\delta^{18}O$ in soil, stem and leaf water vary from 0.1‰ to 0.2‰ depending on the sites and involved stable isotope laboratory (Table 2).

Meteorological, turbulent fluxes and soil moisture measurements: At most of the sites, meteorological parameters (radiation, air temperature and humidity, soil temperature and moisture) are continuously measured and are used to construct the meteorological forcing for ORCHIDEE.

Fluxes of latent and sensible energy are measured using the eddy co-variance technique and are used for evaluating the hydrological simulation. Gaps are filled using ERA-Interim reanalyses [94]. Soil moisture observations are available at most sites.

Simulation set-up

To evaluate in detail the isotope composition of different water pools, stand-alone ORCHIDEE simulations on the ten MIBA and Carbo-Europe sites were performed. We prescribe the vegetation type and properties and the bare soil fraction based on local knowledge at each site (Table 3).

ORCHIDEE offline simulations require as forcing several meteorological variables: near-surface temperature, humidity and winds, surface pressure, precipitation, downward longwave and shortwave radiation fluxes. At Le Bray and Yatir, we use local meteorological measurements available at hourly time scale. At other

Site name	Biome	Dominant Species	Annual-mean temperature (°C)	Annual-mean precipitation (mm/year)	Elevation (m)
Le Bray	Temperate coniferous forest	Maritime pine	12	1022	60
Yatir	semi-arid forest	Aleppo pine	15.3	270	650
Morgan-Monroe	Temperate deciduous forest	Liriodendron tulipifera	12.4	1094	275
Donaldson Forest	Tropical pine plantation	Pinus palustris	21.7	1330	50
Anchorage	Boreal coniferous forest	Picea glauca	2.3	408	35
Mitra	Mediteranean forest	Sparse holm oak trees with patches of cork trees	13.9	480	230
Bily Kriz	Temperate coniferous forest	Pine forest	3.4	1024	936
Brloh	Temperate deciduous forest	Beech forest	7.6	832	630
Hainich	Temperate deciduous forest	Fagus Sylvatica	8	800	440
Tharandt	Temperate deciduous forest	Pine forest	8.1	1000	380

Table 2: Vegetation and climtological information on the 10 sites used in this study: biome, dominant species, annual-mean temperature and precipitation, elevation.

Site name	Prescribe vegetation in ORCHIDEE	Meteoro-logical forcing	Isotopic forcing for precipitation and vapor	local, GNIP, USNIP or Carbo-Europe stations used to calculate isotopic forcing
Le Bray	70% temperate needleleaf evergreen (LAI=0.4), 30% C3 grass (LAI=0.4)	obs	obs_iso	Le Bray local data for both precipitation and water vapor
Yatir	100% temperate needleleaf evergreen (LAI=4)	obs	obs_iso	Rehovot for water vapor and Beit Dagan GNIP station for precipitation
Morgan-Monroe	100% temperate broad-leaved summergreen (LAI=4.5)	obs_ERA	NIP_LMDZ	USNIP_IN22, USNIP_KY03
Donaldson Forest	100% temperate needleleaf evergreen (LAI=2.85)	obs_ERA	NIP_LMDZ	USNIP_FL14, USNIP_FL99
Anchorage	40% boreal needle-leaved evergreen (LAI=4), 60% boreal broad-leaved summergreen (LAI=4.5)	ERA	NIP_LMDZ	Bethel, USNIP_SOGR_10, USNIP_CA45
Mitra	50% temperate broad-leaved evergreen (LAI=2), 50% C3 grass (LAI=0.4)	obs_ERA	NIP_LMDZ	Beja, Faro, Penhas, Mitra, Portoallegre
Bily Kriz	100% temperate needleleaf evergreen (LAI=7.5)	obs_ERA	NIP_LMDZ	Vienna, Podersdorf, Apetlon, Liptovsky, Krakow
Brloh	100% temperate broad-leaved summergreen (LAI=4.5)	ERA	NIP_LMDZ	Leipzig, Hohhohensaas, Regensburg, Vienna, Petzenkirchen
Hainich	80% temperate broad-leaved summergreen (LAI=4.5), 20% C3 grass (LAI=0.4)	obs_ERA	NIP_LMDZ	Leipzig, Hohhohensaas, Braunschweig, BadSalzuflen, Wuerzburg, Wasserkuppe
Tharandt	80% temperate needleleaf evergreen (LAI=4), 20% C3 grass (LAI=0.4)	obs_ERA	NIP_LMDZ	Leipzig, Berlin, Hohhohensaas, Regensburg

Table 3: Information on the offline simulations performed on the 10 sites listed in Table 1: meteorological forcing (6 hourly observations of temperature, humidity, winds, precipitation and radiative fluxes), isotopic forcing (monthly isotopic composition of the precipitation and near-surface water vapor), and prescribed vegetation type and LAI (leaf area index) properties. We give proportions (in %) of the total vegetated area, excluding bare soil. For example, if a given vegetation type covers 100% of the vegetated area and the bare soil fraction is 30%, then the vegetation type covers only 70% of the total area. Three kinds of meteorological forcing are possible: meteorological observations only (obs), meteorological observations filled with ERA-Interim for missing variables (obs_ERA) or ERA-Interim (ERA). Two kinds of isotopic forcing are possible: isotopic composition of precipitation and water vapor observed on the site (obs_iso), or interpolation between GNIP, USNIP or Carbo-Europe stations using the LMDZ atmospheric general circulation model. In the former case, the datasets used for prescribing the water vapor and precipitation isotopic composition forcing are mentionned. In the latter case, GNIP, USNIP or Carbo-Europe stations used to construct the interpolated precipitation isotopic composition forcing are listed.

sites, we use local meteorological measurements when available and combine them with ERA-Interim reanalyses at 6-hourly time scale for missing variables. At other sites, no nearby meteorological measurements are available and only ERA-Interim reanalyses [94] are used (Table 3).

At each site, we run the model three times over the first year of isotopic measurement (e.g., 2007 at Le Bray). These three years are discarded as spin-up. Then we run the model over the full period of isotopic measurements (e.g., 2007-2008 at Le Bray). We checked that at all sites, the seasonal distribution of $\delta^{18}O$, which is the slowest variable to spin-up, is identical between the last year of spin-up and the following year.

We force ORCHIDEE with monthly isotopic composition of precipitation and near-surface water vapor. Since we evaluate the results at the monthly time scale, we assume that monthly isotopic forcing is sufficient. At Le Bray and Yatir, monthly observations of isotopic composition of precipitation and near-surface water vapor are available to construct the forcing. Unfortunately, these observations

are not available on the other sites. Therefore, we create isotopic forcing using isotopic measurements in the precipitation performed on nearby GNIP or USNIP stations. To interpolate between the nearby stations, we take into account spatial gradients and altitude effects by exploiting outputs from an LMDZ simulation.

Model-data comparison methods

Simulated isotopic composition in soil, stem and leaf water: The soil profile option is activated in all our stand-alone ORCHIDEE simulations. We compare the soil water samples collected in the first 15 cm of the soil (in the first 5-10 cm at many sites) to the soil water composition simulated in the uppermost layer.

The observed composition of stem water is compared to the simulated composition of the transpiration flux.

When comparing observed and simulated composition of leaf water, the Peclet effect, which mixes stomatal water with xylem water (Equation 8), is deactivated. Neglecting the Peclet effect may lead to overestimate of $\delta^{18}O_{leaf}$ values.

Impact of the temporal sampling: Over the ten sites, samples were collected during specific days and hours. This temporal sampling may induce artifacts when comparing observations to monthly-mean simulated ORCHIDEE values. For soil and stem water, the effect of temporal sampling can be neglected because simulated soil and stem water composition vary at a very low frequency. For leaf water however, there are large diurnal variations [95]. For example, if leaf water is sampled every day at noon when $\delta^{18}O_{leaf}$ is maximum, then observed $\delta^{18}O_{leaf}$ will be more enriched than monthly-mean $\delta^{18}O_{leaf}$. The exact sampling time is available for Le Bray site only, where we will estimate the effect of temporal sampling.

Spatial heterogeneities: We are aware of the scale mismatch between punctual in-situ measurements and an LSM designed for large scales (a typical GCM grid box is more than 100 km wide). However, for soil moisture it has been shown that local measurements represent a combination of small scale (10-100 m) variability [20,21] and a large-scale (100-1000 km) signal [96] that a large-scale model should capture [97]. The sampling protocol allows us to evaluate the spatial heterogeneities. For example at Le Bray, two samples were systematically taken a few meters apart, allowing us to calculate the difference between these two samples. On average over all months, the difference between the two samples is 3.5‰ for $\delta^{18}O_s$, 4.8‰ for $\delta^{18}O_{stem}$ and 1.3‰ for $\delta^{18}O_{leaf}$. At Yatir, samples were taken several days every month, allowing us to calculate a standard deviation between the different samples for every month. On average of all months, the standard deviation is 0.9‰ for $\delta^{18}O_s$, 0.4‰ for $\delta^{18}O_{stem}$ and 1.2‰ for $\delta^{18}O_{leaf}$. These error bars need to be kept in mind when assessing model-data agreement.

Soil moisture: Soil moisture have a different physical meaning in observations and model. Soil moisture is measured as volumetric soil water content (SWC) and expressed in %. In ORCHIDEE, the soil moisture is expressed in mm and cannot be easily converted to volumetric soil water content: the maximum soil water holding capacity of 300 mm and soil depth of 2 m are arbitrary choices and do not reflect realistic values at all sites. In LSMs, soil moisture is more an index than an actual soil moisture content [3]. In this version of ORCHIDEE in particular, it is an index to compute soil water stress, but it was not meant to be compared with soil water content measurements. Therefore, to compare soil moisture between model and observations, we normalize values to ensure that they remains between 0 and 1. The observed normalized SWC is calculated as $\frac{SWC - SWC_{min}}{SWC_{max} - SWC_{min}}$ where SWC_{min} and SWC_{max} are the minimum and maximum observed values of monthly SWC at each site. Similarly, simulated normalized SWC is calculated as $\frac{SWC - SWC_{min}}{SWC_{max} - SWC_{min}}$ where SWC_{min} and SWC_{max} are the minimum and maximum simulated values of monthly SWC at each site.

Evaluation at measurement sites

In this section, we evaluate the simulated isotopic composition in different water reservoirs of the soil-vegetation-atmosphere continuum at the seasonal scale.

Hydrological simulation: Before evaluating the isotopic composition of the different water reservoirs, we check whether the simulations are reasonable from a hydrological point of view. ORCHIDEE captures reasonably well the magnitude and seasonality of the latent and sensible heat fluxes at most sites (Figures 3 and 4). At Le Bray for example, the correlation between monthly values of evapo-transpiration is 0.98 and simulated and observed annual mean evapo-transpiration rates are 2.4 mm/d and 2.0 mm/d respectively. However, the model tends to overestimate the latent heat flux at the expense

of the sensible heat flux at several sites. This is especially the case at the dry sites Mitra and Yatir: the observed evapo-transpiration is at its maximum in spring and then declines in summer due to soil water stress. ORCHIDEE underestimates the effect of soil water stress on evapo-transpiration and maintains the evapo-transpiration too strong throughout the summer.

The soil moisture seasonality is very well simulated at all sites where data is available (Figures 3 and 4), except for a two-month offset at Yatir (Figure 3f).

Water isotopes in the soil water: The evaluation of the isotopic composition of soil water is crucial before using ORCHIDEE to investigate the sensitivity to the evapo-transpiration partitioning or to infiltration processes, or in the future to simulate the isotopic composition of paleo-proxies such as speleothems [98].

In observations, at all sites, $\delta^{18}O_s$ remains close to $\delta^{18}O_p$, within the relatively large month-to-month noise and spatial heterogeneities (Figures 3 and 4) At most sites (Le Bray, Donaldson Forest, Anchorage, Bily Kriz and Hainich), observed $\delta^{18}O_s$ exhibits no clear seasonal variations distinguishable from month-to-month noise. At Morgan-Monroe and Mitra, and to a lesser extent at Brloh and Tharandt, $\delta^{18}O_s$ progressively increases throughout the spring, summer and early fall, by up to 5‰ at Morgan-Monroe. The increase in $\delta^{18}O_s$ in spring can be due to the increase in $\delta^{18}O_p$. The increase in $\delta^{18}O_s$ in late summer and early fall, while $\delta^{18}O_p$ starts to decrease, is probably due to the enriching effect of bare soil evaporation. At Yatir, $\delta^{18}O_s$ increases by 10‰ from January to June, probably due to the strong evaporative enrichment on this dry site. Then, the $\delta^{18}O_s$ starts to decline again in July. This could be due to the diffusion of depleted atmospheric water vapor in the very dry soil.

ORCHIDEE captures the order of magnitude of annual-mean $\delta^{18}O_s$ on most sites, and captures the fact that it remains close to $\delta^{18}O_p$. ORCHIDEE captures the typical $\delta^{18}O_s$ seasonality, with an increase in $\delta^{18}O_s$ in spring-summer at Morgan-Monroe, Donaldson Forest, Mitra and Bily Kriz. However, the sites with a spring-summer enrichment in ORCHIDEE are not necessarily those with a spring-summer enrichment in observations. This means that ORCHIDEE misses what controls the inter-site variations in the amplitude of the $\delta^{18}O_s$ seasonality. The seasonality is not well simulated at Yatir. This could be due to the missed seasonality in soil moisture and evapo-transpiration. This could be due also to the fact that at Yatir ORCHIDEE underestimates the proportion of bare soil evaporation to total evapo-transpiration: less than 10% in ORCHIDEE versus 38% observed [89], which could explain why the spring enrichment is underestimated. Besides, ORCHIDEE does not represent the diffusion of water vapor in the soil, which could explain why the observed $\delta^{18}O_s$ decrease at Yatir in fall is missed.

When comparing the different sites, annual-mean $\delta^{18}O_s$ follows annual-mean $\delta^{18}O_p$, with an inter-site correlation of 0.99 in observations. Therefore, it is easy for ORCHIDEE to capture the inter-site variations in annual-mean $\delta^{18}O_s$. A more stringent test is whether ORCHIDEE is able to capture the inter-site variations in annual-mean $\delta^{18}O_s$ - $\delta^{18}O_p$. This is the case, with a correlation of 0.85 (Figure 5a) between ORCHIDEE and observations. In ORCHIDEE (and probably in observations), spatial variations in $\delta^{18}O_s$ - $\delta^{18}O_p$ are associated with the relative importance of bare soil evaporation.

Water isotopes in the stem water: In observations, observed $\delta^{18}O_{stem}$ exhibits no seasonal variations distinguishable from month-to-month noise (Figures 3 and 4). At Le Bray, Yatir, Mitra, Brloh, Hainich, observed $\delta^{18}O_{stem}$ is more depleted than the surface soil water. It likely

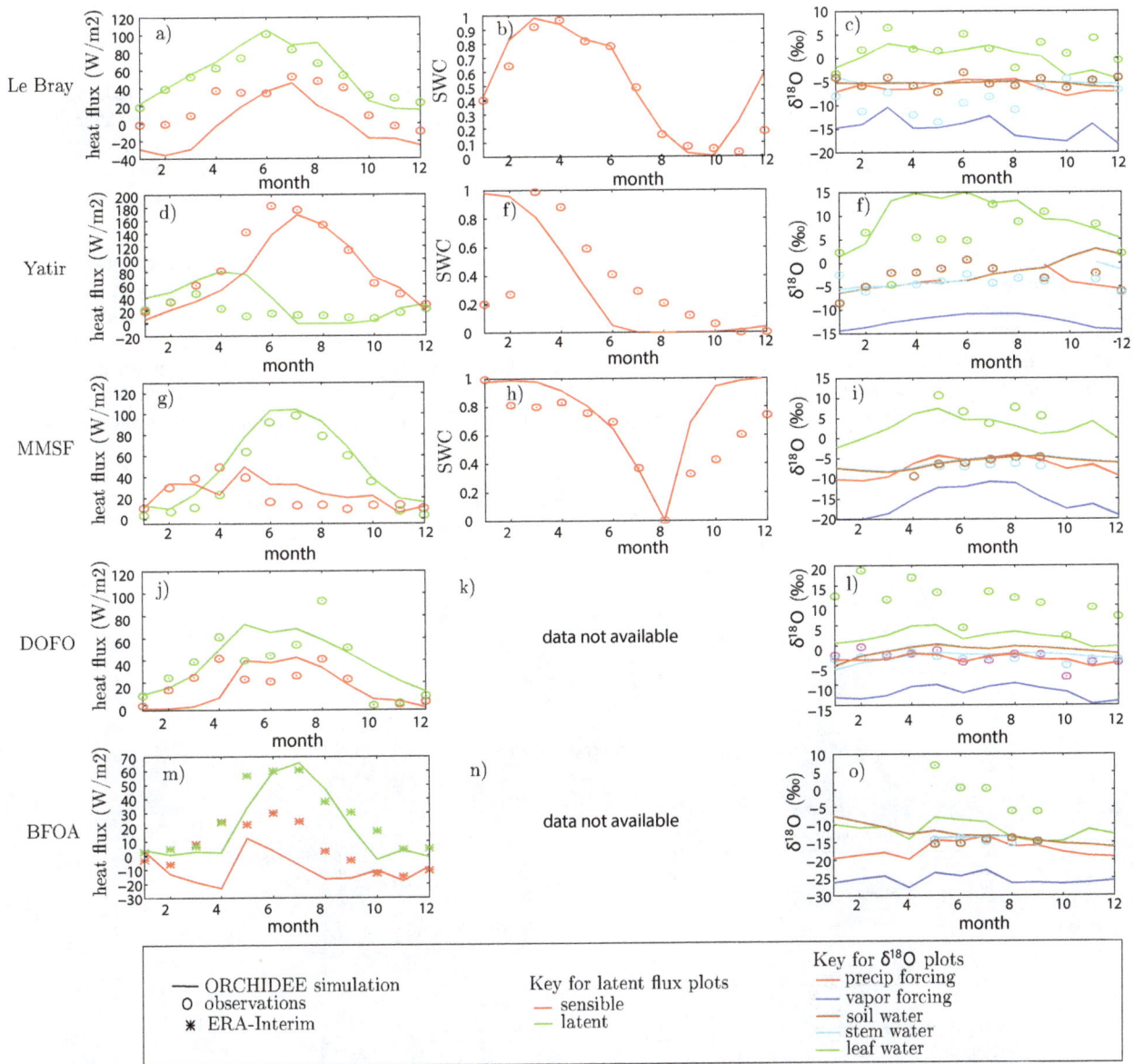

Figure 3: Evaluation of hydrological and isotopic variables simulated by ORCHIDEE on different MIBA or Carbo-Europe sites. a, d, g, j, m: latent (green) and sensible (red) heat fluxes observed locally when available (circles), simulated in the ERA-Interim reanalyses (stars) and simulated by ORCHIDEE (lines). b, e, h, k, n: normalized soil moisture content (SWC, without unit) observed locally (circles) and simulated by ORCHIDEE (lines). c, f, i, l, o: $\delta^{18}O$ of the surface soil (brown) and stems (green) simulated by ORCHIDEE in the control offline simulations (thin curves) and observed (circles). Observed $\delta^{18}O$ in precipitation (thick dashed red) and vapor (thick dashed blue) used as forcing are also shown. a-c: Le Bray, d-f: Yatir, g-i: Morgan-Monroe, j-l: Donaldson Forest, m-o: Anchorage. The normalized SWC (soil water content) is calculated.

corresponds to the $\delta^{18}O$ values in deeper soil layers, suggesting that the rooting system is quite deep. For example, at Mitra, the root system reaches least 6 m deep, and could at some places reach as deep as 13 m where it could use depleted ground water. At Donaldson Forest, Morgan-Monroe, Anchorage and Tharandt, $\delta^{18}O_{stem}$ is very close to $\delta^{18}O_s$, maybe reflecting small vertical variations in isotopic composition within the soil or shallow root profiles.

At Bily Kriz, observed $\delta^{18}O_{stem}$ is surprisingly more enriched than surface soil water. Several hypotheses could explain this result: (1) the surface soil water could be depleted by dew or frost at this mountainous, foggy site; (2) spruce has shallow roots and therefore sample soil water

that is not so depleted; (3) the twigs that were sampled were relatively young so that evaporation from their surface could have occurred when they were still at tree; (4) twigs were sampled in sun-exposed part of the spruce crowns during sunny conditions, which could favor some evaporative enrichment. Additional measurements show a lower Deuterium excess in the stem water compared to the soil water, supporting evaporative enrichment of stems.

ORCHIDEE captures the fact that $\delta^{18}O_{stem}$ is nearly uniform throughout the year. As for soil water, it is easy for ORCHIDEE to capture the inter-site variations in annual-mean $\delta^{18}O_{stem}$ (inter-site correlation between ORCHIDEE and observations of 0.90). ORCHIDEE is able to capture some of the inter-site variations in annual-mean $\delta^{18}O_{stem}$ - $\delta^{18}O_p$, with a inter-site correlation between ORCHIDEE and observations of 0.60. However, ORCHIDEE simulates $\delta^{18}O_{stem}$ values that are very close to $\delta^{18}O_s$ values (Figure 5b). It is not able to capture $\delta^{18}O_{stem}$ values that are either more enriched or more depleted than $\delta^{18}O_s$. This could be due to the fact that ORCHIDEE underestimates vertical variations in soil isotopic composition. Also, ORCHIDEE is not designed to represent deep ground water sources or photosynthesizing twigs.

Vertical profiles of soil water isotope composition: At Le Bray, we compare our offline simulation for 2007 with soil profiles collected from 1993 to 1997 and in 2007 (Figure 6a-6b). The year mismatch adds a source of uncertainty to the comparison. In summer (profiles of August 1993 and September 1997), the data exhibits an isotopic enrichment at the soil surface of about 2.5‰ compared to the soil at 1 m depth (Figure 6a), likely due to surface evaporation [99]. Then, by the end of September 1994, the surface becomes depleted, likely due to the input of depleted rainfall. Previously enriched water remains

between 20 and 60 cm below the ground, suggesting an infiltration through piston-flow [100]. ORCHIDEE predicts the summer isotopic enrichment at the surface, but slightly later in the season (maximum in September rather than August) and underestimates it compared to the data (1.5‰ enrichment compared to 2.5‰ observed, Figure 6b). The model also captures the surface depletion observed after the summer, as well as the imprint of the previous summer enrichment at depth. However, ORCHIDEE simulates the surface depletion in December, whereas the surface depletion can be observed sooner in the data, at the end of September 1994.

At Yatir, observed profiles exhibit a strong isotopic enrichment from deep to shallow soil layers in May-June by up to 10‰ (Figure 6c). As for Le Bray, the model captures but underestimates this isotopic enrichment in spring and summer by about 3‰ (Figure 6d). This discrepancy could be the result of underestimated bare soil evaporation. Observed profiles also feature a depletion at the surface in winter that the model does not reproduce. This depletion could be due

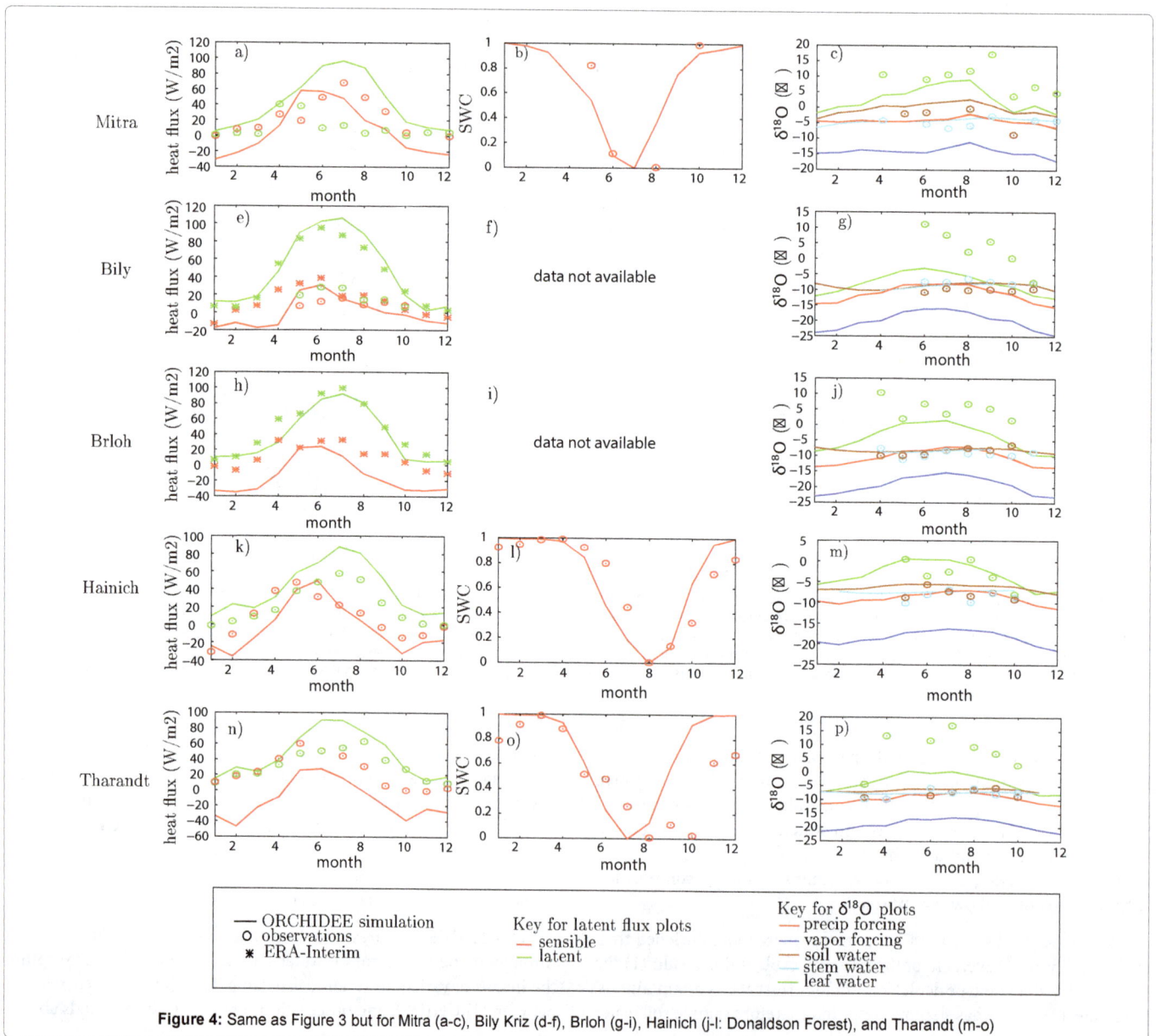

Figure 4: Same as Figure 3 but for Mitra (a-c), Bily Kriz (d-f), Brloh (g-i), Hainich (j-l: Donaldson Forest), and Tharandt (m-o)

to back-diffusion of depleted vapor in dry soils [99,101-103], a process that is not represented in ORCHIDEE but likely to be significant in this region. Soil evaporation fluxes measured with a soil chamber at Yatir shows that when soils are dry, there is adsorption of vapor from the atmosphere to the dry soil pores before sunrise and after sunset [104].

Water isotopes in leaf water: It is important to evaluate the simulation of the isotopic composition of leaf water by ORCHIDEE if we want to use this model in the future for the simulation of paleo-climate proxies such tree-ring cellulose [105,106], for the simulation of the isotopic composition of atmospheric CO_2 which may be used to partition CO_2 fluxes into respiration from vegetation and soil [107,108] or for the simulation of the isotopic composition of atmospheric O_2 which may be used to infer biological productivity [109,110].

In the observations, $\delta^{18}O_{leaf}$ exhibits a large temporal variability reflecting a response to changes in environmental conditions (e.g., relative humidity and the isotopic composition of atmospheric water vapor). At all sites except at Yatir, $\delta^{18}O_{leaf}$ is most enriched in summer than in winter, by up to 15‰ (Figures 3 and 4). This is because the evaporative enrichment is maximum in summer due to drier and warmer conditions.

ORCHIDEE captures the maximum enrichment in summer. However, ORCHIDEE underestimates the annual-mean $\delta^{18}O_{leaf}$ at most sites (Figure 5). This could be due to the fact that most leaf samples were collected during the day, when the evaporative enrichment is at its maximum, while for ORCHIDEE we plot the daily-mean $\delta^{18}O_{leaf}$. At Le Bray, if we sample the simulated $\delta^{18}O_{leaf}$ during the correct days and hours, simulated $\delta^{18}O_{leaf}$ increases by 4‰ in winter and by 10‰ in summer. Such an effect can thus quantitatively explain the model-data mismatch. After taking this effect into account, simulated $\delta^{18}O_{leaf}$ may even become more enriched than observed. This is the case at Le Bray, especially in summer. The overestimation of summer $\delta^{18}O_{leaf}$ could be due to neglecting diffusion in leaves or non-steady state effects.

Again, Yatir is a particular case. Minimum $\delta^{18}O_{leaf}$ occurs in spring-summer while the soil evaporative enrichment is maximum. In arid regions and seasons, leaves may close stomata during the most stressful periods of the day, inhibiting transpiration, and thus retain the depleted isotopic signal associated with the moister conditions of the morning [111,112]. ORCHIDEE does not represent this process and thus simulates too enriched $\delta^{18}O_{leaf}$.

Summary: Overall, ORCHIDEE is able to reproduce the main features of the seasonal and vertical variations in soil water isotope content, and seasonal variations in stem and leaf water content. Discrepancies can be explained by some sampling protocols, by shortcomings in the hydrological simulation or by neglected processes in ORCHIDEE (e.g., fractionation in the vapor phase).

The strong spatial heterogeneity of the land surface at small scales does not prevent ORCHIDEE from performing reasonably well. This suggests that in spite of some small-scale spatial heterogeneities at each site, local isotope measurements contain large-scale information and are relevant for the evaluation of large-scale LSMs.

Sensitivity analysis

Sensitivity to evapo-transpiration partitioning: Several studies have attempted to partition evapo-transpiration into the transpiration and bare soil evaporation terms at the local scale [29-31,113]. Estimating E/ET, where E is the bare soil evaporation and ET is the evapo-transpiration, requires measuring the isotopic composition of soil water, stem water and of the evapo-transpiration flux. The isotopic composition of the evapo-transpiration can be estimated through "Keeling plots" approach [114], but this is costly [29] and the assumptions underlying this approach are not always valid [115].

Considering a simple soil water budget at steady state and with vertically-uniform isotopic distribution, we show that although estimating E/ET requires measuring the isotopic composition of the

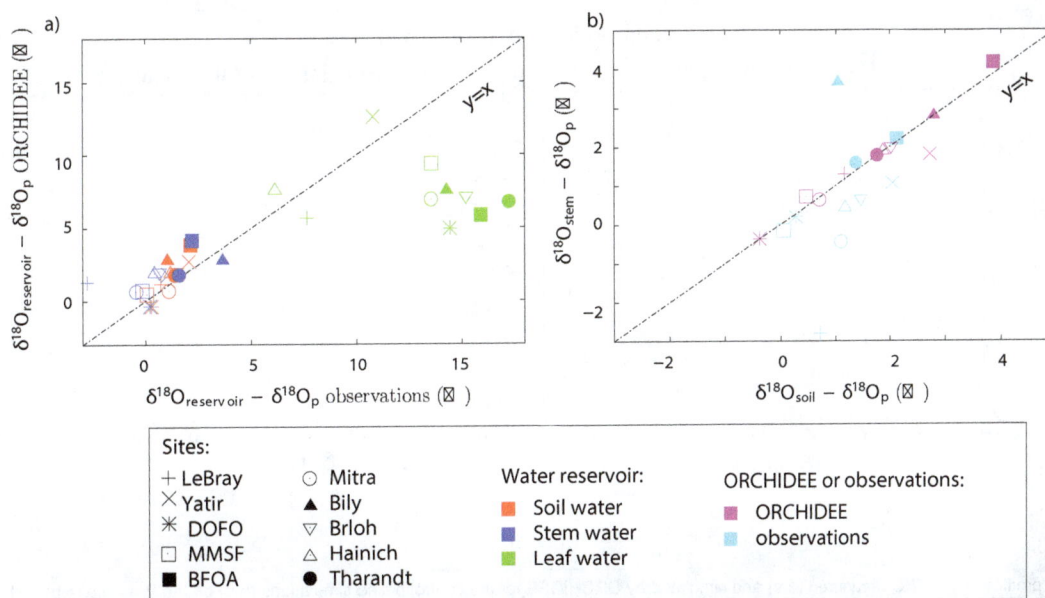

Figure 5: a) Relationship between simulated and observed annual-mean $\delta^{18}O$ in the soil water (red), stem water (blue) and leaf water (green), to which the precipitation-weighted annual-mean precipitation $\delta^{18}O$ is subtracted. In the case of perfect model-data agreement, markers should fall on the y=x line. b) Relationship between the annual-mean $\delta^{18}O$ in the soil water and in stem water, to which the precipitation-weighted annual-mean precipitation $\delta^{18}O$ is subtracted, for both ORCHIDEE (magenta) and observations (cyan). When soil and stem water share the same $\delta^{18}O$, they fall on the y=x line.

evapo-transpiration flux, estimating E/I (where I is the precipitation that infiltrates into the soil) requires measuring temperature, relative humidity (h) and the isotopic composition of the soil water ($\delta^{18}O_s$), water vapor ($\delta^{18}O_v$) and precipitation ($\delta^{18}O_p$) only. Such variables are available from several MIBA and Carbo-Europe sites. More specifically, E/I is proportional to $\delta^{18}O_p - \delta^{18}O_s$:

$$E/I = \frac{\alpha_{eq} \cdot \alpha_k \cdot (1-h) \cdot \left(\delta^{18}O_p - \delta^{18}O_s\right)}{\left(\delta^{18}O_s + 10^3\right) \cdot \left(1 - \alpha_{eq} \cdot \alpha_k \cdot (1-h)\right) - \alpha_{eq} \cdot h \cdot \left(\delta^{18}O_v + 10^3\right)} \quad (1)$$

where α_{eq} and α_k are the equilibrium and kinetic fractionation coefficients respectively.

Below, we show that this equation can apply to annual-mean quantities, neglecting effects associated with daily or monthly co-variations between different variables. We investigate to what extent this equation allows us to estimate the magnitude of E/I at local sites.

At the Yatir site, all the necessary data for equation 1 is available. An independent study has estimated E/I =38% [89]. Using annually averaged observed values ($\delta^{18}O_p$=-5.1‰ and $\delta^{18}O_s$=-3.7‰ in the the surface soil), we obtain E/I=46%. However, in ORCHIDEE, the annually averaged surface $\delta^{18}O_s$ is 0.8 lower when sampled at the same days as in the data. When correcting for this bias, we obtain E/I =28%. Observed E/I lies between these two estimates. This shows the applicability of this estimation method, keeping in mind that estimating E/I is the most accurate where E/I is lower.

When we perform sensitivity tests to ORCHIDEE parameters at the various sites, the main factor controlling $\delta^{18}O_s$ is the E/I fraction.

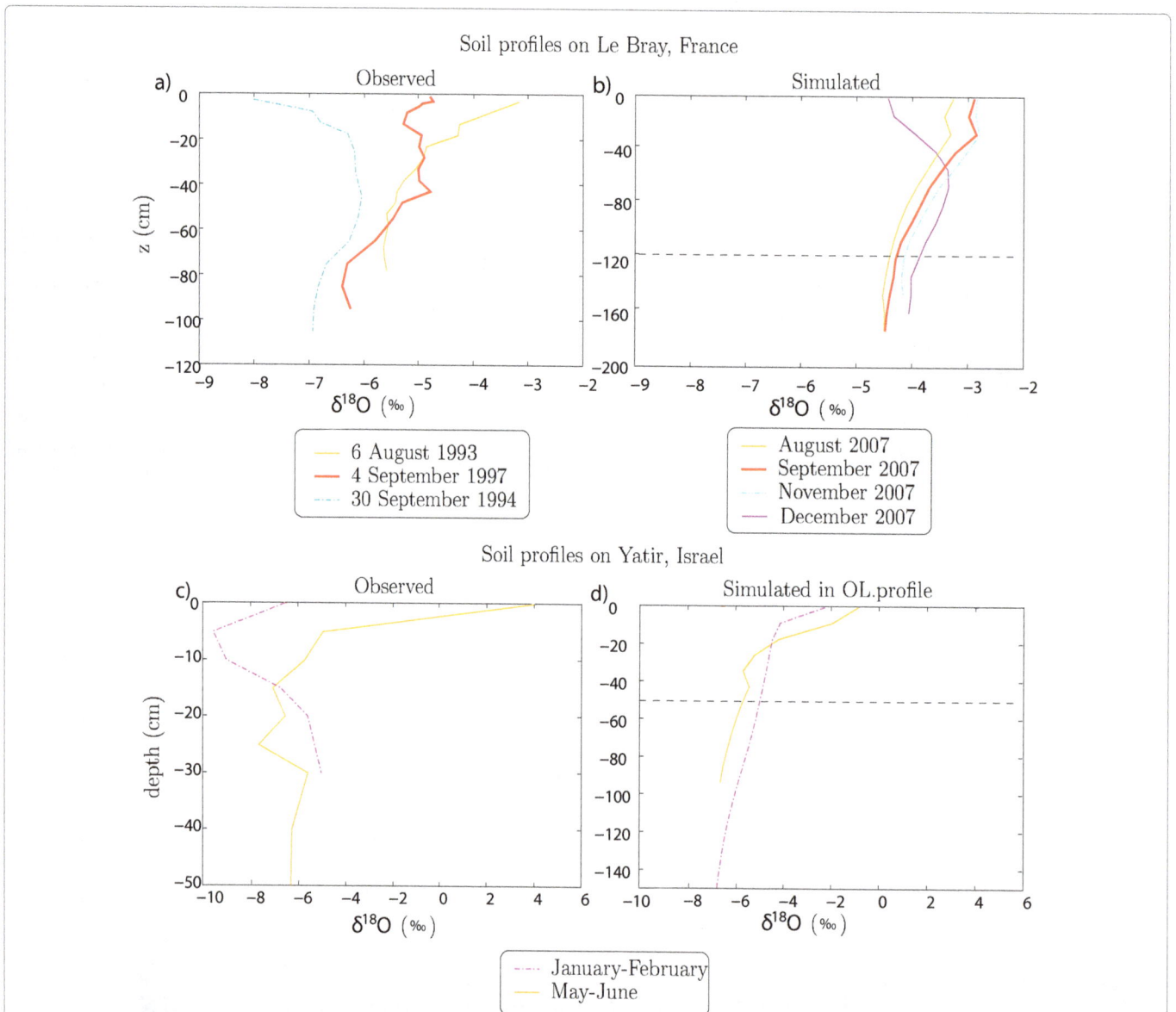

Figure 6: Vertical profiles of soil $\delta^{18}O$ measured (a,c) and simulated by ORCHIDEE for the control offline simulations (b,d) on the Bray site (a,b) and the Yatir sites (b,d). Beware that the y-scales for observations and simulations are different. This is because the representation of the soil water content is very rudimentary in the ORCHIDEE model, preventing any quantitative comparison of measured and simulated soil depth. The horizontal black dashed line represents the bottom of the observed profiles. Model outputs are sampled at the same time as the data. For the Yatir sites, frequent soil sampling for the same year allowed us plot representative bi-monthly averages for both measured and simulated profiles. This could not be the case for Le Bray. Some soil profiles were observed at Le Bray in 2007, but we do not show them because they are limited to the top 24 cm of the soil only.

This is illustrated as an example at Le Bray and Mitra sites (Figure 7). Sensitivity tests to parameters as diverse as the rooting depth or the stomatal resistance lead to changes in $\delta^{18}O_s$ - $\delta^{18}O_p$ and in E/I that are very well correlated, as qualitatively predicted by equation 13. This means that whatever the reason for a change in E/I, the effect on $\delta^{18}O_s$ - $\delta^{18}O_p$ is very robust.

Quantitatively, the slope of $\delta^{18}O_s$ - $\delta^{18}O_p$ as a function of E/I among the ORCHIDEE tests is of 0.78‰/% (r=0.94, n=6) at Le Bray and of 0.25‰/% (r=0.999, n=5) at Mitra, compared to about 0.25-0.3‰/% predicted by equation 13. The agreement is thus very good at Mitra. The better agreement at Mitra is because it is a dry site where E/I varies greatly depending on sensitivity tests. In contrast, Le Bray is a moist site where E/I values remains small for all the sensitivity tests, so numerous effects other than E/I and neglected in equation 13 can impact $\delta^{18}O_s$ - $\delta^{18}O_p$.

To summarize, local observations of $\delta^{18}O_s$ - $\delta^{18}O_p$ could help constrain the simulation of E/I in models. This would be useful since the evapo-transpiration partitioning has a strong impact on how an LSMs represents land-atmosphere interactions [116].

Sensitivity to soil infiltration processes: Partitioning between evapo-transpiration, surface runoff and drainage depends critically on how precipitation water infiltrates the soil [5,8,10], which is a key uncertainty even in multi-layer soil models where infiltration processes are represented explicitly [62]. It has been suggested that observed isotopic profiles could help understand infiltration processes at the local scale [100]. The capacity of ORCHIDEE to simulate soil profile allows us to investigate whether measured isotope profiles in the soil could help evaluate the representation of these processes also in large-scale LSMs.

With this aim, we performed sensitivity tests at Le Bray. The simulated profiles are sensitive to vertical water fluxes in the soil. When the diffusivity of water in the soil column is decreased by a factor 10 from 0.1 to 0.01 compared to the control simulation, the deep soil layer becomes more depleted by about 0.7‰ (Figure 8) and the isotopic gradient from soil bottom to top becomes 30% steeper in summer, because the enriched soil water diffuses slower through the soil column.

Simulated profiles are also sensitive to the way precipitation infiltrates the soil. When precipitation is added only to the top layer (piston-flow infiltration) the summer enrichment is reduced by mixing of the surface soil water with rainfall, and it propagates more easily to lower layers during fall and winter. Conversely, when rainfall is evenly spread throughout the soil column (a crude representation of preferential pathway infiltration), the surface enrichment is slightly

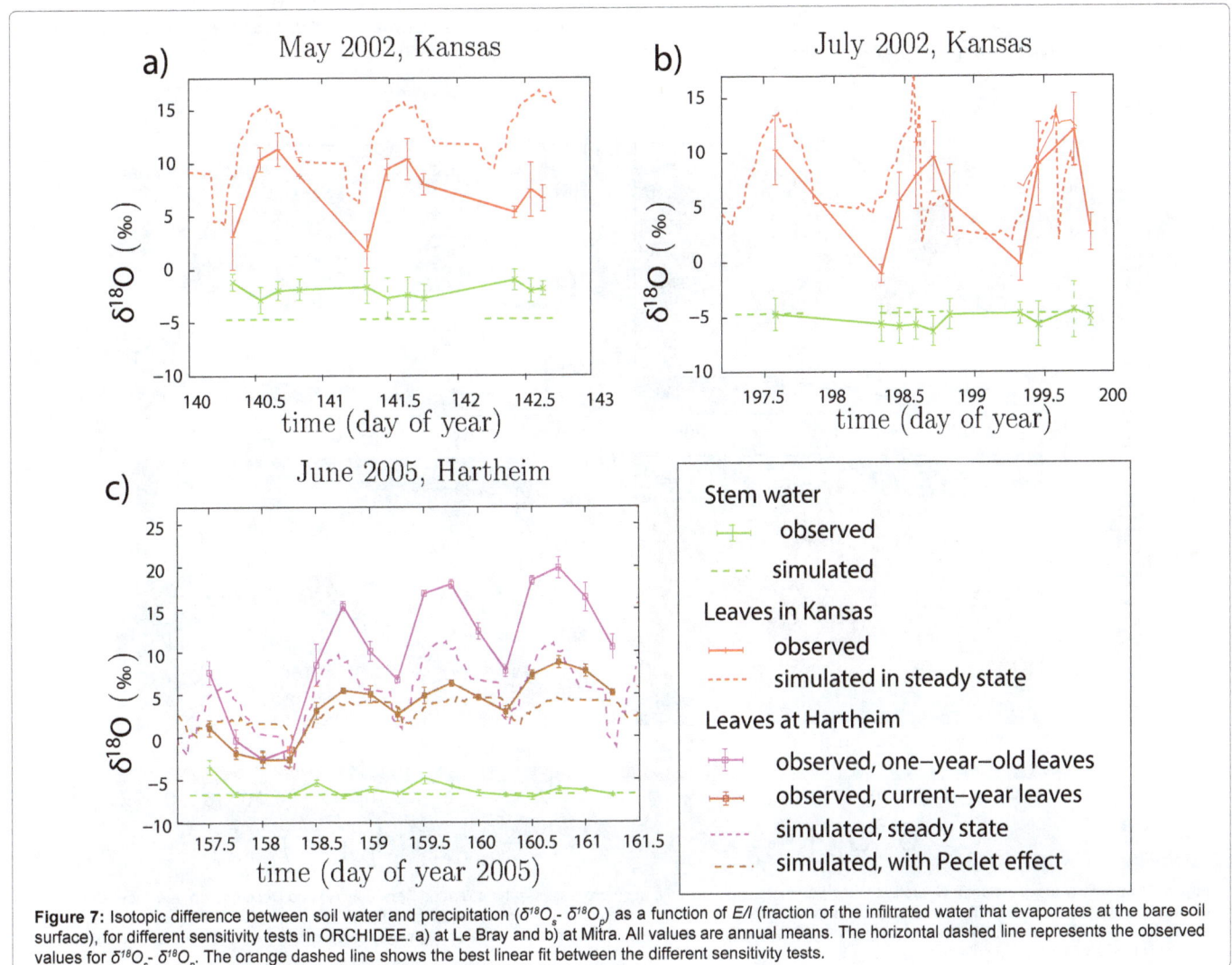

Figure 7: Isotopic difference between soil water and precipitation ($\delta^{18}O_s$ - $\delta^{18}O_p$) as a function of E/I (fraction of the infiltrated water that evaporates at the bare soil surface), for different sensitivity tests in ORCHIDEE. a) at Le Bray and b) at Mitra. All values are annual means. The horizontal dashed line represents the observed values for $\delta^{18}O_s$ - $\delta^{18}O_p$. The orange dashed line shows the best linear fit between the different sensitivity tests.

more pronounced and the deep soil water is more depleted by up to 0.8‰ in winter (Figure 8). However, the observed surface depletion occurs in February with preferential pathways, compared to December in the piston-like in infiltration. The quick surface depletion observed after the summer suggests that infiltration is dominated by the piston-like mechanisms.

To summarize, we show that vertical and seasonal variations of $\delta^{18}O_s$ are very sensitive to infiltration processes, and are a powerful tool to evaluate the representation of these processes in LSMs.

Global-scale Simulations Using the Coupled LMDZ-ORCHIDEE Model

Simulation set-up

To compare with global datasets, we performed LMDZ-ORCHIDEE coupled simulations. In all our experiments, LMDZ three-dimensional fields of horizontal winds are nudged towards ECMWF (European Center for Medium range Weather Forecast) reanalyses [117]. This ensures a realistic simulation of the large-scale atmospheric circulation and allows us to perform a day-to-day comparison with field campaign data [60,118]. At each time step, the simulated horizontal wind field $\vec{u}$ is relaxed towards the reanalysis following this equation:

$$\frac{\partial \vec{u}}{\partial t} = \vec{F} + \frac{\vec{u}_{obs} - \vec{u}}{\tau}$$

where $\vec{u}_{obs}$ is the reanalysis horizontal wind field, $\vec{F}$ is the effect of all simulated dynamical and physical processes on $\vec{u}$, and τ is a time constant set to 1 h in our simulations [119].

To compare with global datasets, LMDZ-ORCHIDEE simulations are performed for the year 2006, chosen arbitrarily. We are not interested in inter-annual variations and focus on signals that are much larger. To ensure that the water balance is closed at the annual scale, we performed iteratively 10 times the year 2006 as spin-up. In these simulations, the Peclet and non-steady state effects are de-activated.

To compare with field campaign observations in 2002 and 2005, we use simulations performed for these specific years, initialized from the 2006 simulation. In these simulations, we test activating or de-activating the Peclet effect.

In all LMDZ-ORCHIDEE simulations, canopy-interception was de-activated (consistent with simulations that our modeling group performed for the Fourth Assessment Report).

Evaluation of water isotopes in leaf water at the diel scale during campaign cases

Daily data from field campaigns: Two field campaigns are used to evaluate the representation of $\delta^{18}O_{leaf}$ diurnal variability. The first campaign covers six diurnal cycles in May and July 2002 in a grassland prairie in Kansas (39.20 ° N 96.58 ° W [120]). The second campaign covers four diurnal cycles in June 2005 in a pine plantation in Hartheim, Germany (7.93 ° N, 7.60 ° E [121]).

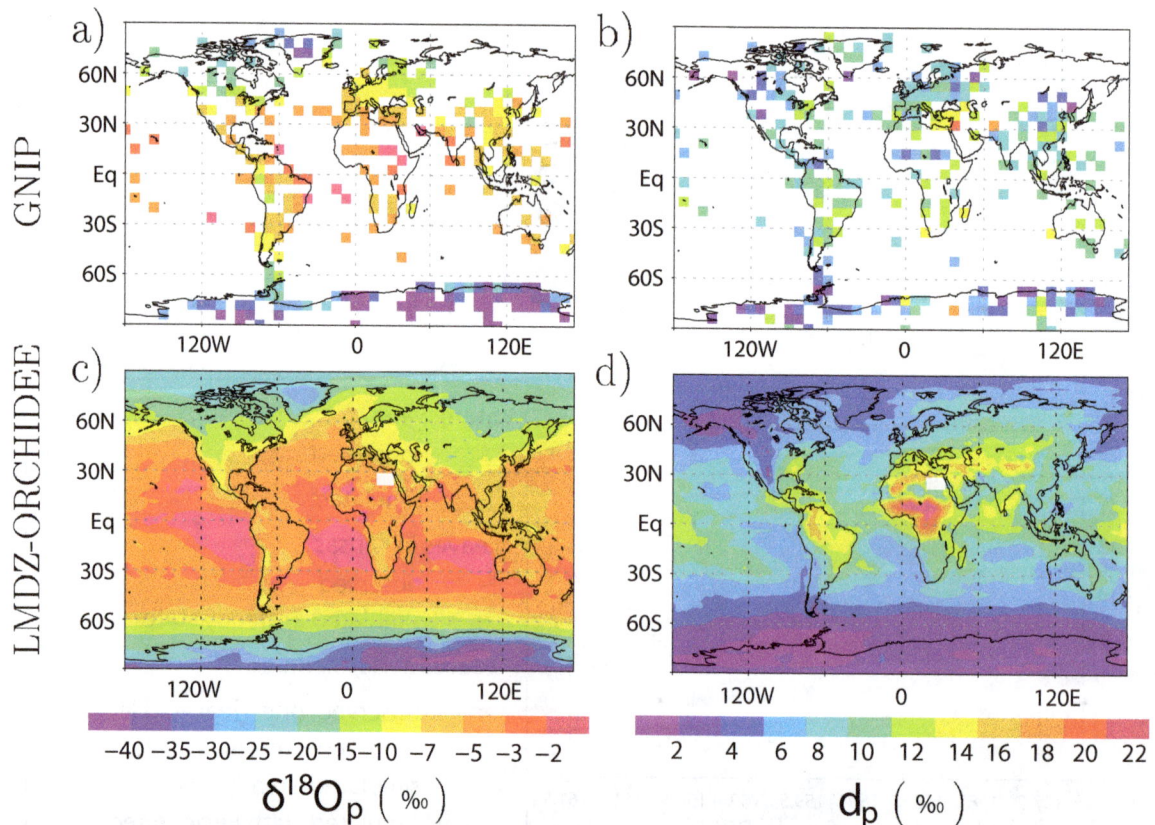

Figure 8: Sensitivity of simulated $\delta^{18}O_s$ profiles to the parameterization of infiltration processes in the soil at Le Bray. July (a) and December (b) are shown for three different parameterizations in offline simulations: control simulation (solid red), a simulation in which the soil water diffusivity was divided by 10 (dashed blue) and a simulation is which the water infiltrates the soil uniformly in the vertical (crude representation of preferential pathways, dash-dotted green) rather than in a piston-like way as is the case for other simulations.

Because meteorological and isotopic forcing are not available for the entire year, we prefer to compare these measurements with LMDZ-ORCHIDEE simulations. At both sites, the simulated $\delta^{18}O_v$ and $\delta^{18}O_{stem}$ are consistent with those observed (model-data mean difference lower than 1.4‰ in Kansas and 0.4‰ at Hartheim), allowing us to focus on the evaluation of leaf processes.

Evaluation results: At the Kansas grassland site, $\delta^{18}O_{leaf}$ exhibits a diel cycle with an amplitude of about 10‰ [120]. LMDZ-ORCHIDEE captures this diel variability, both in terms of phasing and amplitude (Figure 9). The model systematically overestimates $\delta^{18}O_{leaf}$ by about 4‰, in spite of the underestimation of the stem water by 1.4‰ on average. This may be due to a bias in the simulated relative humidity (LMDZ is on average 13% too dry at the surface, which translates into an expected enrichment bias of 3.9‰ on the leaf water assuming steady state based on Equation 7) or to uncertainties in the kinetic fractionation during leaf water evaporation.

At the Hartheim pine plantation, $\delta^{18}O_{leaf}$ is on average 8‰ more depleted for current-year needles than for 1-year-old needles. Also, the observed diel amplitude is weaker for current-year needles (5 to 8‰) than for 1-year-old needles (10 to 15‰). These observations are consistent with a longer diffusion length for current-year needles (15 cm) than for 1-year-old needles (5 cm) [121] and with a larger transpiration rate, leading to a stronger Peclet effect. When neglecting Peclet and non-steady state effects, ORCHIDEE simulates an average $\delta^{18}O_{leaf}$ close to that of 1-year-old needles, consistent with the small diffusion length and evaporation rate of these leaves. ORCHIDEE captures the phasing of the diurnal cycle, but underestimates the diel amplitude by about 4‰. This is probably due to the underestimate of the simulated diel amplitude of relative humidity by 20%. Accounting for Peclet and non-steady state effects strongly reduces both the average $\delta^{18}O_{leaf}$ and its diel amplitude (Figure 9), in closer agreement with current-year needles.

To summarize, ORCHIDEE simulates well the leaf water isotopic composition. The leaf water isotope calculation based on Craig et al. [75] simulates the right phasing and amplitude for leaves that have short diffusive lengths or low transpiration rates. Non-steady state and diffusion effects need to be considered in other cases. By activating or de-activating these effects, ORCHIDEE can simulate all cases.

Evaluation of water isotopes in precipitation

Precipitation datasets: To evaluate the spatial distribution of precipitation isotopic composition simulated by the LMDZ-ORCHIDEE coupled model, we use data from the Global Network for Isotopes in Precipitation (GNIP [122]), further complemented by data from Antarctica [123] and Greenland [124]. We also use this network to construct isotopic forcing at sites where the precipitation was not sampled, complemented with the USNIP (United States Network for Isotopes in Precipitation [125]) network.

Evaluation results: At the global scale, the LMDZ-ORCHIDEE coupled model reproduces the annual mean distribution in $\delta^{18}O_p$ and d_p observed by the GNIP network reasonably well (Figure 10), with correlations of 0.98 and 0.46 and Root Mean Square Errors (RMSE) of 3.3‰ and 3.5‰ respectively.

This good model-data agreement can be obtained even when we de-activate ORCHIDEE. When we use LMDZ in a stand-alone mode, in which the isotope fractionation at the land surface is neglected [60], the model-data agreement is as good as when we use LMDZ-ORCHIDEE. Therefore, fractionating processes at the land surface have a second

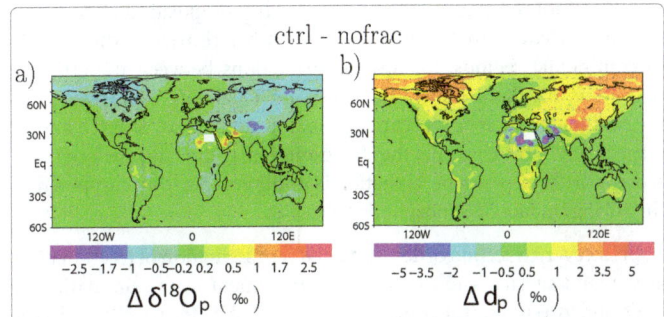

ctrl - nofrac

Figure 9: $\delta^{18}O$ of stem and grass leaves measured during two series of 3 diurnal cycles in May and July 2002 over the plains of Kansas [120] and simulated by LMDZ-ORCHIDEE for the same year in the grid box containing the observation site. $\delta^{18}O$ of vapor (blue), pine leaves (pink and red) and stems (green) measured during four diurnal cycles in June 2005 in Hartheim, Germany [121] and simulated by LMDZ-ORCHIDEE for the same year in the grid box containing the observation site. Simulated values are dashed, observed values solid. Two kinds of leaves were sampled during this campaign: one-year-old leaves (solid pink) and current-year leaves (solid brown). Two leaf water diagnostics were computed for in LMDZ-ORCHIDEE: stationary state at the evaporative site (dashed red, equation 7) or non-stationary state in the lamina, taking into account the Peclet effect (dashed brown, equation 9, using an effective length scale of 25 mm).

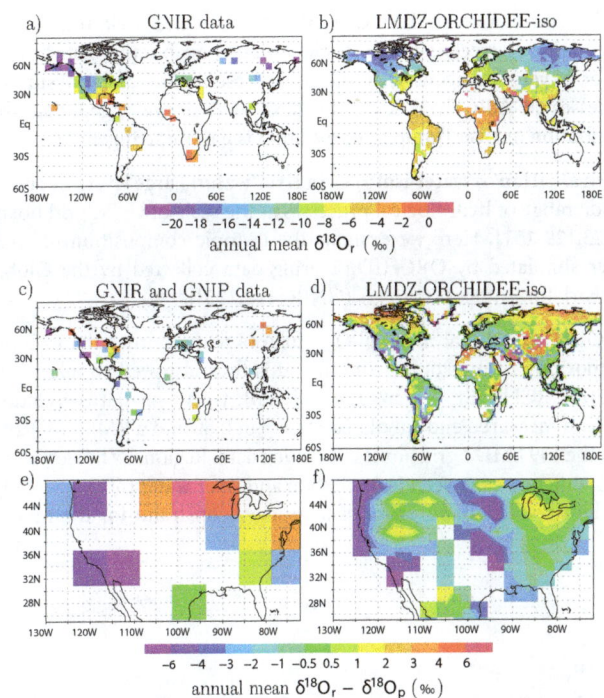

Figure 10: a) Annual mean $\delta^{18}O_p$ from GNIP [122], Antarctica [123] and Greenland [124] data. The data is gridded over a coarse 7.5 × 6.5° grid for visualization purposes. b) Same as a) but for annual mean d_p. c) Annual mean $\delta^{18}O_p$ simulated by coupled LMDZ-ORCHIDEE model for the control simulation. d) same as c) but for annual mean d_p.

order effect on precipitation isotopic composition, consistent with [28,71-73].

To quantify in more detail the effect of fractionation at the land surface, we performed additional coupled simulations with LMDZ-ORCHIDEE. We compare the control simulation described above (ctrl) to a simulation in which fractionation at the land surface was de-activated (nofrac) (Figure 11). In nofrac, the composition of bare soil evaporation equals that of soil water. Even when restricting the analysis to continental regions, the spatial correlations between the ctrl and nofrac simulations are 0.999 and 0.95 for $\delta^{18}O_p$ and d_p respectively, and the root mean square differences are 0.27‰ and 1.1‰ for $\delta^{18}O_p$ and d_p respectively. This confirms that fractionation at the land surface has a second-order effect on precipitation isotopic composition compared to the strong impact of atmospheric processes.

However, to second order, a detailed representation of fractionation at the land surface lead to a slight improvement in the simulation of $\delta^{18}O_p$ and to a significant improvement in that of d_p. In ctrl, $\delta^{18}O_p$ is lower by up to 1.5‰ and d_p higher by up to 5‰ than in nofrac over boreal continental regions such as Siberia, Canada and central Asia, consistent with the expected effect of fractionation at surface evaporation [42]. Taking into account fractionation at the land surface leads to a better agreement with the GNIP data over these regions, where $\delta^{18}O_p$ is overestimated by about 4‰ and d_p underestimated by 4 to 7‰ when neglecting fractionation at the land surface. The effect of fractionation is maximal over these boreal regions because (1) the fraction of bare soil evaporation is maximal, (2) a significant proportion of evaporatively-enriched soil water is lost by drainage and (3) a larger proportion of the moisture comes from land surface recycling [48,126,127]. Similar results were obtained with other models [128].

To summarize, LMDZ-ORCHIDEE simulates well the spatial distribution of precipitation isotopic composition, but this distribution is not a very stringent test for the representation of land surface processes in ORCHIDEE. In the next section, we argue that the distribution of river isotopic composition is a more stringent test.

Evaluation of water isotopes in river water: Large rivers integrate a wide range of hydrological processes at the scale of GCM grid boxes [22,23,129-131]. Here we evaluate the isotopic composition of river water simulated by ORCHIDEE using data collected by the Global Network for isotopes in Rivers (GNIR [132,133]).

Observed annual mean $\delta^{18}O_{river}$ follows to first order the isotopic composition of precipitation [79], and is thus also well simulated by LMDZ-ORCHIDEE (Figure 12a and 12b), with a spatial correlation between measured and simulated $\delta^{18}O_{river}$ of 0.80 and a RMSE of 3.2‰ over the 149 LMDZ grid boxes containing data. Regionally however, the $\delta^{18}O$ difference between precipitation and river water ($\delta^{18}O_{river} - \delta^{18}O_p$) can be substantial and provides a stronger constraint for the model.

Over South America, Europe and some parts of the US, the river water is typically 1‰ to 4‰ more depleted than the precipitation (Figure 12a), because precipitation contributes more to rivers during seasons when it is the most depleted [134]. In contrast, over central Asia or northern America, river water is more enriched than precipitation, due to evaporative enrichment of soil water [79,134,135]. This is further confirmed by a simulation where fractionation at the land surface was neglected (not shown), for which the river water is in global average 5‰ more depleted.

ORCHIDEE reproduces moderately well the magnitude and patterns of $\delta^{18}O_{river} - \delta^{18}O_p$, with a spatial correlation of 0.39 and a RMSE of 2.7‰ over the 22 LMDZ grid boxes that contain $\delta^{18}O_{river}$ observations. It simulates the negative values over the western US, Europe and South America and the positive value over Mongolia. However, the model does not capture the positive $\delta^{18}O_{river} - \delta^{18}O_p$ in Eastern US, though positive values are simulated further North. This suggests that such a diagnostic may help identify biases in the representation of the soil water budget, as discussed in the following section.

Sensitivity to the representation of pathways from precipitation to rivers

At the local scale, water isotopes have already been used to partition river discharge peaks into the contributions from recent rainfall and soil water [37-39]. Given the property of rivers to integrate hydrological processes at the basin scales [22,23,129-131], we now explore to what extent $\delta^{18}O_{river}$ could help evaluate pathways from precipitation to rivers in LSMs. We illustrate this using seasonal variations in $\delta^{18}O_{river}$ on two well established GNIR and GNIP stations in Vienna (Danube river) and Manaus (the Amazon) (Figure 13). The seasonal cycle in $\delta^{18}O_{river}$ is attenuated compared to that in $\delta^{18}O_p$, and $\delta^{18}O_{river}$ lags $\delta^{18}O_p$ (by 5 month at Vienna and 1-3 months at Manaus).

LMDZ-ORCHIDEE (control simulation) simulates qualitatively well the amplitude and the phasing observed in $\delta^{18}O_p$ and $\delta^{18}O_{river}$. To understand better what determines the attenuation and lag of the seasonality in $\delta^{18}O_{river}$ compared to that in $\delta^{18}O_p$, we perform sensitivity tests to ORCHIDEE parameters. Parameters tested include the partitioning of excess rainfall into surface runoff and drainage and the residence time scale of different reservoirs (slow, fast and stream) in the routing scheme. River discharge is extremely sensitive to these parameters [136].

If all the runoff occurs as surface runoff (Figure 13), then the seasonal cycle of $\delta^{18}O_{river}$ is similar to that of $\delta^{18}O_p$. This shows that the attenuation and lag of the seasonality in $\delta^{18}O_{river}$ compared to that in $\delta^{18}O_p$ are caused by the storage of water into the slow reservoir, which accumulates drainage water.

When the residence time scale of the slow reservoir is multiplied by 2 (i.e., the water from the slow reservoir is poured twice faster into the streams, Figure 12), the simulated lag of $\delta^{18}O_{river}$ at Vienna increases from 4 to 5 months (in closer agreement with the data). In contrast, the seasonal cycle in $\delta^{18}O_{river}$ is not sensitive to residence time scales in the stream and fast reservoirs, which are too short to have any impact at the seasonal scale.

To summarize, ORCHIDEE performs well in simulating the seasonal variations in $\delta^{18}O_{river}$. In turn, $\delta^{18}O_{river}$ observations could help estimate the proportion of surface runoff versus drainage and calibrate empirical residence time constants in the routing scheme, offering a mean to enhance model performance.

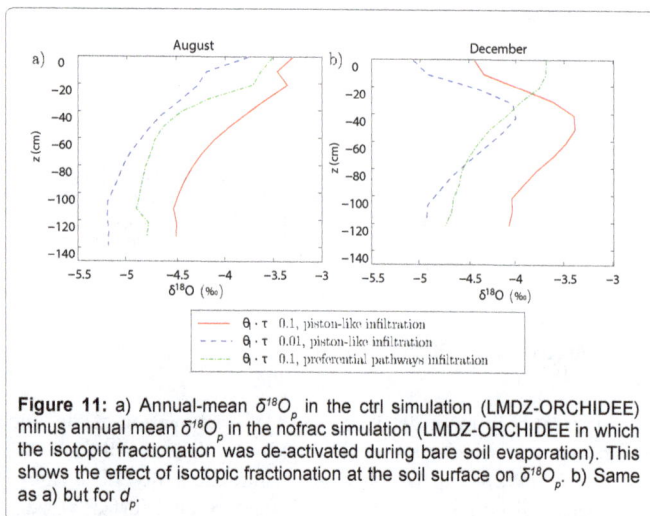

Figure 11: a) Annual-mean $\delta^{18}O_p$ in the ctrl simulation (LMDZ-ORCHIDEE) minus annual mean $\delta^{18}O_p$ in the nofrac simulation (LMDZ-ORCHIDEE in which the isotopic fractionation was de-activated during bare soil evaporation). This shows the effect of isotopic fractionation at the soil surface on $\delta^{18}O_p$. b) Same as a) but for d_p.

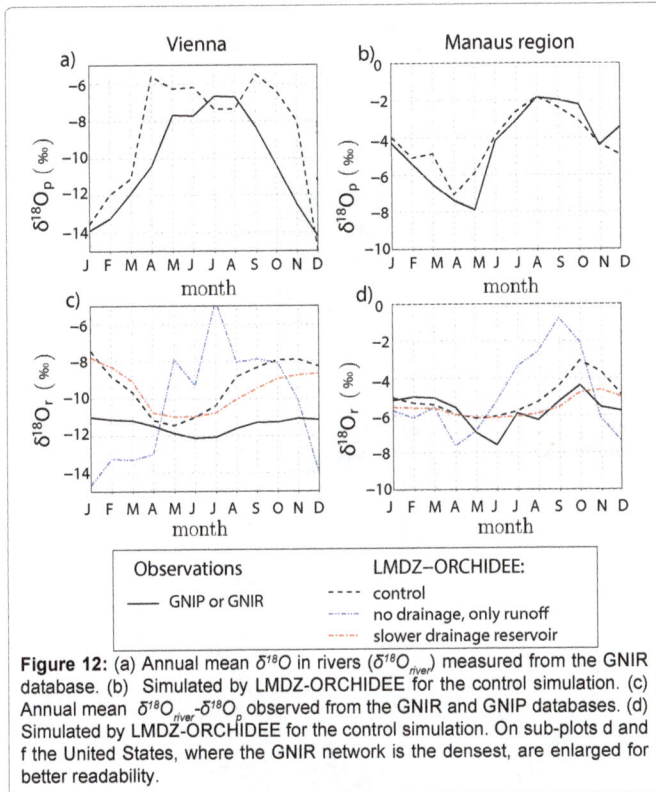

Figure 12: (a) Annual mean $\delta^{18}O$ in rivers ($\delta^{18}O_{river}$) measured from the GNIR database. (b) Simulated by LMDZ-ORCHIDEE for the control simulation. (c) Annual mean $\delta^{18}O_{river}$-$\delta^{18}O_p$ observed from the GNIR and GNIP databases. (d) Simulated by LMDZ-ORCHIDEE for the control simulation. On sub-plots d and f the United States, where the GNIR network is the densest, are enlarged for better readability.

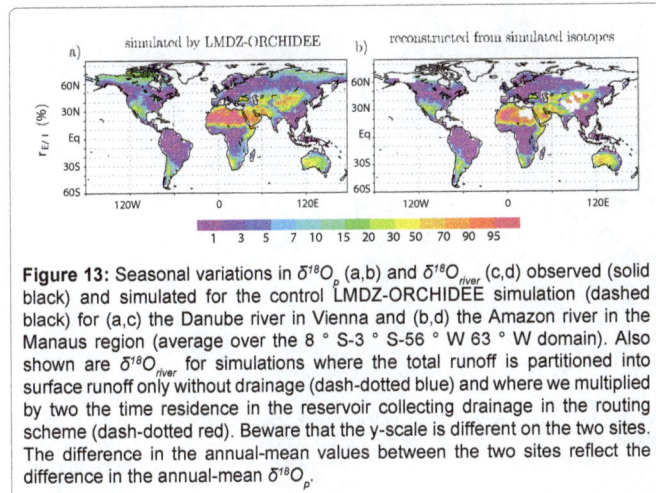

Figure 13: Seasonal variations in $\delta^{18}O_p$ (a,b) and $\delta^{18}O_{river}$ (c,d) observed (solid black) and simulated for the control LMDZ-ORCHIDEE simulation (dashed black) for (a,c) the Danube river in Vienna and (b,d) the Amazon river in the Manaus region (average over the 8 ° S-3 ° S-56 ° W 63 ° W domain). Also shown are $\delta^{18}O_{river}$ for simulations where the total runoff is partitioned into surface runoff only without drainage (dash-dotted blue) and where we multiplied by two the time residence in the reservoir collecting drainage in the routing scheme (dash-dotted red). Beware that the y-scale is different on the two sites. The difference in the annual-mean values between the two sites reflect the difference in the annual-mean $\delta^{18}O_p$.

Evapo-transpiration partitioning

In this section, we generalize at the global scale our results on evapo-transpiration partitioning estimates.

We apply equation 1 to annual-mean outputs from a LMDZ-ORCHIDEE simulation. We compare E/I estimated from Equation 1 to E/I directly simulated by LMDZ-ORCHIDEE. The spatial pattern of E/I is remarkably well estimated by Equation 1 (Figure 14). The equation captures the maximum over the Sahara, Southern South America, Australia, central Asia, Siberia and Northern America. The isotope-derived spatial distribution of E/I correlates well with the simulated distribution (r=0.91). Average errors are lower than 50% of the standard deviation at the global scale. This confirms that co-variation between the different variables at sub-annual time scales has

a negligible effect, so that the equation can be applied to annual-mean quantities. Generally, E/I estimates are best where E/I is relatively small.

To test the effect of the assumption that the soil water isotopic composition is vertically constant, we applied Equation 1 using $\delta^{18}O_s$-$\delta^{18}O_p$ from a simulation with soil profiles activated. This assumption is a significant source of uncertainty on estimating E/I (Table 4). We also analyzed the effect of potential measurement errors in $\delta^{18}O_s, \delta^{18}O_p, \delta^{18}O_v$ temperature or relative humidity on the E/I reconstruction. Results are relatively insensitive to small errors in these measurements (Table 4). However, results are sensitive to the choice of the n exponent in the calculation of the kinetic fractionation α_k (Table 4): knowing the n exponent with an accuracy of 0.07 (e.g., estimated n ranges from 0.63 to 0.70) is necessary to estimate E/I with an absolute precision of 2%.

Finally, estimating E/I using equation 1 bears additional sources of uncertainty in that we cannot estimate using the ORCHIDEE model. These are related to all processes that ORCHIDEE does not simulate. For example, ORCHIDEE underestimates or mis-represents the vertical isotopic gradients in soil water at some sites and does not represent the effect of water vapor diffusion in the soil. These effects may disturb the proportionality between E/I and $\delta^{18}O_s$ - $\delta^{18}O_p$ in practical applications.

To summarize, co-located isotope measurements in precipitation, vapor and soil water could provide an accurate constrain on the proportion of bare soil evaporation to precipitation infiltration.

Conclusion and Perspectives

The ORCHIDEE LSM, in which we have implemented water stable isotopes, reproduces the isotopic compositions of the different water pools of the land surface reasonably well compared to local data from MIBA and Carbo-Europe and to global observations from the GNIP and GNIR networks. Despite the scale mismatch between local measurements and a GCM grid box, and despite the strong spatial heterogeneity in the land surface, the capacity of ORCHIDEE to reproduce the seasonal and vertical variations in the soil isotope composition suggests that even local measurements can yield relevant information to evaluate LSMs at the large scale.

We show that the simulated isotope soil profiles are sensitive to infiltration pathways and diffusion rates in the soil. The spatial and seasonal distribution of the isotope composition of rivers is sensitive to the partitioning of total runoff into surface runoff and drainage and to the residence time scales in underground reservoirs. The isotopic composition of soil water is strongly tied to the fraction of infiltrated water that evaporates through the bare soil. These sensitivity tests suggest that isotope measurements, combined with more conventional measurements, could help evaluate the parameterization of infiltration processes, runoff parameterizations and the representation of surface water budgets in LSMs.

Evaluating an isotopic LSM requires co-located observations of the isotope composition in precipitation, vapor and soil at least at the monthly scale. However, such co-located measurements are still very scarce, and most MIBA and Carbo-Europe sites are missing one of the components. Therefore, for LSM evaluation purpose, we advocate for the development of co-located isotope measurements in the different water pools at each site, together with meteorological variables. Our results suggest that isotope measurements are spatially relatively well representative and that even monthly values are already valuable to identify model bias or to estimate soil water budgets. Therefore, in the perspective of LSM evaluation, if a compromise should be made with sampling frequency and spatial coverage, we favor co-located

Absolute or relative error	RMS absolute error on $r_{s/l}$	RMS relative error on $r_{s/l}$, when $r_{s/l} > 4\%$ (37% of total land aread)
soil profiles	12%	50%
$\Delta T = 1°C$	0.2%	1%
$\Delta rh = 1\%$	0.5%	1%
$\Delta \delta_p = 1$	3%	35%
$\Delta \delta_v = 1$	1%	8%
$\Delta \delta_s = 1$	5%	49%
$\Delta n = 0.5$	14%	52%

Table 4: Uncertainties in the estimation of *E/I* related to measurement errors and assumptions necessary in the simple conceptual model. Values give absolute (in ratio) and relative variations (in %) in estimated *E/I* when temperature *T* is modified by 1 ° C (line 4), when relative humidity *rh* is modified by 1% (line 5), when $\delta^{18}O_v$, $\delta^{18}O_p$ and $\delta^{18}O_s$ are modified by 1, when *n* in the kinetic fractionation is varied from 0.5 to 1, and when the soil $\delta^{18}O$ is not homogeneous vertically. The resulting variations in estimated *E/I* are averaged over all land grid points where the estimation could be performed.

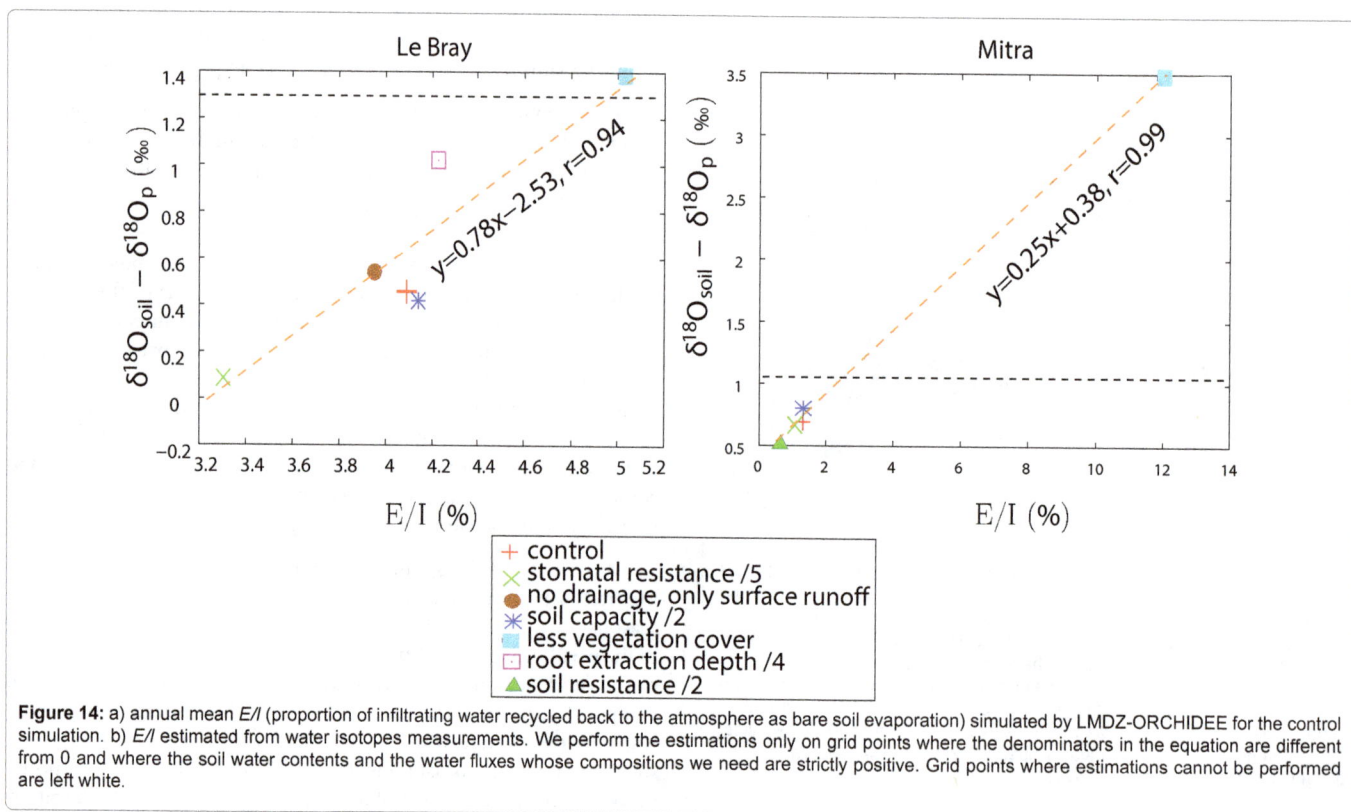

Figure 14: a) annual mean *E/I* (proportion of infiltrating water recycled back to the atmosphere as bare soil evaporation) simulated by LMDZ-ORCHIDEE for the control simulation. b) *E/I* estimated from water isotopes measurements. We perform the estimations only on grid points where the denominators in the equation are different from 0 and where the soil water contents and the water fluxes whose compositions we need are strictly positive. Grid points where estimations cannot be performed are left white.

measurements of all the different water pools at the monthly scale on a few sites representative of different climatic conditions, rather than multiplying sites where water pools are not all sampled. Additionally, at each observation site, collecting different soil samples a few meters apart is helpful to check that they are spatial representative. In the future, development in laser technology [137,138] will allow the generalization of water vapor isotope monitoring at the different sampling sites, which has long been a very tedious activity [90].

From the modeling point of view, kinetic fractionation processes during bare soil evaporation are a source of uncertainty, and a better understanding and quantification of this fractionation is necessary [103,139]. In addition, the accuracy of isotopic simulations by LSM is expected to improve as the representation of hydrological processes

improves. In particular, given the importance of vertical water exchanges for the isotopic simulation, implementing water isotopes in a multi-layer hydrological parameterization with sufficient vertical resolution [69] is crucial. In the future, we plan to implement water isotopes in the latest version of ORCHIDEE, which is multi-layer and more sophisticated [140-142]. Finally, latest findings largely based on water isotopic measurements suggest that different water pools co-exist within a soil column and that evaporation, transpiration, runoff and drainage tap from these different pools [77,143,144]. These effects are not yet represented explicitly in global LSMs. These effects were mainly evidenced based on isotope measurements, and in turn, their representation expected to significantly impact isotopic simulations. Such feedbacks between isotopic research and hydrological parameterization improvements should lead to LSM improvements in the future. With this in mind, LSM inter-comparison projects would strongly benefit from including water isotopes as part of their diagnostics, in the lines of iPILSP (isotope counterpart of the Project for Intercomparison of Land-surface Parameterization Schemes [27]).

Representation of isotope fractionation during evaporation from land surface water pools

Processes for which we neglect fractionation: Snow sublimation is associated with a slight fractionation due to exchanges between snow and vapor in snow pores [115,145,146]. However, we assume that these effects are small enough to be neglected, as in other GCMs [58].

Water uptake by roots has been shown to be a non-fractionating process [147,148], but fractionation at the leaf surface during transpiration impacts the composition of transpired fluxes at scales shorter than daily [95,137]. As the application of ORCHIDEE in the context of our study focuses mainly on time scales of a month or longer, we assume here that the transpiration and stem water have the composition of soil water extracted by the roots.

Evaporation from bare soils and canopy-intercepted water: We represent isotope fractionation during evaporation of soil and canopy-intercepted water using the model of Craig [75]: at any time t, the isotopic composition of evaporation R_E is given by:

$$R_E(t) = \frac{R_l(t) - \alpha_{eq} \cdot h \cdot R_v(t)}{\alpha_K \cdot \alpha_{eq} \cdot (1-h)} \quad (2)$$

where R_l and R_v are the isotopic compositions of liquid water at the evaporative site and of water vapor respectively, h is the relative humidity normalized to surface temperature, α_{eq} is the isotopic fractionation during liquid-vapor equilibrium [149] and α_k is the kinetic fractionation during water vapor diffusion. The kinetic fractionation during soil evaporation is still very uncertain [103,150]. We use the very widespread formulation of [99,151]:

$$\alpha_K = \left(\frac{D}{D_i}\right)^n \quad (3)$$

where D and D_i are the molecular diffusivities of light and heavy water vapor in air, respectively, and n is an exponent that depends on the flow regime (0.5, 0.67 and 1 for turbulent, laminar and stagnant regimes respectively) but remains difficult to estimate [103,150]. In this study, we take $n = 0.67$ for both evaporation of soil and canopy-intercepted water, corresponding to moist conditions in the case of soils [99]. However, we also tried 0.5 and 1.0 to estimate the range of uncertainty related to this parameter. The isotopic composition of precipitation is only slightly sensitive to the formulation of the kinetic fractionation: when n varies from 0.5 to 1, significant changes in $\delta^{18}O_p$ and d_p are restricted to areas where bare soil covers more than 70%. Even in those case, changes in $\delta^{18}O_p$ and d_p never exceed 2‰ and 7‰ respectively. The impact is slightly stronger on soils. Varying n from 0.5 to 1 leads to $\delta^{18}O_s$ variations of 2‰ in offline simulations on the Bray site, of the order of the observed average difference between two samples collected on the same day (2.2‰). In coupled simulations, the impact on $\delta^{18}O_s$ and d_s reaches 8‰ and 20‰ respectively on very arid regions such as the Sahara.

To calculate the temporal mean isotopic composition of evaporation over the time step Δt, $\overline{R_E}$, we assume R_v and h are constant throughout each time step. On the other hand, we allow the isotopic ratio of liquid water to vary over the simulation time step Δt following [151]. While assuming constant R_l is a valid assumption for models with very short time steps [152], it is not the case in ORCHIDEE (Δt =30min). We then calculate $\overline{R_E}$ as:

$$\overline{R_E} = \frac{R_{l0} \cdot \left(1 - f^{\beta+1}\right) - \gamma \cdot R_v \cdot f \cdot \left(1 - f^\beta\right)}{1 - f} \quad (4)$$

where R_{l0} is the initial isotopic ratio of liquid water, f is the remaining liquid fraction in the water reservoir affected by isotopic enrichment, and β and γ are parameters defined by Stewart [151]:

$$\beta = \frac{1 - \alpha_{eq} \cdot \alpha_K \cdot (1-h)}{\alpha_{eq} \cdot \alpha_K \cdot (1-h)}$$

and

$$\gamma = \frac{\alpha_{eq} \cdot h}{1 - \alpha_{eq} \cdot \alpha_K \cdot (1-h)}$$

For canopy-intercepted water, the water reservoir is sufficiently small to assume that the water reservoir affected by isotopic enrichment is the total canopy-intercepted water. For soil evaporation on the other hand, we assume that the depth of the water reservoir affected by isotopic enrichment equals the average distance traveled by water molecules in the soil:

$$L = \sqrt{K_D \cdot \Delta t} \quad (5)$$

where K_D is the effective self-diffusivity of liquid water in the soil column. Neglecting the dispersion term, K_D is given by Munnich et al. [100,147,151-153]:

$$K_D = D_m \cdot \tau \cdot \theta_l \quad (6)$$

where $D_m = 2.5 \cdot 10^{-9} m^2/s$ is the molecular liquid water self-diffusivity [154,155], τ is the soil tortuosity and θ_l is the volumetric soil water content. In the control simulation, we assume $\theta_l \tau = 0.1$ leading to $L = 0.67$ mm. This choice is consistent with a τ of 0.67 [151] and an average θ_l of about 15%. At the Bray, measurements along profiles show θ_l varying from about 5 to 30%. Since these values are difficult to constrain observationally and very variable spatially and temporally, sensitivity tests to $\theta_l \tau$ are performed and described. We neglect the vapor phase in the soil and associated fractionation and diffusion processes [153].

Dew formation: We assume fractionation during dew and frost formation following a Rayleigh distillation of the vapor in the lowest 10 hPa (~80 m) of the atmosphere. Since the atmospheric water vapor condenses in small proportion during frost and dew, this choice of the depth of atmosphere involved in the condensation has almost no impact on the composition of the dew and frost formed. Following common practice, we use equilibrium fractionation coefficient from Merlivat et al. [148,156,157] and the kinetic fractionation formation of [158] with $\lambda = 0.004$, whose choice has very little impact on the results.

Leaf water evaporation: At isotopic steady state, the composition of water transpired by the vegetation is equal to that of the soil water extracted by the roots. In default simulations, we assume that isotopic steady state for plant water is established at any time and we diagnose the composition of the leaf water at the evaporation site, R_e^{ss}, by inverting the Craig and Gordon equation [75]:

$$R_e^{SS} = \alpha_{eq} \cdot \left(\alpha_K \cdot (1-h) \cdot R_s + h \cdot R_v\right) \quad (7)$$

where R_s and R_v are the isotopic ratio in soil water and water vapor respectively, h is the relative humidity normalized to surface temperature, α_{eq} is the isotopic fractionation during liquid-vapor equilibrium [148] and α_k is the kinetic fractionation during water vapor diffusion. We take the same kinetic fractionation formulation as for the soil evaporation [150], with $n = 0.67$ [31,69]. Leaf water compositions are significantly sensitive to parameter n, with variations of the order of 10‰ as n varies from 0.5 to 1. We assume that the leaf temperature used to calculate α_{eq} is equal to the soil temperature, but results are very little sensitive to this assumption.

The isotopic composition of leaf water has been the subject of many observational and numerical modeling studies [86,159-161]. Several studies have shown that the composition of the leaves is affected by mixing with xylem water and by non-stationary effects [161,162]. Non-steady state effects are also incorporated in ORCHIDEE following [159]. The isotopic ratio in the leaf mesophyll R_L^{SS} is the result of the mixing between leaf water at the evaporative site and xylem water (Peclet effect):

$$R_L^{SS} = R_e^{SS} \cdot f + R_s(1-f) \tag{8}$$

where f is a coefficient decreasing as the Peclet effect increases:

$$f = \frac{1-e^{-P}}{P}$$

and p is the Peclet parameter [120,160]:

$$P = \frac{E \cdot L_{eff}}{W \cdot D_m}$$

E is the transpiration rate per leaf area, L_{eff} is the effective diffusion length and W is the leaf water content per leaf volume (assumed equal to 10^3 kg/m^3, order of magnitude in [121]). The Peclet number P can be tuned by changing L_{eff} that depends on leaf geometry and drought intensity (e.g., 7 to 12 mm in Cuntz et al. [161], 50 to 150 mm in Barnard et al. [121]). We take L_{eff} =8 mm to optimize our simulation on Hartheim.

For some simulations, we account for the effect of water storage in leaves (leading to some memory in the leaf water isotopic composition) following Dongmann [163]. Assuming that W is constant, we calculate the leaf lamina composition R_L as Farquhar [159]:

$$R_L(t) = R_L(t - dt) \cdot e^{-dt/\tau} + R_L^{SS}(t) \cdot \left(1 - e^{-dt/\tau}\right) \tag{9}$$

where

$$\tau = \frac{W \cdot \alpha_K \cdot \alpha_{eq} \cdot f}{g}$$

and g is the sum of the total (stomatic and boundary layer) conductances. The isotopic composition of transpiration is then calculated so as to conserve isotope mass.

Representation of the vertical distribution of soil water isotopic composition

Principle: In control simulations, we assume that the isotopic composition of soil water is homogeneous vertically and equals the weighted average of the two soil layers. In addition, to test this assumption, we implemented a representation of the vertical distribution of the soil water isotopic composition: the soil water is spread vertically between several layers. The first layer contains a water height $L = \sqrt{K_D \cdot \Delta t}$, where K_D is the diffusivity of water molecules in water and Δt is the time step of the simulation, and the other layers contain a water height $resol.L$. The parameter $resol$ can be tuned to find a compromise between vertical resolution and computational time. Layers are created from the top to bottom until all layers are full with water except the deepest one. For example, with L =0.67 mm, up to 16 layers can thus be created if the soil is saturated. Bare soil evaporation is extracted from the first layer. Transpiration is extracted from the different layers following a root extraction profile that reflects the sensitivity of transpiration to soil moisture [164]. Drainage takes water from the deepest layer. In the control simulation, rain and snow melt are added to the first layer (piston-like flow). In a sensitivity test, that can also be homogeneously

distributed in the different layers, to crudely represent preferential pathways through fractures or pores in the soil.

At each time step, the soil water isotopic composition in each layer is re-calculated by taking into account the sources and sinks for each layer and ensuring that each layer remains full except the deepest one. Isotopic diffusion between adjacent layers is applied at each time step (Equation 6). The water budget of the total soil remains exactly the same as without vertical discretization.

Evaluation for an idealized case: The module representing vertical distribution of water isotopes in the soil is first evaluated for an idealized case when it is not yet embedded into ORCHIDEE.

First, we use a case in which the soil column evaporates at its top and is permanently refilled at the bottom by a water with $\delta^{18}O$ of -8‰ [152]. The soil remains saturated, and we focus on the steady state reached after a few hundreds of days [152]. An analytical solution is available for this case [100,165]. The analytical solution and a much more sophisticated model of soil water isotopes (MuSICA [166]) yield very similar results (Figure 15a): the bottom of the soil is at -8‰ while the top of the soil is enriched up to 15‰. The soil module of ORCHIDEE is able to reproduce these results when the value of $\theta_t\tau$ is set to be very low (0.001) and when the vertical resolution is sufficiently high (layers of 0.75 mm). Whatever the value for $\theta_t\tau$, ORCHIDEE results become less sensitive to the vertical discretization when layers are thinner than about 2 mm.

Second, we use a case in which the soil column, initially with a soil water of -8‰, evaporates at its top until the soil water content is only 20% [99]. The atmosphere has a relative humidity of 20% and a vapor $\delta^{18}O$ of -15‰. The sophisticated models MuSICA and SiSPAT [152] feature a typical evaporative enrichment profile, with $\delta^{18}O$ increasing from its initial value of -8‰ at the bottom to a maximum $\delta^{18}O$ of 13‰ about 10 mm below the surface (Figure 15b). In the uppermost 10 mm, there is a slight depletion due to diffusion of water vapor into the soil column [101]. ORCHIDEE is not able to reproduce this vertical profile. First, since diffusion of water vapor in the soil is neglected, it is not able to simulate the depletion near the surface. Second, since $\theta_t\tau$ is temporally and vertically constant in ORCHIDEE, it is not able to adapt to the drying of the soil. In the sophisticated model, as the soil dries, the soil water content θ_t decrease, thus inhibiting vertical mixing of soil water and favoring strong isotopic gradients. In contrast in ORCHIDEE, $\theta_t\tau$ remains constant at a value representative of a moister soil, thus favoring vertical mixing of soil water and leading to a nearly uniform enrichment with depth [167-170].

To summarize, our representation of isotopic vertical profiles in ORCHIDEE is probably most suited when soil moisture remains high and does not vary too strongly.

Calculation of isotopic forcing from LMDZ outputs and nearby GNIP or USNIP stations

When precipitation and water vapor isotopic observations are not available at a given site, we create isotopic forcing using isotopic measurements in the precipitation performed on nearby GNIP (Global Network for Isotopes in Precipitation [122]) or USNIP (United States Network for Isotopes in Precipitation [125]) precipitation stations. To interpolate between the nearby stations, taking into account spatial gradients and altitude effects, we use outputs from an LMDZ simulation.

Let's assume there are n GNIP or USNIP stations around the

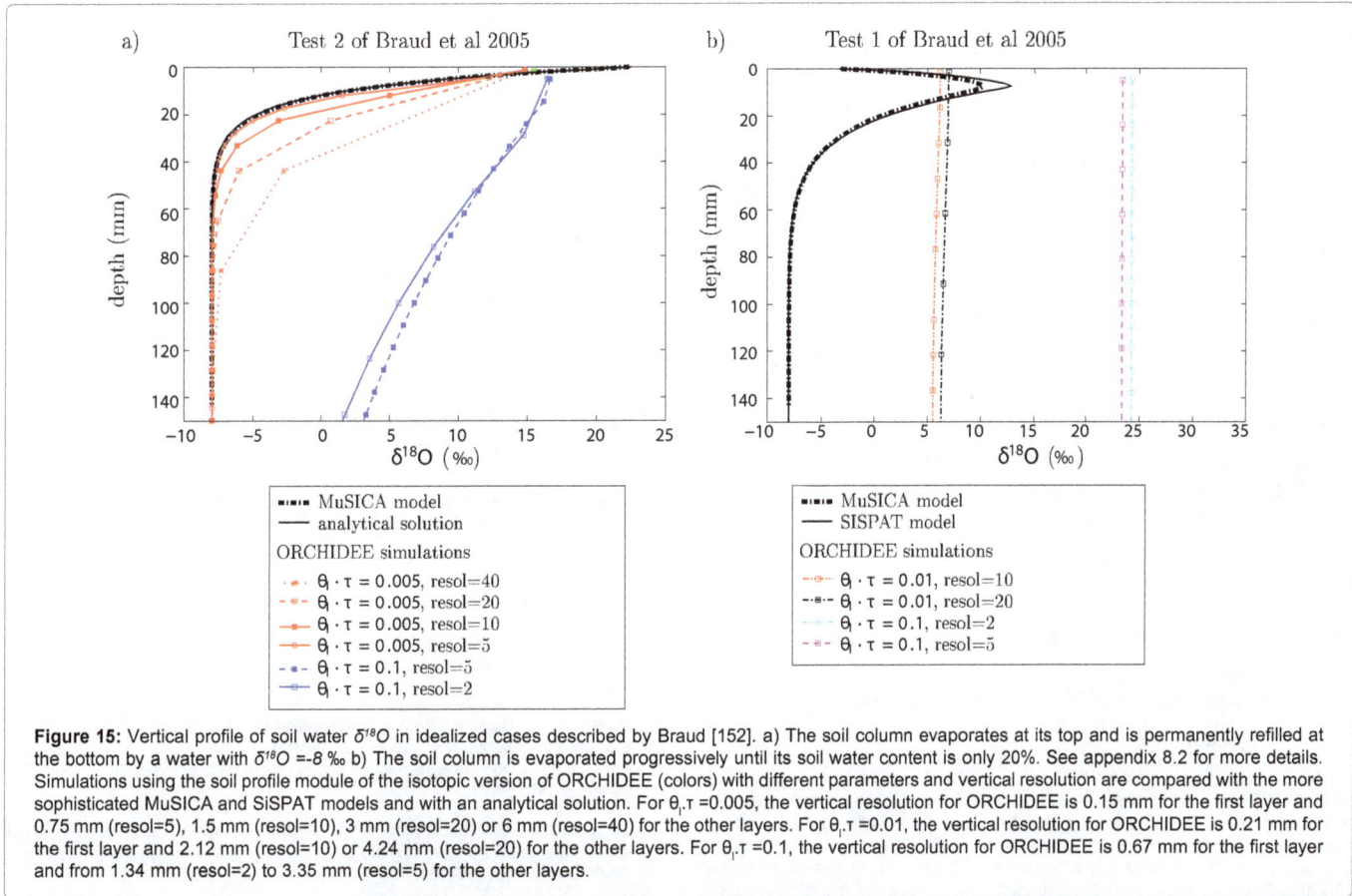

Figure 15: Vertical profile of soil water $\delta^{18}O$ in idealized cases described by Braud [152]. a) The soil column evaporates at its top and is permanently refilled at the bottom by a water with $\delta^{18}O = -8$ ‰ b) The soil column is evaporated progressively until its soil water content is only 20%. See appendix 8.2 for more details. Simulations using the soil profile module of the isotopic version of ORCHIDEE (colors) with different parameters and vertical resolution are compared with the more sophisticated MuSICA and SiSPAT models and with an analytical solution. For $\theta_l.\tau = 0.005$, the vertical resolution for ORCHIDEE is 0.15 mm for the first layer and 0.75 mm (resol=5), 1.5 mm (resol=10), 3 mm (resol=20) or 6 mm (resol=40) for the other layers. For $\theta_l.\tau = 0.01$, the vertical resolution for ORCHIDEE is 0.21 mm for the first layer and 2.12 mm (resol=10) or 4.24 mm (resol=20) for the other layers. For $\theta_l.\tau = 0.1$, the vertical resolution for ORCHIDEE is 0.67 mm for the first layer and from 1.34 mm (resol=2) to 3.35 mm (resol=5) for the other layers.

site of interest (MIBA or Carbo-Europe). The isotopic composition of precipitation at the site of interest and for a given month, $\delta_{p,site}$, is calculated as:

$$\delta_{p,site} = \delta_{p,lmdz}(s) + a_s \cdot (z_{site} - z_{lmdz}(s)) + \sum_{i=1}^{n} r_i \cdot (\delta_{p,NIP}(i) - \delta_{p,lmdz}(i))$$

where

$$r_i = \frac{1/d_i}{\sum_{j=1}^{n} 1/d_j}$$

and where d_i is the geographical distance between the site of interest and the GNIP or USNIP station, $\delta_{p,lmdz}(s)$ is the precipitation isotopic composition simulated by LMDZ in the grid box containing the site s, $\delta_{p,lmdz}(i)$ is the precipitation isotopic composition simulated by LMDZ in the grid box containing the GNIP or USNIP station, $\delta_{p,NIP}(i)$ is the precipitation isotopic composition observed at the GNIP or USNIP station, Z_{site} is the altitude of the site of interest, $Z_{lmdz}(s)$ is the altitude of the LMDZ grid box containing the site of interest and a_s is the slope of the isotopic composition as a function of altitude simulated by LMDZ in the grid boxes containing and surrounding the site of interest [171]. The first term on the right hand side corresponds to the raw LMDZ output for the site of interest. The second term allows us to correct for the altitude effect. Since LMDZ is run at a 2.5 ° latitude * 3.75 ° longitude resolution, we cannot expect the average grid box size to be representative of the local altitude at the site. The third term allows us to correct for possible biases in LMDZ compared to GNIP and USNIP observations. Table 3 lists the GNIP and USNIP stations used to construct the forcing at each site of interest.

To calculate the isotopic composition of the water vapor, we assume that although LMDZ might have biases for simulating the absolute values of precipitation and water vapor composition, it simulates properly the precipitation-vapor difference [47,60]. Therefore, the isotopic composition of water vapor at the site of interest, $\delta_{v,site}$, is calculated as:

$$\delta_{v,site} = \delta_{p,site} + \delta_{v,lmdz}(s) - \delta_{p,lmdz}$$

where $\delta_{v,lmdz}(s)$ is the isotopic composition of water vapor simulated by LMDZ in the grid box containing the site of interest.

A simple equation to relate the soil water isotopic composition to the surface soil water budget

To explore how the isotopic composition of soil water can help estimate terms of the soil water budget, we derive here a very simple theoretical framework.

We assume that the water mass balance is:

$$P = E + T + D + R \tag{10}$$

where P is the precipitation, R the surface runoff, E is the bare soil evaporation, T the transpiration and D the drainage. Similarly, the isotopic mass balance is:

$$P.R_p = E.R_E + T.R_T + D.R_D + R.R_R \tag{11}$$

Where R_p, R_E, R_T, R_D and R_R are the isotopic ratios of incoming water at the soil surface, bare soil evaporation, transpiration, drainage and surface runoff respectively.

We assume that the bare soil evaporation isotope ratio depends on that of the soil (R_s) following the Craig [75] relationship (Equation 2) and that the transpiration composition is equal to that of the soil ($R_T = R_s$), implying little vertical variations in soil water isotope ratios [172]. We assume that the isotopic composition of surface runoff is that of the incoming water ($R_R = R_p$) and that the isotopic composition of drainage is that of the soil water ($R_D = R_s$). In doing so, we neglect again vertical isotope variations in the soil and the temporal co-variation between R_s, D and T. Combining equations for the mass balance of water (Equation 11) and of water isotopes (Equation 10) then yields:

$$R_p = E/I.R_E + (1-E/I).R_S \qquad (12)$$

where $I = P - R$ represents the incoming water that infiltrates into the soil. E/I represents the proportion of the infiltrated water which is evaporated at the soil surface.

The composition of the bare soil evaporation flux, R_E, is a function of R_s following the Craig [75] formulation (Equation 2). Replacing R_E by its function of R_s in Equation 12 allows us to deduce E/I:

$$E/I = \frac{\alpha_{eq} \cdot \alpha_K \cdot (1-h) \cdot (R_p - R_s)}{R_s \cdot (1 - \alpha_{eq} \cdot \alpha_K \cdot (1-h)) - \alpha_{eq} \cdot h \cdot R_v} \qquad (13)$$

Therefore, E/I is a function of the isotopic difference between the soil water and the precipitation water, which is easy to observe on instrumented sites such as MIBA or Carbo-Europe sites.

Acknowledgments

We thank a reviewer for his thorough review and detailed comments. LMDZ and ORCHIDEE simulations were performed on IDRIS machines to which access was granted by GENCI under project 0292. We thank Katia Laval for fruitful discussion and comments on an earlier version of this manuscript. We thank Matthias Cuntz for discussions. We thank Arthur Gessler and Romain Barnard for providing their data from Hartheim, and thank Chun-Ta Lai for providing his data from the Kansas prairie. We thank Danilo Dragoni, Kim Novick and Rich Phillips for providing information and data on the Morgan-Monroe site. We thank Marion Devaux, Cathy Lambrot (Inra-Ephyse, France), Rolf Siegwolf (Paul Scherrer Institute, Switzerland), Glyn Jones and Howard Griffiths (University of Cambridge, UK) for sampling and analysis of the isotopic data on the Bray and Mitra sites. We thank Eyal Rotenberg and Jean-Marc Bonnefond for providing the meteorological forcing over Yatir and the Bray respectively. We thank Dan Yakir for the isotopic and meteorological data collection in Yatir, his role in the MIBA initiative and comments on the manuscript. Part of the work was done while Camille Risi was a post-doc advised by David Noone, who I thank as well. This work benefited from financial support of the LEFE project MISSTERRE. Cathy Kurz-Besson was supported by the Fundação para a Ciência e Tecnologia (PTDC/AAG-REC/7046/2014). Lisa Wingate was supported by a Marie Curie Career Development Fellowship, thus some of the research leading to these results has received funding from the [European Community's] Seventh Framework Programme ([FP7/2007-2013] under grant agreement n ° [237582]. The research was supported partly by the Czech Science Foundation project to JS (14-12262S) and by the Czech research infrastructure for systems biology C4SYS project (LM2015055).

References

1. Henderson-Sellers A, Irannejad P, McGuffie K, Pitman AJ (2003) Predicting land-surface climates-better skill or moving targets? Geophy Res Lett 30: 1777-1780.

2. Qu W, Henderson-Sellers A (1998) Comparing the scatter in pilps off-line experiments with that in amip i coupled experiments. Global and Planetary Change 19: 209-223.

3. Koster RD, Milly PCD (1996) The Interplay between Transpiration and Runoff Formulations in Land Surface Schemes Used with Atmospheric Models. J Clim 10: 1578-1591.

4. Polcher J, Laval K, Dfimenil L, Lean J, Rowntree P (1996) Comparing three land surface schemes used in general circulation models. J Hydrol 180: 373-394.

5. Wetzel PJ, Liang X, Irannejad P, Boone A, Noilhane J, et al. (1996) Modeling vadose zone liquid water fluxes: Infiltration, runoff, drainage, interflow. Global and Planetary Change 13: 57-71.

6. Desborough C, Pitman A, Irannejad P (1996) Glob Planet Change 13: 47-56.

7. Mahfouf JF, Ciret C, Ducharne A, Irannejad P, Noilhan J, et al. (1996) Analysis of transpiration results from the RICE and PILPS Workshop. Glob Planet Change 13: 73-88.

8. Ducharne A, Laval K, Polcher J (1998) Sensitivity of the hydrological cycle to the parametrization of soil hydrology in a gcm. Clim Dyn 14: 307-327.

9. Boone A (2004) The Rhone-Aggregation Land Surface Scheme Intercomparison Project: An Overview. J Clim 17: 187-208.

10. Boone A, deRosnay P, Balsamo G, Beljaars A, Chopin F et al. (2009) The AMMA Land Surface Model intercomparison Project (ALMIP). Bull Am Meteor Soc 90: 1865-1880.

11. Crossley JF, Polcher J, Cox PM, Gedney N, Planton S (2000) Uncertainties linked to land-surface processes in climate change simulations. Clim Dyn 16: 949-961.

12. Gedney N, Cox PM, Douville H, Polcher J, Valdes P (2000) Characterizing gcm land surface schemes to understand their responses to climate change. J Clim 13: 3066-3079.

13. Milly PCD, Dunne KA, Vecchia AV (2005) Global pattern of trends in streamflow and water availability in a changing climate. Nature 17.

14. Lean, Rowntree P (1997) Understanding the sensitivity of a GCM simulation of Amazonian deforestation to the specification of vegetation and soil characteristics. J Clim 10: 1216-1235.

15. Pitman AJ, deNoblet-Ducoudre N, Cruz FT, Davin EL, Bonan GB, et al. (2009) Uncertainties in climate responses to past land cover change: First results from the LUCID intercomparison study. Geophy Res Lett 36: L14814.

16. Moran M, Scotta R, Keefera T, Emmericha W, Hernandeza M, et al. (2009) Partitioning evapotranspiration in semiarid grassland and shrubland ecosystems using time series of soil surface temperature. Agric and For Meteorol 149: 59-72.

17. Seneviratne SI, Corti T, Davin EL, Hirschi M, Jaeger EB, et al. (2010) Investigating soil moisture-climate interactions in a changing climate: a review. Earth-Sci Rev 99: 125-161.

18. Baldocchi D, Falge E, Gu L, Olson R, Hollinger D, et al. (2001) FLUXNET: A New Tool to Study the Temporal and Spatial Variability of Ecosystem-Scale Carbon Dioxide, Water Vapor, and Energy Flux Densities. Bull Am Meteor Soc 82: 2415-2434.

19. Robock A, Vinnikov KY, Srinivasan G, Entin JK, Hollinger SE, et al. (2000) The global soil moisture data bank. Bull Am Meteor Soc 81: 1281-1299.

20. Vachaud G, Passeratde Silans A, Balabanis P, Vauclin M (1985) Temporal stability of spatially measured soil water probability density function. Soil Sci Soc Am 49: 822-828.

21. Rodriguez-Iturbe I, Vogel G, Rigon R, Entekhabi D, Castelli F, et al. (1995) On the spatial organization of soil moisture fields. Geophys Res Lett 22: 2757-2760.

22. Nijssen B, Lettenmaier DP, XuLiang S, Wetzel W, Wood EF (1997) Streamflow simulation for continental-scale river basins. Water Resour Res 33: 711-724.

23. Oki T, Sud YC (1998) Design of Total Runoff Integrating Pathways (TRIP) - A Global River Channel Network. Earth Interactions 2: 1-36.

24. Gat JR (1996) Oxygen and hydrogen isotopes in the hydrologic cycle. Annual Review of Earth and Planetary Sciences 24: 225-262.

25. Henderson-Sellers A, McGuffie K, Noone D, Irannejad P (2004) Using Stable Water Isotopes to Evaluate Basin-Scale Simulations of Surface Water Budgets. J Hydromet 5: 805-822.

26. Henderson-Sellers A, McGuffie K, Zhang H (2001) Stable Isotopes as Validation Tools for Global Climate Model Predictions of the Impact of Amazonian Deforestation. J Clim 15: 2664-2677.

27. Henderson-Sellers A (2006) Improving land-surface parameterization schemes using stable water isotopes: Introducing the 'iPILPS' initiative. Glob Planet Change 51: 3-24.

28. Wong T (2016) The Impact of Stable Water Isotopic Information on Parameter Calibration in a Land Surface Model. PhD thesis, University of Colorado at Boulder.

29. Moreira M, Sternberg L, Martinelli L, Victoria R, Barbosa E, et al. (1997)

Contribution of transpiration to forest ambient vapor based on isotopic measurements. Global Change Biol 3: 439-450.

30. Yepez E, Williams S, Scott R, Lin G (2003) Partitioning overstory and understory evapotranspiration in a semiarid savanna woodland from the isotopic composition of water vapor. Agricultural and Forest Meteorology 119: 53-68.

31. Williams DG, Cable W, Hultine K et al. (2004) Evapotranspiration components determined by stable isotope, sap flow and eddy covariance techniques. Agricult Forest Meteor 125: 241-258.

32. Rothfuss Y, Biron P, Braud I, Canale L, Durand JL, et al. (2010) Partitioning evapotranspiration fluxes into soil evaporation and plant transpiration using water stable isotopes under controlled conditions. Hydrological processes 24: 3177-3194.

33. Brunel J, Walker G, Dighton J, Montenya B (1997) Use of stable isotopes of water to determine the origin of water used by the vegetation and to partition evapotranspiration. A case study from HAPEX-Sahel. J Hydrol 188-189: 466-481.

34. Krabbenhoft DP (1990) Estimating groundwater exchange with lakes 1. the stable isotope mass balance method. Water Resour Res 26: 2445-2453.

35. Gibson J (2002) Short-term evaporation and water budget comparisons in shallow Arctic lakes using non-steady isotope mass balance. J Hydrol 264: 242-261.

36. Gibson JJ, Edwards TWD (2002) Regional water balance trends and evaporation-transpiration partitioning from a stable isotope survey of lakes in northern Canada. Glob Biogeochem Cycles 16: 1026.

37. Wels C, Cornett J, Lazerte BD (1991) Hydrograph separation: a comparison of geochmical and isotopic tracers. J Hydrol 122: 253-274.

38. Millet A, Bariac T, Ladouche B, Mathieu R, Grimaldi C, et al. (1997) Influence of deforestation on the hydrological behavior of small tropical watersheds. Revue des Sciences de leau 1: 61-84.

39. Weiler M, McGlynn BL, McGuire KJ, McDonnell JJ (2003) How does rainfall become runoff? A combined tracer and runoff transfer function approach. Water Resources Research 39.

40. Ladouche B, Probst A, Viville D, Idir S, Baque D, et al. (2001) Hydrograph separation using isotopic, chemical and hydrological approaches (strengbach catchment, france). Journal of hydrology 242: 255-274.

41. Salati E, DallOlio A, Matsui E, Gat J (1979) Recycling of water in the Amazon basin: An isotopic study. Water Resources Research 15: 1250-1258.

42. Gat JR, Matsui E (1991) Atmospheric water balance in the Amazon basin: An isotopic evapotranspiration model. J Geophys Res 96: 13179-13188.

43. Jasechko S, Sharp WD, Sharp JJ, Birks SJ, Yi Y, et al. (2013) Terrestrial water fluxes dominated by transpiration. Nature 496: 347-350.

44. Ducoudre N, Laval K, Perrier A (1993) SECHIBA, a new set of parametrizations of the hydrological exchanges at the land-atmosphere interface within the LMD atmospheric general circulation model. J Clim 6: 248-273.

45. Krinner G, Viovy N, deNoblet-Ducoudre N, Ogee J, Polcher J et al. (2005) A dynamic global vegetation model for studies of the coupled atmosphere-biosphere system. Glob Biogeochem Cycles 19.

46. Shi C, Masson-Delmotte V, Risi C, Eglin T, Stievenard M, et al. (2011) Sampling Strategy and Climatic Implications of Tree-Ring Stable isotopes in Southeast Tibetan Plateau. Earth Planet Sci Lett 301: 307-316.

47. Risi C, Bony S, Vimeux F, Frankenberg C, Noone D (2010) Understanding the Sahelian water budget through the isotopic composition of water vapor and precipitation. J Geophys Res 115: D24110.

48. Risi C, Noone D, Frankenberg C, Worden J (2013) Role of continental recycling in intraseasonal variations of continental moisture as deduced from model simulations and water vapor isotopic measurements. Water Resour Res 49: 4136-4156.

49. Hourdin F, Musat I, Bony S, Braconnot P, Codron F, et al. (2006) The LMDZ4 general circulation model: climate performance and sensitivity to parametrized physics with emphasis on tropical convection. Clim Dyn 27: 787-813.

50. Craig H (1961) Isotopic variations in meteoric waters. Science 133: 1702-1703.

51. Gonfiantini R (1978) Standards for stable isotope measurements in natural compounds. Nature 271: 534-536.

52. Dansgaard (1964) Stable isotopes in precipitation. Tellus 16: 436-468.

53. Marti O, Braconnot P, Bellier J, Benshila R, Bony S, et al. (2005) The new IPSL climate system model: IPSL-CM4. Technical report, IPSL, Note du pole de modelisation de lIPSL, 26: 1-86.

54. Dufresne JL, Foujols MA, Denvil S, Caubel A, Marti O, et al. (2012) Climate change projections using the IPSL-CM5 Earth System Model: from CMIP3 to CMIP5. Clim Dyn 40: 1-43.

55. Solomon S (2007) Climate change 2007-the physical science basis: Working group I contribution to the fourth assessment report of the IPCC, volume 4, Cambridge University Press.

56. Meehl GA, Covey K, Delworth T, Latif M, McAvaney B, et al. (2007) The WCRP CMIP3 multimodel dataset: A new era in climate change research. Bull Am Meteor Soc 7: 1383-1394.

57. Joussaume S, Jouzel J, Sadourny R (1984) A general circulation model of water isotope cycles in the atmosphere. Nature 311: 24-29.

58. Hoffmann G, Werner M, Heimann M (1998) Water isotope module of the ECHAM atmospheric general circulation model: A study on timescales from days to several years. J Geophys Res 103: 16871-16896.

59. Bony S, Risi C, Vimeux F (2008) Influence of convective processes on the isotopic composition (deltaO18 and deltaD) of precipitation and water vapor in the Tropics. Part 1: Radiative-convective equilibrium and TOGA-COARE simulations. J Geophys Res 113: D19305.

60. Risi C, Bony S, Vimeux F, Jouzel J (2010) Water stable isotopes in the LMDZ4 General Circulation Model: model evaluation for present day and past climates and applications to climatic interpretation of tropical isotopic records. J Geophys Res 115: D12118.

61. Risi C, Noone D, Worden J, Frankenberg C, Stiller G, et al. (2012) Process-evaluation of tropical and subtropical tropospheric humidity simulated by general circulation models using water vapor isotopic observations. Part 1: model-data intercomparison. J Geophy Res 117: D05303.

62. DeRosnay P. (1999) Representation de linteraction sol-v\'eg\'etation-atmosph\'ere dans le Mod\'ele de Circulation G\'en\'erale du Laboratoire de M\'et\'eorologie Dynamique. PhD thesis, Universit\'e de Paris 06.

63. Sitch S (2003) Evaluation of ecosystem dynamics, plant geography and terrestrial carbon cycling in the LPJ dynamic vegetation model. Global Change Biol 9: 161-185.

64. Choisnel E (1977) Le bilan d'\'energie et hydrique du sol. La M'et'eorologie 6: 103-133.

65. Choisnel E, Jourdain SV, Jaquart CJ (1995) Climatological evaluation of some fluxes of the surface energy and soil water balances over France. Annales Geophysicae 13: 666-674.

66. Ngo-Duc T (2005) Modelisation des bilans hydrologiques continentaux: variabilite interannuelle et tendances. Comparaison aux observations. PhD thesis, Universite Pierre et Marie Curie.

67. Manabe S, Smagorinsky J, Strickler R (1965) Simulated climatology of a general circulation model with a hydrologic cycle. Mon Weath Rev 93: 769-798.

68. Polcher J (2003) Les processus de surface a lechelle globale et leurs interactions avec latmosphere. In These dhabilitation a diriger des recherches, Universite Paris 6.

69. Riley WJ, Still J, Torn MS, Berry JA (2002) A mechanistic model of H218O and C18OO fluxes between ecosystems and the atmosphere: Model description and sensitivity analyses. Global Biogeochem Cycles 16: 1095

70. Cuntz M, Ciais PandHoffmann G, Knorr W (2003) A comprehensive global three-dimensional model of D18O in atmospheric CO2: 1. Validation of surface processes. J Geophys Res 108.

71. Aleinov I, Schmidt GA (2006) Water isotopes in the GISS ModelE land surface scheme. Global and Planet Change 51: 108-120.

72. Yoshimura K, Miyazaki S, Kanae S, Oki T (2006) Iso-MATSIRO, a land surface model that incorporates stable water isotopes. Glob Planet Change 51: 90-107.

73. Haese B, Werner M, Lohmann G (2013) Stable water isotopes in the coupled atmosphere-land surface model ECHAM5-JSBACH. Geoscientific Model Development 6: 1463-1480.

74. Risi C (2009) Les isotopes stables de leau: applications a letude du cycle de leau et des variations du climat. PhD thesis, Universite Pierre et Marie Curie.

75. Craig H, Gordon LI (1965) Deuterium and oxygen-18 variations in the ocean and marine atmosphere. Stable Isotope in Oceanographic Studies and Paleotemperatures, Laboratorio di Geologia Nucleate, Pisa, Italy, pp: 9-130.

76. Brooks JR, Barnard HR, Coulombe R, McDonnell JJ (2010) Ecohydrologic separation of water between trees and streams in a mediterranean climate. Nature Geoscience 3: 100-104.

77. Bowen G (2015) Hydrology: The diversified economics of soil water. Nature 525: 43-44.

78. Good SP, Noone D, Bowen G (2015) Hydrologic connectivity constrains partitioning of global terrestrial water fluxes. Science 349: 175-177.

79. Kendall C, Coplen TB (2001) Distribution of oxygen-18 and deuterium in river waters across the United States. Hydrol Processes 15: 1363-1393.

80. Twining J, Stone D, Tadros C, Henderson-Sellers A, A W (2006) Moisture Isotopes in the Biosphere and Atmosphere (MIBA) in Australia: A priori estimates and preliminary observations of stable water isotopes in soil, plant and vapour for the Tumbarumba Field Campaign. Global and Planetary Change 51: 59-72.

81. Knohl A, Tu KP, Boukili V, Brooks PD, Mambelli S, et al. (2007) MIBA-US: Temporal and Spatial Variation of Water Isotopes in Terrestrial Ecosystems Across the United States. Eos Trans AGU 88.

82. Hemming D, Griffiths H, Loader A, Robertson I, Wingate L, et al. (2007) The Moisture Isotopes in Biosphere and Atmosphere network (MIBA): initial results from the UK. Eos Trans AGU 88.

83. Valentini R, Matteucci G, Dolman A, Schulze ED, Rebmann C, et al. (2000) Respiration as the main determinant of carbon balance in european forests. Nature 404: 861-865.

84. Hemming D, Yakir D, Ambus P, Aurela M, Besson C, et al. (2005) Pan-european δ13c values of air and organic matter from forest ecosystems. Global Change Biology 11: 1065-1093.

85. Stella P, Lamaud E, Brunet Y, Bonnefond JM, Loustau D, et al. (2009) Simultaneous measurements of CO2 and water exchanges over three agroecosystems in South-West France. Biogeosciences Discuss 6: 2489-2522.

86. Wingate L, Ogee J, Burlett R, Bosc A (2010) Strong seasonal disequilibrium measured between the oxygen isotope signals of leaf and soil CO2 exchange. Glob Change Biology.

87. Wingate L, Ogee J, Cuntz M, Genty B, andUlli Seibtf IR, et al. (2009) The impact of soil microorganisms on the global budget of deltaO18 in atmospheric CO2. PNAS.

88. Grunzweig JM, Hemming D, Maseyk K, Lin T, Rotenberg E, et al. (2009) Water limitation to soil co2 efflux in a pine forest at the semiarid ?timberline? Journal of Geophysical Research: Biogeosciences 114.

89. Raz-Yaseef N, Yakir D, Rotenberg E, Schiller G, Cohen S (2009) Ecohydrology of a semi-arid forest: partitionning among water balance components and its implications for predicted precipitation changes. Ecohydrolohy pp: 10.1002/eco.65.

90. Angert A, Lee JE, Yakir D (2008) Seasonal variations in the isotopic composition of near surface water vapour in the eastern Mediterranean. Tellus 60: 674-684.

91. Zhang G, Leclerc1 MY, Karipot A (2010) Local flux-profile relationships of wind speed and temperature in a canopy layer in atmospheric stable conditions. Biogeosciences 7: 3625-3636.

92. Kratochvilova I, Janous D, Marek M, Bartak M, Riha L (1989) Production activity of mountain cultivated norway spruce stands under the impact of air pollution. i. general description of problems. EKOLOGIA(CSSR)/ECOLOGY(CSSR) 8: 407-419.

93. Voelker S, Brooks J, Meinzer F, Roden J, Pazdur A, et al. (2014) Isolating relative humidity: dual isotopes delta18o and deltad as deuterium deviations from the global meteoric water line. Ecological Applications 24: 960-975.

94. Dee D, Uppala S, Simmons A, Berrisford P, Poli P, Kobayashi S, et al. (2011) The era-interim reanalysis: Configuration and performance of the data assimilation system. Quarterly Journal of the royal meteorological society 137: 553-597.

95. Lai CT, Ehleringer J, Bond B, U KP (2006) Contributions of evaporation, isotopic non-steady state transpiration, and atmospheric mixing on the deltaO18 of water vapor in Pacific Northwest coniferous forests. Plant Cell and Environment 29: 77-94.

96. Vinnikov K, Robock A, Speranskaya N, Schlosser CA (1996) Scales of temporal and spatial variability of midlatitude soil moisture. J Geophys Res 101: 7163-7174.

97. Robock A, Schlossera CA, Vinnikova KY, Speranskayad NA, Entina JK, et al. (1998) Evaluation of the AMIP soil moisture simulations. Glob Planet Change 19: 181-208.

98. McDermott F (2004) Palaeo-climate reconstruction from stable isotope variations in speleothems: a review. Quaternary Science Reviews 23: 901-918.

99. Mathieu R, Bariac T (1996) A numerical model for the simulation of stable isotope profiles in drying soils. J Geophys Res 101: 12685-12696.

100. Gazis C, Geng X (2004) A stable isotope study of soil water: evidence for mixing and preferential flow paths. Geoderma 119: 97-111.

101. Barnes CJ, Allison GB (1983) The distribution of deuterium and oxygen 18 in dry soils: I Theory. J Hydrol 60: 141-156.

102. Allison GB, Barnes CJ, Hughes MW (1983) The distribution of deuterium and oxygen 18 in dry soils: II. Experimental. J Hydrol 64: 377-397.

103. Braud I, Biron P, Bariac T, Richard P, Canale L, et al. (2009) Isotopic composition of bare soil evaporated water vapor. Part I: RUBIC IV experimental setup and results. J Hydrol 369: 1-16.

104. Raz-Yaseef N, Yakir D, Schill, Cohen S (2012) Dynamics of evapotranspiration partitioning in a semi-arid forest as affected by temporal rainfall patterns. Agr Forest Meteorol 157: 77-85.

105. McCarroll D, Loader N (2004) Stable isotopes in tree rings. Quat Sci Rev 23: 771-801.

106. Shi C, Daux V, Risi C, Hou SG, Stievenard M, et al. (2011) Reconstruction of southeast Tibetan Plateau summer cloud cover over the past two centuries using tree ring delta18O. Clim Past.

107. Yakir D, Wang XF (1996) Fluxe of CO2 and water between terrestrial vegetation and the atmosphere estimated from isotope measurements. Nature 380: 515-517.

108. Yakir D, Sternberg LdSL (2000) The use of stable isotopes to study ecosystem gas exchange. Oecologia 123: 297-311.

109. Bender M, Sowerms T, Labeyrie L (1994) The Dole Effect and Its Variations During the Last 130,000 Years as Measured in the Vostok Ice Core. Glob Biogeochem Cycles 8: 363-376.

110. Blunier T, Barnett B, Bender ML, Hendricks MB (2002) Biological oxygen productivity during the last 60,000 years from triple oxygen isotope measurements. Glob Biogeochem Cycles 16.

111. Yakir D, Yechieli Y (1995) Plant invasion of newly exposed hypersaline Dead Sea shore. Nature 374: 803-805.

112. Gat JR, Yakir D, Goodfriend G, Fritz P, Trimborn P, et al. (2007) Stable isotope composition of water in desert plants. Plant Soil 298: 31-45.

113. Wang L, Caylor KK, Villegas JC, Barron-Gafford GA, Breshears DD, et al. (2010) Partitioning evapotranspiration across gradients of woody plant cover: Assessment of a stable isotope technique. Geophy Res Lett 37: L09401.

114. Keeling C (1961) The concentration and isotopic abundances of carbon dioxide and marine air. Geochim Cosmochim Acta 24: 277-298.

115. Noone D, Risi C, Bailey A, Brown D, Buenning N, et al. (2012) Factors controlling moisture in the boundary layer derived from tall tower profiles of water vapor isotopic composition following a snowstorm in colorado. Atmos Chem Phys Discuss 12: 16327-16375.

116. Lawrence DM, Thornton PE, Oleson KW, Bonan GB (2007) The partitioning of evapotranspiration into transpiration, soil evaporation, and canopy evaporation in a gcm: Impacts on land-atmosphere interaction. J Hydrometeor 8: 862-880.

117. Uppala S, Kallberg P, Simmons A, Andrae U, daCostaBechtold V, et al. (2005) The ERA-40 re-analysis. Quart J Roy Meteor Soc 131: 2961-3012.

118. Yoshimura K, Kanamitsu M, Noone D, Oki T (2008) Historical isotope simulation using reanalysis atmospheric data. J Geophys Res 113: D19108.

119. Coindreau O, Hourdin F, Haeffelin M, Mathieu A, Rio C (2007) Assessment of physical parameterizations using a global climate model with stretchable grid and nudging. Mon Wea Rev 135: 1474.

120. Lai CT, Riley W, Owensby C, Ham J, Schauer A, et al. (2006) Seasonal and interannual variations of carbon and oxygen isotopes of respired CO_2 in a tallgrass prairie: Measurements and modeling results from 3 years with contrasting water availability. J Geophys Res 111: D08S06.

121. Barnard RL, Salmon Y, Kodama N, Sorgel K, Holst J, et al. (2007) Evaporative enrichment and time lags between d18O of leaf water and organic pools in a pine stand. Plant Cell and Environment 30: 539-550.

122. Rozanski K, Araguas-Araguas L, Gonfiantini R (1993) Isotopic patterns in modern global precipitation. Geophys Monogr Seri, AGU, Climate Change in Continental Isotopic records.

123. Masson-Delmotte V, Hou S, Ekaykin A, Jouzel J, Aristarain A, et al. (2008) A review of Antarctic surface snow isotopic composition: observations, atmospheric circulation and isotopic modelling. J Climate 21: 3359-3387.

124. Masson-Delmotte V, Landais A, Stievenard M, Cattani O, Falourd S, et al. (2005) Holocene climatic changes in Greenland: Different deuterium excess signals at Greenland Ice Core Project (GRIP) and NorthGRIP. J Geophys Res 110.

125. Vachon RW, White JWC, Gutmann E, Welker JM (2007) Amount-weighted annual isotopic (δ18O) values are affected by the seasonality of precipitation: A sensitivity study. Geophy Res Lett 34: L21707.

126. Yoshimura K, Oki T, Ohte N, Kanae S (2004) Colored moisture analysis estimates of variations in 1998 asian monsoon water sources. J Meteor Soc Japan 82: 1315-1329.

127. Vander Ent RJ, Savenje HHG, Schaefli B, Steele-Dunne SC (2010) Origin and fate of atmopheric moisture over continents. Water Resour Res 46: W09525.

128. Kanner LC, Buenning NH, Stott LD, Timmermann A (2013) The role of soil evaporation in delao18 terrestrial climate proxies. Glob Biogeochem Cycles.

129. Abdulla FA, Lettenmaier DP, Wood EF, Smith JA (1996) Application of a macroscale hydrological model to estimate the water balance of the Arkansas-Red River Basin. J Geophys Res 101: 7449-7459.

130. Bosilovich MG, Yang R, Houser PR (1999) River basin hydrology in a global offline land-surface model. J Geophys Res 104: 19661-19673.

131. Ducharne A, Golazb C, Leblois E, Lavala K, Polcher J, et al. (2003) Development of a high resolution runoff routing model, calibration and application to assess runoff from the LMD GCM. J Hydrol 280: 207-228.

132. Vitvar T, Aggarwal P, Herczeg A (2006) Towards a global network for monitoring isotopes in rivers. Geophys Res Abstracts, EGU, 8.

133. Vitvar T, Aggarwal PK, Herczeg AL (2007) Global network is launched to monitor isotopes in rivers. Eos Trans AGU 88: 325-332.

134. Dutton AL, Wilkinson B, Welker JM, Lohmann KC (2005) Comparison of river water and precipitation delta18O across the 48 contiguous United States. Hydrol Processes 19: 3551-3572.

135. Gibson JJ, Edwards TWD, Birks SJ, Amour NAS, Buhay WM, et al. (2005) Progress in isotope tracer hydrology in Canada. Hydrol Processes 19: 303-327.

136. Guimberteau M, Laval K, Perrier A, Polcher J (2008) Streamflow Simulations by the Land Surface Model ORCHIDEE Over the Mississippi River Basin: Impact of Resolution and Data Source on the Model. In American Geophysical Union, Fall Meeting.

137. Lee X, Kim K, Smith R (2007) Temporal variations of the 18O/16O signal of the whole-canopy transpiration in a temperate forest. Global Biogeochem Cycles 21:GB3013.

138. Gupta P, Noone D, Galewsky J, Sweeney C, Vaughn BH (2009) Demonstration of high-precision continuous measurements of water vapor isotopologues in laboratory and remote field deployments using wavelength-scanned cavity ring-down spectroscopy (WS-CRDS) technology. Rapid Commun Mass Spectrom. 23: 2534-2542.

139. Nusbaumer J (2016) An examination of atmospheric river moisture transport and hydrology using isotope-enabled CAM5. PhD thesis, University of Colorado at Boulder.

140. deRosnay P, Bruen M, Polcher J (2000) Sensitivity of the surface fluxes to the number of layers in the soil model used in GCMs. Geophys Res Let 27: 3329-3332.

141. Zhu D, Peng S, Ciais P, Viovy N, Druel A, et al. (2015) Improving the dynamics of northern hemisphere high-latitude vegetation in the orchidee ecosystem model. Geoscientific Model Development 8: 2263-2283.

142. Ryder J, Polcher J, Peylin P, Ottle C, Chen Y, et al. (2016) A multi-layer land surface energy budget model for implicit coupling with global atmospheric simulations. Geoscientific Model Development 9: 223-245.

143. Botter G, Bertuzzo E, Rinaldo A (2011) Catchment residence and travel time distributions: The master equation. Geophysical Research Letters 38.

144. Evaristo J, Jasechko S, McDonnell JJ (2015) Global separation of plant transpiration from groundwater and streamflow. Nature 525: 91-94.

145. Sokratov SA, Golubev VN (2009) Snow isotopic content change by sublimation. Journal of Glaciology 55: 823-828.

146. Ekaykin AA, Hondoh T, Lipenkov VY, Miyamoto A (2009) Post-depositional changes in snow isotope content: preliminary results of laboratory experiments. Clim Past Discuss 5: 2239-2267.

147. Washburn E, Smith E (1934) The isotopie fractionation of water by physiological processes. Science 79: 188-189.

148. Barnes C, Allison G (1988) Tracing of water movement in the unsaturated zone using stable isotopes of hydrogen and oxygen. J Hydrol 100: 143-176.

149. Majoube M (1971) Fractionnement en Oxygene 18 et en Deuterium entre leau et sa vapeur. Journal de Chimie Physique 10: 1423-1436.

150. Braud I, Bariac T, Biron P, Vauclin M (2009) Isotopic composition of bare soil evaporated water vapor. Part II: Modeling of RUBIC IV experimental results. J Hydrol 369: 17-29.

151. Stewart MK (1975) Stable isotope fractionation due to evaporation and isotopic exchange of falling waterdrops: Applications to atmospheric processes and evaporation of lakes. J Geophys Res 80: 1133-1146.

152. Braud I, Bariac T, Gaudet JP Vauclin M (2005) SiSPAT-Isotope, a coupled heat, water and stable isotope (HDO and H218O) transport model for bare soil. Part I. Model description and first verifications. J Hydrol 309: 301-320.

153. Munnich KO, Sonntag C, Christmann D, Thoma G (1980) Isotope fractionation due to evaporation from sand dunes. Z Mitt Zentralinst Isot Stralenforsch 29: 319-332.

154. Melayah A, Bruckler L, Bariac T (1996) Modeling the transport of water stable isotopes in unsaturated soils under natural conditions 1. theory. water resources res 32: 2047-2054.

155. Mills R (1973) Self diffusion in normal and heavy water in the range 1-45C. J Phys Chem 77: 685-688.

156. Harris KA, Woolf LA (1980) Pressure and temperature dependence of the self-diffusion coefficient of water and oxygen-18 water. J Chem Soc Faraday Trans 76: 377-385.

157. Merlivat L, Nief G (1967) Fractionnement isotopique lors des changements d'état solide-vapeur et liquide-vapeur de l'eau à des températures inférieures à 0°C. Tellus 19: 122-127.

158. Majoube M (1971) Fractionnement en O18 entre la glace et la vapeur deau. Journal de Chimie Physique 68: 625-636.

159. Jouzel J, Merlivat L (1984) Deuterium and oxygen 18 in precipitation: modeling of the isotopic effects during snow formation. J Geophys Res 89: 749.

160. Farquhar G, Cernusak L (2005) On the isotopic composition of leaf water in the non-steady state. Functional Plant Biology 32: 293-303.

161. Cuntz M, Ogee J, Farquhar G, Peylin P, Cernuzak L (2007) Modelling advection and diffusion of water isotopologues in leaves. Plant cell and environment 30: 892-909.

162. Ogee J, Cuntz M, Peylin P, Bariac T (2007) Non-steady-state, non-uniform transpiration rate and leaf anatomy effects on the progressive stable isotope enrichment of leaf water along monocot leaves. Plant Cell and Environment 30: 367-387.

163. Dongmann G, Nurnberg H, Forstel H, Wagener K (1974) On the enrichment of H2018 in the leaves of transpiring plants. Rad and Environm Biophys 11: 41-52.

164. Rosnay PD, Polcher J (1998) Modelling root water uptake in a complex land surface scheme coupled to a GCM. Hydrol Earth Sci 2: 239-255.

165. Zimmermann U, Ehhalt E, Munnich K (1967) Soil-water movement and evapotranspiration: changes in the isotopic composition of the water. Proceedings of the symposium on isotopes in hydrology, 14-18 November, IAEA, Vienna, pp: 567-585.

166. Ogee J, Brunet Y, Loustau D, Berbigier P, Delzon S (2003) MuSICA, a CO_2, water and energy multilayer, multileaf pine forest model: evaluation from hourly to yearly time scales and sensitivity analysis. Global Change Biology 9: 697-717.

167. Dragoni D, Schmid HP, Wayson CA, Potter H, Grimmond CSB, et al. (2011) Evidence of increased net ecosystem productivity associated with a longer vegetated season in a deciduous forest in south?central Indiana, USA. Global Change Biology 17: 886-897.

168. Dubbert M, Cuntz M, Piayda A, Werner C (2014) Oxygen isotope signatures of transpired water vapor: the role of isotopic non-steady-state transpiration under natural conditions. New Phytologist 203: 1242-1252.

169. Gholz HL, Clark KL (2002) Energy exchange across a chronosequence of slash pine forests in Florida. Agricultural and Forest Meteorology 112: 87-102.

170. Knohl A, Schulze ED, Kolle O, Buchmann N (2003) Large carbon uptake by an unmanaged 250-year-old deciduous forest in Central Germany. Agricultural and Forest Meteorology 118: 151-167.

171. Kurz-Besson C, Otieno D, Lobodo Vale R, Siegwolf R, Schmidt M, et al. (2006) Hydraulic Lift in Cork Oak Trees in a Savannah-Type Mediterranean Ecosystem and its Contribution to the Local Water Balance. Plant and Soil 282: 361-378.

172. Schmid HP, Grimmond CSB, Cropley F, Offerle B, Su HB (2000) Measurements of Co2 and energy fluxes over a mixed hardwood forest in the mid-western united states. Agricultural and Forest Meteorology 103: 357-374.

10

Baseline Study of Drinking Water Quality – A Case of Leh Town, Ladakh (J&K), India

Konchok Dolma*, Madhuri S Rishi and Herojeet R

Department of Environment Studies, Punjab University, Chandigarh-160014, India

Abstract

Water is the vital resource on which life sustains and water becomes more valuable in this cold desert part of the northernmost region of India, called Ladakh which comprises of two districts, Leh and Kargil. Groundwater since ancient times in the form of springs provided ample water for the region and its contribution has increased manifold in the wake of recent spurt in bore well installations, especially, in Leh-Town. Due to increasing urbanization, with surge in a huge floating population in the absence of a sewerage link in summer tourism boom season, puts extra stress on the limited water resources of the area and with the rising living standards, grey and black water is being disposed off in the ground-pit or in septic tanks without any treatment. This may lead to pollution of groundwater resources especially, in the densely populated residential areas. For insuring sustainable development of groundwater, in the absence of any observation wells for constant monitoring of quality or quantity of groundwater and the unregulated installation of bore-wells makes this quality characterization very significant and helps in future management. The physico-chemical parameters like pH, electrical conductivity, turbidity, total dissolved solids, hardness, alkalinity, nitrates, fluoride, and chlorides were analyzed to meet the objective of the study. The results revealed that in general, the present status of groundwater quality is suitable for drinking purposes and out of 20 total samples evaluated, 75% of samples had NTU above desirable limit while 10% samples each recorded TDS and EC above desirable limits.

Keywords: Drinking water; Cold desert; Sustainable development; Turbidity; Sewerage

Introduction

"Hydrology" is the science to know the properties, distribution and behavior of water in nature. Among the various needs of water, the most essential need is drinking. Surface water and ground water are two major sources for the supply of drinking water. Surface water comes from lakes, reservoirs and rivers. Groundwater comes from wells that the water supplier drills into aquifers. Demand for freshwater is increasing with rising population, especially groundwater, since its quality is better than the surface water, which is getting polluted with time. Groundwater is about 20% of the world's fresh water resource and widely used by industries, for irrigation and domestic purposes [1]. The quality of groundwater depends upon overall proportional amount of different chemical constituents present in groundwater. Groundwater quality is more significant in the case of Leh Town as this town is facing rapid urbanization, propelled by the tourism boom, in recent years. The town's water supply network is dependent wholly-solely on groundwater supplied by the extensive spring network in the region. Along with the region's Public Health Engineering Department (PHE) which supplies water to the town's 21 wards from groundwater sources through pipelines and Public Stand Posts (PSPs) in summers and in winters by water tankers as the pipes freezes and explodes during winters. There is a wide presence of private bore wells in the town and adjoining areas through which the residents fulfill their water requirements. The water quotient is very important in this town as it lies in a cold desert and if we see the whole surrounding areas, we can definitely say this town is an oasis in a desert as it is surrounded by icy cold mountain ranges of the Himalayas which are one of the highest in the world. Water since time immemorial has played an important role in the historical development of this area as popular culture revolves around water conservation aspects. Whether it is traditional irrigation system, sanitation system or folkways and mores, judicious use of water is at the heart of this town's history and culture.

Description of the study area

Ladakh Region which is part of the northernmost Indian state, Jammu and Kashmir comprises of two districts Leh and Kargil. Topographically the whole district is mountainous with parallel ranges of the Himalayas, mainly Zanskar, Ladakh and the Karakoram ranges. The Shayok, Indus and Zanskar rivers flow between these ranges and most of the population is inhabited in these valleys. The elevation of Ladakh Region ranges from 2300 m to 5000 m above mean sea level. District Leh with an area of 45100 km² is the 2nd largest district in India after Kutch in Gujarat with an area of 45652 km². As Ladakh is a cold desert precipitation averages around 61 mm/annum and occurs in the form of snow and rain both in winter and summer months respectively [2]. Ladakh is a semi-autonomous region of India, governed by the Ladakh Autonomous Hill Development Council of Leh and Kargil simultaneously for Leh and Kargil District. The district is divided into 9 Community Development Blocks namely Leh, Khaltsi, Nyoma, Durbuk, Kharu, Nubra, Saspol, Panamic and Chuchot which is further divided into 03 tehsil namely Leh, Sumoor and Khaltsi. Leh is the district headquarter and the only township in the district.

Leh town: Leh town is located in District Leh between North latitude 34°13'00" to 34°8'00" and East longitude 77°38'00" to 77°32'00". Leh is about 434 kilometers from Srinagar (NH 1D) and 474 kilometers

---

*Corresponding author: Konchok Dolma, Department of Environment Studies, Punjab University, Chandigarh-160014, India
E-mail: kono20jan@gmail.com

from Manali (NH 21). The town is divided into 21 administrative wards and the relief ranges from highest 4500 m (amsl) to lowest 3250 m (amsl). The town is classified as Class III Urban Agglomeration (UA) and has a Notified Area Council (NAC). The population as per 2011 census is 30,870 which have risen from 28,639 according to 2001 census thus recording a growth rate of 7.8%. The town caters to a huge population of army personnel permanently deployed here with an ever surging tourist's population. The tourism industry and army presence is the pull factor for a large floating population of tourists and migrants labourers particularly in the short summer seasons from June to September.

Historical background: Historical background of Leh Town dates back to the 17th century when this became the capital of Ladakh region replacing Shey, located 15 km away from Leh city centre by the great king Singhe Namgyal. Ladakh being an important stopover in the trans Himalayan trade especially the pashm trade so was invaded by the Dogra rulers and the Britishers in order to control the Silk route connecting with central asia [3].

Geology and hydrogeology: Geology and hydrogeology of a particular area determines the groundwater flow and its properties. Rocks of the district is constituted by igneous, metamorphic and sedimentary rocks that are sandwiched between tertiary granitoid batholiths of Ladakh and Karakoram ranges. Leh Town shows terraces and valleys show a plethora, both erosional (amphitheaters) and depositional land forms, belonging to glacial (moraines), fluvio-glacial (glacial out wash), mass wasting (alluvial fans). Terraces forms the high ground between the hill ranges and the valleys, sloping southward and are composed of boulders and cobbles of moraine sediments with sand, silt, clay and gravel of fluvio glacial origin, derived from Ladakh range. The movement of groundwater is affected by the unconsolidated formations like alluvium, scree and talus formations [4]. The upper part of this valley is laden with well-preserved repository of lateral and terminal moraines also the glaciers melt/rain water drains the lower reaches, gently sloping valley southwards. The valley fill deposits are mainly boulders and gravel mixed with salt and sand material. Groundwater occurs as unconfined condition in this formation and aquifers in this region are made of boulders and clastic material in clay, silt, matrix of sand, silt and clay (Figure 1).

Materials and Methods

A total of 20 groundwater samples were collected from different locations in Leh Town which spans an area of 1893 hectares/18.93 km² and is divided into 21 administrative wards [5]. The samples were collected in the month of May (Pre monsoon) and October (post monsoon), 2013 (Figure 2). Prior to sample collection, all the plastic bottles were thoroughly washed and sun-dried and before sample collection the plastic bottles were rinsed twice with the water sample to be collected. The bottles were then labeled and the co-ordinates of the sampling sites were duly noted. Parameters like Temperature, pH, and EC were analyzed on the spot using potable water and soil analysis kit. For the analysis of other parameters, the bottles were taken to the laboratory and stored at 4°C and further analysis completed as per standard procedures. Water samples were analyzed in the geochemical laboratory of the Department of Geology and Water Resources Department, Chandigarh according to the standard methodology given by American Public Health Association [6], Trivedy and Goel and Central Pollution Control Board, New Delhi [7]. For map making survey of India toposheet no. 52F/12 was used for digitization and GPS (Global Positioning System) device was used for identifying sampling location.

Results and Discussions

Physico-chemical analysis of water samples collected from different groundwater locations around Leh Town was analyzed and the distribution of water samples showing various parameters against maximum permissible and desirable limits are shown in Table 1.

pH and electrical conductivity (EC)

The pH (hydrogen ion concentration) of water is very important indicator of its quality as it depends on the presence of phosphates, silicates, borates, fluorides and some other salts in dissociated form. In general waters having pH between 6.5 and 8.5 are categorized as suitable, whereas waters with pH 7.0 to 8.0 are highly suitable for all purposes.

The pH value of the surface water of the study area during pre-monsoon varies from 7 to 8 with mean value of 7.55 and varied from 7.00 to 7.9 with mean value of 7.24 during post monsoon which indicated that water is slightly alkaline in nature but suitable for domestic purposes [8]. Electrical conductivity of water is also an important parameter for determining the water quality. It is a measurement of water's capacity for carrying electrical current and is directly related to the concentration of ionized substance in the water. In the present study, EC values of surface water ranged between 205 µmhos/cm to 849 µmhos/cm with mean value of 441.59 µmhos/cm in pre monsoon and between 201 µmhos/cm to 826 µmhos/cm with mean value of 404.54 µmhos/cm during post monsoon. Distribution of pH and EC in samples is shown in Figure 3a and 3b.

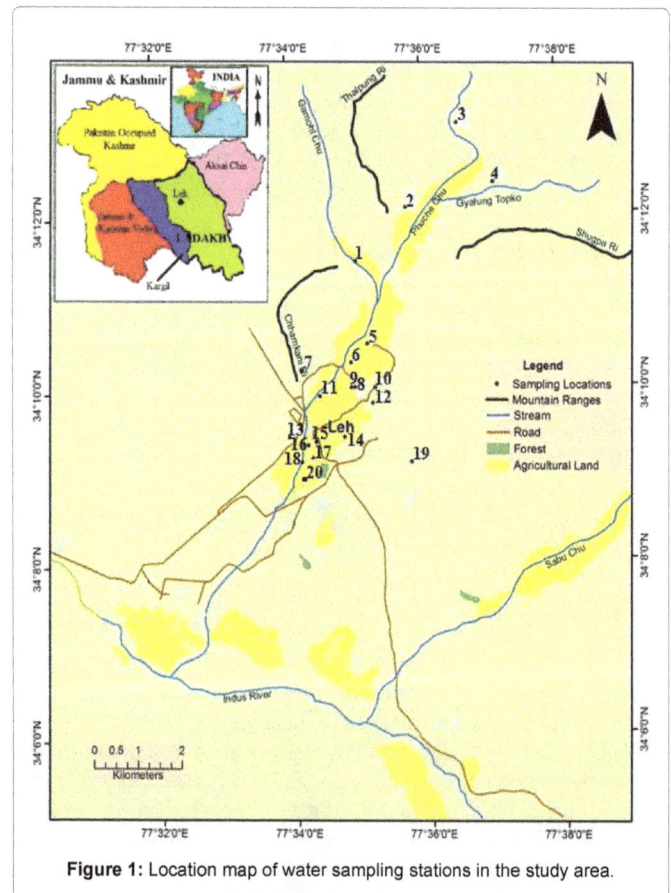

Figure 1: Location map of water sampling stations in the study area.

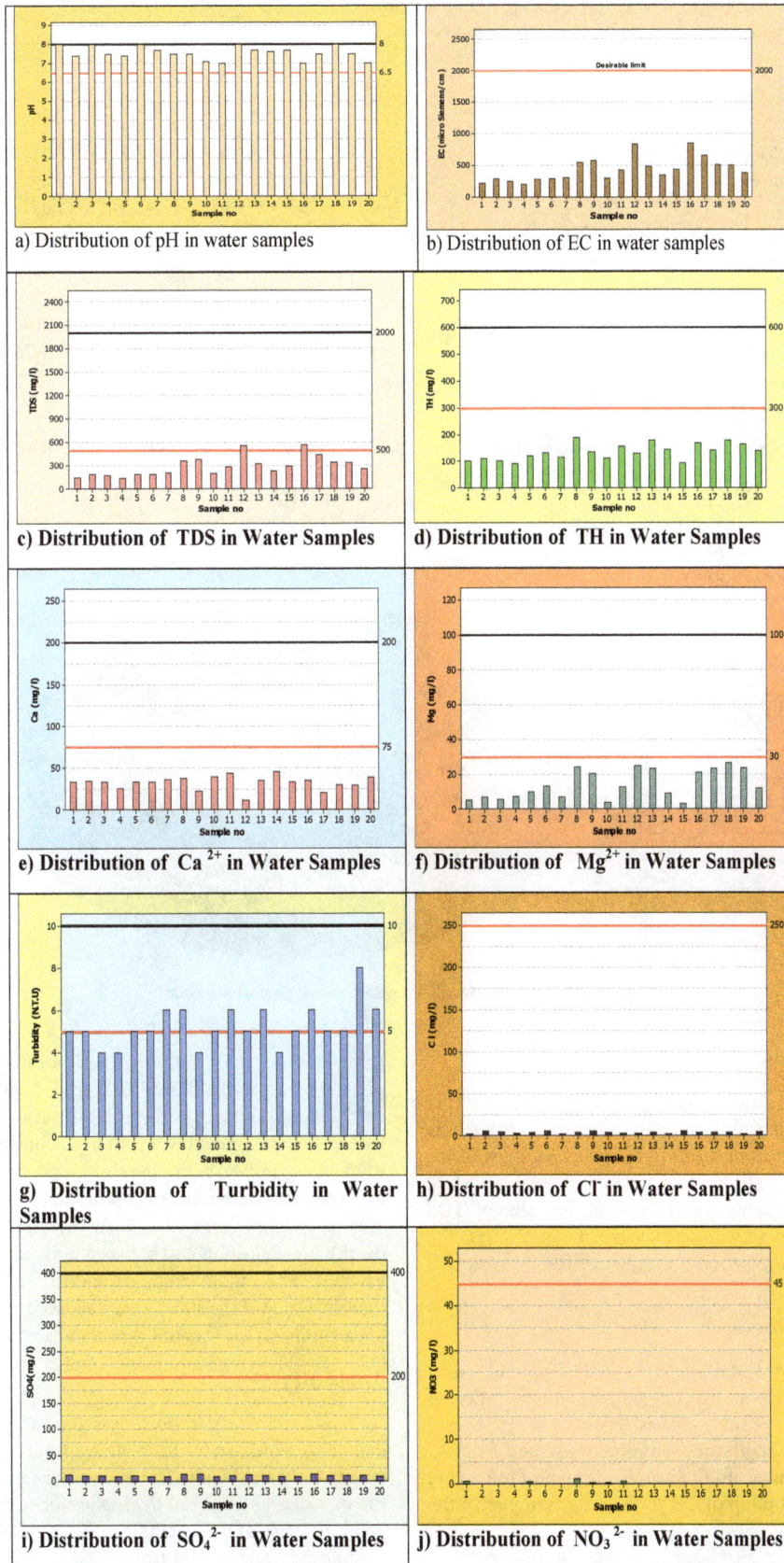

Figure 2: Distribution of various parameters in analysed water samples in the study area.

S. No	Parameter	Max. permissible limit for drinking water	Desirable limit for drinking water	No.of water samples analyzed	No. of Samples above *DL	No. of Samples above *PL	Range Min	Range Max	Mean Value	Std Deviation
1.	pH	No relaxation	6.5-8.5	20	Nil	Nil	7	8	7.55	0.34
2.	EC	0-2000 µS/cm	750 µS/cm	20	2	Nil	205	849	441.51	188.87
3.	TDS	2000 mg/l	500 mg/l	20	2	Nil	134	565	291.22	125.31
4.	TH	600 mg/l	300 mg/l	20	Nil	Nil	90	188	135	30.29
5.	Cl^-	1000 mg/l	250 mg/l	20	Nil	Nil	2.13	4.97	3.51	0.99
6.	Ca^{2+}	200 mg/l	75 mg/l	20	Nil	Nil	11.77	45.11	32.15	8.00
7.	Mg^{2+}	100 mg/l	30 mg/l	20	Nil	Nil	2.92	26.35	14.04	8.22
8.	NO_3^{2-}	No relaxation	45 mg/l	20	Nil	Nil	0	1.00	0.16	0.26
9.	Na^+			20	-	-	0.1	31.9	7.54	8.35
10.	K^+			20	-	-	0.8	3.2	1.6	0.62
11.	SO_4^{2-}	400 mg/l	200 mg/l	20	Nil	Nil	9.39	13.70	10.69	1.53
12.	PO_4^{2-}	0.1 mg/l	-	20	Nil	Nil	0	0.47	0.06	0.10

*DL- Desirable Limit; *PL- Permissible Limit

Table 1: Distribution of groundwater samples showing various parameters against maximum permissible and desirable limits.

Parameter	Percentage of Samples above *DL %	Percentage of Samples above *PL %
EC	10	Nil
TDS	10	Nil
Turbidity	75	Nil
pH	Nil	Nil
Ca^{2+}	Nil	Nil
Mg^{2+}	Nil	Nil
Na^+	Nil	-
K^+	-	-
Cl^-	Nil	Nil
F^-	Nil	Nil
SO_4^{2-}	Nil	Nil
NO_3^{2-}	Nil	Nil

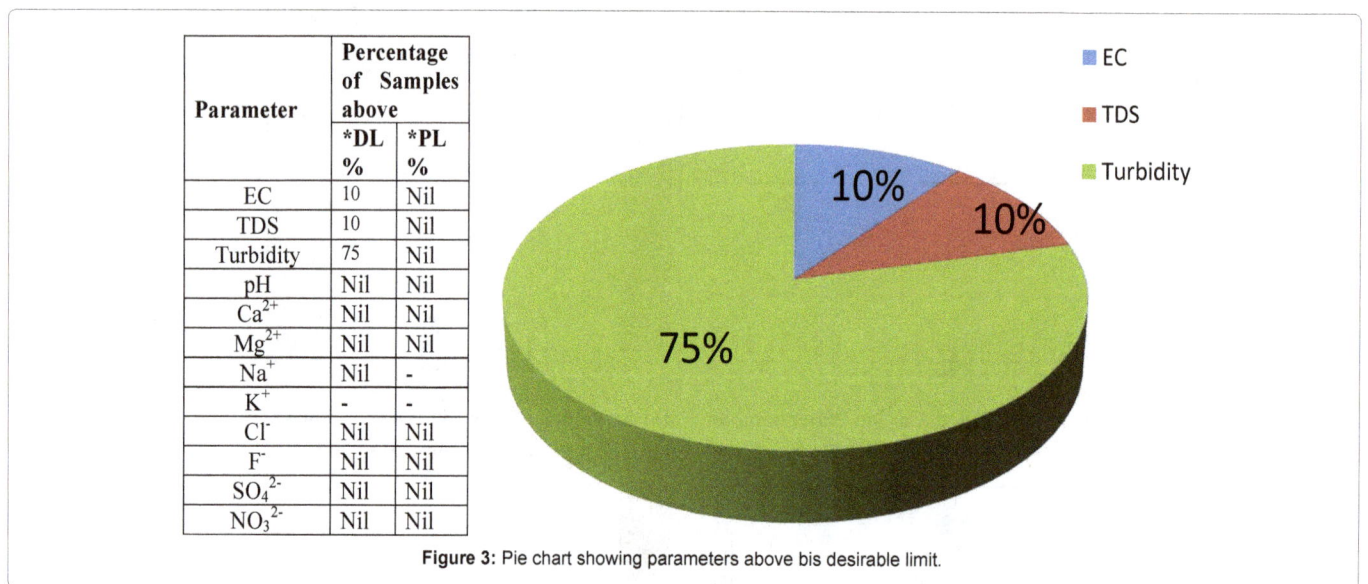

Figure 3: Pie chart showing parameters above bis desirable limit.

Total dissolved solids (TDS)

Total dissolved salt concentrations is the primary indicator of the total mineral content in water and are related to problems such as excessive hardness. Total dissolved solids in the water samples of the study area varied from 134 mg/l to 565 mg/l with mean value of 291.22 mg/l during pre-monsoon. During post monsoon, the value of TDS varied from 130 mg/l to 561 mg/l with mean value of 279.22 mg/l. The distribution of TDS in the surface water samples of the study area is depicted in Figure 1.3 (c).

Total hardness (TH)

It results from the presence of divalent metallic cations of which calcium and magnesium are the most abundant. The concentration of total hardness in the water of the study area varied from 90 mg/l to 188 mg/l with mean value of 135 mg/l during pre-monsoon and from 72 mg/l to 172 mg/l with mean value of 120.63 mg/l during post monsoon. Figure 1.3 d shows the distribution of TH in the water samples of the study area.

Calcium (Ca^{2+}) and Magnesium (Mg^{2+})

The amount of calcium in the surface water of the study area varied from 11.77 mg/l to 45.41 mg/l with mean value of 32.15 mg/l during pre-monsoon and between 10.09 mg/l to 40.37 mg/l with mean value of 27.29 mg/l during post monsoon. Magnesium concentration in the surface water is generally less than calcium due to the slow dissolution of magnesium bearing minerals and greater abundance of calcium in earth crust. Magnesium concentration in the surface water of the study area ranged between 2.92 to 26.35 mg/l with mean value of 14.04 mg/l during pre-monsoon and between 3.41 mg/l to 24.88 mg/l with mean value of 13.66 mg/l during post monsoon. Figure 1.3 e and f indicates that the water of the study area were well within the permissible limits of calcium and magnesium thus, it is safe for drinking purposes.

Turbidity

Turbidity refers to how clear the water is. It is a measure of the degree to which water loses its transparency due to the presence of suspended particulates. The more total suspended solids in the water, the murkier it seems and the higher the turbidity. It is considered as a good measure of the quality of water. Turbidity in the water samples of the study area varied from 3 NTU to 8 NTU with mean value of 5.8 NTU during pre-monsoon. During post monsoon, the value of turbidity varied from 5 NTU to 8 NTU with mean value of 5.93 NTU.

All the samples analyzed were within desirable limit of 5 NTU [9] and are fit for human consumption. The distribution of turbidity at different sampling locations is shown in Figure 3 g.

Chloride (Cl⁻) and Fluoride (F⁻)

Chloride in drinking water is not generally harmful to human beings until high concentration is present. The chloride ion in the surface water of the study area in pre monsoon season varied between 2.13 mg/l to 4.91 mg/l with mean value of 3.51 mg/l. In post monsoon season, it varied between 1.42 mg/l to 6.39 mg/l with mean value of 2.90 mg/l. Figure 3 h Clearly indicates that all the water samples of the study area were within the desirable limit and hence fit for consumption. Fluoride (F) is essential in trace amounts for all human beings and is one of the normal constituents of all diets. The desirable limit of fluoride in drinking water is 1 mg/l [9]. Fluoride concentrations in all the water samples were found to be within the permissible limit of 1.5 mg/l Figure 3 i. In pre monsoon season the fluoride concentration of surface water varied between 0.02 mg/l to 1.30 mg/l with mean value of 0.59 mg/l. In post monsoon season the concentration of fluoride varied between 0.04 mg/l to 1.30 mg/l with mean value of 0.62 mg/l. Only three samples crossed the desirable limit of 1.0 mg/l.

Nitrates (NO₃²⁻)

Excess of nitrates consumed by humans particularly infants is likely to cause health hazards and may lead to Methaemoglobinemia (Blue baby) disease. Distribution of nitrate in water samples are shown in Figure 3 j. The nitrate content of water samples in the study area was varied from 0 mg/l to 1.00 mg/l with mean value of 0.16 mg/l during pre-monsoon and between 0 mg/l to 0.97 mg/l with mean value of 0.16 during post monsoon.

Sodium (Na⁺) and Potassium (K⁺)

Sodium is the most abundant element of the alkali-earth group in the earth crust with average value 2.5%. In igneous rock, sodium is slightly more abundant than potassium, but in sediment, sodium is less abundant. BIS (1991) & WHO (2006) have not given any guideline limit for sodium and potassium in drinking water. Sodium concentration ranged between 0.1 mg/l to 31.9 mg/l with mean value 7.54 mg/l during pre-monsoon. During post monsoon, the sodium content varies from 0.07 mg/l to 27.8 mg/l with mean value 6.56 mg/l. While potassium is

an essential element for both plants and animals. Very high potassium concentration may be harmful to human nervous and digestive system. Potassium concentration ranged between 0.8 mg/l to 3.2 mg/l with mean value 1.6 mg/l during pre-monsoon and between 0.6 to 2.7 mg/l with mean value 1.26 mg/l during post monsoon. The concentration of potassium in the study area is very low. It is not feasible to assess the suitability of water for drinking purpose as no agency have given any standard with respect to potassium.

Hydrochemical facies of surface water

The hydrochemical facies of a particular place are influenced by geology of the area and distribution of facies by the hydro-geological controls. In the present study, the water samples of the study area have been classified as per Chadha's diagram [10]. The diagram is a modified version of Piper trilinear diagram and the expanded Durov diagram [11].

The chemical analyses data of all the water samples collected from the study area have been plotted on Chadha's diagram (Figure 4) and results have been summarized in Table 2. It is evident from the results that the water samples collected from the study area fall in Group 1 (Ca²⁺- Mg²⁺-Na⁺- K⁺), Group 5 (Ca²⁺-Mg²⁺- HCO₃⁻) and Group 6 (Ca²⁺-Mg²⁺- Cl⁻-SO₄²⁻) type. However majority of the surface water sample fall in Group 5 (Ca²⁺-Mg²⁺- HCO₃⁻) which means alkaline earths and weak acidic anions exceed both alkali metals and strong acidic anions, respectively. Such waters have temporary hardness.

Conclusion

The complete and systemic study of water quality in the study area revealed that apart from domestic sources, there were no other major sources of pollution observed in the project area. Turbidity of 75% of samples was above desirable limit and the majority of samples which recorded a high level of NTU could be attributed to the loose friable and weak sandy loamy soil. Terraces form the main topography of the valley area. The unconsolidated formations like alluvium, scree and talus formations present along the terrace plays a vital role in the recharge of ground water. This fluvo glacial structure bring clastic materials in clay, silt and sand matrix along with glacial melt water in the aquifer system and is responsible for the flow of groundwater. Also 10% of the samples were having EC and TDS above desirable limit but they were well within permissible limit. Eventually it is concluded that

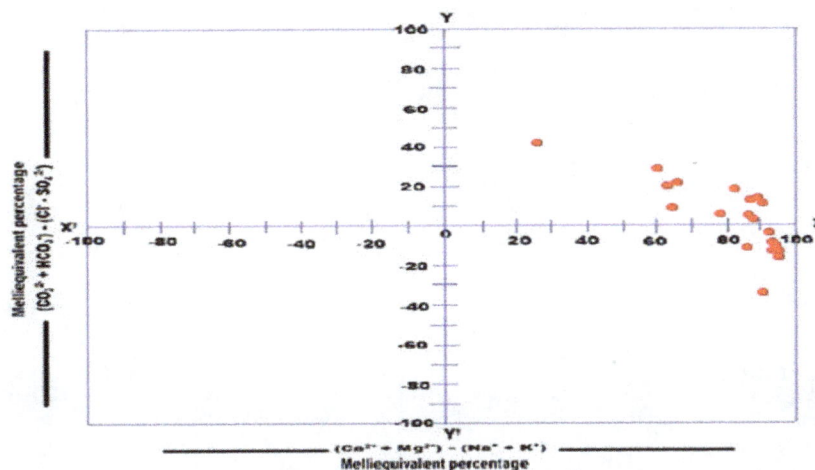

Figure 4: Chadha diagram for groundwater samples.

Classification/ Type	Surface water
Group 1 (Ca^{2+}- Mg^{2+}-Na$^+$- K$^+$)	5
Group 2 (Na$^+$- K$^+$- Ca^{2+}- Mg^{2+})	--------
Group 3 (HCO$_3$- Cl-SO$_4^{2-}$)	---------
Group 4 (SO$_4^{2-}$- HCO$_3$-Cl-)	---------
Group 5 (Ca^{2+}- Mg^{2+}- HCO$^-$)	10
Group 6 (Ca^{2+}- Mg^{2+}- Cl-SO$_4^{2-}$)	5
Group 7 (Na$^+$- K$^+$- Cl-SO$_4^{2-}$)	----------
Group 8 (Na$^+$- K$^+$- HCO$^-$)	----------

Table 2: Summarized results of chadha's classification.

the groundwater quality in the study area in general, can be designated as good for domestic and drinking purposes but enforcement of precautionary and preventive approaches required. Instead of unregulated rampant bore well development a well-managed plan should be taken into consideration to avoid further contamination, degradation and overexploitation of valuable groundwater resource.

References

1. Usha O, Vasavi A, Spoorthy, Swamy PM (2011) The Physcio Chemical and a Bacteriological Analysis of Groundwater in and around Tirupati, Pollution Research, 30: 339-343.

2. Raksha A, Aswathanarayana LG (2013) Selection of Sustainable Sources for Reliable Water Supply Schemes, In the Localities of Abnormal Climatic Conditions, Research and Reviews: Journal of Engineering and Technology 2: 256-264.

3. Kimua M (2013) Past Forward: Understanding Change in Old Leh Town, Ladakh, North India, Norwegian University of Life Sciences.

4. Central Ground Water Board (2009) Ground Water Information Booklet of Leh District, Jammu and Kashmir State.

5. Gondhalekar D, Akhtar A, Kebschull J, Kilmann P, Dawa S, et al. (2013) Water-related health risks in rapidly developing towns: the potential of integrated GIS-based urban planning, Water International, 38: 902-920.

6. APHA (2002) Standard methods for the examination of water and wastewater (20nd ed.). Washington D.C.: American Public and Health Association.

7. Central Pollution Control Board (2001) Pollution Control Acts, Rules, and Notifications, (4thedn), New Delhi: Central Pollution Control Board, Ministry of Environment and Forests, Government of India.

8. Herojeet RK, Rishi S, Madhuri, Sidhu N (2013) Hydrochemical Characterization, Classification and Evaluation of Groundwater Regime in Sirsa Watershed, Nalagarh Valley, Himachal Pradesh, India, Civil and Environmental Research 3: 47-57.

9. Bureau of Indian Standard (1991) Indian Standard Specification for Drinking Water IS: 10500, Indian Standard Institute, New Delhi, 1-31.

10. Chadha DK (1999) A Proposed new Diagram for Geochemical Classification of Natural Waters and Interpretation of Chemical Data, Hydrogeology Journal 7: 431-439.

11. Durov SA (1948) Natural Waters and Graphic Representation of their Composition Dokl, Akad, Nauk SSSR, 59: 87-90.

Dam Site Selection Using Remote Sensing Techniques and Geographical Information System to Control Flood Events in Tabuk City

Eyad Abushandi[1]* and Saleh Alatawi[2]

[1]Civil Engineering Department, Faculty of Engineering, University of Tabuk, Saudi Arabia
[2]Vice Rector Office for Graduate Studies and Scientific Research, University of Tabuk, Kingdom of Saudi Arabia

Abstract

Constructing dams around the city of Tabuk is an important solution to controlling flood events, as well as increasing surface water budget and creating ground water recharge spots. The success of this effort is mainly based on locating the best site for a dam in the area. The aim of this study is to demonstrate the use of remote sensing, and geographic information system in dam site selection within the context of a catchment scale. Digital Elevation Model was used from the Advanced Spaceborne Thermal Emission and Reflection Radiometer (ASTER 30m) to characterise the catchment area. Enhanced Thematic Mapper Plus images (ETM+) from LandSat 7 were used to classify the land cover in the study area. Several software packages such as ERDAS 11, Global Mapper 15.2 and ArcGIS 10.1 were used to construct and process the basic database. In addition, model builder from ArcGIS 10.1 was employed to construct a simplified model and integratesraster and vector datasets. The parameters of this model were: catchment slope (less 3%), delineation network order more than or equal to 7 and Runoff Coefficient (0.4). Six suitable locations were chosen: Wadi Dam in connection with Wadi Al Baqqr, Wadi Na'am, Wadi Atanah, Wadi Abu Nishayfah A, Wadi Abu Nishayfah B and Qa'a Sharawra.

Keywords: Information technology; Azrid area; Flash flood; Remote sensing; DEM; Earth dams

Introduction

Flash flood affects both civilian and agricultural activities as the negative impact of flooding lasts a long time after the event. In addition to the unexpected flash flood, the morphological characteristics of an area may make the flash flood much worse, even catastrophic. To avoid this negative impact many projects are proposed by different authorities such as dams, tracking canals, or storm drainage systems. Ancient dams were certainly built to meet a single purpose, which is water harvesting and irrigation, e.g. Ma'rib dam of 1750 and 1700 BCE [1]. Nowadays, there are multi-reasons for building modern dams and water structures (Figure 1):

i. Regulation of excess water and flood control

ii. Soil erosion and sediment control

iii. Drought control

iv. Irrigation

v. Generation of hydro-electric power

However, 60% of total stream flow is regulated by dams and reservoirs for flood control [2].

In general, there are different classifications of dams mainly based on engineering and study area aspects such as hydraulic design, structural design, usage of dams, construction material used and/or capacity. Based on construction material, earth dams account for 62% of all reported dams [2]. In addition, there are several considerations of dam construction:

i. Site topography and valley shape

ii. Geological structure and foundation conditions

iii. Availability of construction materials

iv. Overall cost

v. Spillway size and location

vi. Roadway

vii. Earthquake hazards

viii. Climatic conditions

ix. Environmental considerations

x. Length and height of dam

xi. Life of dam

However, ground investigation is important before planning to construct an earth dam especially site selection and earthworks considerations. Generally, dams in arid regions, particularly, in Saudi Arabia, are built to guarantee water supply and reduce the negative impact of flood events. In fact, Tabuk city is suffering surface water scarcity compensated by flash flooding, which has been affecting the entire city for many years during the winter season. The catchment experienced major floods during the years 1981, 1988, 2010 and 2012. Excess water management is a key aspect of sustaining the water sector in the Kingdom of Saudi Arabia. Unexpected flash floods and strong population growth rates associated with high water needs for agriculture will increase pressure on water resources in Tabuk especially groundwater. In addition, flash flooding destroys the infrastructure of the city. However, water resources are limited and construction

*Corresponding author: Eyad Abushandi, Civil Engineering Department, Faculty of Engineering, University of Tabuk, Saudi Arabia, E-mail: eabushandi@gmail.com

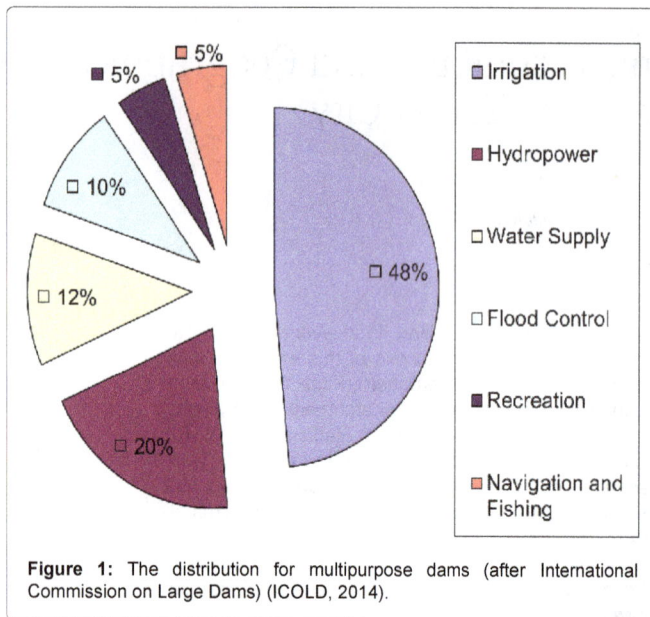

Figure 1: The distribution for multipurpose dams (after International Commission on Large Dams) (ICOLD, 2014).

of new dams will be an important part of the solution. At the initial stage of dam construction, full information on a selected site should be available for better planning and design. A tool for selecting the best site at Tabuk city catchment is the subject of this paper. After selecting suitable places for the dams, calculations of the inflow to the dam reservoirs is very important in order to forecast the amount of water received and design structures to control the inflow. In this case many hydrological and statistical models can be used such as IHACRES [3], ARMA, or ARIMA [4].

GIS and remote sensing techniques in dam site selection

Application of remote sensing and GIS techniques in hydrology is today one of the most effective approaches. Recently, remote sensing has provided valuable datasets to examine hydrological variables and morphological changes over large regions at different spatial and temporal scales (e.g. MSR-E, TRMM, GSMAP, ASTER, SAR and several others). Many researchers over the past 20 years have focused on satellite imagery applications in hydrology [5-9].

GIS and remote sensing can provide a huge amount of valuable data in spatial and temporal resolutions for areas where ground data are not easily available. In 2008, GIS and remote sensing techniques were applied by Forzieri et al. to assess the suitability of sites for the installation of small dams for the purpose of water harvesting in arid areas. The selection criteria are defined both in a qualitative and quantitative way. Qualitative criteria imply the identification of suitable valleys, wadi physical characteristics, based on satellite images interpretation – mainly DEM – and large-scale available cartography; other qualitative selection criteria concern the distance from settlements and infrastructures. Quantitative criteria were based on indexes that synthesise the effectiveness and feasibility of the possible interventions such as the alluvial plan index (α) and the hydrologic index (P).

In a similar study, Singh et al. [10] used several remote sensing imageries to select dam sites. They used some parameters to select the sites including:

i) Slope (less than 10%)

ii) Soil infiltration rate (moderate)

iii) Land use (shrubs and river beds)

iv) Soil type (sandy clay loam)

The results proposed 14 dam sites which could be used for water harvesting and cultivation.

Information, such as terrain surface, wadis network, land use and catchment boundaries are important for dam site selection, which can be gathered from remotely sensed images. In particular, DEM can provide slope data which is the most influential factor on flood behaviour [11,12] described in detail a workflow for the Digital Elevation Model (DEM) preprocessing and extraction from ASTER images including accuracy estimation. In addition to the low spatial and temporal resolutions in the majority of remote sensing imagery, data accuracy in most related research is still a dilemma. A review by Sanyal and Lu [13] on the application of Remote Sensing in flood management noted that DEM model is the main part of flood hazard mapping. In particular, slopes data from DEM are useful for many hydrological studies and can be employed for dam location selection. Furthermore, a limited effort has been devoted in recent years to determine the capability of these techniques in assisting engineering dam design by allowing efficient, quick and economic data collection [14,15]. However, the application of remote sensing in ephemeral streams is limited compared with permanent rivers [16].

Kumar [17] used remote sensing and GIS techniques to assign the location of small water harvesting structures across streams/watersheds. Various thematic layers such as Landuse/Landcover, geomorphology and lineaments were used. These layers along with geology and drainage were integrated using GIS techniques to derive suitable water harvesting sites. In addition to the suitable site selection of the dams, they calculated the storage and transmittance of groundwater in the study area. Youssef et al. [18] furthermore, proposed three dam site locations for Jeddah City in Saudi Arabia based on topographical analysis using different data sets such as topographic maps, remote sensing images, a digital elevation model with 90-m resolution (STRM 2000) and geological map. In addition, these selected locations were in the outlet areas where large tracts of land could be temporarily inundated by water as a result of water being held back by the proposed dam.

Martire et al. applied most recently satellite data (Synthetic Aperture Radar (SAR)) for the precise monitoring of earth dams ground deformation. They found high agreement between final SAR and in-situ instrumental data, which demonstrated the reliability of such a technique for future use.

Besides ground application, remote sensing can be used in the case of atmospheric variables such as cloud thickness, relative humidity and Evapotranspiration (ET) vital for water resource management such as MODIS application in ET [19] and GSMaP [7].

Catchment characteristics

The study area is located in the north-western part of Saudi Arabia (Figure 2) about 96 Km southwest of the Jordanian border. The study area covers about 20892.6 km². In terms of aridity, the catchment is classified as hyper-arid, which means high evaporation rate, low vegetative cover and flash flood due to unexpected rainfall storms. There are two main climatic seasons in Tabuk: winter season with limited rainfall storms and a hot and dry summer. The annual rainfall in Tabuk is around 33 mm; the main rainfall occurs in the wet season between October and April.

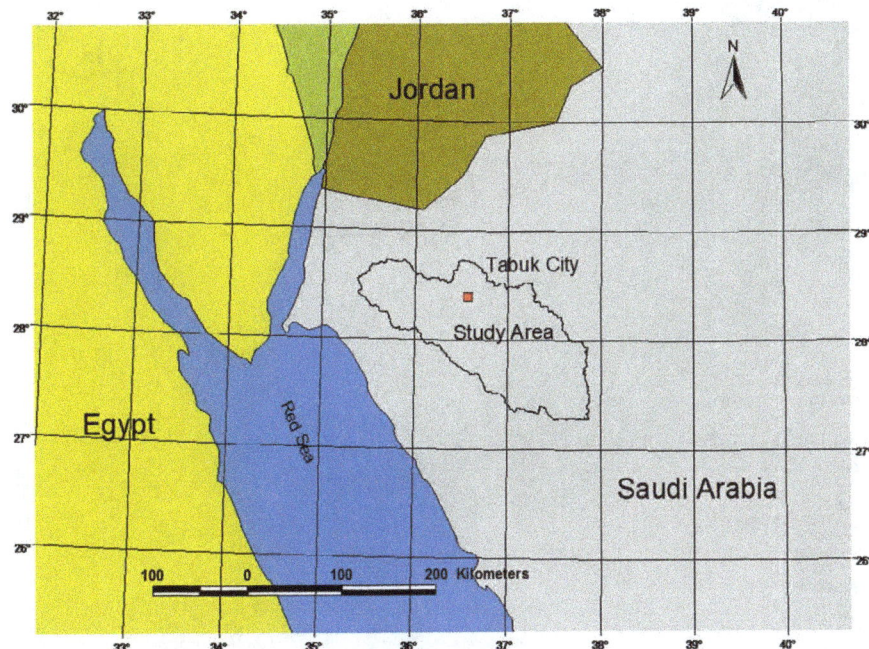

Figure 2: Study area location.

Materials and Method

Digital elevation data from ASTER, 30 m resolution, were used to derive the geomorphological structure. The boundaries and wadis network for Tabuk catchment were derived from the digital elevation data using River Tools V. 2.4. Further processes including modelling, stratification of different layers and orders were conducted using ArcGIS V. 10.0. The path profiles and wadis tracking were conducted using Global Mapper V.15.2. There are five main wadis crossing Tabuk City from east to west (Figure 3):

   i. Wadi Al-Akhdar

   ii. Wadi Abu Nishayfah

   iii. Wadi Na'am

   iv. Wadi Al Baqqar (extension from Wadi Ayrin and Wadi Al Hadarah)

   v. Wadi Damm

The names of the wadis are common but might be changed from one area to another and all ended at a single outlet called Qa'a Sharawra. The area of Qa'a Sharawra is around 253 Km^2. The wadis are mainly generated from three main mountain series surrounding the city of Tabuk from the east and west parts: Ghawanim Mountains, Rays Mountains and Asafir Mountains (Figure 4). However, these mountain series create slope zones which have potential for flood and soil erosion.

Wadi Al-Akhdar is generated from Ghawanim Mountains in the east and is 68 km long (Figure 5). The wadi is connected to Wadi Mishash Bani Atiyah and Tuus Al Arqanah in the upland. However, the wadi does not directly cross the city of Tabuk.

Wadi Abu Nishayfah is the longest wadi (226 km) in the catchment and crosses directly the city of Tabuk. Therefore, the flood behaviour in this wadi is important for any flood protection projects. The wadi is generated from Nahamah Mountain in the southern part of the catchment. The wadi has many feeding spots and sub-streams along the way from both sides east and west (Figure 6). Geomorphologically, at the distance of 207 km from its generation point and before crossing the city of Tabuk, Wadi Abu Nishayfah is connected to Wadi Ar Radwan.

Wadi Na'am is around 60 km long and generated from a series of mountains from the western part of the catchment, namely: Jibal Al Asafir (1257 m above sea level) and Jibal As Salitiyat (1190 above sea level) (Figure 7). The wadi also has many feeding spots and sub-streams along the way from the west and south parts especially in the first 15 km from the generation site. In addition, there are some sharing streams between Wadi Na'am and Wadi Atanah.

Wadi Al Baqqar is the extension of Wadi Al Hadarah and slightly connected to Wadi Ayrin. They are all generated from Jibal Al Asafir and Jibal As Salitiyat (Figure 8).

Wadi Damm crosses many mountain series from the northwestern part of the catchment, but relatively lower elevations such as Al Baydi Mountain (elevation 961 m), Umm Jiba Mountains (elevation 1008m), Shuka'ah Mountains (elevation 978 m) and many others. This means the wadi has many feeding sub-streams. Wadi Damm is around 137 km long (Figure 9).

The figures from 5 to 9 show the longitudinal profiles of the wadis which have a gradually lower gradient as they movie from west or south to the north. The figures show marked variations in wadis -floor slope over relatively short distances. However, the longitudinal profiles show the linear features in order to define the channel sections to the direction of flow. This concept will provide useful information concerning flood plains and possible over spilling. These information are important to model parameters of selected control sites.

Statistical rational method

The Statistical Rational Method was used to compute runoff coefficient. The Rational Method equation is:

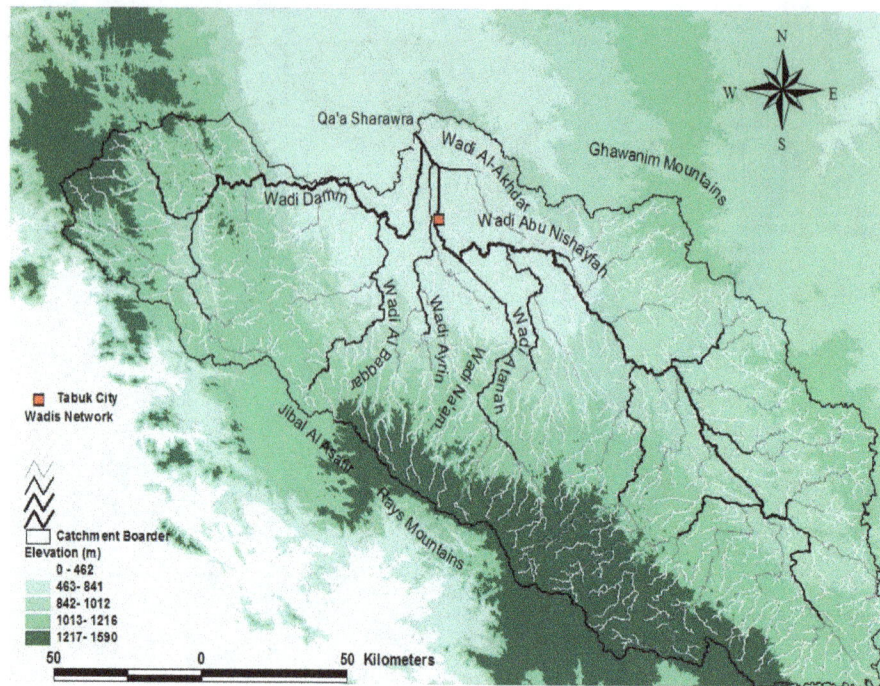

Figure 3: Altitude and catchment network of Tabuk.

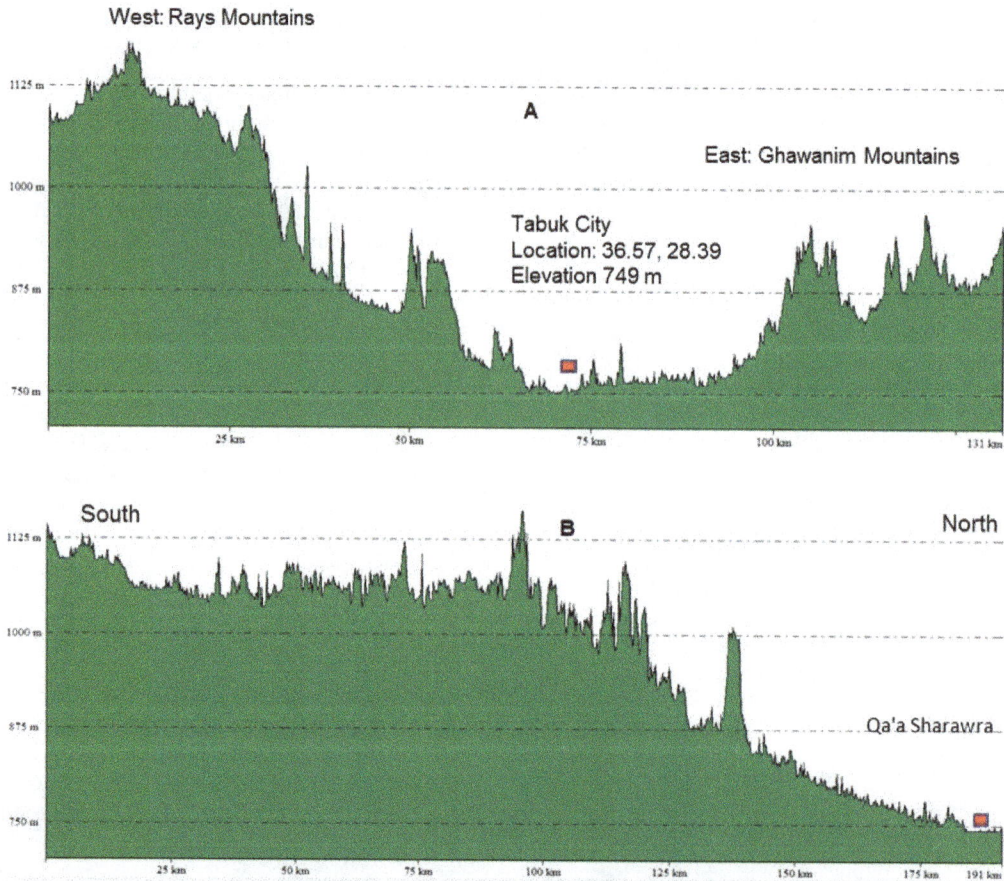

Figure 4: A: Cross section of Tabuk Catchment from West to East, B: longitudinal section from South to North.

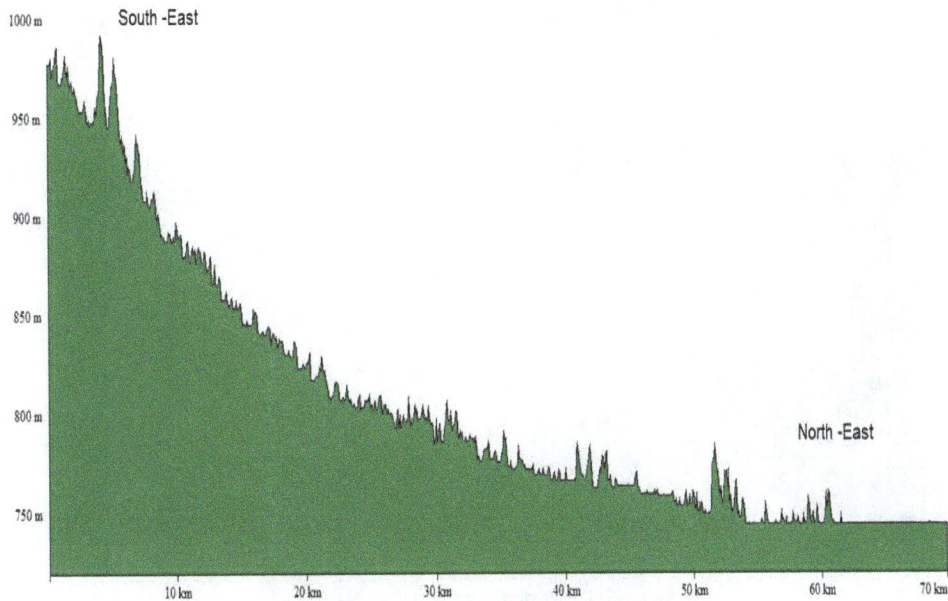

Figure 5: Longitudinal section of Wadi Al-Akhdar.

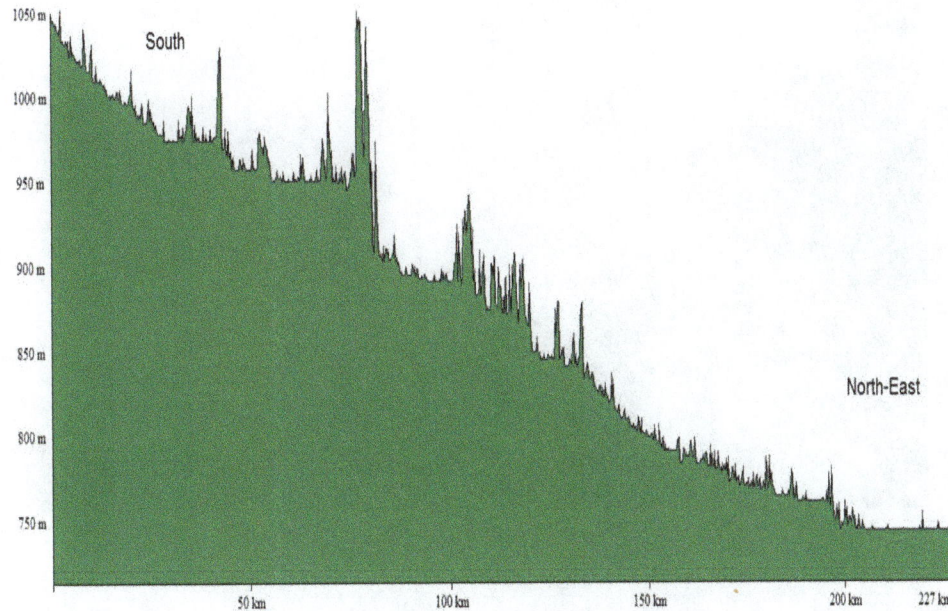

Figure 6: Longitudinal section of Wadi Abu Nishayfah.

Q=CIA　　　*Equation 1*

Where Q=Runoff in cubic feet per second (cfs), C=Runoff coefficient, I=Rainfall intensity (inches per hour) and A=Drainage area (acres). The main parameters that determine Runoff Coefficient (C) are: i) Land cover type, ii) Slope, iii) Soil type. Spatial analysis of the land cover types was successfully applied using ETM+ to classify land cover types for Tabuk Catchment. Because of available ground knowledge about land cover identity and distribution, supervised classification was successfully applied to estimate these types with high accuracy referring to ground truth data. In addition, the extracted information on land use types and their spatial distribution was used to select Runoff Coefficient (C). Finally, ArcGIS Model builder model was used to run the best location of the dams based on the following parameters (Figure 10):

i. Catchment Slope (less than 3%).

ii. Delineation network order more than or equal to 7

iii. Runoff Coefficient (0.4)

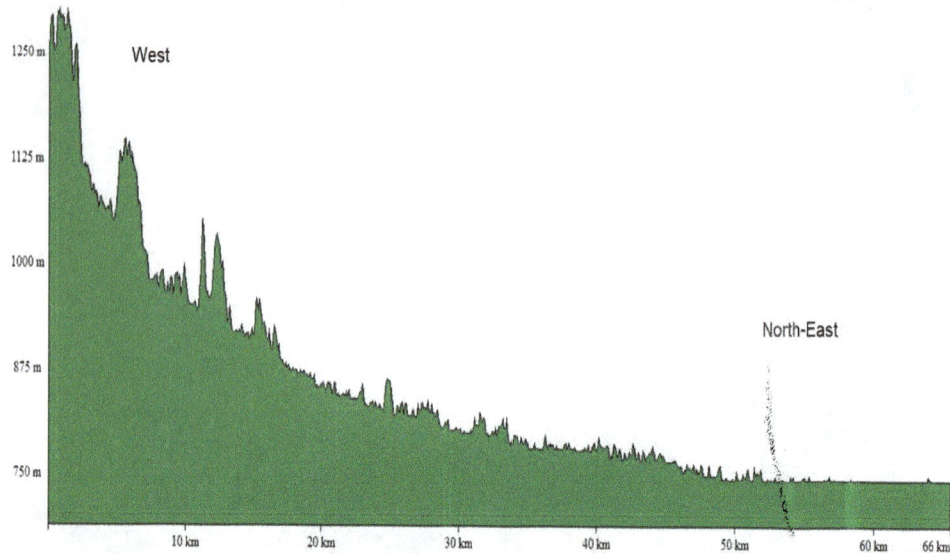

Figure 7: Longitudinal section of Wadi Na'am.

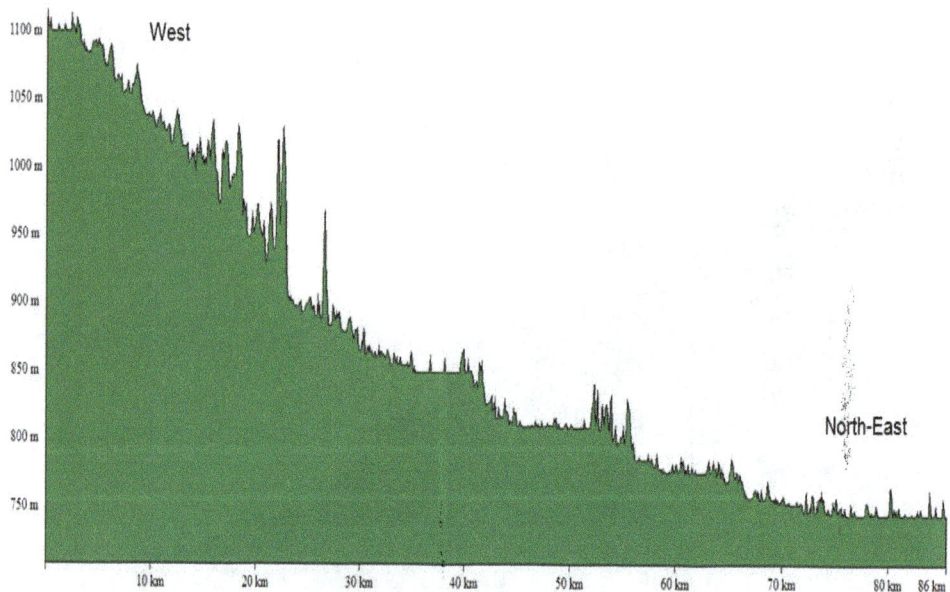

Figure 8: Longitudinal section of Wadi Al Baqqr.

Results

Land cover types are determined for Tabuk Catchment by applying supervised classification on Landsat TM images. The ETM+ data were processed using ERDAS software and further analysis of the land cover map was conducted with ArcGIS 10.0. A set of seven land use/land cover classes were chosen for Tabuk catchment. The land cover types map and distribution (Figures 11 and 12) show that the major land cover types include: i) Weathered Eroded Rocky Land, ii) Eroded Flooded Land, iii) Low Land. Because Tabuk area is a hyper-arid regime, land cover is characterised by fragmented landscapes with low availability of vegetation cover.

According to the land cover map and Rational Method empirical table [20], Runoff Coefficient (C) was calculated (Table 1).

Different geospatial data sets and analyses have been extracted and conducted from ASTER image including network delineation, slope map, side flow, sub-basin and wadis longitudinal and cross sections. Based on the sub-basins areas from ASTER analysis around 95% of water contributing flood events in Tabuk City runs from outside (Figure 13). In addition, the sub-basins within area greater than 130 km² are mainly in the higher altitudes located 50-220 km from the city.

Basically, ASTER and ETM+ data sets are the main inputs into the ArcGIS Model builder. To find suitable dam location, the constructed

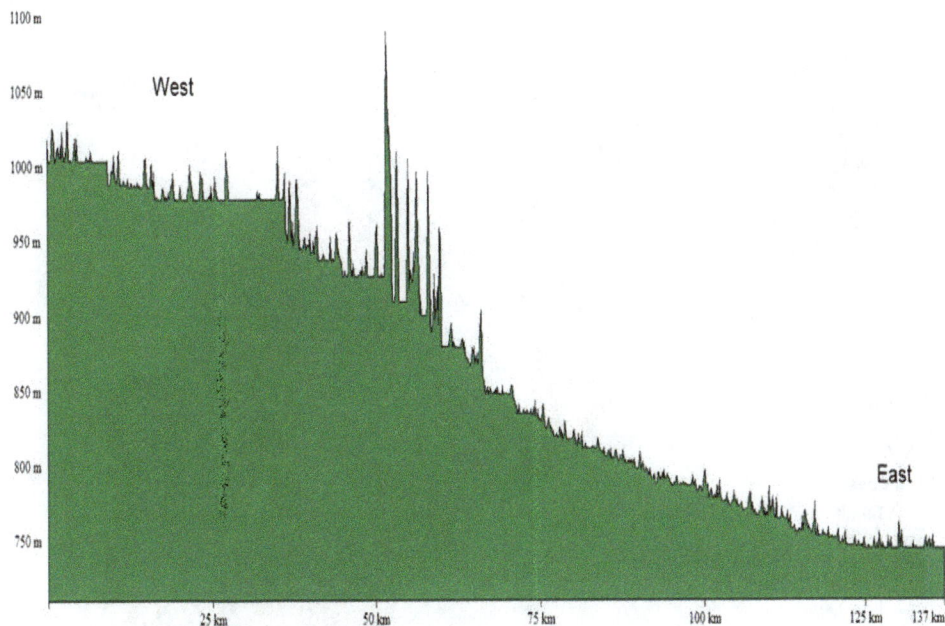

Figure 9: Longitudinal section of Wadi Damm.

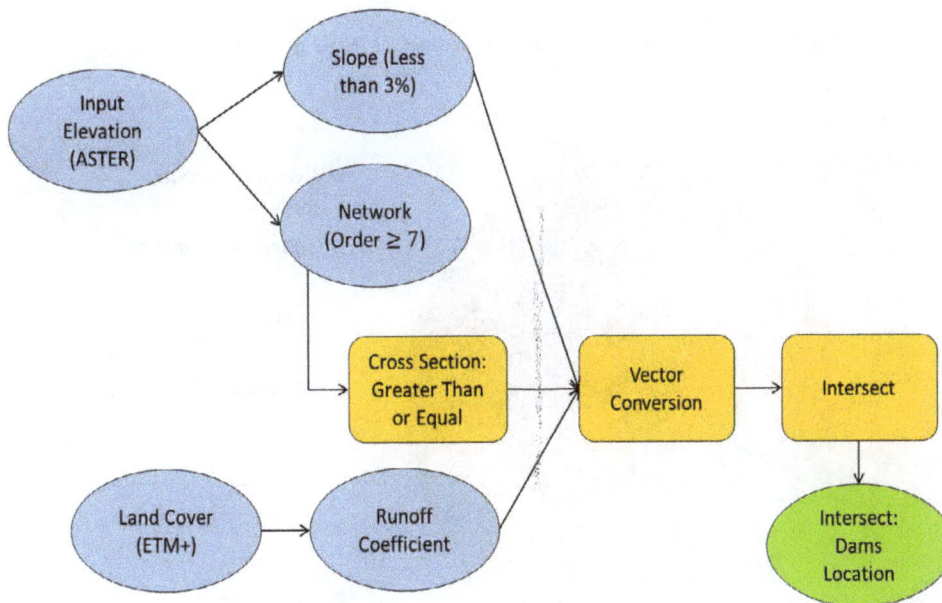

Figure 10: Dam Location Simplified Model Flowchart.

model in the ArcGIS Model builder used different parameters: slope less than 3%, delineation network order more than or equal to 7 with cross section (Net Diameter) greater than or equal to 500 m, continuity of at least 20 km and 40 km distance between two dams proposal on each wadi. As a cross factor the model used Low Land class because it has the lowest Runoff Coefficient which is equal to 0.4 and connects the major wadis. According to the geomorphological analyses, six sites were selected to build dams (Figure 14). The locations of the dams are as follows:

i.   Wadi Damm in connection with Wadi Al Baqqr

ii.   Wadi Na'am

iii.   Wadi Atanah

iv.   Wadi Abu Nishayfah A

v.   Wadi Abu Nishayfah B

vi.   Qa'a Sharawra

Discussion

Digital elevation models (DEMs) as well as layers produced by

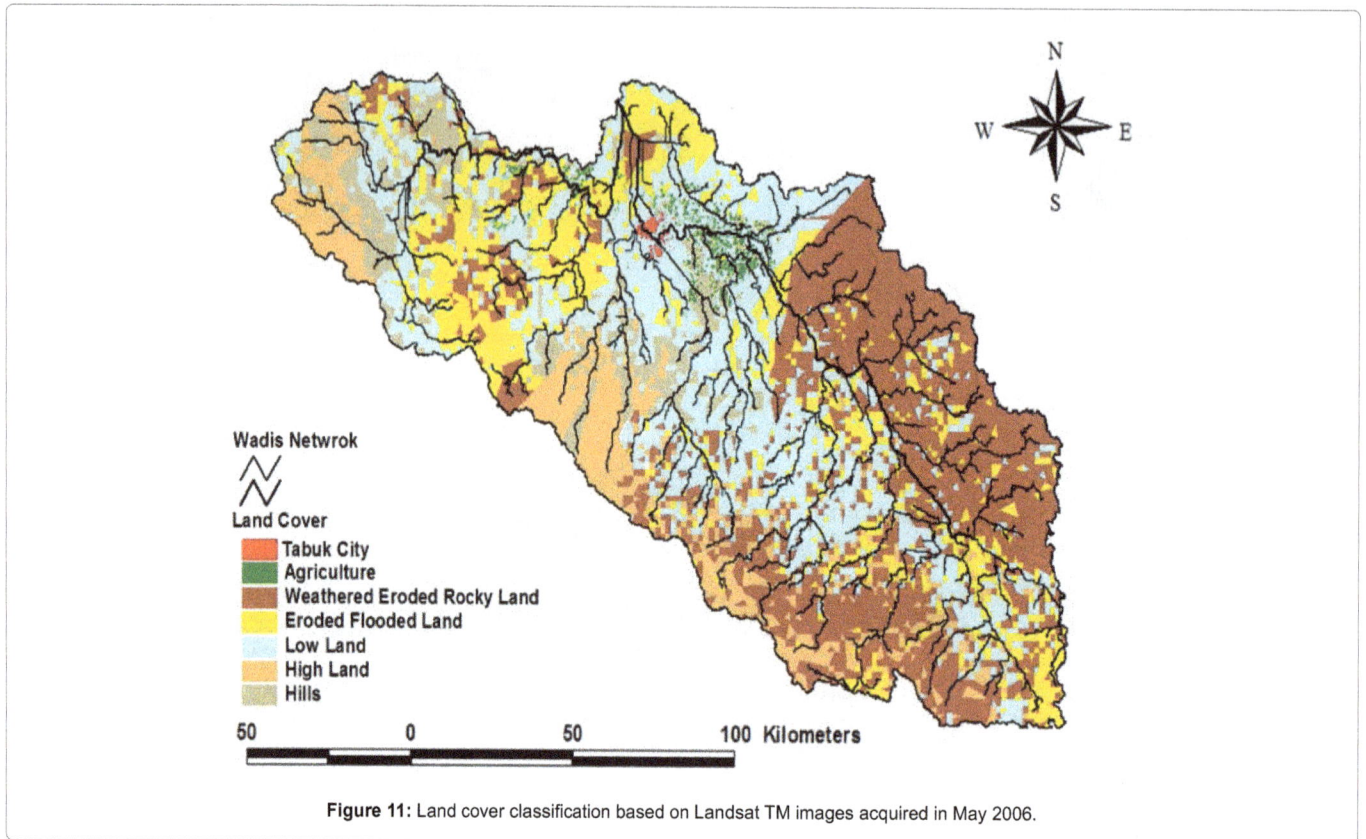

Figure 11: Land cover classification based on Landsat TM images acquired in May 2006.

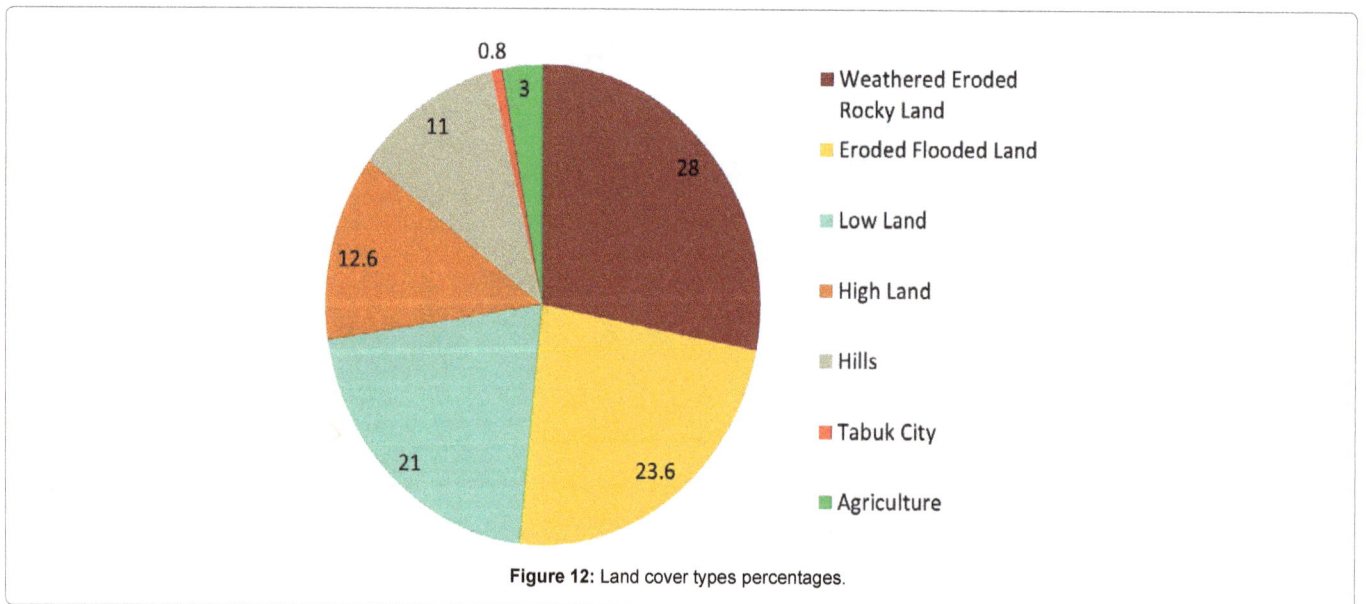

Figure 12: Land cover types percentages.

Class Name	C Value	A (Km²)
Weathered Eroded Rocky Land	0.7	5849.928
Eroded Flooded Land	0.6	4930.6536
Low Land	0.4	4387.446
Hills	0.7	2298.186
High Land	0.7	2632.4676
Tabuk City	0.9	167.1408
Agriculture	0.5	626.778

Table 1: Runoff Coefficient (C) values.

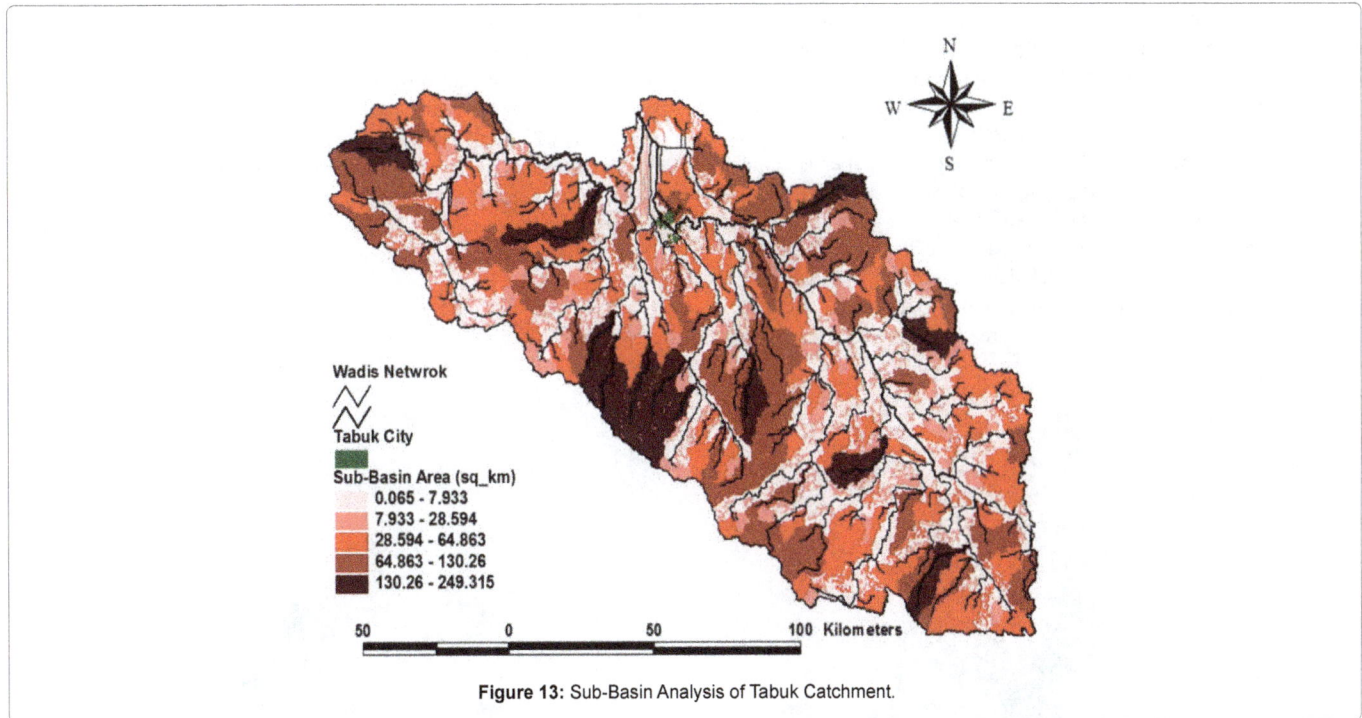

Figure 13: Sub-Basin Analysis of Tabuk Catchment.

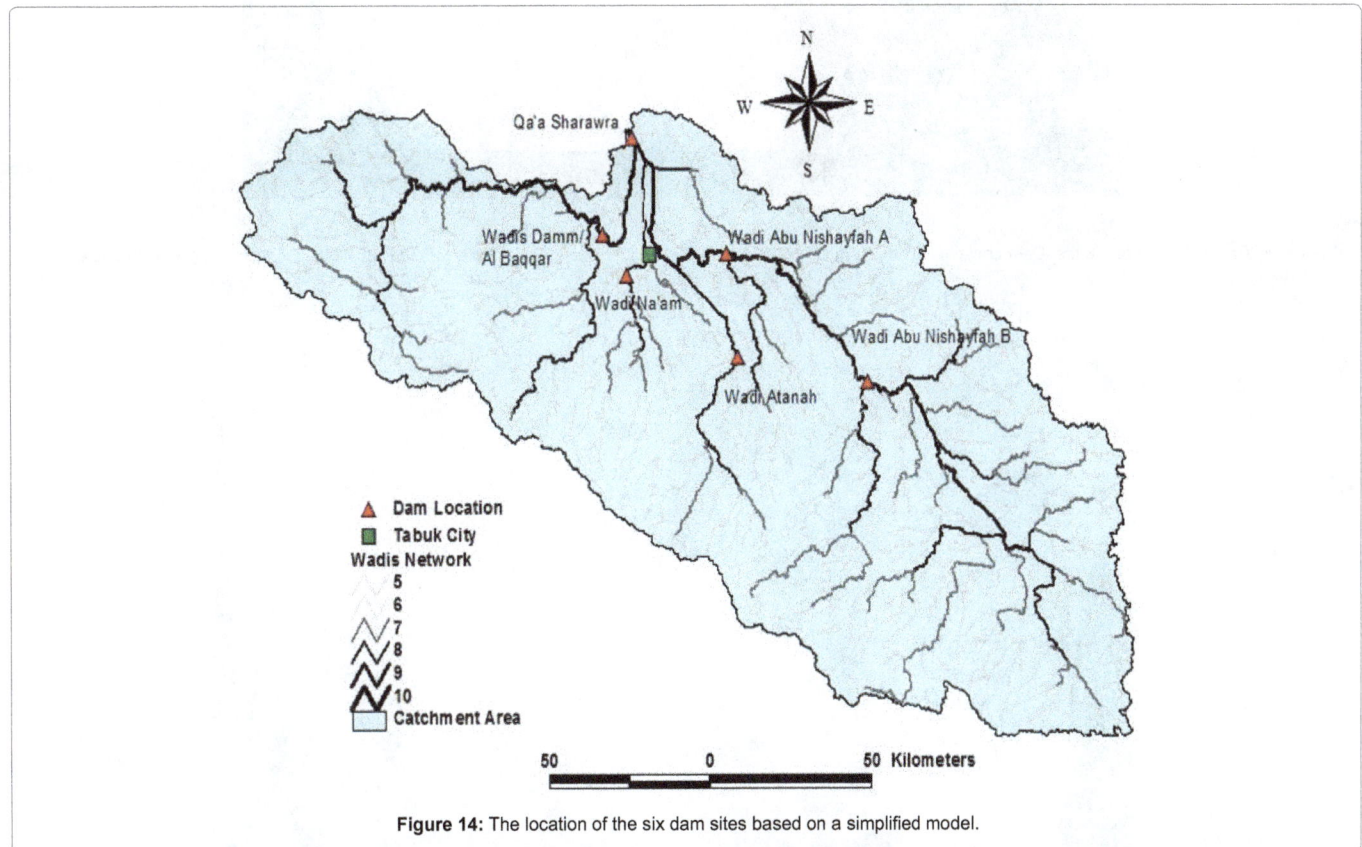

Figure 14: The location of the six dam sites based on a simplified model.

RS and GIS provide a solid geospatial data foundation, which is suitable for addressing questions on choosing the best sites of dams. To accomplish this, there is a need to understand geospatial processes of gathering, organising and analysing data using RS and GIS. On the other hand, theoretical and practical backgrounds are needed to understand linkages, relationships and thresholds that allow faster identification of the study area. DEMs are essential for topographic characterisation by representing land surface, hydrological boundaries and terrain attributes, such as slope and aspect. In addition, DEMs are useful to model the distribution of flood site potential and topographic moisture with a resolution not more than 30 m (cell size). Concerning the accuracy of DEMs research has indicated that 10-20 m resolution is required to depict accurately topographic features and other biophysical processes [21]. The scope of this paper is achieved by developing a method needed to address the complex morphological relationships for selecting suitable dam location in an arid catchment (Tabuk/Northwest). Software such as ArcGIS (ESRI) provides modelling tools to achieve spatial data visualisation in geographic space and time. The intersection function in ArcGIS is used to compute the common parameters in the study area including catchment slope,

delineation network and Runoff Coefficient, taking into account the continuity of the stream and the distance between two dams proposal on each wadi. The values of these parameters were logically chosen, e.g. slope less than 3% indicates the flattest area, wadi network order greater than 7 has all sub stream and feeding points clustered, Runoff Coefficient is less or equal to 4 meaning the water collection will be in the Low Land class. In addition, the length and the stream width play an important role in dam selection. The missing important parameter is the spatial distribution of soil type to achieve higher credibility of the selected sites.

In terms of dam type, embankment (earth) dams can be used

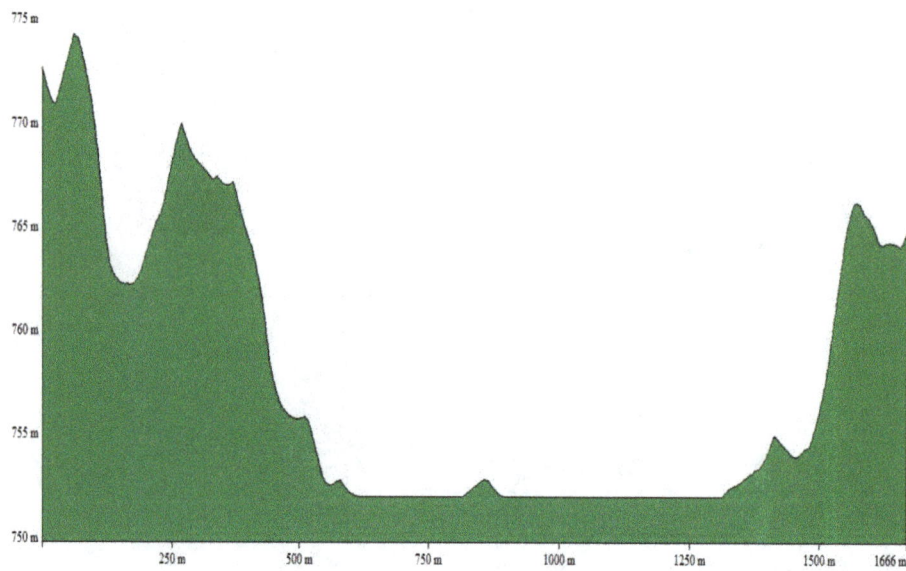

Figure 15: Cross Section of the first Dam Location: Wadi Damm in connection with Wadi Al Baqqr Coordinate (36.38E, 28.44N), Elevation 752 m above sea level, order No. 9.

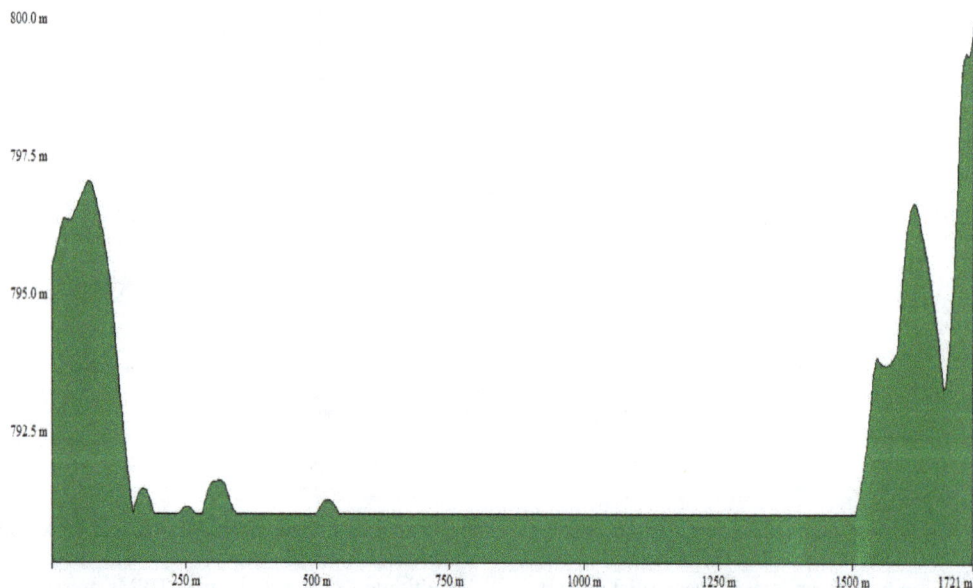

Figure 16: Cross Section of the first Dam Location: Wadi Abu Nishayfah A, Coordinate (36.86E, 28.28N), Elevation 790 m above sea level, order No. 9.

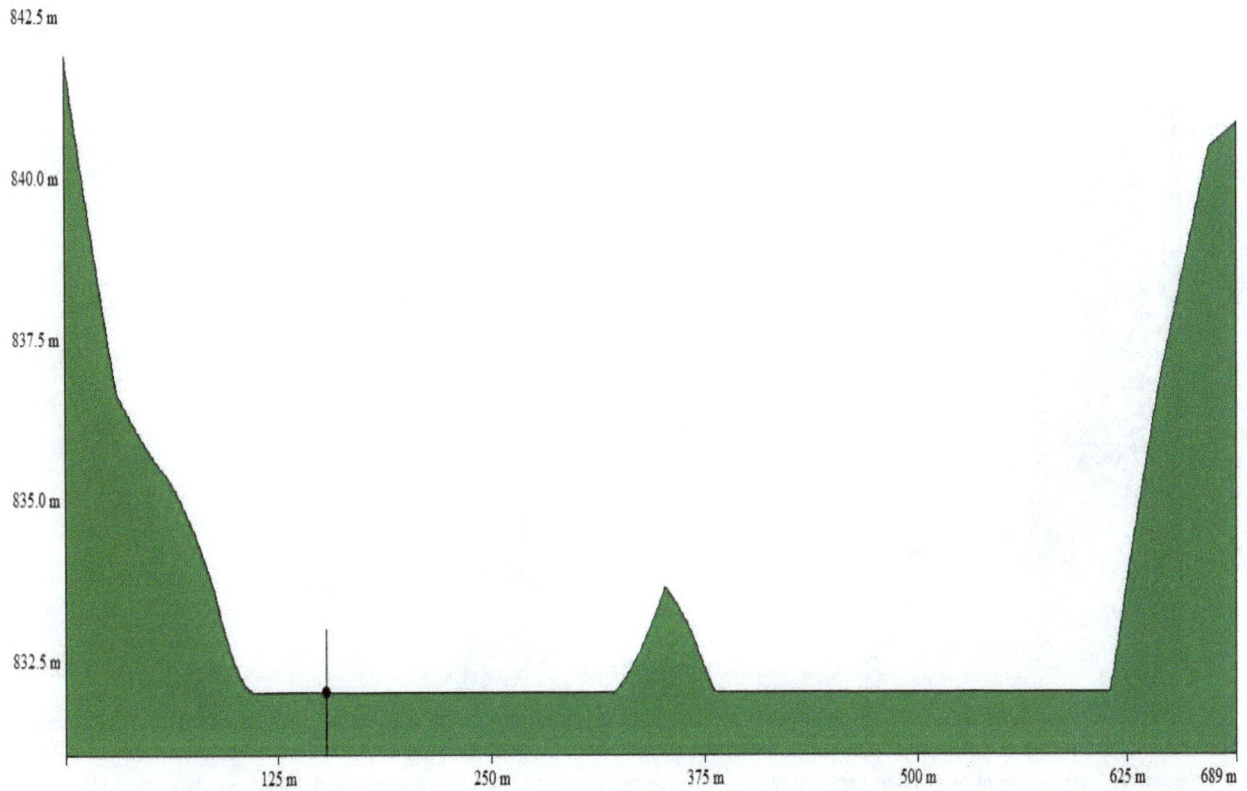

Figure 17: Cross Section of the first Dam Location: Wadi Atanah, Coordinate (36.75E, 28.17N), Elevation 832 m above sea level, order No. 8.

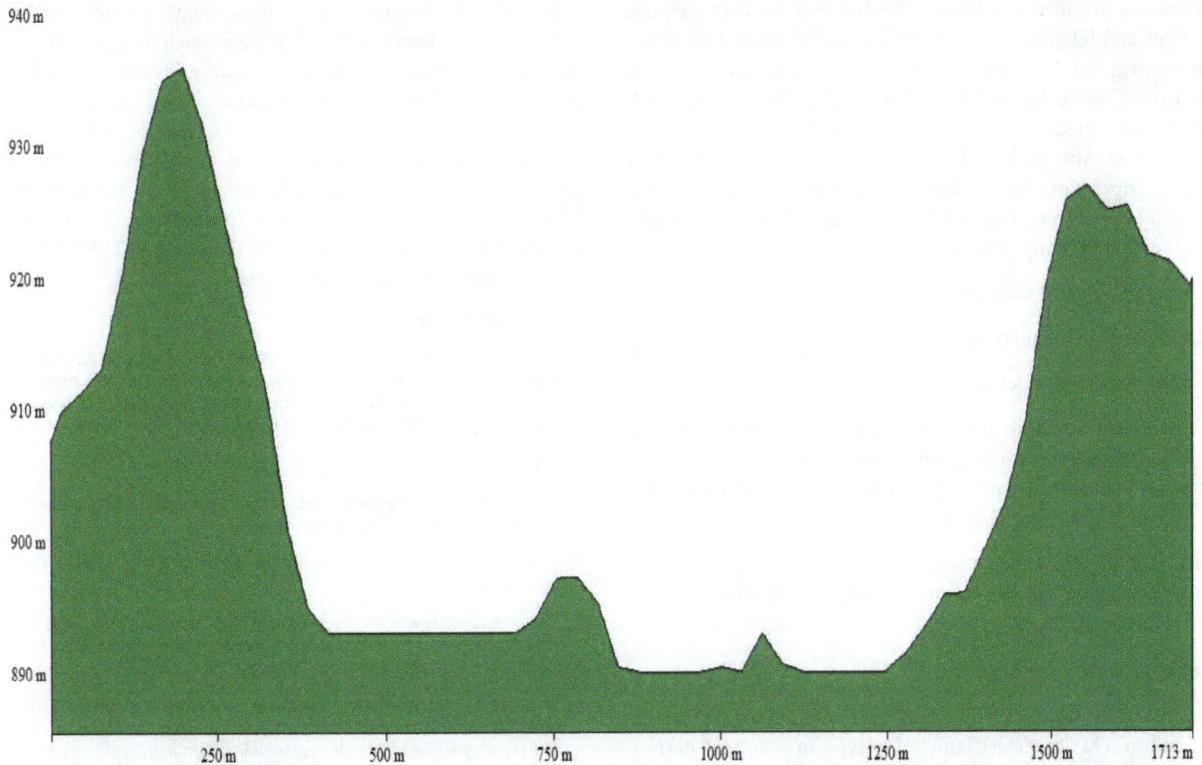

Figure 18: Cross Section of the first Dam Location: Wadi Abu Nishayfah B, Coordinate (37.22E, 28.08N), Elevation 790 m above sea level, order No. 9.

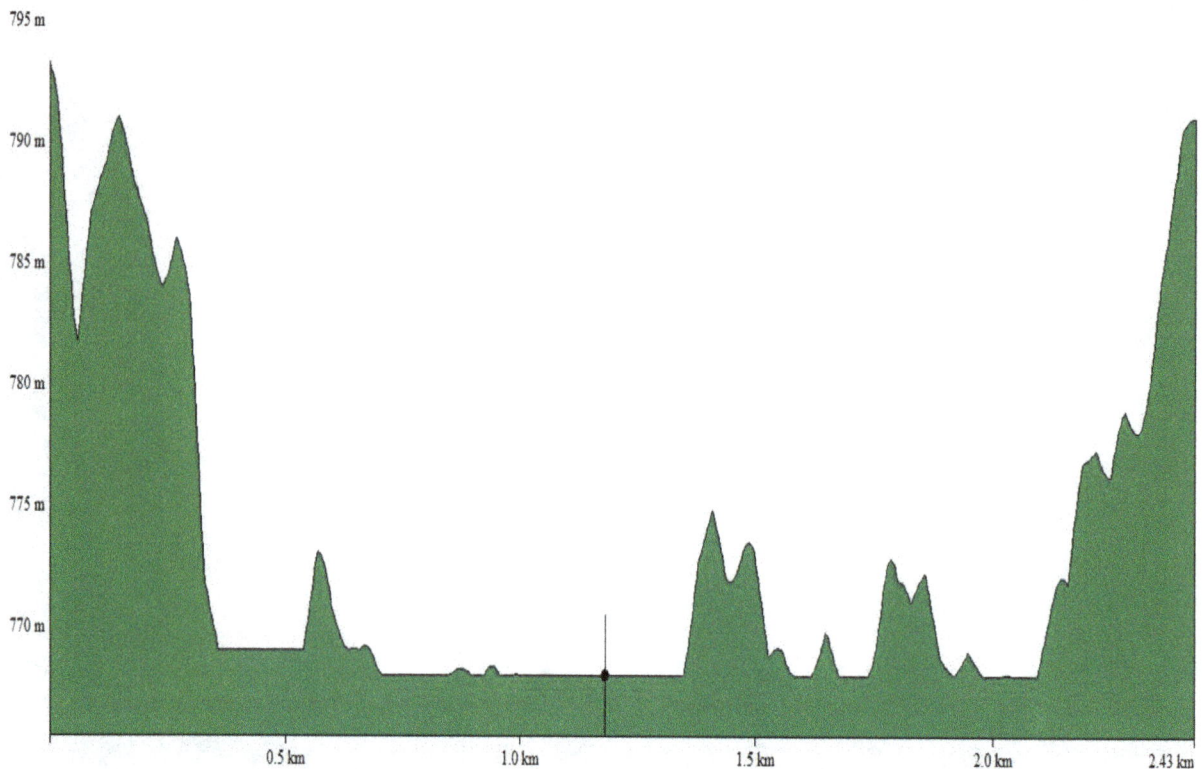

Figure 19: Cross Section of the first Dam Location: Wadi Na'am, Coordinate (36.44E, 28.33N), Elevation 768.5 m above sea level, order No. 8.

to protect the city of Tabuk from flash flood event, at the same time harvesting water and creating ground water recharge spots. Furthermore, the six sites which were selected to build dams require a range of cut and fill earthworks operations from low to high range of operations, e.g. the locations of Wadi Damm in connection with Wadi Al Baqqr (Figure 15) and Wadi Abu Nishayfah A (Figure 16), and Wadi Atanah (Figure 17) require low range operations, while the locations of Wadi Abu Nishayfah B (Figure 18) and Qa'a Sharawra need moderate operation. Finally, the location of Wadi Na'am requires full cut and fill operations (Figure 19). The cross section of the wadis are very important in many aspects:

i. Construction cost calculations

ii. Dams storage calculation

iii. Stability and risk assessment

There are many methods based on information technology used in suitable dam location such as image differencing, vegetation index differencing, selective principal components analysis, direct multi-date classification, post-classification analysis, etc.

Finally, it is useful to recognize that anthropogenic changes can be the most typical and serious factors of land cover change, through direct and indirect effects.

Conclusion

Spatial analysis was applied using Landsat Thematic Mapper images (Landsat TM) to detect land cover type in a certain period. Land cover types have been classified successfully using supervised classification referring to ground truth data. DEM information (slope,

network delineation, boundaries) was extracted from ASTER images. Analysis of these datasets using different software packages accurately reflects the development of the technology and software in order to construct a net between modelling approach and software. Finally, all information extracted from satellite imageries is imported into ArcGIS (Model Builder), and spatial analysis is carried out to identify the best dam location. According to the intersection of different parameters and analyses, six sites were selected to build earth dams. All sites require earthworks operations in the range from low to high range, which accordingly affect the cost of the operation. The achieved results demonstrate the great potential of ETM+ and ASTER satellite imagery for hydrology applications at acceptable scales.

Acknowledgement

The authors would like to acknowledge financial support for this work from the Deanship of Scientific Research (DSR), University of Tabuk, Tabuk Saudi Arabia, under grant no. 2/052/1435, and ongoing project entitled "Topographic Study of Tabuk Wadis: Changes and Behavior During Flash Flood Events".

References

1. Hill R (1996) Irrigation and water supply, A History of Engineering in Classical and Medieval Times. Routledge 17-46.

2. International Commission on Large Dams (ICOLD) (2014) Dams general synthesis.

3. Jakeman J, Littlewood G, Whitehead G (1990) Computation of the instantaneous unit hydrograph and identifiable component flows with application to two small upland catchments. Journal of Hydrology 117: 275-300.

4. Valipour M, Banihabib M, Behbahani S (2013) Comparison of the ARMA, ARIMA, and the autoregressive artificial neural network models in forecasting the monthly inflow of Dez dam reservoir. Journal of Hydrology 476: 433-441.

5. Kite W, Pietroniro A (1996) Remote sensing applications in hydrological modelling. Hydrol Sci 41: 563-591.

6. Liu J, Chen JM, Cihlar J (2003) Mapping evapotranspiration based on remote sensing: An application to Canada's landmass. Water Resour Res 39: 1189.

7. Forzieri G, Gardenti M, Caparrini F, Castelli F (2008) A methodology for the pre-selection of suitable sites for surface and underground small dams in arid areas: A case study in the region of Kidal. Physics and Chemistry of the Earth 33: 74-85.

8. Abushandi E, Merkel B (2011) Rainfall estimation over the Wadi Dhuliel arid catchment, Jordan from GSMaP_MVK+. Hydrol. and Earth Syst Sci 8: 1665-1704.

9. Mahmoud S (2014) Investigation of rainfall-runoff modeling for Egypt by using remote sensing and GIS integration. CATENA 120: 111-121.

10. Singh P, Singh D, Litoria P (2009) Selection of suitable sites for water harvesting structures in Soankhad watershed, Punjab using remote sensing and geographical information system (RS&GIS) approach-A case study. Journal of the Indian Society of Remote Sensing 37: 21-35.

11. McCabe M, Kustas W, Kongoli C, Ershadi A, Hain C (2013) Global-scale estimation of land surface heat fluxes from space: current status, opportunities, and future directions, in: Petropoulos. Remote Sensing of Energy Fluxes and Soil Moisture Content, E-Publishing Inc., Pp 447-462.

12. Shafapour M, Pradhan B, Mansor S, Ahmad N (2015) Flood susceptibility assessment using GIS-based support vector machine model with different kernel types. Catena 125: 91-101.

13. Sanyal J, Lu X (2004) Application of remote sensing in flood management with special reference to monsoon Asia: A Review. Nat. Hazard 33: 283-301.

14. Salih SA, Al-Tarif ASM (2012) Using of GIS spatial analyses to study the selected location for dam reservoir on Wadi Al-Jirnaf, West of Shirqat area. Iraq J Geogr Inf Syst 4: 117-127.

15. Ramakrishnan D, Bandyopadhyay A, Kusuma KN (2009) SCS-CN and GIS-based approach for identifying potential water harvesting sites in the Kali Watershed, Mahi River Basin. India J Earth Syst Sci 118: 355-368.

16. Clark J (1998) Putting water in its place: A perspective on GIS in hydrology and water. Hydrol Process 12: 823-834.

17. Kumar M, Agarwal A, Bali R (2009) Delineation of potential sites for water harvesting structures using remote sensing and GIS. Journal of the Indian Society of Remote Sensing 36: 323-334.

18. Youssef A, Pradhan B, Sefry S (2014) Remote sensing-based studies coupled with field data reveal urgent solutions to avert the risk of flash floods in the Wadi Qus (east of Jeddah) Kingdom of Saudi Arabia. Natural Hazards.

19. Rahimi S, Gholami Sefidkouhi MA, Raeini-Sarjaz M, Valipour M (2014) Estimation of actual evapotranspiration by using MODIS images (a case study: Tajan catchment). Arch Agron Soil Sci.

20. David T (2006) The Rational Method. Civil Engineering Department, Texas Tech University, 1-7.

21. Deng Y, Wilsonj J, Bauer O (2007) DEM resolution dependencies of terrain attributes across a landscape. Int J Georg Info Sci 21: 187-213.

12

Farmers Participatory Evaluation of Chickpea Varieties in Mirab Badwacho and Damot Fullasa Districts of Southern Ethiopia

Yasin Goa[1]*, Demelash Bassa[1], Genene Gezahagn[1] and Mekasha Chichaybelew[2]

[1]*Areka Agricultural Research Center, SARI, Ethiopia*
[2]*Debrezeit Agricultural Research Center, EIAR, Ethiopia*

Abstract

Chickpea is one of the grain food legumes contributing an enormous amount of protein to the human diet in Southern Ethiopia. Though a lot of improved varieties were released by research centers farmers depend on low yield and local varieties. Hence, participatory variety selection is one of the methods used to evaluate varieties through involvement of users. Participatory Variety Selection (PVS) were conducted during 2015/2016 in Mirab Badwacho and Damot Fullasa districts of South region, Ethiopia to assess the performance of chickpea (*Cicer arietinum* L.) varieties and to evaluate farmers' selection criteria for chickpea. Six improved varieties with local check were laid out in a randomized complete block design with four replications. Significant variation among chickpea varieties were observed for most the agronomic traits collected except for number of pod per plant which was not significant. Concerning location, the majority of the traits were showed significant difference indicating dissimilarity in agro ecologies of the two districts. The study also revealed that in some cases the researchers' selection match with farmers' preferences. However, in general farmers have shown their own way of selecting a variety for their localities. These parameters include earliness, diseases and pest resistance, seed colour, branch number and length and seed size. Hence, including farmers' preferences in a variety selection process is a paramount important. Therefore, based on attentively measured parameters, farmers' favourites and the agro ecologies of the site the varieties Natoli, Dalota and Arerti are selected for the area. The varieties Habru and Ejere should also be given due consideration by farmers for its earlier maturity in the study area.

Keywords: Chickpea; Variety selection; Participatory; Farmers' preferences; Grain yield

Introduction

Chickpea (*Cicer arietinum* L.) is the third most important pulse crop with a total annual global production of 9.7 M tones from 11.5 Mha [1]. According to FAOSTAT [2], the global chickpea area was 12.0 million ha, production was 10.9 million metric tons and yield was 913 kg.ha⁻¹ in 2010. India is the largest chickpea producing country, with 68% of world chickpea production. The other major chickpea producing countries include Ethiopia, Australia, Pakistan, Turkey, Myanmar, Iran, Mexico, Canada, and USA. In Ethiopia, chickpea is mainly grown at an altitude of 1400-2300 m.a.s.l., where annual rainfall ranges between 700 and 2000 mm [3]. It is the major cool season food legumes ranked second next to Faba bean, which occupies about 239,747.51 hectares of land annually with estimated production of 4,586, 822.55 quintals. The national average seed yield is 19.13 qt/ha [4]. Two types of chickpea; Kabuli type is grown in temperate regions while the desi type chickpea is grown in the semi-arid tropics. Ethiopia has suitable agro-climatic conditions for the production of both types of chickpeas. Whereas, in the country Desi type chickpeas has traditionally grown for both home consumption and sale. Currently, the Kabuli types are just beginning to expand in Ethiopia as well as in southern region.

Chickpea seeds are eaten fresh as green vegetables, parched, fried, roasted, and boiled and it is valued for its nutritive seeds with high protein content, 25.3-28.9%, after dehulling [5]. Chickpea seed has 38-59% carbohydrate, 3% fiber, 4.8-5.5% oil, 3% ash, 0.2% calcium, and 0.3% phosphorus [5].

In Ethiopia smallholder farmers grow chickpea at the end of the main rainy season using residual soil moisture because of its ability to withstand drought stress. Through efficient use of the residual moisture, chickpea also allows farmers to harvest two crops in a growing season (cereal followed by chickpea), improving their food supply, and

secure an additional source of income. Similarly, in southern Nations, Nationalities, and Peoples Regional State (SNNPRS), chick pea is occupying about 5,662.23 hectares of land annually with estimated production of 93,892.80 quintals [4].

Despite its importance, the national (19.13 qt/ha) as well as regional average yields (16.58 qt/ha) of chickpea are low due to various production constraints including: Low yield potential of landraces, lack of superior varieties, their susceptibility to biotic and a biotic stresses and poor cultural practices are some the serious constraints in chickpea production in Ethiopia [6-8]. Chickpea varieties were released by the various national and regional research centers of the country. Farmers have no ample information about the released varieties because they were released with poor involvement of farmers and the released varieties had not yet tested in the study area. In the country, efforts have been made through PVS to develop and popularize improved varieties of some crops. Participatory approach is being carried out in many crops like bread wheat [9], common bean [10] and maize [11]. Danial et al. [12] reported that farmer's preferences vary with environmental conditions, traits of interest, ease of cultural practice, processing, use and marketability of the product, ceremonial and religious values. However, the farmers' selection criteria for improved

*Corresponding author: Yasin Goa, Areka Agricultural Research center, SARI, Ethiopia, E-mail: yasingoac76@yahoo.com chickpea varieties were not adequately assessed and well documented especially in the Southern parts of Ethiopia. Therefore, the objectives of this study were to evaluate the performance of the released chickpea varieties through PVS and to assess farmers' selection criteria for future chickpea improvement work with the participation of farmers in Southern Ethiopia [13].

Materials and Methods

The trial was conducted during 2015/2016 main cropping season at two districts (Mirab Badewacho and Damot Fullasa) in Southern Ethiopia. Geographically, Mirab Badewacho (Woybra kebele) is found at 37° 51' E Longitude and N 7° 08 Latitude and at 1899, whereas Damot Fullasa (Shanto Zuria kebele) is found at 37° 51' E Longitude and 07° 00' N Latitude and at an altitude of 1922 m above sea level. The chickpea varieties used in the trials at two districts were Habru, Arerti, Natoli, Ejere, Dalota and farmers' varieties. Six varieties including the local checks were tested in the study. The trials were planted by farmers in collaboration with researchers and the agricultural extension workers in RCBD with four replications. Each farmer acted as a replicate. Each experimental plot had an area of 10 × 10 m (100 m²). Each plot comprised of thirty three rows which were 10 m long. Spacing of 30 cm between rows and 10 cm between plants was used.

Farmers participated in the entire crop cultivation operations from sowing till harvesting. Technical advice on management practice of the improved chickpea varieties was given by the researchers. Weeding, spraying such as karate/malathion against insect pests (pod borer) was done as required. Trials were managed according to recommended agronomic practices. Farmers were participated in evaluation and selection of improved chickpea varieties at pod filling and maturity stage through organizing field days. Farmers, higher officials, researchers, and extension personnel were participated in field day. Such field days were beneficial for participating farmers in comparing the varieties when they are in the field and an opportunity for research and extension work to point out the differences. Conducting field day at the maturity and harvest time were more vital in evaluating the yield potential and other yield contributing attributes. Farmers set their selection criteria and ranking of varieties according to their setting criteria. The rank sum method each trait for each variety was used to rank varieties based on farmers' selection criteria. The value of each trait has equal weight. Data were collected on plant height, number of pod per plant, number of days to maturity, biomass yield and grain yield per hectare. The data was subjected to Analysis of Variance (ANOVA) using the Statistical Analysis System (SAS). Means were separated using the Least Significant Difference (LSD) at P=0.05.

Results and Discussions

The Participatory Variety Selection (PVS) ensured farmers to be participant in selection of improved chickpea crop varieties in comparison with local check based on their preference criteria.

Researchers' evaluation

Combined analyses of data from the two trial sites (Mirab Badwacho and Damot Fullasa) shown very highly significant varietal differences (P<0.01) in grain yield, days to 95% maturity and biomass yield (Table 1). Table 1 shows mean square-values of researchers' evaluation of agronomic trait for the varieties, locations, and error. Researchers evaluated the varieties based on yield and other agronomic traits. The varieties revealed highly significant to significant variation in all agronomic traits recorded except number of pod per plant. With regard to locations, most of the agronomic traits recorded shows statistically significant. This indicates that all the varieties responded not similarly to the tested locations.

Table 2 indicates researchers' evaluation of the average values of the different agronomic traits. The average data showed that chickpea varieties differs in days to maturity, biomass yield (kg/ha) and grain yield (kg/ha). However, the varieties did not vary in number of pod per plant.

The study revealed that chickpea varieties highly significantly differ in biomass yield (kg/ha) and grain yield (kg/ha) of plant (Table 2). The variety Natoli had the highest grain yield (1837.5 kg.ha⁻¹), while local check had the lowest grain yield (1075.5 kg.ha⁻¹). Variation in environment and in their genetic makeup could be the possible reason for the observed differences in chickpea varieties. These findings are in line with those of Biru et al. [7], who tested different improved chickpea varieties and reported that average grain yield over environments varied from 520-2010 kg.ha⁻¹. From the varieties evaluated, Ejere had maximum biomass yield (5425 kg/ha) next by varieties Arerti and Natoli with the biomass yield of 5337.5 kg/ha and 5118.8 kg/ha, respectively.

Mean values were differing significantly for plant height of test varieties and local check (Table 2). The highest plant was observed in Ejere (58.68 cm) followed by Dalota and Natoli with height of 58.3 cm and 57.97 cm respectively. The past research work reported by Biru et al. [7] regarding plant height in chickpea is agreement with the present investigation.

The data analysis of revealed significant variations among the evaluated varieties of chickpea for pod per plant (Table 2). The variety Arerti (86.33) had larger pod per plant, while local check (65.62) had smaller pod per plant. This is in line with Biru et al. [7] who stated that average pod per plant among four chickpea varieties varied from 37.5 to 83.6. According to the mean values Habru was earliest for days to maturity (102.75 days) followed by Ejere and local check with the values of 110.5 and 114.63 days, respectively. These results are similar to Biru et al. [7,8], who also suggested significant genetic differences for this trait among chickpea genotypes.

Farmers chickpea variety evaluation

Representative farmers to the area and having long experience in farming of chickpea were selected to participate on trial evaluation. Before beginning of the selection process, selected farmers from the two villages were asked to set their priority selection criteria. Accordingly, earliness, Disease and pest resistance, Seed colour, Branch number and length, Seed size Ground cover, Emergence, Pod number, Seed number and yield were identified as the most important farmers' preference criteria.

Scale of 1-5 were used for ranking of varieties, whereby 1 being very good and 5 being very poor. Table 3 showed the Damot Fullas districts farmers' selection of the varieties based on the criteria they fixed. Shanto Zuria village farmers (Table 3) varietal assessment showed that variety Arerti was ranked highest (1.4) followed by Ejere and Dalota with the values of 1.7 and 1.8. Accordingly, chickpea varieties such as Arerti, Ejere, Dalota, Habru, Natoli, and local were preferred at Damot Fullasa site as 1st, 2nd, 3rd, 4th, 5th and 6th respectively. Their grain yields were 2200 kg/ha, 2175 kg/ha, 1900 kg/ha, 1875 kg/ha, 1852.5 kg/ha and 1325 kg/ha of Natoli, Dalota, Arerti, Ejere, Habru and local check, respectively (Table 2). It can be said that improved chickpea varieties evaluated in this site were superior to local check and best adapted to specific environment and similar agro ecology provided that other

Source of variation	DF	Mean square				
		DM	PH (cm)	PP	BY (kg/ha)	GY (kg/ha)
Replication	3	19.81ns	195.12*	2547.82**	14158524.31**	1289863.19**
Varieties	5	279.90**	295.77*	572.79ns	6039302.08**	617327.08**
Location	1	1752.08**	3512.34**	26245.45**	34256302.08**	5076502.08**
Loc' variety	5	69.433ns	21.67ns	890.08ns	1339302.08ns	16277.08ns
Rep (loc)	3	141.25ns	133.75ns	2058.31**	12966857.64**	1215002.08**
Error	30	19.98	50.74	174.94	804691	69940.97
CV (%)		3.94	12.92	17.05	19.45	16.92
LSD (0.05)		4.56	7.27	13.51	916.01	270.1

Where, **=Highly significant at P ≤ 0.01; *=Significant at P ≤ 0.05; ns=Not significant at P=0.05; DM=Days to maturity; PH=Plant height; PP=Pod per plant; BY=Biomass yield;
GY=Grain yield (kg/ha)

Table 1: Yield and yield components of chickpea varieties mean square values of combined over location.

Varieties	DM	PH(cm)	PP	BY(kg/ha)	GY(kg/ha)	MGY(kg/ha)	DFGY(kg/ha)
Habru	102.75^c	57.400^a	78.9^{abc}	3812.5^{bc}	1488.8^c	1125.00^b	1852.5^{ab}
Arerti	118.0^a	55.425^a	86.338^a	5337.5^a	1600.0^{abc}	1300.00^{ab}	1900.0^a
Natoli	118.1^a	57.975^a	85.400^a	5118.8^a	1837.5^a	1475.00^a	2200.0^a
Ejere	110.5^b	58.688^a	69.2^{bc}	5425.0^a	1556.3^{bc}	1237.50^b	1875.0^a
Dalota	116.3^a	58.325^a	79.9^{ab}	4662.5^{ab}	1818.8^{ab}	1462.50^a	2175.0^a
Local	114.6^{ab}	42.938^b	65.6^c	3312.5^c	1075.0^d	825.00^c	1325.0^b
CV(%)	3.94	12.92	17.05	19.45	16.92	9.504258	18.6
LSD(0.05)	4.56	7.27	13.51	916.01	270.1	177.3	431.9

*Means with the same letter are not significantly different; DM=Days to maturity; PH=Plant height; PP=Pod per plant; BY=Biomass yield; GY=Grain yield (kg/ha) of combined;
MGY=Seed yield of Mirab Badwacho; DFGY=Grain yield of damot Fullasa

Table 2: Average values of yield and yield related attributes of chickpea varieties across location and individual districts.

Preference Criteria	Varieties					
	Habru	Arerti	Natoli	Dalota	Ejere	Local
Earliness	1	3	5	3	2	2
Disease and pest resistance	5	1	2	1	3	4
Seed colour	1	1	3	3	1	4
Branch number and length	2	1	2	2	2	5
Seed size	1	1	2	1	1	4
Grain yield	3	2	2	2	2	5
Ground cover	2	1	1	1	1	3
Emergence	2	1	1	1	2	3
Pod number	1	1	2	2	2	5
Seed number	2	1	2	2	2	2
Overall score	21	14	22	18	18	37
Average score	2.1	1.4	2.2	1.8	1.8	3.7
Rank	4	1	5	3	2	6

NB: "1" means very good and "5" means very poor

Table 3: Shanto Zuria village farmers' variety preference result in Damot Fullassa district of Southern Ethiopia in 2015/2016.

Preference Criteria	Varieties					
	Habru	Arerti	Natoli	Dalota	Ejere	Local
Earliness	1	5	4	3	2	2
Disease and pest resistance	5	1	1	1	3	4
Seed colour	3	3	1	2	3	4
Branch number and length	4	1	2	2	3	5
Seed size	1	1	1	1	1	4
Grain yield	4	3	1	2	3	5
Ground cover	2	1	1	1	1	3
Pod number	1	1	1	2	3	5
Seed number	2	1	1	1	2	2
Overall score	23	17	13	15	21	34
Average score	2.6	1.9	1.4	1.7	2.3	3.8
Rank	5	3	1	2	4	6

NB: "1" means very good and "5" means very poor

Table 4: Wobara village Farmers' variety preference result in Mirab Badewacho district of Southern Ethiopia in 2015/2016.

factors kept constant. Similarly, Woybara village farmers' preferred varieties are Natoli, Dalota and Arerti with the mean values of 1.4, 1.7 and 1.9 respectively (Table 4). In the same talken, the yield of Natoli, Dalota, Arerti, Ejere, Habru, and local were 1475 kg/ha, 1462.5 kg/ha, 1300 kg/ha, 1237.50 kg/ha, 1125.00 kg/ha and 825 kg/ha in Mihrab Badiwacho woreda of wobera kebele (Table 2). In view of that, chickpea varieties such as Natoli, Dalota, Arerti, Ejere, Habru, and local check were preferred at Mirab Badewacho site (woybra kebele) as 1st, 2nd, 3rd, 4th, 5th and 6th respectively.

According to CSA 2014/2015, area in hectares, production in Quintals and yield per hectare of chickpea was 239, 755.25, 4, 586, 822.55, and 19.13 during Meher season in Ethiopia and distribution of area (hectare) under chickpea production in country was shown on Figure 1. Similarly, area in hectares, production in Quintals and yield per hectare of chickpea was 5,662.23, 93, 892.80, and 16.58 during Meher (the main season) in SNNPRE [4]. Photos taken during field day farmers PVS demonstration plot evaluation (Figure 2).

Table 5 presented average value of the two districts farmers' preferences for the studied varieties. Accordingly, the two districts farmers' interest of traits combined result indicated that varieties Arerti, Dalota and Natoli are the three best varieties with the average values of 1.65, 1.75 and 1.8, respectively. Researchers and farmers rank comparison are given below (Table 6). Table 6 below revealed that some of farmers' rank did match with researchers rank except for the varieties Natoli and Arerti which was ranked 3rd and 1st by farmers, 1st and 3rd by researchers. The present study confirms farmers' perception about crop

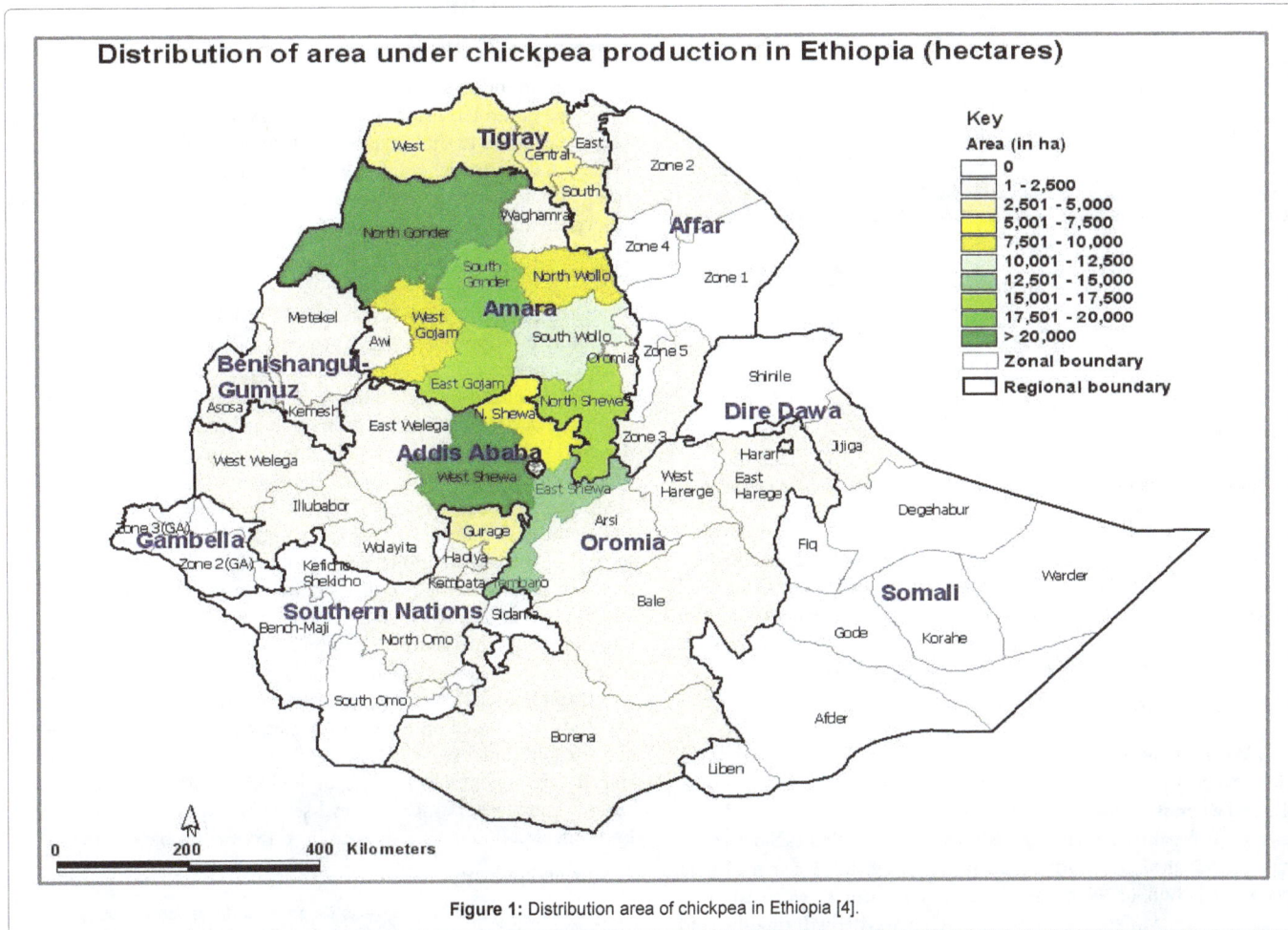

Figure 1: Distribution area of chickpea in Ethiopia [4].

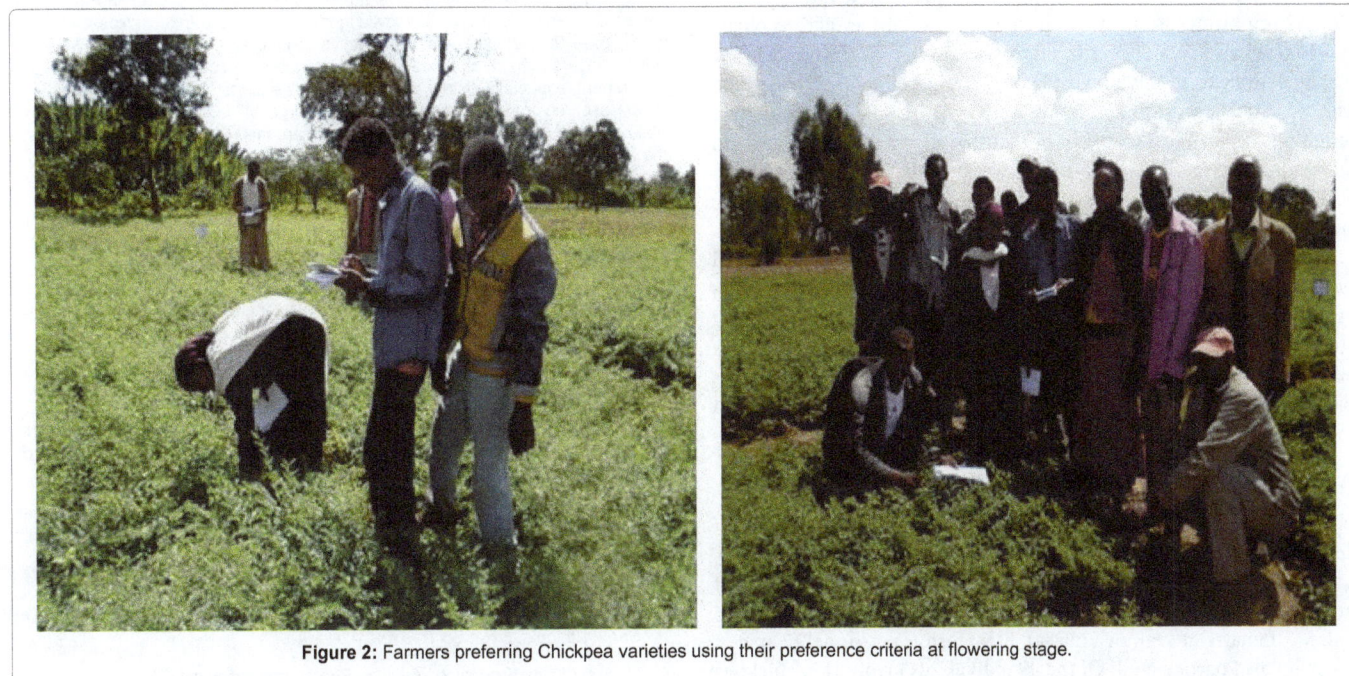

Figure 2: Farmers preferring Chickpea varieties using their preference criteria at flowering stage.

Varieties	Damot Fullasa	Mirab Badwacho	Average	Rank
Habru	2.1	2.6	2.35	5
Arerti	1.4	1.9	1.65	1
Natoli	2.2	1.4	1.8	3
Dalota	1.8	1.7	1.75	2
Ejere	1.8	2.3	2.05	4
Local	3.7	3.8	3.75	6

Table 5: Farmers average varietal assessment result in two target districts of Southern Ethiopia (Mirab Badwacho and Damot Fullasa) in Meher season of 2015/2016.

Varities	Researchers rank	Farmers rank
Habru	5	5
Arerti	3	1
Natoli	1	3
Dalota	2	2
Ejere	4	4
Local	6	6

Table 6: Chickpea varieties ranking according to farmers and researchers.

varieties are not similar to researchers and if given the chance, farmers are able to express their preferences differently. Though the researchers rank did match with farmers for varieties Habru and Ejere farmers had preferred them as of their early maturity. This is in agreement with Biru et al. [7,8] who stated that there were growing interests among farmers in the use of early chickpea varieties in short rain fall season.

Farmers' field days

PVS trials by involving 5 released varieties were involved along with farmer' variety as a check, were carried out in Mirab Badewacho (4), and Damot Fullasa (4), and 8 farmers. In total, 2 farmers' field days were organized in target districts of Mirab Badwacho (33; 31 males, 2 females) and Damot Fullasa (62; 56 males, 6 females) with the participation of 99 farmers. Participant farmers were asked to select preferred varieties along with preference criteria during the field days. The choice of parameters that farmers prefer in a variety differs widely from farmer to farmer and area to area. Besides agro ecology, social issues also influence a farmer's selection of crop varieties. The overall analysis from this activity enabled the selection of high yielding released chick pea varieties in respective district and facilitated in designing for further demonstration and scale up strategy. The field day feedback target village farmers indicated that old local varieties were used by farmers because of the absence of alternatives. These varieties are not only low yielding but are highly susceptible to insect-pest (pod borer attacks). During the discussions, it was suggested that owing to consecutive years of drought and current Elino event, farmers are looking for early maturing crop varieties which the existing agricultural office and other public agencies are unable to provide. In general farmers' favorite criteria also delivered constructive comments to researchers and development personnel involved in chickpea improvement to devise the research strategy.

Training

Training on available technologies and socio-economic aspects of chickpea in Ethiopia and southern regions was provided to 49 farmers. A training program was prepared to improve the attentiveness of farmers on chickpea available technology in which 41 farmers (36 males; 5 females at Damot Fullasa), 8 extension personnel (7 males, 1 female) also participated. Of the 49 individuals involved in chickpea training, 12% were women. An information manual was prepared on improved chickpea technologies in Damot Fullasa district in Amharic for development agents.

Awareness activities

Creation of awareness activities were conducted through FM radio and South television. PVS village training, demonstrations, and farmer field days to share experience were used in awareness creation. In Damot Fullasa, higher officials were involved in awareness creation. Events of all the field days were transmitted on public media (South Television and FM Radio) in Amharic.

Conclusions

Participatory Varietal Selection (PVS) on chickpea indicated variability of improved varieties preferences among farmers as well as from district to districts. Continuous evaluation of diverse chickpea varieties to substitute local varieties might accelerate the adoption of improved varieties and at the same time maintain genetic diversity of the chickpea. Farmers may require multiple traits from one key crop such as chickpea. However, researchers may not know the traits that are important to farmers and vice versa. Participatory varietal selection has significant role in technology adaptation and dissemination in short time than conventional approach. In these investigation farmers' selection criteria in the two districts were earliness, earliness, Disease and pest resistance, Seed colour, Branch number and length, Seed size Ground cover, Emergence, Pod number, Seed number and grain yield. Based on the criteria they set, their preferred varieties were Natoli, Arerti and Dalota. Researchers also recommend these three varieties for the study area based on the data analysis, agro ecologically suitability and the additional two early maturing chickpea varieties Ejere and Habru were preferred by farmers for double cropping and short rainfall season. Therefore, farmers' varietal selection criteria should be taken into consideration during chickpea improvement programme.

Acknowledgement

We are grateful to TL-III projects for providing funds for this study. We are very grate to farmers of the Woybara and Shanto Zuria village farmers who made this participatory variety evaluation study possible. Debrezeit Agricultural Research Centre is highly acknowledged for providing the chickpea varieties seeds used for the experiment and active collaborator of our center. We would like to sincerely thank all those who contributed to successful completion of this experiment, especially district agricultural office experts and development agents, Crop Research process owner and Mr Tadel Hirgo, W/rt Bogalech Utta, Mr Deneke Makae, Waza Morgito and Filimon Uliso who are the key role player in conducting the PVS activity and served as technical assistant.

References

1. http://faostat.fao.org/

2. http://faostat.fao.org/site/567/DesktopDefault.aspx

3. Anbessa Y, Bejiga G (2002) Evaluation of Ethiopian chickpea landraces for tolerance to drought. Genetic Resources and Crop Evolution 49: 557-564.

4. CSA (Central Statistical Agency) (2014/15) Agricultural Sample Survey Report on Area and Production of Crops Private Peasant Holdings, Meher Season.

5. Hulse JH (1991) Nature, composition, and utilization of grain legumes, In: Uses of tropical Legumes: Proceedings of a Consultants' Meeting, 27-30 March 1989, ICRISAT Center, Patancheru, India p: 11-27.

6. Dadi L, Regassa S, Fikre A, Mitiku D, Gaur PM, et al. (2005) Adoption Studies on Improved Chickpea Varieties in Ethiopia. Ethiopian Agricultural Research Organization, Addis Ababa, Ethiopia.

7. Alemu B, Abera D, Adugna A, Terefe M (2014) Adaptation Study of Improved Kabuli Chickpea (Cicer Arietinum L.) Varieties at Kellem Wollega Zone, Haro Sabu, Ethiopia. J Nat Sci Res 4: 21-24.

8. Goa Y (2014) Evaluation of Chick Pea (Cicerarietinum L.) Varieties for Yield Performance and Adaptability to Southern Ethiopia. J Biol Agric Healthc 4: 34-38.

9. Demelash A, Desalegn T, Alemayehu G (2013) Participatory Varietal Selection of Bread Wheat (*Triticum aestivum* L.) Genotypes at Marwold Kebele, Womberma Woreda, West Gojam, Ethiopia. Int J Agron Plant Prod 4: 3543-3550.

10. Gurmu F (2013) Assessment of Farmers' Criteria for Common Bean Variety Selection: The case of Umbullo Watershed in Sidama Zone of the Southern Region of Ethiopia. Ethiopian E-Journal for Research and Innovation Foresight 5: 4-13.

11. Tadesse D, Medhin ZG, Ayalew A (2014) Participatory on Farm Evaluation of Improved Maize Varieties in Chilga District of North Western Ethiopia. Int J Agric Forest 2014 4: 402-407.

12. Danial D, Parlevliet J, Almekinders C, Thiele G (2007) Farmers participation and breeding for durable disease resistance in the Andean region. Euphytica 153: 385-396.

13. Yadaw RB, Khatiwada SP, Chandhary B, Adhikari NP, Baniya B, et al. (2006) Participatory varietal selection (PVS) of rice varieties for rainfed conditions. Rice Fact Sheet, International Rice Research Institute (IRRI).

13

Preliminary Performance Evaluation of the Gold Nanoparticle Method for Quantification of Residual Poly-(Diallyldimethyl Ammonium Chloride) in Treated Waters in the Umgeni Water Catchment, Kwazulu-Natal (South Africa)

Manickum T[1]*, John W[1], Toolsee N[2] and Rachi Rajagopaul[1]

[1]*Scientific Services Laboratories: Chemical Sciences, Engineering and Scientific Services Division, Umgeni Water, 310 Burger Street, Pietermaritzburg 3201, KwaZulu-Natal, South Africa*
[2]*Wiggins Process Evaluation Facility, Wiggins Water Works, 251 Wiggins Road, Mayville, Engineering & Scientific Services Division, Umgeni Water, KwaZulu-Natal, South Africa*

Abstract

A "real world" study to assess the performance characteristics (precision, accuracy) of the citrate-capped, gold nanoparticle, Ultraviolet-Visible colorimetric method, for quantifying residual poly-diallyl dimethylammonium chloride (poly-DADMAC) in four raw dam and treated potable waters, was undertaken. Using three calibration methods, the method was found to be sensitive (LOQ=2 µg/L), over the linear range 10-30 µg/L. The overall mean within-batch precision (%RSD) was: 7.42 (±7.07) for Method 1, and 7.66 (±7.37) for Method 2; between-batch (reproducibility) (%RSD) was 54.37 ± 30.03) and 35.89 ± 34.89). Statistical data analysis indicated fairly good agreement (no significant difference) for poly-DADMAC levels in 30 samples analyzed by the two methods Method 1 and 2. The residual poly-DADMAC potable water levels (range: <2-8 µg/L), were: on average (±SD) (µg/L), 1.21 (±1.31) for Hazelmere Dam, 1.22 (±0.55) for Midmar Dam, 3.40 ± 3.89) for Inanda Dam, and 3.64 (±3.83) for Nagel Dam. The observed, apparent poly-DADMAC levels, obtained by Method 1, (range: 6-16 µg/L) were, on average (±SD) (µg/L), for the raw water samples: 3.73 (±0.46) for Inanda Dam, 5.73 (±6.57) for Nagle Dam, 6.82 (±9.03) for Hazelmere Dam and 10.12 (±6.94) for Midmar Dam. The study indicated compliance of all treated, potable water for residual poly-DADMAC, to the current international limit of ≤50 µg/L. The relatively high apparent concentration (range: <2-24 µg/L) of poly-DADMAC observed on the raw dam waters was attributed to the presence of Natural Organic Matter (NOM).

Keywords: Citrate-capped gold nanoparticle; Water treatment polymer; Poly-(diallyldimethylammonium chloride); Residual polyelectrolyte; Colorimetry; Ultraviolet-visible spectroscopy; Natural organic matter; Disinfection by-product; Toxicity; N-nitrosodimethylamine

Introduction

Poly-diallyldimethylammonium chloride (poly-DADMAC) is one of the most commonly used organic polyelectrolytes in wastewater and potable water treatment plants, as a coagulant and as a flocculent aid, for floc formation and for improved settling of larger particles [1-5]. Due to its potential to form N-nitrosodimethylamine (NDMA) [6-8], there has been, in recent years, a growing concern over the fate of poly-DADMAC within the water treatment process. Some early work has demonstrated that NDMA is a disinfection by-product formed during chlorination steps within the water treatment process [9].

Furthermore, NDMA is a suspected carcinogen [1,6,8,10,11]. The presence of residual poly-DADMAC depends on its reactivity during the disinfection processes, and whether it degrades into toxic compounds, or other by-products, that pass through the various stages in the water treatment process. Due to the highly charged nature the main assumption is that it will be removed together with the sludge during flocculation in the water treatment process.

Personal care products are another source of polyelectrolytes that can enter the environment and water treatment facilities, where they may not be adequately removed in the water treatment process [12,13]. Residual amounts may persist if the incorrect dose is used. The American Water Works Association (AWWA), American Society for Testing Materials, The European Committee for Standardization, the National Sanitation Foundation International, and the American National Standards Institute, provide standards for the maximum dosage of polyelectrolytes (10-100 mg/L) that can be used in water treatment. They have set the residual amount of poly-DADMAC in drinking water at 50 µg/L [3,14,15]. Recent work has shown that polyelectrolytes, like poly-DADMAC, can be toxic to aquatic organisms at levels above 50 µg/L [12,13].

Thus, for water treatment plants using poly-DADMAC as coagulant, and from an environmental, human health perspective, there is a strong requirement to determine the amount of residual polyelectrolytes, like poly-DADMAC, in the drinking water. To monitor residual concentrations in water, down to the required limit of ≤50 µg/L, sensitive analytical methods are therefore required.

Colloid titration has been used to determine residual poly-DADMAC in water samples [1,2,16]. However, the sensitivity of such

---

*Corresponding author: Manickum T, Scientific Services Laboratories: Chemical Sciences, Engineering and Scientific Services, Umgeni Water, P.O Box 9, Pietermaritzburg-3200 KwaZulu-Natal, South Africa
E-mail: thavrin.manickum@umgeni.co.za

techniques is 0.5-1.0 mg/L. The AWWA standard for poly-DADMAC [17] uses a gravimetric method; the method is long, labor-intensive and cannot be applied to analysis of residues in treated water. The other challenge is that the cationic polymer is ultraviolet (UV) inactive, and it is therefore not possible to employ UV-Visible (Vis) spectrophotometry for its analysis [18,19]. Pre- and post-fluorescent tagging of poly-DADMAC, with 10-40 µg/L detection limits in water, was developed by Elridge [4]. However, these methods are complicated, require several pre-treatment steps, can be expensive, are very time-consuming, and may not be suitable for routine analysis. A novel gel permeation chromatography (GPC) method, using RI detection, was developed by W John [18,19], for poly-DADMAC analysis in water, with a detection limit of 50 mg/L, and 1% precision. A spectrophotometric detection method, of poly-DADMAC, exploiting the flocculation properties using 4-hydroxy-1-napthylazo-benzene-sulfonic acid, which forms a coloured colloid ion pair, was reported by Ndungu et al. [20]. The method had a linear range of 0.1-1.8 mg/L, with a limit of detection (LOD) of 0.07 mg/L.

Umgeni Water, a bulk potable water supplier, in KwaZulu-Natal (KZN), makes use of this organic polymeric poly-DADMAC as a coagulant in some of its water treatment plants. However, to date the residual amount of poly-DADMAC in the final drinking water from any of the plants using poly-DADMAC, has not been fully investigated or accurately determined. Although we have a state-of-art water testing laboratory at the head office in Pietermaritzburg (KZN), which is ISO/IEC 17025-accredited, and beside the earlier work on analytical method development for quantification of residual poly-DADMAC [2,18,19], there are no automated, rapid, simple "in-house" test methods available, at national, and international, level, for accurate, low level, residual poly-DADMAC analysis in water.

Gumbi et al. [21], from the University of KwaZulu-Natal (South Africa), developed a novel, sensitive spectroscopic technique for poly-DADMAC analysis, using citrate-capped gold nanoparticles (Au-NPs), which was applied to analysis of river water samples, with a reported 1-100 µg/L detection range [21]. We hypothesized that this newly developed gold-nanoparticle analytical method would be suitable for the accurate, precise low level quantification (≤50 µg/L) of residual poly-DADMAC in treated water. The aim of this study was thus: to evaluate the suitability of the recently reported gold-nanoparticle method [21] for determination of residual poly-DADMAC in typical/real potable, water samples treated with polyelectrolyte-based coagulant: poly-DADMAC; to assess the precision and accuracy of this analytical test method.

We now report on the preliminary performance of this analytical test method. To our knowledge, this is the first report of a "real world" study application of the citrate-capped, Au-NP colorimetric method for quantitation of residual poly-DADMAC in treated, and raw, dam water for potable use.

Materials and Methodology

Reagents and chemicals

The three organic polyelectrolyte-based coagulants, containing poly-DADMAC, (with the water works at which it is used) were: Z553D (DV Harris, Wiggins), obtained from Zetchem (KwaZulu-Natal, South Africa); SF3456 (Hazelmere water works) and SF3435 (Durban Heights), were obtained from Improchem (KwaZulu-Natal). The composition of the coagulant blends (Aluminium chlorhydrin-DADMAC) was unknown due to it being proprietary information.

The Acrodisc premium 25 mm syringe filter with GxF/0.45 µm GHP membrane (HPLC certified - Glass fiber prefilter (GHP: hydrophilic polypropylene, Part Number: AP-4559T) was obtained from Pall life Sciences. Gold (III) chloride tri-hydrate (HAuCl$_4$.3H$_2$O), tri-sodium citrate (99%) (Na$_3$C$_6$H5O$_7$.2H$_2$O) and poly diallyl dimethyl ammonium chloride (poly-DADMAC) 35% weight (average molecular weight 100,000) (C$_8$H$_{16}$ClN) were obtained from Sigma Aldrich and were of analytical grade. All chemicals were used without further purification. All glassware used was salinized to prevent the adsorption of poly-DADMAC and other charged species. Plastic containers were used to store all solutions.

Instrumentation for poly-(DADMAC) analysis

The Ultraviolet (UV)-Visible (Vis) spectra were measured with an Ocean Optics spectrometer (model HR2000+), equipped with a tungsten halogen (Ocean Optics) based module, and two fiber optic cables (QP 600-2-vis-BX model 727-733-2447, suitable for 400-2100 nm range, from Narich Ltd (Milnerton, South Africa, agents for Ocean Optics); raw data were captured and analyzed with the spectrometer SpectroSuite^R software. Samples were transferred to a 1.0 ml quartz cuvette and placed in the cuvette holder (Ocean Optics CUV-UV with a 1 cm path length). The light was passed through a fiber optic cable, then the cuvette holder and finally via a second fiber optic cable to the spectrometer.

Instrumentation for physical tests

For the 51 study samples, the pH, salinity, conductivity, Redox, TDS and temperature, were determined at UKZN. The Redox and pH were measured with an 827 pH lab meter equipped with a probe (6.0220.100) bought from Metrohm (Switzerland). Salinity, conductivity and total dissolved solids (TDS) were measured with an InoLab® Cond level 1 (8F93) instrument, equipped with a probe (WTW Tetracon 325), bought from Germany, through Merck. Both probes were conditioned with standards (as per the manufacturer's recommendations) before use every day. Similar physical tests were performed at the Umgeni Water Chemistry laboratory. Turbidity, in Nephelometric Turbidity Units (NTU), was measured with a HACH 2100 turbidimeter. The total organic carbon (TOC) was measured with a Tekmar Torch analyser, from LabHouse (Midrand, South Africa), agents for Tekmar. The total dissolved solids (TDS) were determined by gravimetry. A JOEL 1010 transmission electron microscope was used for transmission electron microscopy (TEM) analysis of the gold nanoparticles [20]. Samples were initially prepared by dipping a 200 mesh copper grid (Formvar support film) in the sample solution, air-drying on filter paper, followed by TEM analysis.

Preparation of stock solution of gold nanoparticles

As per the previous report [21], Au-NPs were prepared by the citrate reduction method [22]. Gold(III) chloride tri-hydrate (0.4768 g) was added to 400 ml of ultrapure water. The gold solution was then heated on a hotplate, 10 ml of 0.2746M tri-sodium citrate was added to the boiling gold solution. The solution was stirred (300 rpm) and carefully observed for the color change, from yellow to colorless and finally to deep red. The red solution was immediately taken off the hotplate and allowed to cool to 25°C. The Au-NP solution (400 ml) was then transferred to a 2.0 L volumetric flask, and diluted with ultrapure water to volume. The solution was thoroughly mixed by inverting the flask for 20 times; no precipitate was observed. The Au-NPs were characterized by UV-Vis spectroscopy and TEM.

Jar test procedure

The purpose of Jar tests are used to predict clarification at water works. This method may be used to determine optimum dose of a polyelectrolyte for use as a primary coagulant and for comparing the performance of different polyelectrolytes. The Jar tests were performed at Wiggins Water Works, Umgeni Water. The standard procedure [23] is described in the Supplementary Material Text A. The optimum dose and most suitable coagulant for a particular site can be deduced from the Jar test results.

Sample collection

The four selected raw water sources (dams) (and respective water works (WW)) were: Inanda Dam (Wiggins WW), Nagle Dam (Durban Heights WW), Hazelmere Dam (Hazelmere WW) and Midmar Dam (DV Harris WW). Grab, raw water samples, and treated water samples, were collected from the designated water works sampling points that each dam supplies; each sample site has a unique sample point code. Samples were collected into 1 L plastic bottles, during the three-month study period: May, June and July 2014. The raw, potable (treated) water, and processed samples from the Jar Test procedure, were then submitted, in 1 L plastic bottles, to UKZN for subsequent analysis for poly-DADMAC. All collected and processed samples were assigned a unique identification number; the composition of the samples was not disclosed to the testing laboratory (UKZN) for the purpose of establishing accuracy and precision of the analytical method for residual poly-DADMAC.

Determination of poly-DADMAC by colorimetry-Au-NP

Poly-DADMAC in the various water samples were analysed by the standard addition method [21]. The UV-Vis data obtained on each water sample was analysed by using three techniques: Absorbance of the peak at 690 nm (Method 1) (M1), Area of the peak at 690 nm (Method 2) (M2), and Ratio of the peak absorbances at 690 nm and 520 nm (A_{690nm}/A_{520nm}) (Method 3) (M3).

Calibration standards: All glassware was silanized before use. A 1.185 g sample of poly-DADMAC (35% weight) was weighed into a vial (plastic) and transferred into a 1 L volumetric flask to make a 400 mg/L of poly-DADMAC stock solution. A 50 mg/L working stock solution of poly-DADMAC was prepared by transferring 12.5 ml of poly-DADMAC stock solution (400 mg/L) into a 100 mL volumetric flask. All flasks were then diluted to volume with double distilled water. Three calibration standards: 0, 10 and 20 µg/L, in 50 ml flasks, were used for Method 1 and Method 2; for Method 3, the 3 calibration standards were between 10-50 µg/L.

Determination of poly-DADMAC by colorimetry-Au-NP: The real water samples were initially filtered through the 0.45 µm GHP pre-filter membrane. Approximately 25 ml of ultrapure water (or water sample) was added into a 50 ml volumetric flask. For the calibration standards, the required amount (e.g., 10 µL for a 50 µg/L standard) of a 50 mg/L poly-DADMAC solution was added to the contents of the flask. A volume of 20 ml of the Au-NP solution was then added to the flask. The blue solution was made up the 50 mL mark, mixed (by inverting the flask 20 times) and analyzed within 20 minutes. This was done to avoid the coagulant effect of poly-DADMAC.

Statistical data analysis

The comparison of observed poly-DADMAC levels for each sample determined by all three calibration methods were determined by one-way analysis of variance, and Bonferroni adjustment was performed afterwards to investigate significant pairwise differences. Tests for correlation and significance (p-values less than 5%) were determined using STATA12, by analysis of Scatter plots and determination of Pearson's correlation coefficient.

Other Physico-chemical water quality data

Other physico-chemical data, like TOC and turbidity, were, as required, obtained from the Umgeni Water Intranet, via the Labware Information Management System (LIMS).

Results

Physico-chemical water quality

After data analysis, the average turbidity, conductivity, TDS and TOC values are summarized in Table 1. The raw data is appended in the Electronic Supplementary Material Table A, Figures A and B.

For the 3-month study period, the average raw water turbidity

Dam (raw water source)	Water works: Raw water Sample point	n	Turb NTU Mean (±SD) (NTU)	Turb NTU% RSD	Turb NTU Range (NTU)	TOC Mean (±SD) (mg/L) (%RSD)	Con	Con	TDS	TDS	WW sample point	(NTU) Mean ± SD (%RSD)	TOC Mean ± SD (mg/L) (%RSD)
							\multicolumn{4}{c	}{Mean ± SD (% RSD)}			
							Raw	Potable	Raw	Potable			
Inanda	Wiggins: TWG001 (0.08-1.01)[a]	92	1.04 (±0.30)	28.60	0.50-1.99	2.59 ± 0.20 (7.81)	197.5 ± 65 (32.92)	229.8 ± 5 6.4 (24.52)	232.4 ± 12.1 (0.05)	251.2 ± 29.4 (11.71)	TWG010	0.26 ± 0.08 (32.10)	2.33 ± 0.27 (11.67)
Nagel	Durban Heights: TDH001 (2.4)	92	4.96 (±2.12)	42.70	2.11-14.80	2.31 ± 0.12 (5.07)	73.7 ± 23.2 (31.55)	95 ± 50.9 (53.59)	123.2 ± 22 (17.87)	128.5 ± 15.9 (12.39)	TDH010	0.22 ± 0.06 (28.15)	2.16 ± 0.26 (12.20)
Hazelmere	Hazelmere: THM001 (2-5)	92	6.71 (±1.54)	22.92	4.00-14.50	2.48 ± 0.25 (9.96)	112 ± 62.2 (55.56)	159.3 ± 30.1 (18.87)	151.4 ± 11.4 (7.52)	208.5 ± 75.1 (36.04)	THM008	0.66 ± 0.21 (32.46)	1.91 ± 0.35 (18.17)
Midmar	DV Harris: TMM001 (1.63-5)	92	9.38 (±4.54)	48.46	0.90-43.10	3.17 ± 0.62 (19.44)	59.3 ± 23.7 (39.93)	77 ± 33.2 (43.05)	70.3 ± 5 (7.13)	90.4 ± 8.5 (9.37)	TMM007	0.25 ± 0.07 (28.08)	2.04 ± 0.39 (19.14)

[a]Coagulant dosing level (mg/L)

[b]Con: Conductivity; TOC: Total Organic Carbon; TDS: Total Dissolved Solids; Turb: Turbidity; NTU: Nephelometric Turbidity Units; WW: Water Works.

Table 1: Physico-chemical water quality data.

(NTU) (±Standard Deviation (SD)) was: 1.04 (±0.30) for Inanda, 4.936 (±2.12) for Nagle, 6.71 (±1.54) for Hazelmere and 9.38 (±4.54) for Midmar. For the conductivity, the average levels (mS/m) were: 59.3 (±23.7) for Midmar, 73.7 (±23.2) for Nagle, 112.0 (±62.2) for Hazelmere and 197.5 (±65) for Inanda. The average TDS levels (mg/L) were: 70.3 (±5) for Midmar, 123.2 (±22) for Nagle, 151.4 (±11.4) for Hazelmere and 232.4 (±12.1) for Inanda. The average TOC levels (mg/L) were: 2.31 (±0.12) for Nagle, 2.48 (±0.25) for Hazelmere, 2.59 (±0.2) for Inanda and 3.17 (±0.62) for Midmar. The overall Redox potential values (mV) were (mean ± SD) (median) (range): -30 ± (29) (-25) (range=+7 to -80) for all four dams. Individual average values were: -41 ± (49) (-41) (-76 to -6) for Midmar, -32 ± (43) (-25) (-79 to 7) for Hazelmere, -25 ± (1) (-25) (-25 to -24) for Nagle, -20 ± (18) (-27) (-34 to 12) for Inanda. The average values for pH were (mean ± SD): 6.82 (±0.37) for Inanda, 7.09 (±0.16) for Nagle, 7.15 (±0.66) for Hazelmere and 7.17 (±0.6) for Midmar.

For the potable water, the average raw water turbidity (NTU) was: 0.22 (±0.06) for Nagle, 0.25 (±0.07) for Midmar, 0.26 (±0.08) for Inanda and 0.66 (±0.21) for Hazelmere. For conductivity, the average levels (mS/m) were: 77 (±33.2) for Midmar, 95.0 (±50.9) for Nagle, 159.3 (±30.1) for Hazelmere and 229.8 (±56.4) for Inanda. The average TDS levels (mg/L) were: 90.4 (±8.5) for Midmar, 128.5 (±15.9) for Nagle, 208.5 (±75.1) for Hazelmere and 251.2 (±29.4) for Inanda. The average TOC levels (mg/L) were: 1.91 (±0.35) for Hazelmere, 2.04 (± 0.39) for Midmar, 2.16 (±0.26), for Nagle and 2.33 (±0.27) for Inanda. The overall Redox potential values (mV) were (mean ± SD) (median) (range): -14 ± (12) (-11) (range=-34 to 2) for all four dams. Individual values were: -18 ± (14) (-11) (-34 to -11) for Midmar, -13 ± (14) (-5) (-29 to -4) for Hazelmere, -11 ± (15) (-5) (-29 to 2) for Inanda, -9 ± (6) (-9) (-13 to -52) for Nagle. The average values for pH were (mean ± SD): 6.77 (±0.10) for Nagle, 6.79 (±0.22) for Inanda, 6.86 (±0.24) for Hazelmere and 6.94 (±0.20) for Midmar.

Sample ID[a]	Water works/ Sample description/ Coagulant dose concentration	Month	Calibration method								
			M1[a] Peak absorbance @ 690 nm		M2 Peak area @ 690 nm		M3 Ratio of peak absorbances (A_{690nm}/A_{520nm})		"M1 and M2" combined data		
			Observed conc[a] (μg/L)	Observed% RSD	Observed conc (μg/L)	Observed% RSD	Observed conc (μg/L)	Observed% RSD	Calculated mean conc (μg/L)	Calculated SD (μg/L)	Calculated %RSD
4R	Wiggins: raw	May	3.16	7.09	4.21	4.83	38.2	1.03	3.69	0.74	20.15
5R	Wiggins: raw	May	4.23	10.8	3.73	15.1	nd[a]	0.92	3.98	0.35	8.88
Mean			3.70		3.97		38.20				
SD			0.76		0.34						
%RSD			20.48		8.55						
6F	Wiggins: potable	May	9.14	3.90	10.30	3.82	65.30	1.26	9.72	0.82	8.44
7F	Wiggins: potable	May	1.38	2.35	1.11	26.10	64.60	1.76	1.25	0.19	15.33
8F	Wiggins: potable	May	2.38	2.03	2.58	7.37	55.60	0.40	2.48	0.14	5.70
Mean			4.30		4.66		61.83				
SD			4.22		4.94		5.41				
%RSD			98.17		105.86		8.75				
9B	Wiggins: raw: Jar test	May	12.70	6.62	11.00	6.46	nd	3.01	11.85	1.20	10.14
10B	Wiggins: raw: Jar test	May	7.84	5.25	8.61	3.01	28.20	0.58	8.23	0.54	6.62
Mean			10.27		9.81		28.20				
SD			3.44		1.69						
%RSD			33.46		17.24						
11O	Wiggins raw: Jar test: optimal dose	May	22.70	7.86	21.30	7.34	nd	3.61	22.00	0.99	4.50
12O	Wiggins raw: Jar test: optimal dose	May	17.21						17.21		
13O	Wiggins/ optimal dose	May	14.10	3.01	19.10	1.50	nd	3.01	16.60	3.54	21.30
Mean			18.00		20.20		3.31				
SD			4.35		1.56		0.42				
%RSD			24.19		7.70		12.82				
14	Wiggins raw: Jar test; 30% overdose	May	1.90						1.90		
15	Wiggins raw: Jar test: 60% overdose	May	0.29	16.00	0.47	31.30	54.30	0.37	0.38	0.12	32.92
16	Wiggins raw: Jar test: 100% overdose	May	2.79	11.70	4.17	21.70	nd	2.39	3.48	0.98	28.04
20R	Wiggins: raw	June	3.60	4.34	4.11	9.75	48.20	0.98	3.86	0.36	9.35
20F	Wiggins: potable	June	0.713	3.39	1.53	2.04	15.3	0.38	1.12	0.58	51.51
21B	Wiggins: raw: Jar test	June	5.12	12.60	5.59	5.60	40.50	3.45	5.36	0.33	6.21
22B	Wiggins: raw: Jar test	June	1.49	4.90	1.95	5.20	22.70	0.17	1.72	0.33	18.91

ID	Sample	Month									
Mean			3.31		3.77	5.40	31.60				
SD			2.57		2.57	0.28	12.59				
%RSD			77.66		68.27	5.24	39.83				
230	Wiggins: raw: Jar test: optimal	June	0.53	8.90	0.64	15.00	13.20	0.44	0.58	0.08	14.15
24	Wiggins: raw: Jar test: 30% overdose	June	3.64	8.10	4.41	2.72	42.80	1.01	4.03	0.54	13.53
25	Wiggins: raw: Jar test: 30% overdose	June	4.77	1.20	5.88	1.17	47.30	0.65	5.33	0.78	14.74
26	Wiggins: raw: Jar test: 30% overdose	June	1.78	5.09	2.16	8.85	30.90	5.46	1.97	0.27	13.64
Mean			3.40		4.15	4.25	40.33				
SD			1.51		1.87	4.06	8.47				
%RSD			44.45		45.15	95.63	21.01				
27	Wiggins: raw: Jar test: 60% overdose	June	5.65	0.48	7.35	0.07	35.60	0.50	6.50	1.20	18.49
28	Wiggins: raw: Jar test: 100% overdose	June	1.61	5.82	0.72	1.70	11.90	1.03	1.17	0.63	53.77
32R		July	3.93	7.19	4.00	7.57	47.20	0.97	3.97	0.05	1.25
32F		July	1.27	4.27	1.80	0.22	9.44	0.73	1.54	0.37	24.41
33B	Wiggins: raw: Jar test	July	5.51	0.63	7.21	1.79	54.20	0.80	6.36	1.20	18.90
34B	Wiggins: raw: Jar test	July	12.90	4.84	12.00	7.20	44.10	2.73	12.45	0.64	5.11
Mean			9.21		9.61		49.15				
SD			5.23		3.39		7.14				
%RSD			56.77		35.26		14.53				
350	Wiggins optimal	July	1.08	8.09	0.41	19.50	7.85	1.12	0.75	0.47	63.32
36	Wiggins 30% overdose	July	1.12	4.15	1.19	24.40	6.44	1.89	1.16	0.05	4.29
37	Wiggins 60% overdose	July	2.48	2.51	3.53	1.24	9.58	0.39	3.01	0.74	24.71
38	Wiggins 60% overdose	July	0.19	12.00	0.74	4.01	9.71	0.17	0.46	0.39	83.45
39	Wiggins 60% overdose	July	3.11	4.46	5.18	2.99	8.03	1.20	4.15	1.46	35.31
Mean			1.93		3.15		9.11				
SD			1.54		2.25		0.93				
%RSD			79.76		71.32		10.26				
40	Wiggins 100% overdose	July	2.70	2.52	2.16	6.52	10.20	1.26	2.43	0.38	15.71
1R	DV Harris: raw	May	6.79	7.15	6.61	7.72	nd	2.08	6.70	0.13	1.90
1F	DV Harris: potable	May	1.85	8.84	2.34	12.00	17.20	0.95	2.10	0.35	16.54
17R	DV Harris: raw	June	5.48	2.93	8.14	0.79	5.32	2.39	6.81	1.88	27.62
17F	DV Harris: potable	June	0.99	33.00	2.71	2.43	12.20	0.28	1.85	1.22	65.85
29R	DV Harris: raw	July	18.10	9.70	18.90	8.92	62.00	6.20	18.50	0.57	3.06
29F	DV Harris: potable	July	0.82	5.52	0.40	6.57	13.40	0.40	0.61	0.30	49.07
2R	Hazelmere: raw	May	1.43	2.06	2.30	0.41	23.30	0.33	1.87	0.62	32.99
2F	Hazelmere: potable	May	2.70	7.91	4.20	7.67	15.80	1.36	3.45	1.06	30.74
18R	Hazelmere: raw	June	17.25	0.95	23.80	0.69	52.20	0.83	20.53	4.63	22.57
18F	Hazelmere potable	June	0.22	34.10	0.67	16.50	10.70	0.40	0.44	0.31	70.79
30R	Hazelmere: raw	July	1.78	4.33	2.31	0.65	16.80	0.69	2.05	0.37	18.33
30F	Hazelmere: potable	July	0.72	5.33	1.25	4.99	8.50	0.47	0.98	0.38	38.23
3R	Durban Heights: raw	May	13.30	4.70	13.40	6.36	nd	0.31	13.35	0.07	0.53
3F	Durban Heights: potable	May	2.01	23.80	2.87	15.10	57.00	2.12	2.44	0.61	24.92
19R	Durban Heights: potable	June	2.36	12.70	3.50	11.70	16.50	1.65	2.93	0.81	27.51
19F	Durban Heights: potable	June	0.89	8.68	0.95	9.66	13.80	0.40	0.92	0.04	4.61
31R	Durban Heights: raw	July	1.54	4.33	2.31	0.65	16.80	0.69	1.93	0.54	28.28
31F	Durban Heights: potable	July	8.02	1.54	11.20	0.99	26.00	0.29	9.61	2.25	23.40

[a]conc.: Concentration; ID: Identity; nd: Not Detected.

Table 2: The observed levels of poly-DADMAC for all the samples.

Water works/ dam	Coag.[a]	Dosing level: May-July (mg/L)	n	Poly-DADMAC concentration (µg/L)										Mean M3/M1	Mean M3/ M2
				Calibration technique											
				Method 1 (M1)		Method 2 (M2)		Method 3 (M3)		Redox potential (mV)					
				Mean ± SD	%RSD	Mean ± SD	%RSD	Mean ± SD	%RSD	Mean ± SD	Median	Range			
				Raw water											
Wiggins/ Inanda			4	3.73 ± 0.46	12.30	4.01 ± 0.21	5.16	44.53 ± 5.51	12.37	-20 ± (18)	-27	-34 to +12		11.9	11.1
DV Harris/ Midmar			3	10.12 ± 6.94	68.54	11.22 ± 6.70	59.71	33.66 ± 40.08	119.07	-41 ± (49)	-41	-76 to -6		3.3	3.0
Hazelmere/ Hazelmere			3	6.82 ± 9.03	132.47	9.47 ± 12.41	131.05	30.77 ± 18.84	61.25	-32 ± (43)	-25	-79 to +7		4.5	3.2
Durban Heights/ Nagel			3	5.73 ± 6.57	114.52	6.40 ± 6.09	95.08	16.65 ± 0.21	1.27	-25 ± (1)	-25	-25 to -24		2.9	2.6
				Potable water											
Wiggins/ Inanda	Z553D	0.08-1.01	4	3.40 ± 3.89	114.21	3.88 ± 4.32	111.45	50.20 ± 23.68	47.18	-11 ± (15)	-5	-29 to 1		14.8	12.9
DV Harris/ Midmar	Z553D	1.63-5	3	1.22 ± 0.55	45.19	1.82 ± 1.24	68.34	14.27 ± 2.61	18.30	-18 (14)	-11	-34 to -10		11.7	7.8
Hazelmere/ Hazelmere	SF3456	2-5	3	1.21 ± 1.31	108.06	2.04 ± 1.89	92.89	11.67 ± 3.74	32.10	-13 ± (14)	-5	-29 to -4		9.6	5.7
Durban Heights/ Nagle	SF3435	2.4	3	3.64 ± 3.83	105.36	5.01 ± 5.45	108.89	32.27 ± 22.27	69.02	-9 ± (6)	-9	-13 to -5		8.9	6.4

Table 3: The computed mean residual poly-DADMAC values for the raw dam waters and corresponding potable water samples.

Poly-DADMAC levels on raw and potable water

The observed levels of poly-DADMAC for all water samples is listed in Table 2; computed averages for the four raw dam water and associated potable waters are listed in Table 3 (Supplementary Figure A). The observed, residual poly-DADMAC levels, obtained by Method 1, (range: 6-16 µg/L) were, on average (±SD) (µg/L), for the raw water samples: 3.73 (±0.46) for Inanda Dam, 5.73 (±6.57) for Nagle Dam, 6.82 (±9.03) for Hazelmere Dam, and 10.12 (±6.94) for Midmar Dam (Table 3). The corresponding potable water levels (range: <2-8 µg/L), were: on average (±SD) (µg/L), 1.21 (±1.31) for Hazelmere Dam, 1.22 (±0.55) for Midmar Dam, 3.40 (±3.89) for Inanda Dam, and 3.64 (±3.83) for Nagle Dam. The observed, residual poly-DADMAC levels, obtained by Method 2, (range: 0-22 µg/L) were, on average (±SD) (µg/L), for the raw water samples: 4.01 (±0.21) for Inanda Dam, 6.40 (±6.09) for Nagle Dam, 9.47 (±12.41) for Hazelmere Dam, and 11.22 (±6.70) for Midmar Dam. The corresponding potable water levels (range: <2-11 µg/L), were: on average (±SD) (µg/L), 1.82 (±1.24) for Midmar Dam, 2.04 (±1.89) for Hazelmere Dam, 3.88 (±4.32) for Inanda Dam, and 5.01 (±5.45) for Nagle Dam. The observed, residual poly-DADMAC levels, obtained by Method 3, (range: 16.65-44.53 µg/L) were, on average (±SD) (µg/L), for the raw water samples: 16.65 ± 0.21 for Nagle dam, 30.77 ± 18.84 for Hazelmere Dam, 33.66 ± 40.08 for Midmar Dam and 44.53 ± 5.51 for Inanda Dam. The corresponding potable water levels, (range: 16.65-44.53 µg/L), were, on average (±SD) (µg/L), for the raw water samples: 11.67 ± 3.74 for Hazelmere Dam, 14.27 ± 2.61 for Midmar Dam, 32.27 ± 22.27 for Nagle dam, and 50.20 ± 23.68 for Inanda Dam.

Discussion

Physicochemical water quality

Raw dam water: For the 3-month study period, the average raw water turbidity increases in the order: Inanda (1.04) <Nagle<Hazelmere<Midmar Dam (9.38). For the raw water turbidity, the data indicates the lowest average value of 1 NTU for Inanda Dam, which is very much lower than that of the other three raw dam waters (5 for Nagle and 9 NTU for Midmar dam). The national drinking water guide limit, as per the South African National Standards (SANS) 241: 2011, for turbidity is 1 NTU, and the Umgeni Water internal limit for potable water is lower, at ≤0.5 NTU. In general, very turbid waters will be expected to require a higher concentration of coagulant for flocculation during the water treatment process. This requirement is confirmed in the increasing dosage of poly-DADMAC that was used at the respective raw water treatment plants: the lowest dose being 0.08-1.01 mg/L, for Inanda Dam (1 NTU), up to a maximum 1.63-5 mg/L, for Midmar Dam (9 NTU). There is minimal difference in pH, which ranges from 6.8 for Inanda, to 7.2 for Midmar. These values do comply with the national SANS 241 limit of ≥5 to ≤9.7. The average conductivity and TDS levels increase in the order: Midmar <Nagle< Hazelmere< Inanda. Except for Inanda, all other three dam levels comply with the national SANS 241: 2011 potable water limit of ≤170 mS/m for conductivity, and ≤1200 mg/L for TDS. It is also evident that the coagulant blends, containing poly-DADMAC, function effectively as organic, polymeric flocculants, in lowering the raw water turbidity, from 5-9 NTU to the national limit of ≤1, for all four raw dam waters. The average TOC levels increase in the order: Nagle<Hazelmere<Inanda<Midmar. However, levels for all four dams comply with the national potable water guide (SANS 241: 2011) limit of ≤10 mg/L.

Potable water: The potable water turbidity values (NTU), much lower than the raw waters, indicate fairly good similarity for the three dams: Midmar (0.22), Nagle (0.25), and Inanda Dam (0.26), whilst that for Hazelmere is approximately three times higher (0.66 NTU). However, all values do comply with the national SANS 241: 2011 potable water quality limit of being ≤1 NTU. The average conductivity and TDS levels increase in the order: Midmar <Nagle<Hazelmere<Inanda. Except for Inanda (230 mS/m), all other three dam levels comply with the national SANS 241 potable water limit of ≤170 mS/m for conductivity, and ≤1 200 mg/L for TDS. For all four dams, the conductivity and TDS levels on the treated (potable) water exceed that for the corresponding raw dam water. Whilst there is no significant difference in TOC levels for all four dams, there is a noticeable lower TOC content in all the potable waters compared to the corresponding raw dam water.

Method validation for the poly-DADMAC assay method

Detailed data has been previously reported [21]. The linearity range was between 0 and 30 μg/L with r^2=0.99 in all cases. The method LOD and LOQ (μg/ L) was 0.49 and 1.47 for Method 1 (absorbance of peak at 690 nm), 0.31 and 0.94 for Method 2 (area of peak at 690 nm) and 0.54 and 1.64 for Method 3 (ratio of the absorbance of peaks at 690 and 520 nm (A_{690nm}/A_{520nm}). From the raw data supplied (Supplementary Material Text B), the computed instrument precision (% RSD) (±SD), based on signal response, was: 18.05 (±17.65) for Method 1, 18.81 (±18.44) for Method 2, and 3.24 (±3.23) for Method 3. The overall mean within-batch (repeatability) precision (%RSD) for the triplicate assay values were: 7.42 (±7.07) for Method 1, 7.66 (±7.37) for Method 2, and 1.92 (±2.71) for Method 3. The overall mean between-batch (reproducibility)%RSD was: 54.37 (±30.03) for Method 1, 35.89 (±34.89) for Method 2 and 13.50 (±12.64) for Method 3.

Observed residual levels of poly-DADMAC in all water samples

Raw dam water samples: Typical calibration graphs are shown in Figure 1 (Method 1: absorbance of peak at 690 nm), Figure 2 (Method 3: area of peak at 690 nm), and Figure 3 (Method 3: ratio of peak absorbances at 690 nm and 520 nm). The observed levels of poly-DADMAC for all the samples by all three calibration Methods 1, 2 and 3, are listed in Table 2 (Supplementary Figures C and D). After data processing, the observed mean values (±SD) (Table 3) were (μg/L), by Method 1, 2, and 3: 3.73 (±0.46), 4.01 (±0.21), 44.53 (±5.51) for Wiggins WW, 10.12 (±6.94), 11.22 (±6.70), 33.66 (±40.08) for DV Harris WW, 6.82 (±9.03), 9.47 (±12.41), 30.77 (±18.84) for

Hazelmere WW, 5.73 (±6.57), 6.40 (±6.09), 16.65 (±0.21) for Durban Heights WW. The typical UV-Vis spectra, for sample 4R (Wiggins WW: May), with the calibration standards, is shown in Figure 4. For assay values obtained by Methods 1 and 2, the poly-DADMAC levels, in the four dams, increase in the following order: Inanda <Nagle<Hazelmere<Midmar. However, for values obtained by Method 3, the order is: Nagle<Hazelmere<Midmar<Inanda.

Potable (treated) water samples: The observed levels of poly-DADMAC for all samples are listed in Table 2. After data analysis, the observed mean values (±SD) (Table 3) were (μg/L), by Method 1, 2, and 3: 3.40 (±3.89), 3.88 (±4.32), 50.20 (±23.68) for Wiggins WW, 1.22 (±0.55), 1.82 (±1.24), 12.27 (±6.21) for DV Harris WW, 1.21 (±1.31), 2.04 (±1.89), 11.67 (±3.74) for Hazelmere WW, 3.64 (±3.83), 5.01 (±5.45), 32.27 (±22.27) for Durban Heights WW (Table 3). The typical UV-Vis spectra, for sample 6F (Wiggin's potable: May), with the calibration standards, is shown in Figure 5.

For assay values obtained by Methods 1 and 2, the poly-DADMAC levels, for the four dams, increase in the following order: Midmar/ Hazelmere<Inanda<Nagle. However, for values obtained by Method 3, the order is: Hazelmere <Midmar<Nagle<Inanda.

Determination of the UV-Vis spectra

The UV-Data were recorded once only; different mathematical models are applied to calculate the concentration. This one set of UV-data was then used in the l models (absorbance, area and ratio). The calibration was obtained using three techniques. The first approach involves plotting peak absorbance at 690 nm (corresponds to the

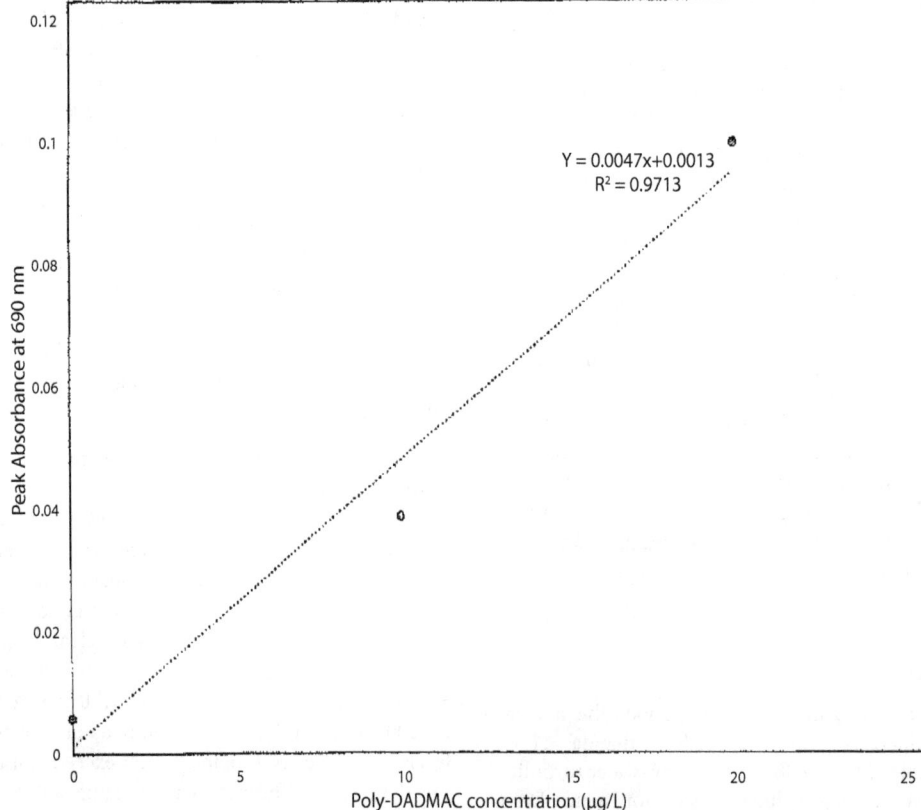

Figure 1: Calibration graph for raw dam water sample 4R by the peak absorbance Method 1.

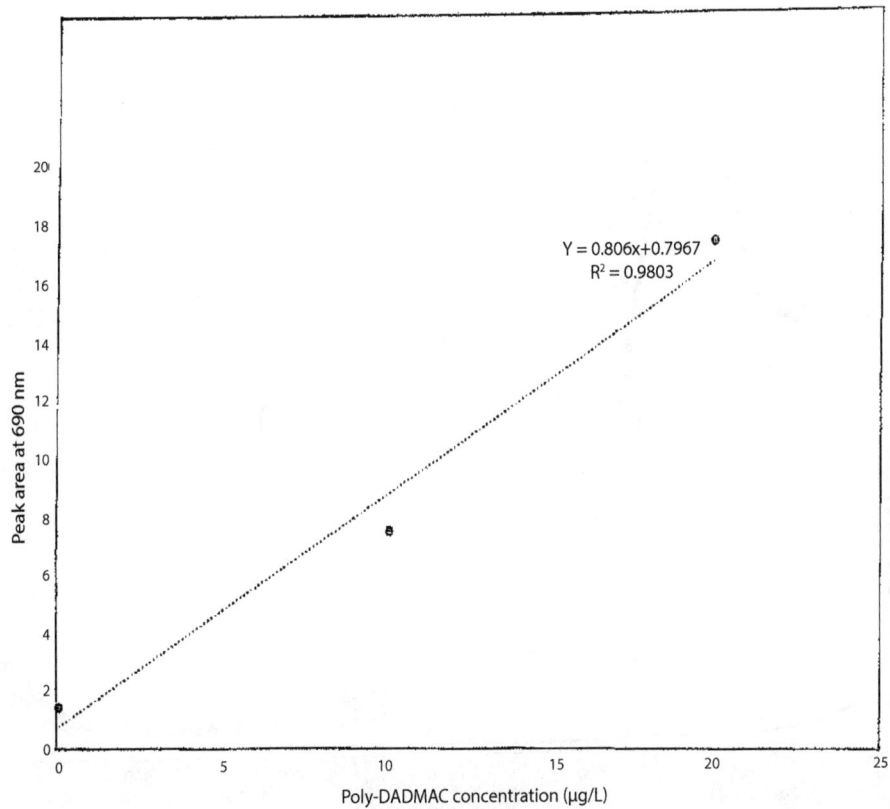

Figure 2: Calibration graph for raw dam water sample 4R by the peak area Method 2.

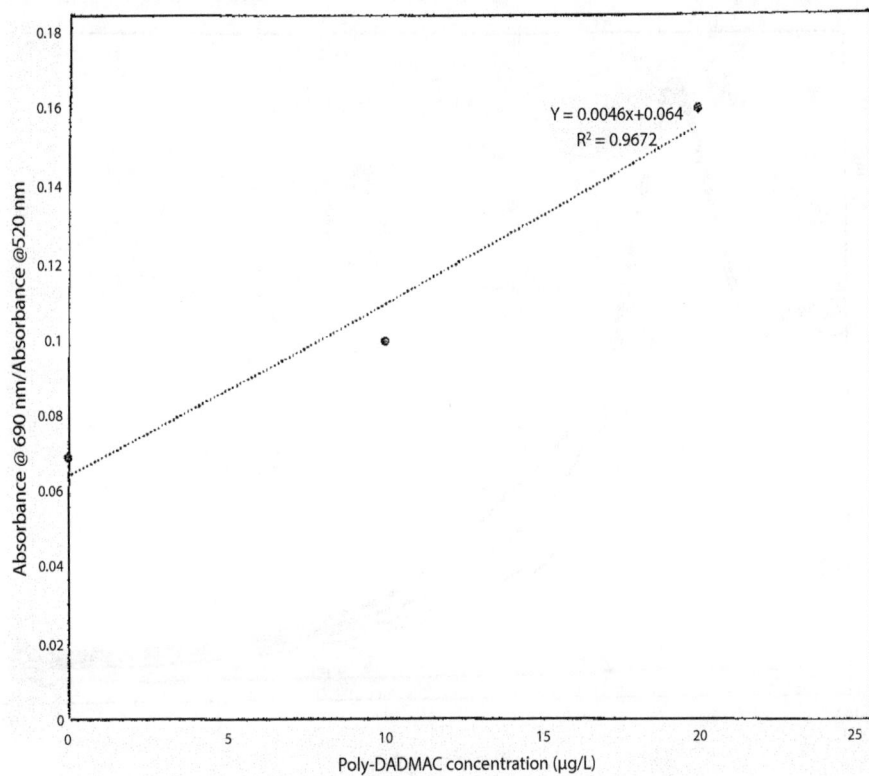

Figure 3: Calibration graph for raw dam water sample 4R by the ratio Method 3.

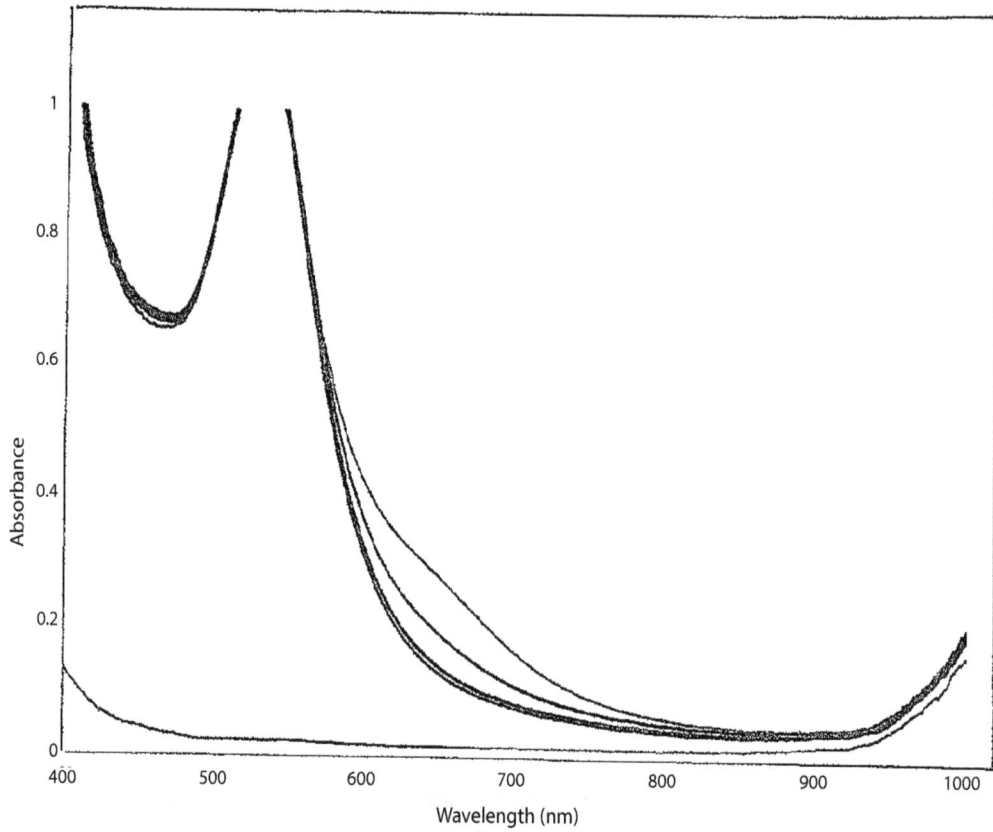

Figure 4: Typical UV-Vis spectra, for raw water sample 4R (Inanda Dam (Wiggins WW), May) with the calibration standards.

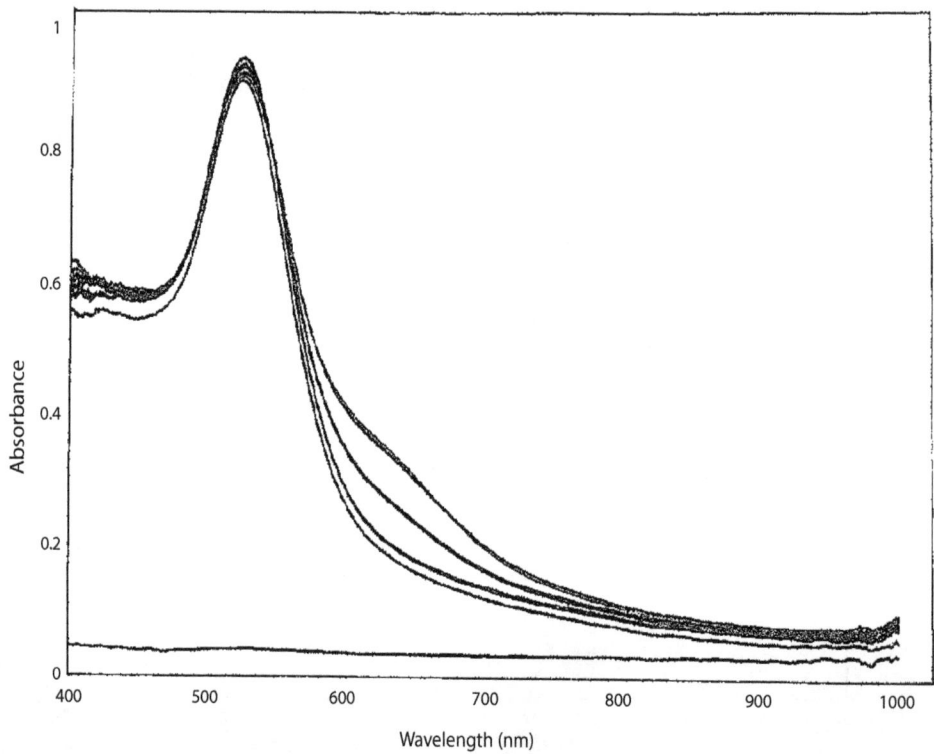

Figure 5: The typical UV-vis spectra, for potable water sample 6F (Inanda Dam (Wiggins), May), with the calibration standards.

aggregate of poly-DADMAC with gold nanoparticles) against poly-DADMAC concentration added to the sample, and then the poly-DADMAC concentration is calculated from the equation of the line where the y-intercept is equal to zero (Method 1). The second approach is similar, but instead the area of the peak at 690 nm is used (Method 2). The rationale for using the area is due to the fact that there is a distribution of poly-DADMAC-gold nanoparticle aggregates, and the area could account for the inherent variations with the aggregates [21]. The third method used a similar approach; however, the response parameter was the ratio of the peak absorbance at 690 nm and at 520 nm (A_{690nm}/A_{520nm}) (Method 3). The ratio method is the preferred method in the literature [24], and as per the earlier work, this method of analysis was also done in order to determine which of the three methods of data analysis provides the most accurate and precise results with the "real world" water samples.

Comparison of assay values for poly-DADMAC obtained by the 3 Methods

Inspection of the observed assay data for poly-DADAMC on all the water samples indicate, in general, fairly good comparison between the Method 1 and Method 2 values, while the values obtained by Method 3 are much higher (Supplementary Figure B). Considering the LOQ (µg/L) for each of the Methods [21], (1.47 for Method 1, 0.94 for Method 2, 1.64 for Method 3) assay values for samples 7F, 17F, 18F, 19F, 20F, 29F, 30F, 32F, 2R, 15, 36, 38, 23O and 35O (fourteen in total) were below the LOQ (1.47 µg/L) for Method 1; assay values for samples 18F, 29F, 15, 28, 38, 23O and 35O (seven in total) were below the LOQ (0.94 µg/L) for Method 2, and assay values for samples 1R, 3R, 5R, 16, 11O, 13O (six in total) were below the LOQ (1.64 µg/L) for Method 3. For both Methods 1 and 3, there was agreement for assay values for six samples: 18F, 29F, 15, 38, 23O and 35O, which were below the respective LOQ of the methods. Simple correlation analysis by linear regression of the assay values by all 3 Methods indicated fairly good correlation for assay values obtained by Method 1 and Method 2: $r^2=0.9415$ (y=1.059x+0.4607) (Supplementary Figure A); the corresponding comparison for Method 1 and Method 3 values gave $r^2=0.3021$(y=2.4848x+18.394), and the Method 2 and Method 3 comparison gave $r^2=0.2504$ (y=1.917x+18.949), indicating no significant correlation. For confirmation, the statistical data analysis indicated that, for the comparison of poly-DADMAC values obtained by all three Methods (1, 2 and 3), there was no significant difference for the poly-DADMAC levels for 31 samples, obtained by Methods 1 and 2 (Supplementary Table B).

Performance evaluation of the analytical test method for poly-DADMAC

Observed precision of poly-DADMAC concentrations: Except for samples 14 and 12O, all samples (n=49) were analyzed in triplicate by the three techniques. The instrument precision data indicates that the precision (%RSD ± SD) for Methods 1 (18.81 ± 18.44%) and Method 2 (18.05 ± 17.65%) exceeds the ≤10% Relative Standard Deviation (RSD) limit used in our internal method validation laboratory procedure. Method 3 (3.24 ± 3.23%) appears to be the most precise. Based on the 49%RSD values, the overall mean within-batch precision (% RSD) (±SD) for the triplicate assay values were: 7.42 (±7.07) for Method 1, 7.66 (±7.37) for Method 2, and 1.92 (2.71) for Method 3. All three methods thus have repeatability RSD ≤10%, and the data indicates that the ratio method (Method 3), with the lowest %RSD, is the most precise. For those 8 samples that were analysed for n=2-4 times, over different days during the three months, the overall mean between-

batch reproducibility (%RSD) (±SD) was: 54.37 (±30.03) for Method 1, 35.89 (±34.89) for Method 2, and 13.50 (±12.64) for Method 3. Again, the ratio method (Method 3) is the most precise. Our internal laboratory procedure for method validation of analytical test methods, for water, wastewater, and soil/sludge matrix, is based on the criteria in the TR reference document (TR 25-02), published by the local ISO/IEC 17025 accrediting body South African National Accreditation Standards (SANAS), and is traceable to the ISO/IEC 17025 guide for method development and validation of test methods for testing and calibration laboratories. Our specified precision limit is RSD ≤10%. The observed reproducibility precision for Methods 1 and 2 is very much greater than 10% (54 and 36%), which exceeds the international limits, whilst that for Method 3 is just over 10% (13%), again indicating that the Ratio Method 3 is the most precise. However, it must be noted that the calculated reproducibility precision is based on data over this lengthy three-month (90 days - May to July) study period. In the initial report [21], intra and inter-day precision ranged from 0.1-0.7% RSD for all three methods. However, this precision was obtained on poly-DADMAC standards (30-90 µg/L) in laboratory water, and not on real environmental sample matrix. It is not clear as to the number of different days the reproducibility precision data was obtained. Furthermore, precision data at the lower, reported LOQ (2 µg/L) level for poly-DADMAC, for all three Methods, was not reported [21].

Accuracy of the poly-DADMAC assay values: The method validation data from the initial report [21] indicated recovery values (%error) of: 87-97% (-3 to -12%) for Method 1, 92-98% (-2 to -8%) for Method 2 and 92-89% (-8 to -11%) on poly-DADMAC solutions (30 µg/L), presumably in ultrapure water matrix, spiked at 10 and 20 µg/L. However, no recovery data is reported for blank matrix, or real samples, spiked at less than 10 µg/L, and/or at the LOQ level (≤2 µg/L). The method was also validated for selectivity where solutions containing DADMAC monomer and choline chloride were spiked with poly-DADMAC [21]. One of the major, and significant, raw materials used in poly-DADMAC synthesis, is the DADMAC monomer. Although the actual percentages are not known, it is a known fact that the commercial blends of the organic coagulants do contain some minor level of the DADMAC monomer. Their relative, residual concentration in the potable water will depend on their removal by the water treatment process at each water treatment works, which in turn is influenced by various factors. Polyelectrolyte applications in potable water production and industrial waste water treatment are in the coagulation and flocculation steps, and dewatering of treatment plant sludge. Polyelectrolytes have strong tendency to adsorb onto surface of particles in aqueous suspension, and is the main reason they are widely used in water treatment processes. The water industries are responsible for producing safe drinking water for people and all organisms in rivers, lakes and oceans. To keep water safe, polyelectrolytes are required to mix with turbid natural water for removing solid waste material before filtration. The main aim of introducing polyelectrolytes in water treatment is to induce flocculation and coagulation processes for the removal of suspended solid particles (colloidal matter). All waters, especially surface water, contain both dissolved and suspended particles, which are often assumed to be negatively charged. In suspension, particles repel each other and they cannot come together (stay stable in solution). As a result, they will remain in suspension. Coagulation is the processes where polyelectrolytes are added to destabilize the suspension or affect the surface of water. In coagulation, polyelectrolytes, overcome the factors that keep particles apart such as repulsion forces, and enable the particles to come together to form micro-flocs (flocs are cluster of small particles). In the flocculation process, polyelectrolytes are further added to induce the agglomeration of micro-flocs to form macroflocs (bigger particles). The macro-flocs, containing poly-DADMAC, settle or precipitate out of water and are removed as sludge. In our study, the observed values indicate significant poly-DADMAC levels (range: 2-24 µg/L) for all the raw dam water samples, by all three Methods. These levels are approximately two times higher compared to the corresponding levels in the treated water (range: 2-11 µg/L). Raw dam water, treated with poly-DADMAC-containing coagulant will be expected to contain some residual polymer present in the potable water, not removed by the sand filtration, or in the final sludge, estimated at ≤50 µg/L. The observation of the lower levels in the potable water indicates efficiency of the raw water treatment process in removing the added poly-DADMAC coagulant. The observed levels of residual poly-DADMAC on the 4 potable waters in this study indicate compliance to the international limit of ≤50 µg/L and accuracy of this gold nanoparticle method. Raw dam water, in the absence of environmental contribution, will not be expected to contain any, or significant levels, of poly-DADMAC. However, the much higher levels noted on our raw water suggests a possible inaccuracy of this gold nanoparticle test method. Alternatively, some other organic compound/s, that are present in the raw dam water, forms an aggregate with the citrate-capped Au-NPs and is subsequently detected by the colorimetric method.

Redox potential: The potential at the boundary (surface of hydrodynamic shear) is the zeta potential. The magnitude of the zeta potential gives an indication of the potential stability of the colloid system. In this study, the zeta potential of the water samples was not determined. The redox potential (or reduction potential) is the tendency of a chemical species to acquire electrons and thereby be reduced. In short, a numerically positive redox potential represents an environment conducive to the oxidation of an introduced substance by reduction of the original media. The redox potential showed a trend of being relatively negative for the raw waters (higher levels of poly-DADMAC) and relatively positive for the potable waters (lower levels of poly-DADMAC) (Supplementary Figures E and F). A decrease in zeta potential, towards negative values, was noted for the citrate-capped Au-NPs (and the other capped Au-NPs) when titrated with TOC as humic acid [25].

Possible sources for the observed levels of apparent poly-DADMAC in the raw water

Environmental contribution of poly-DADMAC: If the observed levels on the raw dam waters is really due to presence of poly-DADMAC, the corresponding catchment and its associated environmental factors, needs to be considered as a potential source: e.g., residence time, closeness of industries, sewage infiltration, agricultural activity, use of poly-DADMAC, the natural, mechanical mechanism of poly-DADMAC-floc aggregate removal by settling to the bottom of the dam by gravity, etc. The four dams supply raw water to the corresponding water treatment plants. The associated environmental data for each dam is summarised in Supplementary Table C. While the process effluent is returned to the Head of Works at Durban Heights (Nagle Dam) (Supplementary Table C), the average poly-(DADMAC) (3.64 µg/L) is not very different from that observed for Wiggins WW (Inanda Dam) (3.40 µg/L) (Table 3). Overall, the data indicates fairly negligible possibility of the environment as a source of poly-DADMAC.

Application of gold nanoparticles: Nanoparticles are one of the most important nanomaterials and they are defined as any material with at least one dimension in the 1-100 nm range. The particle shape may vary and the materials include metals, semiconductors,

polymers and carbon based materials. Gold nanoparticles have unique and very interesting physical-chemical properties, especially their optical properties. Like all other metal nanoparticles, Au-NPs undergo plasmon resonance, whereby the frequency of the incident electromagnetic radiation resonates with the oscillation of the delocalised electron cloud present on the nanoparticle surface. The localised surface plasmon resonance frequency lies in the visible range for Au-NPs, and because it is very sensitive to: diameter of the Au-NPs, the surrounding surface chemistry, the aggregation of Au-NPs, it has found use as a probe, or sensor, for the detection of large and small bio-molecules, various organic molecules and some inorganic ions [25-30].

Characterisation of the Au-NP complex: When gold nanoparticles are introduced into the poly-DADMAC solution, the intense red colour, (which shows an absorbance peak at 526 nm), of the Au-NPs decrease, with a slow appearance of a blue colour. The colour change is due to the shift of the plasmon band to longer wavelengths as the Au-NPs aggregate. The blue colour is attributed to the formation of aggregates between the Au-NPs and poly-DADMAC. Two possible scenarios have been proposed [21] poly-DADMAC has replaced the citrate ions on the surface of the Au-NPs, or, the poly-DADMAC has simply surrounded the citrate ions that are absorbed onto the Au-NPs. The Au-NPs have a high affinity for nitrogen and cationic molecules compared to citrate ions [25-29]. Thus, the former scenario is more likely, where the poly-DADMAC replaces the citrate ion, and destabilises the colloid with aggregate formation. In the current study, the aggregates of Au-NPs and poly-DADMAC are characterised by UV-Vis spectroscopy and TEM. In Figure 5, the peak at 526 nm is due to excess Au-NPs and a second peak at 690 nm, is due to the Au-NP-poly-DADMAC aggregates. The observation of the 164 nm shift to longer wavelengths is due to the formation of Au-NP-poly-DADMAC aggregates, which was confirmed by TEM analysis [21]. Other changes that can occur with NPs, beside aggregation that can cause a shift in the plasmon peak are: refractive index of the surrounding medium, surface chemistry of the Au-NPs, and changes with Au-NP size or shape [27,29,31]. The particle size analysis of the aggregates revealed that the Au-NPs have a similar morphology, before aggregation (27 ± 3 nm) and after aggregation (28 ± 3 nm) [21]. The Au-NP optical properties change with the size. Based on size alone, a shift of colour from red to blue occurs when the particle size changes from 3 to 60 nm [27,29]. Thus the shift in the plasmon resonance peak cannot be due to a change in size or shape of the Au-NPs, but it is due to the aggregation of the Au-NPs with poly-DADMAC [21].

Natural organic matter (NOM): NOM is a mixture of organic compounds, having diverse chemical properties, which occur in all natural water sources as a result of the breakdown of animal and plant material [32]. Since NOM emanates from different sources, it can be assumed that the composition of NOM in various water sources may not be uniform. NOM can be broadly characterised into: humic substances, microbial by-products and colloid natural organic matter. Humic substances constitute the more hydrophobic fraction of NOM and exhibit relatively high specific ultraviolet absorbance (SUVA) values since the humics usually contain a relatively large proportion of aromatic moieties. Huber et al. reported the characterisation of aquatic (river sample in Germany) humic and non-humic matter using size exclusion chromatography (SEC) [33] into biopolymers, humic substances, building blocks, low molecular weight acids, low molecular weight neutrals and hydrophobic organic carbon. The typical organic content (NOM) of raw dam water is biopolymers, which are very high in summer, and moderate in winter (250-800 µg/L), and humics (moderate: ±2,000 µg/L) [Huber S, Personal Communication, 2015].

Natural organic matter in eight South African water treatment plants, including Wiggins Water Works, of Umgeni Water, was characterized using combined techniques [34,35]. The dissolved organic carbon (DOC) results varied from 3.5-22.6 mg/L, indicating the extent of variation of NOM quantities in the different regions where the samples were obtained. The advanced techniques used indicated that the samples contained mainly humic substances, while some had marine humic and non-humic substances.

Reactions of NOM with Au-NPs: Nason et al. [35] studied the interactions between NOM and gold nanoparticles, stabilized with different organic capping agents, like anionic citrate, neutral polyvinylpyrrolidone, and cationic mercapto (trimethylammonium). Another report appeared on the gold(III) and Au-NPs interactions with humic acids [35,36]. A study on effects of NOM type and concentration on the aggregation of citrate-stabilized Au-NPs was further reported by Nason et al. [37]. They showed that four different NOM isolates act to stabilize citrate Au-NPs with respect to aggregation. The resulting stability appears to be due to adsorption of the NOM onto the surfaces of the NPs; the exact nature of the interactions between NOM and the coated Au-NPs is however, unclear. Both the type and concentration of NOM, along with the ionic strength of the system are important factors in determining the colloid stability.

Possible explanation for apparent poly-DADMAC levels in the Umgeni Water catchment (raw dam waters): NOM can be broadly characterized into: humic substances (HS), microbial by-products (composed of acids, with high charge density, polysaccharides, amino sugars, proteins), and colloidal natural organic matter (contain relatively polar amino sugars). Treatment of raw dam water, containing, inter alia, NOM, and other matter, with coagulant, like poly-DADMAC, removes NOM, and other matter, by floc formation, and subsequent filtration. Untreated raw dam water, not treated with coagulant, will contain the same concentration of NOM, and other matter.

The Au-NP colorimetric method, in this study, uses 20 mL of a suspension of about 200 particles per 410 mL, equivalent to ± 10 particles per sample/test. Examination of the typical UV-Vis spectrum of the gold nanoparticle solutions mixed with various concentrations of poly-DADMAC standards (10-100 µg/L) [21] shows the presence of the peak for the Au-NPs, at 520 nm, and the Au-NP-poly-DADMAC aggregate (690 nm), suggesting an excess amount of Au-NPs is still available for possible aggregate formation with other organic compounds, even at 100 µg/L poly-DADMAC concentration. Based on the earlier studies [35-37], we therefore propose that, on addition of Au-NPs to the raw water sample, there is some interaction of the NOM and Au-NPs to form aggregates, which are subsequently detected as "poly-DADMAC", by the colorimetric analytical method employed for determination of residual poly-DADMAC. Presumably, the Au-NP-(NOM) aggregate absorbs at the same wavelength of 690 nm, as the Au-NP-(poly-DADMAC) aggregate.

Influence of sample collection from different sites at different times: This is an important factor that must be considered in the evaluation of the analytical results obtained, and the subsequent conclusions made in this study. The grab, raw and potable, water samples in this study were collected over a 3-month period, from the designated sample sites, but on different days, and at different times. The day-today variation (%RSD) for water quality indices, and for residual poly-DADMAC levels, can therefore be expected to be fairly large. For example, for raw water, the %RSD for TOC ranges from 5-19%, and is 12-19% for potable water. For raw water, the %RSD for the apparent poly-DADMAC levels, by Method 1, ranges from

12-132%, and is fairly similar for potable water: 45-114%. The much higher variation in poly-DADMAC levels, in both raw and potable water, compared to variation for both a raw and potable, water quality index, can be due in part to, amongst others, the following factors: the much smaller number of samples (n=3-4, for poly-DADMAC, Table 3), compared to n=90 (for TOC, Table 1), and the relatively higher inherent imprecision noted for Method 1 (absorbance of peak at 690 nm) and Method 2 (area of peak at 60 nm), variation in the NOM levels in the various water samples taken.

Earlier work on poly-DADMAC quantitation in river water: Gumbi et al. [21] reported relatively lower levels of poly-DADMAC (not detected/2-2.1 µg/L) on samples (n=8) from the Umgeni River, in KwaZulu-Natal, using the same citrate-capped, gold nanoparticle colorimetric method. Again, one would not expect any, or significant, levels of poly-DADMAC, in natural river water. The corresponding TDS levels for these river samples were correspondingly lower: (mean) (±SD) 41 (±25) mg/L (range=18-69); the TOC levels on these river samples were not, however, reported [21]. For our raw dam waters, we observed an average TOC level of 2.6 (±0.40) mg/L for the raw waters. The apparent mean poly-DADMAC levels measured were 6.6 µg/L (range=<2 to 17 µg/L), with much higher TDS levels of 144 (±68) mg/L (range=65-244). Based on our findings in this study, the possibility exists similarly that the apparent levels of 2 µg/L noted in the earlier report [21] could be likely due to a relatively lower level of NOM (not reported) in river water, which forms an aggregate with the citrate-capped Au-NPs, and is detected by the UV-Vis colorimetric method.

Removal of NOM from water: Krause, et al. [32], in their characterisation of NOM in South Africa, investigated the use of cyclodextrin polyurethanes for NOM removal; the hydrophobic basic fraction and the hydrophilic acid fraction were most efficiently removed (24 and 10% respectively). The use of strong base anion exchange resin in the sample, about 10 g/50 mL, and shaking overnight [Huber S, 2015, Personal Communication] may remove most of the NOM while poly-DADMAC should stay in solution; the resin is cationic in charge and should repulse poly-DADMAC.

Evaluation of the application of the developed Au-NP colorimetric method

Although citrate-Au-NPs were synthesised in the present study, they are commercially available from NanoComposix Inc. and nationally (South Africa) (1.67×10^{11}-1.97×10^{11} particles/mL), but at substantial cost: ±R266 per 20 mL/sample. Beside this reagent cost, there are no other major reagent costs. The capital costs of the required equipment are affordable (±7394 Euros). Regarding the method performance, detailed data has been previously reported [22]. The linearity range was between 0 and 30 µg/ L with r^2=0.99 in all cases. The method LOD and LOQ (µg/ L) was 0.49 and 1.47 for Method 1 (absorbance of peak at 690 nm), 0.31 and 0.94 for Method 2 (area of peak at 690 nm) and 0.54 and 1.64 for Method 3 (ratio of the absorbance of peaks at 690 and 520 nm (A_{690nm}/A_{520nm}), indicating that Method 2 is the most sensitive. However, all three Methods are fairly sensitive, for the quantification of residual poly-DADMAC as the international maximum limit is about 25 times higher. However, the recovery (and percentage error) at this level was not reported [22]. The Method detection level (MDL) has been defined as follows: "the constituent concentration that, when processed through the complete method, produces a signal with a 99% probability that it is different from the blank. For seven replicates of the sample, the mean must be 3.14s above the blank where s is the standard deviation of the seven replicates…The MDL will be larger than the LLD…Recoveries should be between 50 and 150% and %RSD values ≤20%..." [38,39]. The Level of quantitation (LOQ) (Minimum quantitation level (MQL)) has also been defined as follows: "the constituent concentration that produces a signal sufficiently greater than the blank that it can be detected within specified levels…Typically it is the concentration that produces a signal 10s above the reagent water blank signal" [38].

The IUPAC method [39] uses the mean concentration and standard deviation from replicate analysis of a "blank" (ultrapure water) sample matrix, as per following equations: mean+10 SD, for LOQ, and mean+3 SD, for Limit of Detection (LOD), respectively. This statistical approach, however, cannot be applied when a negative value is observed for the signal response for the blank sample. The serial dilution technique, although it results in higher LOD and LOQ, would tend to be more accurate, being based on compliance to actual recovery, and precision, limits; selection of the "noise" region in a chromatogram, using the S/N method, is subjective, due to choice by the analyst. The computed instrument precision (% RSD) (±SD) was 18.05 (±17.65). The overall mean within-batch precision (% RSD) for the triplicate assay values were: 7.42 (±7.07) for Method 1, 7.66 (±7.37) for Method 2, and 1.92 (2.71) for Method 3, which complies with our internal limit of ≤10% RSD for method validation. The overall mean between-batch reproducibility was, however ≥10% RSD for all three calibration methods:% RSD (±SD): 54.37 (±30.03) for Method 1, 35.89 (±34.89) for Method 2, and 13.50 (±12.64) for Method 3. Method 3 was the most precise, and most inaccurate. Compared to other analytical methods, like colloid titration and gravimetry, this gold nanoparticle method is much faster, is far less labour-intensive and is much more sensitive.

Correlation analysis between observed residual poly-DADMAC levels and water quality parameters

The average poly-DADMAC levels, obtained by all three calibration methods (M1, M2, M3) on all raw and potable water, were compared with the corresponding average pH, conductivity, turbidity, TDS and TOC values. The results of all the statistical data analysis is summarised in Supplementary Table D. Due to the fairly good correlation of poly-DADMAC levels obtained by M1 and M2, any comparisons, and their possible significance, between poly-DADMAC levels obtained by M3 and water quality parameters can be ignored.

Raw water: For raw water, there was a strong positive linear relationship between: the apparent poly-DADMAC (M1) level and TOC, poly-DADMAC (M1, M2) level and pH, and poly-DADMAC (M1, M2) level and turbidity. A strong negative relationship between: the apparent poly-DADMCAC (M1, M2) level and Conductivity, poly-DADMAC (M1, M2) level and TDS, poly-DADMAC (M1, M2) level and Redox Potential was observed.

It was subsequently proposed that the observed, apparent poly-DADMAC levels are due to the presence and reaction of NOM with the Au-NPs. In such a case, it can be expected that water quality parameters, like TOC, TDS, conductivity (NOM contains some charged material) and turbidity of water would increase as the apparent poly-DADMAC (NOM, indicated by the TOC level) levels increase. However, a strong negative relationship is noted for the comparison with conductivity and TDS. In the absence of actual NOM levels, there will be obvious uncertainty in these comparisons. Particles that occur in natural waters are almost always negatively charged. Thus, as apparent poly-DADMAC (NOM) levels increase, redox potential will decrease (shift toward negative values). The raw water pH for the four dams ranged from 6.8-7.1. Acidic pH is known to destabilise citrate-

capped Au-NPs [30]. Gumbi et al. [22] showed that varying the pH (6-9) did not have any significant effect on absorbance or area for this colorimetric method for poly-DADMAC analysis. It would appear that NOM behaves similarly to poly-DADMAC, so that an increase in apparent poly-DADMAC level is noted with increasing pH.

Potable water: For potable water, there was a strong positive linear relationship between: the poly-DADMAC (M1, M3) level and TOC, poly-DADMAC (M1, M2) (level) and Redox potential. A strong negative relationship between the poly-DADMCAC (M1, M2, and M3) level and pH was noted. The TOC levels can be expected to increase with an increase in poly-DADMAC (organic material) levels. Poly-DADMAC is a cationic (positively charged) polymer. Hence, an increase in level would be expected to result in increasing (positive) Redox potential. The potable water pH for the four treated dam waters ranged from 6.8-6.9, which is not very different to that of the corresponding raw waters, and it falls within the reported stable range studied [21]. Although the sample numbers in this study is rather small (n=4), it would appear that optimum levels of poly-DADMAC (3.4-3.6 mg/L) are observed at average water pH 6.77 (Nagle Dam) and 6.79 (Inanda Dam).

Possible relationship between NOM and poly-DADMAC: Grab raw and potable water samples in this study were collected over a 3-month period, from the designated sample sites, but on different days, and at different times. The day-to-day variation (%RSD) for water quality indices, and for poly-DADMAC levels, can therefore be expected to be fairly large. For example, for raw water, the %RSD for TOC ranges from 5-19%, and is 12-19% for potable water. For raw water, the %RSD for the apparent poly-DADMAC levels, by Method 1, ranges from 12-132%, and is fairly similar for potable water: 45-114%. The much higher variation in poly-DADMAC levels, in both raw and potable water, compared to variation for both a raw and potable, water quality index, can be due in part to, amongst others, the following factors: the much smaller number of samples (n=3-4, for poly-DADMAC, Table 3), compared to n=90 (for TOC, Table 1), and the relatively higher inherent imprecision noted for Method 1 (absorbance of peak at 690 nm) and Method 2 (area of peak at 60 nm) , variation in the NOM levels in the various water samples taken. The TOC value is approximately equal to the NOM value for natural waters. Humic acid, a component of NOM, has been suggested as a standard for mimicking NOM in the laboratory. In the performance of Total Organic Carbon (TOC) analysis, UV persulfate instrumentation demonstrated 95% recovery of humic acid consistently across a linear range of 1 to 100 ppm C, the range typically found in the NOM of source water. In this study, there is a strong positive linear relationship between the poly-DADMAC level (determined by M1) and TOC, for both raw and potable water. The NOM levels were not determined analytically in this study. We can therefore expect some positive linear relationship between the NOM (not measured here) and poly-DADMAC levels. The observed/measured TOC levels can be used as an approximate indicator of the actual NOM levels. Raw dam water, not treated with coagulant, will contain NOM and other material, whereas potable water, treated with coagulant, will contain a much lower level of NOM and other material, due to the effect of the coagulant during the water treatment process. Subsequently, the apparent poly-DADMAC levels, from reaction of NOM with the Au-NPs, would be expected to be greater for the raw water, compared to the potable water. The TOC levels are, in general, expected to be greater for raw water (contains NOM, various organic matter, etc.) compared to treated, potable water (added coagulant aids in NOM, organic matter, etc. removal via flocculation-coagulation during the water treatment process). This is evident for each of the four dams. The TOC levels are as follows (raw vs. potable – mg/L): Inanda: 2.59 vs. 2.33; Nagle: 2.31 vs. 2.16; Hazelmere: 2.48 vs. 1.91; Wiggins: 3.17 vs. 2.04. We would therefore expect the NOM levels for the raw dam waters to exceed that for the treated potable water. Subsequently, the apparent (false positive) poly-DADMAC levels noted for raw water, from the proposed reaction of NOM in raw water, with the Au-NPs, would be expected to be greater than the actual true level of residual poly-DADMAC in the treated potable water (which contains a relatively lower of level of NOM, organic matter, etc.). The source of residual poly-DADMAC in the treated water is from the initially added coagulant (0.08-5 mg/L) during the water treatment process, and is expected to be ≤50 µg/L, the international limit. The latter is observed in this study. The observed poly-DADMAC levels are as follows (raw vs. potable – µg/L (Method 1): Midmar: 10.12 vs. 1.22; Hazelmere: 6.82 vs. 1.21; Nagle: 5.73 vs. 3.64; Inanda: 3.73 vs. 3.40. Compared to the other 3 dams, the average raw water turbidity for Inanda dam is distinctly the lowest (1.04 ± 0.30 NTU), and so is the corresponding coagulant dosing level (0.08-1.01 mg/L) (Table 3) required. Hence there is no significant difference in the apparent poly-DADMAC level in the raw water (3.73), and the treated water (3.40). A combined plot of TOC (converted to µg/L units) (x-axis) vs. poly-DADMAC concentration (y-axis) indicated a significant positive linear relationship: $r^2=0.8017$. We can therefore expect a significant positive linear relationship between the actual NOM and poly-DADMAC levels.

Degradation of poly-DADMAC

Detailed stability studies on poly-DADMAC were conducted by John [19], under different experimental conditions of exposure to temperature (ambient to 80°C/30 min), pH variations (2-12/1 hr.), UV radiation (365 nm/24 hr.) and ozone. At 80°C there was clear change in polymer structure. At pH 12, there was a noted decrease in the polymer peak area, but peak shape and MWD remained essentially unchanged. The UV radiation study showed evidence of polymer degradation. In essence, the GPC results indicated that poly-DADMAC is a very stable polymer and undergoes change only when subjected to extremes of pH, temperature and UV conditions, which are unlikely to be experienced under environmental conditions, and during the normal course of water treatment processes. The stability data on poly-DADMAC indicate very little or no effect on the validity or accuracy of our observed study results.

Conclusion

The current real world study indicates that the citrate-capped gold nanoparticle colorimetric method, using the calibration of peak absorbance at 690 nm (Method 1), or peak area at 690 nm (Method 2), is suitable for quantification of residual poly-DADMAC in potable water, treated with the poly-DADMAC coagulant. However, raw dam water, containing NOM, and possibly any other organic matter that may be present, apparently forms an aggregate with the citrate-capped Au-NPs, which absorbs at the same 690 nm wavelength as that of the Au-NP-poly-DADMAC aggregate and is subsequently detected by the UV-Vis colorimetric method. The test method was found to be sensitive (LOQ=0.9-1.6 µg/L), linear ($r^2=0.99$) and accurate over the range 0-30 µg/L for quantification of residual poly-DADMAC in treated, potable water. However, the instrument and inter-day method precision exceeded the internal limit of being ≥10% RSD. For potable water, there was a strong positive linear relationship between: poly-DADMAC levels and: TOC, Redox potential, and a strong negative, linear relationship between poly-DADMCAC levels and PH. Future research work must consider (inter alia): (1) improvement of the instrument and inter-day precision of the colorimetric analytical

method: The observed instrument precision (%RSD (±SD), for Method 1 (absorbance of peak at 690 nm) and Method 2 (area of peak at 690 nm), was 18.05 (±17.65) and 18.81 (±18.44), respectively, which exceed the typical ≤10% limit. The overall mean between-batch (reproducibility)% RSD was: 54.37 (±30.03) for Method 1, 35.89 (±34.89) for Method 2, and 13.50 (±12.64) for Method 3; (2) evaluation of the recovery at the observed LOQ: No recovery data for blank matrix or real samples, is reported in the original method [21], at spike levels less than 10 µg/L down to 2 µg/L of poly-DADMAC; (3) use of other organic capping agents, (e.g., tannic acid, polyvinylpyrrolidone): The current method development, and application, is based on the use of only citrate-capped gold nanoparticles; (4) efficient sample preparation methods for NOM removal from raw dam water The use of cyclodextrin polyurethanes for NOM removal was shown to achieve 10-24% NOM removal from raw water, in one South African study. The proposed use of strong base anion exchange resin, to remove most of the NOM, is one possible option; (5) transfer of this analytical to a real raw water treatment plant for application: The determination of residual poly-DADMAC in treated water is useful for at least two reasons: (1) establishment of over-dosing with the coagulant and (2) establishment of water quality compliance-health risk assessment, regarding the international allowable limit of a residual of ≤50 µg/L. The initially reported method development work, and this current study, was undertaken at laboratory scale, in an academic (university), and process evaluation, setting; (6) application of this same Au-NP colorimetric method to quantification of NOM, or other organic matter, present in raw dam water: The current study has shown possible interference by NOM, present in untreated raw dam water, by its reaction with the Au-NPs, in the analysis of poly-DADMAC. NOM in natural water can be quantified using size exclusion chromatography-organic carbon detection-organic nitrogen detection (LC-OCD-OND) [40]; (7) toxicity assessment studies of residual poly-DADMAC, and disinfection by-products (DBPs): poly-DADMAC can be toxic to aquatic organisms at levels above 50 µg/L, and has potential to form N-nitrosodimethylamine (NDMA), which is a disinfection by-product, and a suspected carcinogen.

Acknowledgements

The authors acknowledges: Mr N Dladla and Mr L Mthembu (Process Technicians, Wiggins Process Evaluation Facility, Umgeni Water), Dr P Ndungu, Lecturer, (UKZN) for the poly-DADMAC analyses, analytical test method information and other useful comments related to the test results, Mr S Terry (Scientist, Water Quality & Environmental Services, Umgeni Water), for providing some of the supplementary information, Mr J Ramjith (Biostatistician/Lecturer, University of Cape Town, South Africa), for the statistical data analysis, and the Innovative Research Development Committee (IRDC) (Umgeni Water), for funding the cost of the poly-DADMAC analyses.

References

1. Bolto B, Gregory J (2007) Organic polyelectrolytes in water treatment. Water Res 41: 2301-2324.

2. Majam S, Thompson PA (2006) Polyelectrolyte determination in drinking water. Water SA 32: 705-707.

3. Jin F, Hu J, Yang M, Jin X, He W, et al. (2006) Determination of diallyldimethylammonium chloride in drinking water by reversed-phase ion-pair chromatography-electrospray ionization mass spectrometry. J Chromatogr A 1101: 222-225.

4. Becker NSC, Bennet DM, Bolto BA, Dixon DR, Eldridge RJ, et al. (2004) Detection of polyelectrolytes at trace levels in water by fluorescent tagging. React Funct Poly 60: 183-193.

5. Nozaic DJ, Freese SD, Thompson P (2001) Longterm experience in the use of polymeric coagulants at Umgeni Water. Water Sci Technol: Water Supply 1: 43-50.

6. Bond T, Templeton MR, Graham N (2012) Precursors of nitrogenous

disinfection by-products in drinking water--a critical review and analysis. below J Hazard Mater 235-236: 1-16.

7. Shah AD, Mitch WA (2012) Halonitroalkanes, halonitriles, haloamides, and N-nitrosamines: a critical review of nitrogenous disinfection byproduct formation pathways. Environ Sci Technol 46: 119-131.

8. Sharma VK (2012) "Kinetics and Mechanism of Formation and Destruction of N-Nitrosodimethylamine in Water – A Review". Sep Purif Technol 88: 1-10.

9. Choi J, Valentine RL (2002) Formation of N-nitrosodimethylamine (NDMA) from reaction of monochloramine: a new disinfection by-product. Water Res 36: 817-824.

10. Padhye L, Luzinova Y, Cho M, Mizaikoff B, Kim JH, et al. (2011) PolyDADMAC and dimethylamine as precursors of N-nitrosodimethylamine during ozonation: reaction kinetics and mechanisms. Environ Sci Technol 45: 4353-4359.

11. Park SH, Wei S, Mizaikoff B, Taylor AE, Favero C, et al. (2009) Degradation of amine-based water treatment polymers during chloramination as N-nitrosodimethylamine (NDMA) precursors. Environ Sci Technol 43: 1360-1366.

12. Cumming J, Hawker D, Chapman H, Nugent K (2011) Water, Air, Soil Pollut 216: 441-450.

13. Cumming JL, Hawker DW, Nugent KW, Chapman HF (2008) Ecotoxicities of polyquarterniums and their associated polyelectrolyte-surfactant aggregates (PSA) to Gambusia holbrooki. J Environ Sci Health, Part A: Toxic/Hazard Subst. Environ Eng 43: 113-117.

14. NSF Fact Sheet –Polyelectrolytes and NSF/ANSI Standard 60 (2010) National Sanitation Foundation International: 12.

15. Letterman RD, Pero RW (1990) Contaminants in polyelectrolytes used in water treatment. J Amer Water Works Assoc 82: 87-97.

16. Cumming JL, Hawker DW, Matthews C, Chapman HF, Nugent K (2010) Toxicol Environ Chem 92: 1595-1608.

17. American Water Works Association (1993) Standard for Poly DADMAC, ANSI/AWWA B451-9, revision of ANSI/AWWA B451-87, Colorado.

18. John W (2008) Synthesis, properties and analysis of polydadmac for water purification. Ph D Thesis.

19. John W, Buckley CA, Jacobs EP, Sanderson DR (2015) Analysis of polydadmac by off-line membrane filtration gel permeation chromatography.

20. Mwangi IW, Ngila JC, Ndungu P (2012) A new spectrophotometric method for determination of residual polydiallyldimethylammonium chloride flocculant in treated water based on a diazotization-coupled ion pair. Water SA 38: 707-714.

21. Gumbi B, Ngila JC, Ndungu PG (2014) Gold nanoparticles for the quantification of very low levels of poly-diallyldimethlyammonium chloride in river water. Analytical Methods 6: 6963-6972.

22. Turkevich J, Stevenson PC, Hillier J (1951) A study of the nucleation and growth processes in the synthesis of colloidal gold. Discuss Faraday Soc 11: 55-75.

23. Toolsee N (2010) Jar Test procedure. Umgeni Water (Pietermaritzburg), Intranet; E&SS/PS/Proc/10.

24. Haiss W, Thanh NT, Aveyard J, Fernig DG (2007) Determination of size and concentration of gold nanoparticles from UV-vis spectra. below Anal Chem 79: 4215-4221.

25. Liu W, Zhang D, Tang Y, Wang Y, Yan F, et al. (2012) Highly sensitive and selective colorimetric detection of cartap residue in agricultural products. Talanta 101: 382-387.

26. Menon SK, Mistry BR, Joshi KV, Sutariya PG, Patel RV (2012) Analytical detection and method development of anticancer drug Gemcitabine HCl using gold nanoparticles. Spectrochim Acta A Mol Biomol Spectrosc 94: 235-242.

27. Ray PC (2010) Size and shape dependent second order nonlinear optical properties of nanomaterials and their application in biological and chemical sensing. Chem Rev 110: 5332-5365.

28. Vilela D, Gonzalez MC, Escarpa A (2012) Sensing colorimetric approaches based on gold and silver nanoparticles aggregation: Chemical creativity beyond the assay. A review. Anal Chim Acta 751 : 24-43.

29. Wang Z, Ma L (2009) Gold nanoparticle probes. Coord Chem Rev 253: 1607-1618.

30. Zhou Y, Yang Z, Xu M (2012) Colorimetric detection of lysine using gold nanoparticles aggregation. Anal Methods 4: 2711-2714.

31. Sun L, Liu D, Wang Z (2008) Functional gold nanoparticle-peptide complexes as cell-targeting agents. Langmuir 24: 10293-10297.

32. Nkambule TL, Krause RWM, Mamba BB, Haarhoff J (2009) Characterisation of natural organic matter (NOM) and its removal using cyclodextrin polyurethanes. Water SA 35: 200-203.

33. Huber SA, Balz A, Abert M, Pronk W (2011) Characterisation of aquatic humic and non-humic matter with size-exclusion chromatography--organic carbon detection--organic nitrogen detection (LC-OCD-OND). Water Res 45: 879-885.

34. Nkambule TL, Krause RWM, Haarhoff J, Mamba BB (2012) Natural organic matter in South African waters: NOM characterization using combined assessment techniques, Water SA 38: 1-16.

35. Stankus DP, Lohse SE, Hutchison JE, Nason JA (2010) Interactions between natural organic matter and gold nanoparticles stabilized with different organic capping agents. Environmental Science & Technology 45: 3238-3244.

36. Pefia-Mendez EM, Moreno FJ, Abizanda AIJ, Gonzalez JEC, Leon JJA, et al. (2011) Gold (III) and gold nanoparticles interactions with humic acids.

37. Nason JA, McDowell SA, Callahan TW (2012) Effects of natural organic matter type and concentration on the aggregation of citrate-stabilized gold nanoparticles. J Environ Monit 14: 1885-1892.

38. Eaton AD, Clesceri LS, Rice EW, Greenberg AE (eds) (2005) Standard methods for the examination of water and wastewater, (21st edn) USA. Introduction: Glossary 1010 C and Quality Assurance 1020: 1-, 1-60.

39. Long GL, Winefordner JD (1983) Limit of detection: a closer look at the IUPAC definition. Anal Chem 55: 712A-724A.

40. Huber SA, Balz A, Abert M, Pronk W (2011) Characterisation of aquatic humic and non-humic matter with size-exclusion chromatography--organic carbon detection--organic nitrogen detection (LC-OCD-OND). Water Res 45: 879-885.

14

Biostimulator and Biodegradable Chelator to Pytoextract not Very Toxic Cu and Zn

Yeh TY*

Department of Civil and Environmental Engineering, National University of Kaohsiung, Taiwan

Abstract

Taiwan spent too much expenditure to remove not very toxic metals Cu and Zn. The biosorption mechanism of metal removal (copper, Cu and zinc, Zn) by four phytoremediation macrophytes biomasses including sunflower (*Helianthus annuus*), Chinese cabbage (*Brassica campestris*), cattail (*Typha latifolia*), and reed (*Phragmites communis*) was investigated in this study. The primary objectives were exploring the potential of reusing these bio-wastes after harvesting from phytoremediation operations. Based on the surface area, zeta potential, scanning electron microscopy (SEM), and energy dispersive X-ray (EDX) investigations, Chinese cabbage biomass presented the highest metal adsorption property while both cattail and reed revealed a lower adsorption capability for both metals tested. The equilibrium adsorption rate between biomass and metal occurred very fast during the first 10 min. The metal adsorption data were fitted with the Langmuir and Freundlich isotherms and presented that the Langmuir isotherm was the best fitted model for all biomass tested. All tested biomasses are fast growing plants with fairly high biomass production that are able to accumulate metals. The Langmuir model was used to calculate maximum adsorption capacity and related adsorption parameters in this study. The results revealed that the maximum metal adsorption capacity Q_{max} was in the order of Chinese cabbage (Cu: 2000; Zn: 1111 mg/kg)> sunflower (Cu: 1482; Zn:769 mg/kg)> reed (Cu: 238; Zn: 161 mg/kg) cattail (Cu: 200; Zn: 133 mg/kg). The harvested sunflower, Chinese cabbage, cattail, and reed biomass possess the potential to be employed as biosorbents to remove Cu and Zn from aqueous solutions. Adsorption isotherms derived in this study might be crucial information for practical design and operation of adsorption engineering processes and prediction of relation between reused macrophyte biosorbents and heavy metal adsorbates.

Keywords: Heavy metals; Biosorbent; Macrophyte; Adsorption; Phytoremediation

Introduction

Phytoextraction, the use of plant for extraction the metals from contaminated the soils, has been viewed as a vital green remediation approach and has drawn great attention due to its low energy consumption and high public acceptance. It is an economic and non-invasive alternation to conventional civil engineering techniques for remediation of contaminated the soil. Phytoremediation mechanisms mainly include phytoextraction and phytostailization while phytoextration refers to extract the metals from the soils and concentrate them into the harvestable aerial parts while phytostabilization means the metal tolerant plants to reduce the mobility of the metals by leaching into groundwater. The degree of translocation from roots to aerial tissues mainly depends on the species of plants, types of the metals, or the soil the metal bioavailability. Phytoextraction can be used in areas with medium to low the soil pollution levels where physical-chemical the soil remediation techniques spell to be too costly. The wastewater generated from confined swine operations is one of the primary pollution sources in Taiwan [1,2]. The effluent is discharged in the surrounding waterways containing significant amounts of heavy metals such as copper (Cu) and zinc (Zn). These metals are intentionally added in fodder to prevent diarrhea and to enhance immune systems of swine. Conventional physical-chemical technologies employed for heavy metals removal for contaminated water include chemical precipitation, ion-exchange, however, they are usually quite costly and energy consumed. Phytoremediation using green plants in constructed wetlands and soil decontamination recently has drawn great attention in Taiwan and worldwide [3-5]. The biomass can be harvested and used for various purposes such as biosorbents for metal removal in water treatment [6,7]. The use and evaluation of recycled biosorbents

is very important to compare and analyze the adsorption mechanism and optimize the purification techniques that are based on biosorption. Several studies were published recently using recycled bio-wastes to remove pollutants [8-10]. The use of recycled and dried plants for metal removal as a simple biosorbent material has advantages in its efficiency in detoxifying dilute effluents and has been viewed as a cost-effective and energy-efficient wastewater treatment approach. The reuse of harvested macrophytes in wastewater engineering can also benefit waste disposal management and save waste treatment costs. The adsorption properties of phytoremediation macrophytes have been investigated for the removal of metals in polluted effluent. The results revealed that the extent of metal adsorption onto biomass seems to have important consequences in the capacity of metal removal [11]. Therefore, it is important to investigate the biosorption mechanism and related sorption parameters of harvested macrophytes to facilitate future biosorbent water purification operation. Metal cations in polluted effluent can be adsorbed by the negative charge of the macrophyte biomass surface. The process of metal removal by plants involves a combination of rapid sorption on the cell wall surface and slow accumulation and possibly translocation into the biomass

*Corresponding author: Yeh TY, National University of Kaohsiung, Department of Civil and Environmental Engineering, Kaohsiung 811, Taiwan
E-mail: tyyeh@nuk.edu.tw

[12]. The rapid sorption may include chelation and ion exchange. Carboxylic group, one of the functional groups on the plant biomass surface, provides binding sites with metals [13]. Research results indicated that all plant parts might accumulate heavy metals, and the ability to concentrate metals from the external solution varied between both plant parts and metals. Between 24% and 59% of the metal content was adsorbed onto the cell walls of the [14] The biomasses of plants, both living and dead, were heavy metal accumulators. The mechanisms of metal biosorption included extracellular accumulation, cell surface sorption, and intracellular accumulation. The semechanisms resulted from complexation, ion exchange, precipitation, and adsorption [15]. The main mechanism involved in biosorption was reported as ion exchange between metal cations and counterions presented in the macrophytes biomass. The investigation revealed that no significant difference was observed in the exchange amounts while using mutimetal or individual metal solutions [16]. Sunflower (*Helianthus annuus*) and Chinese cabbage (*Brassica campestris*) are fast-growing crops that have been commonly used for phytoextraction of metal contaminated soils, while reed (*Phragmites communis*) and cattail (*Typha latifolia*) are predominant macrophytes that have been employed for water purification within constructed wetlands. These plants contain high amount of lignin and cellulose which may adsorb heavy metal cations from aqueous solution. After harvesting, these plant biomasses could be used as biosorbents for metal adsorption. Brassica family has been reported for its prominent ability to remove heavy metals from contaminated soils [17]. *B. campestris* and *H. annuus* have the potential as biofuel to become the substitute of fossil fuels, especially the increasing oil prize in recent years. The higher biomass production of these economic crops, namely sunflower and Chinese cabbage, contribute them being the candidates of phytoextration contaminant and then harvested as potential biosorbents. Reed and cattail, commonly used macrophytes in constructed wetlands for water pollution mitigation, have been reported as a very high adsorption affinity value, which assist to predict its high ability to adsorb heavy metals in aqueous solutions [18]. This study focused on the biosorption characteristics of the harvested biomass of plants may provide information for enhancing phytoremediation processes to remove metals both in soil and water. The aim of this study was to investigate the biosorption performance and mechanisms of four macrophyte biomasses. The benefits from this study were two folds: to highlight the metal adsorption capability of plant biomass for environmental decontamination, and to test the possibility to recycle the harvested biomass for biosorbents. Biostimulator has been facilitated the plant growth enhancement and been employed for agricultural operation were tested to evaluate vetiver the metal attenuation enhancement. Properties of tested biostimlator. The stiumulators can be borrowed to enhance the vetiver propagation leading to expected phytoattenuation purpose. GA3 and IAA. The properties of GA3 and IAA the objective of this research is to investigate stable the soil Pb and Cr by employing biostimulator IAA and GA3 and biochelator citric acid and humic acid using high uptake vetiver.

Materials and Methods

Plant, biostimulators, and the soil preparation

Vetiver and sunflower were collected from the University of Kaohsiung campus wetlands (22°73'N, 120°28'E) precultured for 5 days and carefully washed with distilled water. Plant samples were dried at 103°C in an oven until completely dried.

Total the metal content, the soil retained fractionation and plant the metal uptake analysis

Harvested plant tissue and final the soil the metal content analysis

Plant after last session of operation was harvested, careful washed, and air dried for the metal analysis. Plant samples were dried at 103°C in an oven until completely dried. Dried plant samples were divided into root and shoot for the metal accumulation assessment. These pretreated plants were digested in a solution containing 11:1 HNO_3: HCl solution via a microwave digestion apparatus (Mars 230/60, CEM Corporation) and diluted to 100 mL with deionized water. 0.2 g of dried the soil adding aqua regia rending for microwave digestion and 2.5 g of dried for sequential extraction experiments. The metals analyses were conducted via an atomic absorption spectrophotometry (AAS, Perkin Elmer). 2.3. Harvested Plant tissue and final the soil the metal content analysis. Plant was harvested, careful washed, and air dried for the metal analysis. Plant samples were dried at 103°C in an oven until completely dried. Dried plant samples were divided into root and shoot for the metal accumulation assessment. These pretreated plants were digested in a solution containing 11:1 HNO_3: HCl solution via a microwave digestion apparatus and diluted to 100 mL with deionized water. 0.2 g of dried the soil was added aqua region rending for microwave digestion. The metals analyses were conducted via an atomic absorption spectrophotometry (AAS, Perkin Elmer).

Macrophet surface adsorption properties detection using Scanning electron microscopy (SEM) and energy dispersive X-ray (EDX) spectroscopy

Pretreated macrophyte samples were gold-coated for SEM observation with qualitative EDX analysis. Specifically, grinded and dried samples were mounted on carbon tape and sputter coated in gold. A Hitachi S-4300 SEM (Tokyo, Japan) was used to capture micrographs. The elements C, O, Cu, and Zn were detected using a SEM coupled with an EDX spectroscopy at an acceleration voltage of 15 kV.2.5.1 FTIR Fourier Transform Infrared (FT-IR) regards as the preferred method of infrared spectroscopy. In infrared spectroscopy, IR radiation is passed through a sample. Some of the infrared radiation is absorbed by the sample and some of it is passed through (transmitted). The resulting spectrum represents the molecular absorption and transmission, creating a molecular fingerprint of the sample. Like a fingerprint without two unique molecular structures produce the same infrared spectrum. This makes infrared spectroscopy useful for several types of analysis. Infrared spectroscopy can result in a positive identification (qualitative analysis) of every different kind of material. In addition, the size of the peaks in the spectrum is a direct indication of the amount of material present. With modern software algorithms, infrared is an excellent tool for quantitative analysis. 2.4 Data and Statistical analysis. Data were evaluated relative to the control to understand their statistical variation. A triplicate of water and sediment samples were measured and recorded for statistical analyses. Statistical significance was assessed using mean comparison test. Differences between treatment concentration means of parameters were determined by Student's t test. One-way ANOVA was also employed to show the variation among sample groups, level of p<0.05 considered statistically significant was used in all comparisons. Means are reported mean ± standard deviation. All statistical analyses were performed with Microsoft Office EXCEL 2007.

Results and Discussion

The properties of tested macrophytes

The surface areas of four studied biomasses were 2.75 ± 0.48,

3.71 ± 0.13, 2.30 ± 0.03, and 2.43 ± 0.17 m²/g, for sunflower, Chinese cabbage, cattail, and reed, respectively, analyzed by the BET method with liquid N_2. Chinese cabbage, the Brassica family, has the largest surface area in this study rendering for better metal adsorption. The adsorption capacity can be further illustrated via comparing the electro kinetic potential (zeta potential) as shown in Figure 1. The effect of pH on the zeta potential of all tested macrophytes was examined. The zeta potential had negative charge for all studied macrophytes rendering for the potential of metal adsorption. The increase in negative charge of the zeta potential was observed while the pH increased. This result indicated that the degree of metal biosorption may increase as the pH increased. The biomass Chinese cabbage was recorded as the negative zeta potential around neutral pH while the lowest recorded was at pH. This result revealed that Chinese cabbage had better metal adsorption capability compared to other macrophytes tested. The rest of tested plants also presented negative charge of zeta potential following the order sunflower <reed< cattail. The lower negative zeta potential also indicated better metal cations adsorption.

The metal adsorption rate and isotherms

The adsorption rate of Cu and Zn by four studied biomasses is depicted in Figure 2. Most of metal biosorption occurred during first 10 min. This adsorption result revealed that a contact time of 120 min for both Cu and Zn was sufficient to achieve equilibrium for four tested macrophytes. Similar rapid metal biosorption has been reported by other researcher Bunluesin [19]. Several factors including the structure of biosorbent and existence of metal species have also been presented to influence adsorption rates. In order to obtain basic information of tested macrophytes as biosorbents, the equilibrium metal concentration (Ce) and the concentration adsorbed onto the surface of the biomass (Q) were linearized and fitted to the Langmuir and Freundlich equations. The Langmuir and Freundlich isotherm models were calculated to determine the adsorption capacities and related parameters. The sorption process for Cu and Zn by four tested biomasses was better described by the Langmuir equation (R^2=0.90-0.99) compared to the Freundlich model (R^2=0.67-0.97). The linear regression was calculated to demonstrate that the Langmuir equation was best fitted, therefore, the sorption as a monolayer can be assumed. The maximum sorption capacity Qmax of Cu was 1482, 2000, 200, and 238 mg/kg while the Qmax of Zn was 769, 1111, 133, and 161 mg/kg for biomass sunflower, Chinese cabbage, cattail, and reed, respectively, predicted by the Langmuir model. The aforementioned maximum sorption capacity was comparable with that of the activated carbon and less than that of the tested biosorbent peanut hulls [20]. The related adsorption parameters were also calculated through the Langmuir equation. For Cu, the binding constant b was 3.00, 3.80, 0.42, and 0.46

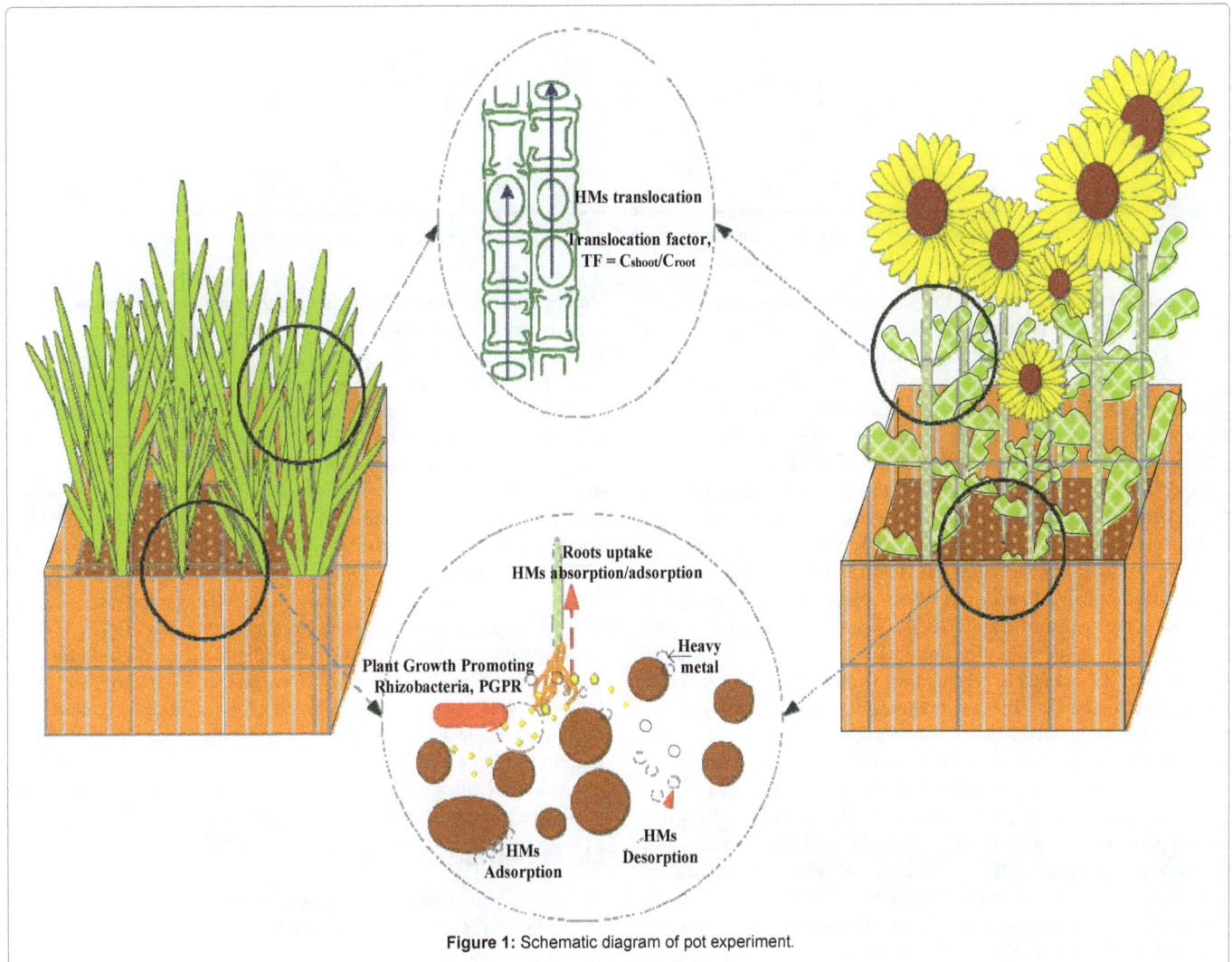

Figure 1: Schematic diagram of pot experiment.

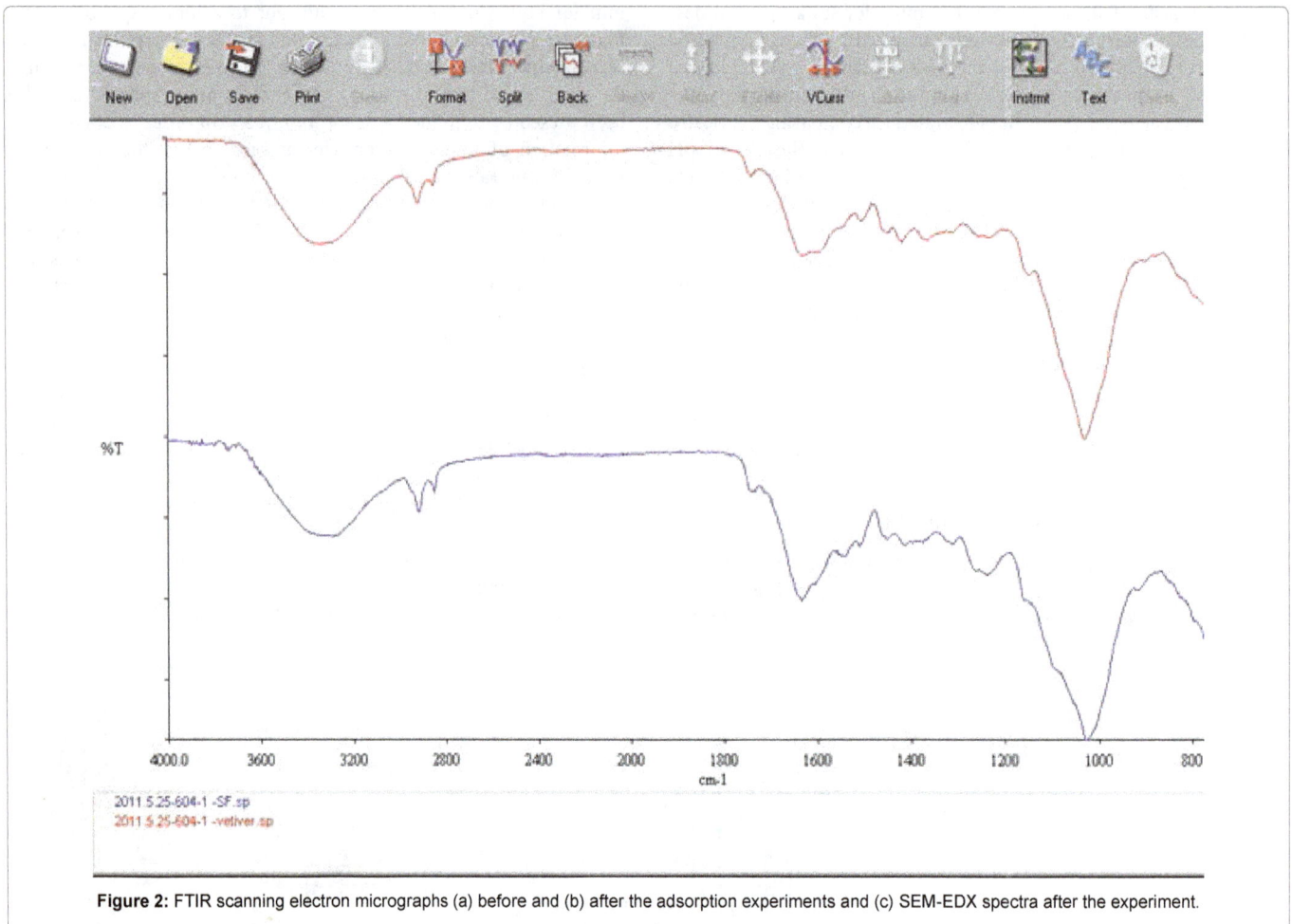

Figure 2: FTIR scanning electron micrographs (a) before and (b) after the adsorption experiments and (c) SEM-EDX spectra after the experiment.

for biomass sunflower, Chinese cabbage, cattail, and reed, respectively. For Zn, the binding parameter b was 2.92, 5.11, 0.51, and 0.54 for biomass sunflower, Chinese cabbage, cattail, and reed, respectively. The high b value of Chinese cabbage biomass is reflected by the steep initial slope of the adsorption isotherm which indicated a high affinity for the adsorbate in dilute metal solutions. Research has presented that wetland macrophyte, Ceratophyllum demersum, was an effective biosorbent for Zn and Cu removal under dilute metal conditions. Batch adsorption experiments showed that the Langmuir isotherm was best fit model and the maximum adsorption capacity was 13.98 mg/g for Zn and 6.17 mg/g for Cu [15]. Similar study was conducted to test the dried free floating macrophyte Lemna minor biomass regarding its adsorption of metals from aqueous solutions. The equilibrium adsorption was reached within 40-60 min. The maximum adsorption capacities of biomass was determined as 83 mg/g for Cu based on the best fitted Langmuir equation [21]. The maximum adsorption capacity might vary with the biomass investigated and adsorption experimental conditions. The equilibrium metal concentration (Ce) after a contact time of 5 h was lower than the initial concentration (Ci). Five hours was assumed to be adequate for the adsorption system to achieve equilibrium which was longer than the time (60 min) to reach equilibrated condition in the aforementioned adsorption rate experiment. The removal efficiency of metals from solutions can be expressed as the fraction of metals adsorbed by studied biomasses which was related to the reciprocal value of the ratio of the metal concentration in the solution at equilibrium to

that in the initial solution. In general, the fraction of metals adsorbed onto biomass decreased as the initial concentration Ci increased. At high initial Cu concentration (10 mg/L), the percentage of Cu that was adsorbed by the biomass decreased to around 20% then gradually leveled off for both cattail and reed while sunflower and Chinese cabbage continued to drop. At high initial Zn concentration (5 mg/L), the percentage of Zn that was adsorbed by the biomass decreased to around 18% then gradually leveled off for both cattail and reed while sunflower and Chinese cabbage gradually decrease. At low initial Cu concentration (1 mg/L), the metals adsorbed by the biomass ranged 72% for sunflower, 73% for Chinese cabbage, 61% for cattail, and 67% for reed, respectively, while at low initial Zn concentration (1 mg/L), the metals adsorbed by the biomass ranged 50% for sunflower, 54% for Chinese cabbage, 35% for cattail, and 40% for reed, respectively. The biosorption efficiency was high at a low metal concentration, especially for Chinese cabbage and sunflower. At a low metal concentration, the ratio of available adsorbent surface area to the metal in solution was high indicating a great metal removal. As metal initial concentration increased, the efficiency was gradually decreased. This result might be attributed to the saturation of the adsorption sites on the biomass.

The microstructure investigation

The microstructures of the tested biosorbents and adsorbed metal determinations onto biomass surface were performed by the scanning electron microscopy (SEM) and energy dispersive X-ray (EDX) (Figures

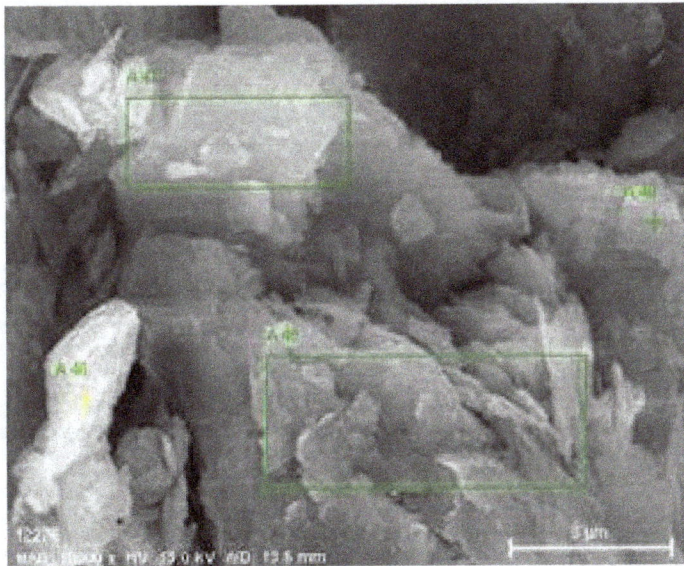

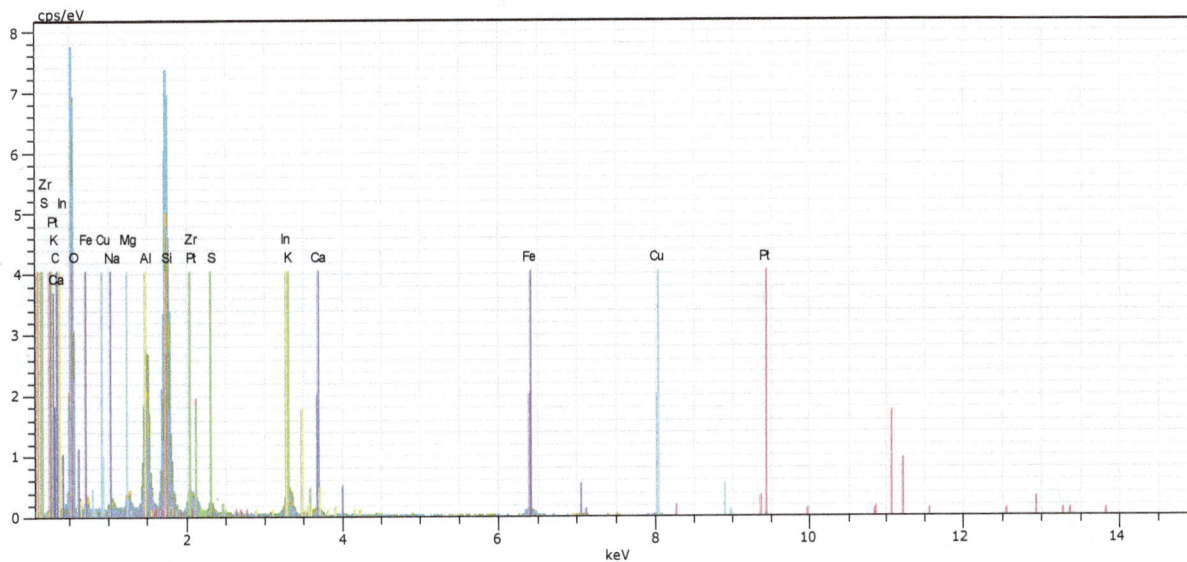

Figure 3: SEM/DEX.

1 and 2). The biomass treated with metals revealed several small bulges that were not observed before the metal sorption experiment. Further EDX observations indicated that small bulges are higher in Cu and Zn. There were more bulges on the surface of Chinese cabbage compared to other three studied macrophytes. The results also suggested that Chinese cabbage might have better metal sorption capacity (Figure 3).

Conclusion

Taiwan spent too much expenditure to remove not very toxic metals Cu and Zn. The harvested biomass of sunflower, Chinese cabbage, cattail, and reed possesses the potential to be used as biosorbents to remove metals from aqueous solutions. Adsorption experiment results showed that Cu and Zn adsorptions were fairly rapid occurring within first 10 min. The adsorption capability of four tested biomasses can be well predicted by the Langmuir adsorption model. The surface area, zeta potential, SEM, and EDX results revealed that Chinese cabbage

biomass presented the highest metal adsorption property while both cattail and reed presented lower adsorption capability for both metals tested. Further study (e.g. FT-IR) might be required to scrutinize the chemical functionalities responsible for the adsorption of the heavy metals. These studied plant biomasses are natural abundant and can be recycled from environmental decontamination operations, namely phytoremediation of metal polluted soil and water purification within constructed wetlands. This research results can benefit adsorption process engineering for mitigation of polluted metal water by reusing harvested macrophytes.

References

1. Lee CY, Lee CC, Lee FY, Tseng SK, Liao CJ (2004) Performance of subsurface flow constructed wetland taking pretreated swine effluent under heavy loads. Bioresour Technol 92: 173-179.

2. Yeh TY, Chou CC, Pan CT (2009) Heavy metal removal within pilot-scale

constructed wetlands receiving river water contaminated by confined swine operations, Desalination 249: 368-373.

3. Dhote S, Dixit S (2008) Water quality improvement through macrophytes--a review. Environ Monit Assess 152: 149-153.

4. Yeh TY, Wu CH (2009) Pollutant removal within hybrid constructed wetland systems in tropical regions. Water Sci Technol 59: 233-240.

5. Yeh TY, Pan CT, Ke TY, Kuo TW (2010) Organic matter and nitrogen removal within field-scale constructed wetlands: Reduction performance and microbial identification studies", Water Environment Research 82: 27-33.

6. Jang A, Seo Y, Bishop PL (2005) The removal of heavy metals in urban runoff by sorption on mulch. Environ Pollut 133: 117-127.

7. Tsui MT, Cheung KC, Tam NF, Wong MH (2006) A comparative study on metal sorption by brown seaweed. Chemosphere 65: 51-57.

8. Bansal M, Singh D, Garg VK (2009) Chromium (VI) uptake from aqueous solution by adsorption onto timber industry waste. Desalination and Water Treatment 12: 238-246.

9. Hannachi Y, Shapovalov NA, Hannachi A (2009) Adsorption of nickel from aqueous solution by the use of low-cost adsorbents. Desalination and Water Treatment 12: 276-283.

10. Okoronkwo AE, Aiyesanmi AF, Olasehinde EF (2009) Biosorption of nickel from aqueous solution by Tithonia diversifolia. Desalination and Water Treatment 12: 352-359.

11. Miretzky P, Saralegui A, Cirelli AF (2004) Aquatic macrophytes potential for the simultaneous removal of heavy metals (Buenos Aires, Argentina). Chemosphere 57: 997-1005.

12. Lesage E, Mundia C, Rousseau DPL, Van de Moortel AMK, Du Laing G, et al. (2007) Sorption of Co, Cu, Ni and Zn from industrial effluents by the submerged aquatic macrophyte Myriophyllum spicatum L. Ecological engineering 30: 320-325.

13. Suñe N, Sánchez G, Caffaratti S, Maine MA (2007) Cadmium and chromium removal kinetics from solution by two aquatic macrophytes. Environ Pollut 145: 467-473.

14. Fritioff A, Greger M (2006) Uptake and distribution of Zn, Cu, Cd, and Pb in an aquatic plant Potamogeton natans. Chemosphere 63: 220-227.

15. Keskinkan O, Goksu MZL, Basibuyuk M, Forster CF (2004) Heavy metal adsorption properties of a submerged aquatic plant (Ceratophyllum demersum). Bioresource Technology 92: 197-200.

16. Miretzky P, Saralegui A, Fernández Cirelli A (2006) Simultaneous heavy metal removal mechanism by dead macrophytes. Chemosphere 62: 247-254.

17. Grispen VM, Nelissen HJ, Verkleij JA (2006) Phytoextraction with Brassica napus L.: a tool for sustainable management of heavy metal contaminated soils. Environ Pollut 144: 77-83.

18. Southichak B, Nakano K, Nomura M, Chiba N, Nishimura O (2006) Phragmites australis: a novel biosorbent for the removal of heavy metals from aqueous solution. Water Res 40: 2295-2302.

19. Bunluesin S, Kruatrachue M, Pokethitiyook P, Upatham S, Lanza GR (2007) Batch and continuous packed column studies of cadmium biosorption by Hydrilla verticillata biomass. J Biosci Bioeng 103: 509-513.

20. Oliverira FD, Paula JH Freitas OM, Figueiredo SA (2009) Copper and lead removal by peanut hulls: Equilibrium and kinetic studies. Desalination 248: 931-940.

21. Saygideger S, Gulnaz O, Istifli ES, Yucel N (2005) Adsorption of Cd(II), Cu(II) and Ni(II) ions by Lemna minor L.: effect of physicochemical environment. J Hazard Mater 126: 96-104.

15

Acid Mine Drainage in Chile: An Opportunity to Apply Bioremediation Technology

Johanna Obreque-Contreras, Danilo Pérez-Flores, Pamela Gutiérrez and Pamela Chávez-Crooker*

Aguamarina SA. Centro Tecnológico SR-97, Las Colonias 580, Antofagasta, Chile

Abstract

The use of micro organisms for heavy metal remediation in water is a technique widely studied. This review describes a number of methods used for acid mine drainage (AMD) remediation, containing high concentration of this type of contaminant. The AMD is a problem generated in abandoned mines and low grade stock of active mines, therefore it is an existing problem in mining countries. In this review it is described the problem in Chile, regulations and the challenge to resolved this problem for a sustainable industrial future.

Keywords: Acid mine drainage; Bioremediation; Chile

Abbreviation: AMD: Acid Mine Drainage

Introduction

The worldwide mining activities of ores are directly associated with acid mine drainage (AMD), which is recognized as one of the most serious environmental problems. Some effluents contain large quantities of toxic substances, such as cyanides and heavy metals, which have serious implications for human and ecological health. AMD is produced when sulfide-bearing material is exposed to oxygen and water.

Copper mining in Chile dates back at least three hundred years. At the beginning, extraction methods were simple because ores contained high grades of gold, silver, and copper. Over time as ore grade decreased, the extraction process has improved through the adoption of more sophisticated technology. At the present time, there is a high demand for treatments in order to recover AMD for further uses. However, the main concern will be focused on mine closure planning.

Many mining operations were abandoned in earlier years, prior to environmental regulations and without proper closure procedure. These abandoned operations are distributed throughout the country. Abandoned gold, copper, silver, poly-metal, carbon, and iron mines all represent significant risks to safety, human health and the environment. Further treatments are needed in order to finally close and recover these areas to give it back to the community.

The abiotic treatment technologies are highly efficient, but high costs due to the high-energy demand and large amounts of chemicals and expensive devices, then bioremediation appears as an attractive alternative; it is economically viable and environmentally friendly.

Bioremediation is an attractive and low cost operation to break down most of these contaminants through different mechanisms according with selected microorganisms, environment and final desired product. In this review we intend to introduce AMD, bioremediation techniques and the opportunity to apply them in the Chilean mining industry.

Acid Mine Drainage

Mining in Chile required great quantities of resources to work, due to the scale of their operations. A solution is to use a close circuit of solutions where every resource is used as much as possible with the positive consequence of reducing the waste production and controlling interactions between the process by-products and the environment [1]. Because acid generation and metals dissolution are the primary problems associated with pollution from mining activities. The chemistry of these processes appeared fairly straightforward, but gradually becomes complicated as geochemistry and physical characteristics can vary greatly from site to site [2].

The principal producer of acidic sulfur wastewaters is the mining industry. Water drainage in active, abandoned mines and mine tailings is often acidic (sometimes extremely so) [3]. Those acidic waters typically pose an additional risk to the environment by the fact that they often contain elevated concentrations of metals like iron, aluminum and manganese, and possibly other heavy metals and metalloid as arsenic, which is generally of greatest concern [3]. In the USA, the area polluted by AMD have been estimated, reaching 180,000 acres of lakes and natural reserves; 12,000 miles of rivers and channels. With an approximated cost of $32-72 billion dollars, while the Canada treating cost would be in the range $2-5 billion dollars [4].

AMD generation

The acidic mine drainage (AMD) is usually produced by the accelerated oxidation of iron pyrite, which is in fact the most common sulfide mineral worldwide [2,3,5].

Acid generations starts with Pyrite (FeS_2) and metals dissolution in coal and hard rock among others types of ore, which explains why AMD is a transversal issue for most of the copper sulfide, coal, and gold operations throughout the world. When pyrite is exposed to oxygen and water gets oxidizes, releasing hydrogen ion, acidity, sulfate ions, and soluble metal cations [2]. There is a bigger group of sulfide mineral which can be referred to as "pyrite type", as pyrrhotite, chalcopyrite, arsenopyrite, spharelite, galena and others that might initiate a similar processes leading to AMD production [5], turning this into a potentially catastrophic issue for countries as Chile, based on the typical mineralogy and the number of running and abandoned

---

*Corresponding author: Pamela Chávez-Crooker, Aguamarina SA, Centro Tecnológico SR-97, Las Colonias 580, Antofagasta, Chile
E-mail: pchavez@aguamarina.cl

Mineral capable of producing AMD			
With oxygen		With ferric iron	
Mineral	Formula	Mineral	Formula
Pyrite	FeS_2	Spharelite	ZnS
Pyrrhotite	$Fe_{1-x}S$	Galena	PbS
Bornite	Cu_5FeS_4	Chalcopyrite	$CuFeS_2$
Arsenopyrite	FeAsS	Covelite	CuS
Enargite	Cu_3AsS_4	Cinnabar	HgS
Luzonite	Cu_3SbS_4		
Sulfarsenide	AsS	Millerite	NiS
Oropimente	As_2S_3	Pentlandite	$(Fe,Ni)_9S_8$
Stibnite	Sb_2S_3	Greenockite	CdS

Table 1: Minerals with AMD producing potential.

operations existing, especially on the north of the country, in case of south Brazil it has become the biggest environmental issue for the mining industry, especially at the south of Santa Catarina state and Rio Grande do Soul [4].

Table 1 shows different ores capable of producing AMD when exposed to oxygen or with ferric iron [1].

AMD is often characterized as low pH water with elevated concentrations of iron, sulfates and heavy metals, the heavy metals profile depend on the originating mineral deposit types [6]. Acid drainage is produced when minerals containing reduced forms of sulfur (S) are exposed to oxygen and water, being oxidized. In coal-mining areas, the most common of these minerals is iron pyrite (FeS_2); also the presences of acidophilic bacteria catalyzed these reactions. In FeS_2 case, AMD formation can be represented by the following equations:

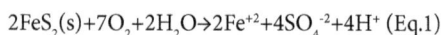

$$2FeS_2(s)+7O_2+2H_2O \rightarrow 2Fe^{+2}+4SO_4^{-2}+4H^+ \ (Eq.1)$$

$$2Fe^{+2}+1/2\ O_2+2H^+ \rightarrow 2Fe^{+3}+H_2O \ (Eq.2)$$

$$2Fe^{+3}+6H_2O \leftrightarrow 2Fe(OH)_3(s)+6\ H^{+2} \ (Eq.3)$$

$$14Fe^{+3}+FeS_2(s)+8H_2O \rightarrow 2SO_4^{-2}+15Fe^{2+}+16H^+ \ (Eq.4)$$

The process is initiated with pyrite oxidation and release of ferrous iron (Fe^{+2}), sulfate (SO_4^{2-}), and hydrogen (H^+) (Eq.1). The sulfuroxidation reactions are catalyzed by bacterial metabolisms; where *Acidithiobacillus ferrooxidans* appears to be the main bacteria, especially because abiotic oxidation of iron is negligible at this pH values [3]. Next, ferrous iron undergoes oxidation forming ferric iron (Fe^{+3}) as equation 2 shows (Eq.2). Ferric ion is hydrolyzed by the reaction with water and formed ferric hydroxide $Fe(OH)_3$, an insoluble organic precipitant, and release additional acidity (Eq.3). The rate of $Fe(OH)_3$ formation is pH-dependent; it occurs rapidly when pH>4. Finally, further pyrite oxidation by ferric iron and another metals (Eq.4) [7,8].

These reactions can be produced in any time during the active mines, nevertheless, after when are closed and abandoned, the pumps turned off and are not anymore keep the water tables artificially low, then the water table can lead to contaminated groundwater being discharged, sometimes in a catastrophic event to human health an environmental [3,9]. A similar scenario occurs on low grade stocks, once the mining operation stops there no new material covering the stock, isolating from oxygen and humidity.

No matter it is an industrial bioleaching operation or AMD production, shows the same behavior, where Initial drainage has higher ionic concentrations and acidity and tend to get lower on time, dissolving any possible salt and metal from the ore. This means that

when water refills the mine and dissolves any acidic salts that have built up on the pore spaces of the exposed walls and ceilings of underground chambers. This initial drainage water tends to be more potentially polluting (in terms of acidity and metal content) than AMD that is subsequently discharged [3,10].

Water with high metal concentration and strong acidity may be formed in mineral tailings and dump heaps, following the same reactions and biological phenomena than mine shafts and adits. Due to the more disaggregated state of ore when leaching ends, or higher concentration on tailings, or high "pyrite type" material on low grade stocks, AMD that flows from them may be more aggressive than discharges from the mine itself. The long term AMD production capacity is another issue to be concerned of, because AMD production may continue for many years after mines are closed and tailing dams are decommissioned. When talking about mine water discharges the terms "acid mine drainage" or "acid rock drainage" are used frequently, even when pH may be almost neutral (above 6), particularly at the point of discharge (where dissolved oxygen concentrations are frequently very low), this allows metals as iron and manganese to be as their reduced form (Fe^{2+} and Mn^{2+}), more stable than the complete oxidized form (Fe^{3+} and Mn^{4+}) at these pH. As water flows, oxygen might dissolved into it and decline its pH (not every AMD does). Net acidity has two components, the proton acidity (hydrogen ion concentration) and mineral acidity, dependent of concentration of soluble metals that produce protons when hydrolyzed. This particularity is used as an ore AMD production potential indicator, useful to create a closure plan for industrial operations [3,11,12]. In another hand, the alkalinity present for the bicarbonate deriving from the dissolution minerals or biological process can fall down the net acidity in AMD [3,4].

The showed reactions occur in a natural way in undisturbed rock, but at a very slow rate, and the water is able to buffer the acid generated. The mining produces an excess acid generation, beyond the water's natural buffering capabilities due to particle size reduction; hence increasing exposed surface area [4].

Characteristics AMD coming from different locations.

Tables 2 and 3 shows AMD characteristic from different localities, main pollutants correspond to high concentrations of iron, sulfate and other metals. In addition acidophilic bacteria that catalyze the generation of AMD are present. No information has been found, for Chile, but one could expect that AMD would have similar characteristics

	King's mine, Norway	Parys mine, United Kingdom	Cantareras, Spain	Richmon mine, USA
pH	2.7	2.5	2.7	0.5-1
Eh (mV)	-	285	425	-
Fe_{total}	171	650	1130	$13-19 \times 10^3$
Fe^{2+}	-	650	915	$13-19 \times 10^3$
Cu	16	40	160	120-650
Zn	25	60	24	700-2600
Sulfate	668	1550	1190	$20-100<10^3$
Moderate Fe-oxidizers	1×10^3	1×10^3	$<10^2$	-
Extreme Fe-oxidizers	6×10^4	3×10^3	1×10^5	-
S-oxidizers	<50	$<10^2$	$<10^2$	-
Heterotrophic acidophiles	2×10^4	2×10^3	$<10^2$	-

Table 2: Physic-chemical characteristics of AMD from various sites worldwide and associated microbial populations (Cupper mine).
All concentrations of metals and sulfate are shown in mg/L. Microbial counts are cells/ml and "-" means that numbers are not determined [13].

	Anna S, USA [14]	North West Province, South Africa [15]	China [16]
Mine Type	Coal	Copper-lead-zinc	Copper
pH	2.8-3.6	1.88	2.75
Al	Mar-36	-	
Fe	Jan-36	98.95	545
Mn	06-Sep	-	190
Zn	-	7.16	
Sulfate	-	4415.51	20800
Lead		2.35	
Copper		3.49	230

Table 3: Physic-chemical characteristics of AMD from various sites worldwide.

of the ore as previously indicated in the Table 1 [13,14,15].

Environmental problems causes by AMD

As we have indicated previously, AMD occurs by exposure of the metals sulphides to water and oxygen, causing major changes to water that receives the AMD. The oxidation of ores promotes the production of sulphuric acid and the release of a wide range of metals. The mixture causes serious environmental problems. AMD is toxic to aquatic organisms, destroys ecosystems, corrodes infrastructure, and taints water in regions where freshwater is already in short supply.

The impact of metals in water for humans as well as animals is the one hand, the persistence in the environment and the other hand, the accumulation in tissues, eventually causing chronic diseases. Usually, the disruption of metabolic functions by the metals is the cause of its toxicity, they accumulate in vital organs and they disrupt their important functions, and they inhibit the absorption, interfere with, or displace the vital nutritional minerals from their original place, thereby hindering their biological functions [16].

In plants experience oxidative stress upon exposure to heavy metals that leads to cellular damage and disturbance of cellular ionic homeostasis, in fact disrupting the physiology and morphology of plants. Plants need a proper balance of macro and micronutrients in the soil and thus the soil pH has an important influence on the availability of nutrients and on the growth of different kinds of plants [5].

It is widely know that aquatic organisms accumulate heavy metals from contaminated water and also by food. The big problems of the heavy metal are highly persistent and toxic even in trace amounts. The pH of water is important to aquatic life because it affects the normal physiological functions of aquatic organisms, including the exchange of ions with the water and respiration. Such important physiological processes operate normally in most aquatic biota under a relatively wide pH range (e.g., 6-9 pH). In fact, most of the freshwater lakes, streams, and ponds have a natural pH in the range of 6-8. When the ambient pH exceeds the range physiologically tolerated by aquatic organisms it can result in numerous sub-lethal effects and even mortality [5].

Chile regulation and legislation

In Chile, as in the rest of South America, the start of mining legislation begins with ordinances issued from Spain, promulgated in 1787, the "Ordinance of New Spain" becoming law in the Republic of Chile in 1833, remained in force until 1874. In 1932, it was performed some modifications, remaining in force for 50 years [17]. These documents lack environmental aspects to protect the surrounding ecosystem at a mine site. This changed in the 1990's, which was promulgated in 1992 Decree Nº185, which seeks to regulate stationary sources of air pollution, in 1994 Law 19.300 of Environmental Framework, where the

principle of promulgating "polluter pays" begins to be implemented and where an environmental impact assessment (EIA) is required for different types of businesses. In 2004, the Mining Safety Regulations required the submission of a closure plan. The Presidential Decree 248 issued in April 2007 ensures that the physical and chemical stability of the tailings deposits are such that they protect "people, property and the environment". In 2012, the new law (20,551) came into effect, which regulates mine closure and mining facilities. The main objective is to protect the life, health and safety of people and the environment, mitigate the negative effects of mining, hold accountable the mining industry job after cessation of operations, ensure the physical and chemical stability of ground where mining was developed, establish economic guarantees to ensure funding for the effective closure and post-closure monitoring chores of mining facilities [18].

AMD remediation

An abandoned mine than already started generating AMD can be considered a perpetual pollution producer. Since both oxygen and water are required to generate AMD, it follows that by excluding either (or both) of these, it should be possible to prevent or minimize AMD production [3,19]. Additionally, bacterial activity control prevents the contamination process from being catalyzed.

One way in which this may be achieved is by flooding and sealing abandoned deep mines. The dissolved oxygen (DO) present in the flooding waters (8-9 mg/l) will be consumed by mineral-oxidizing microorganisms present and replenishment of DO by mass transfer and diffusion will be impeded by sealing off the mine.

A technique widely used is to preventing contact between the mineral and dissolved oxygen, and consequent formation of AMD, is the flooding and sealing abandoned deep mines. The mineral-oxidizing microorganism consumed the dissolved oxygen (DO) present in the flooding waters, and the sealing of mine prevented the replenishment of DO by mass transfer and diffusion. However, this requires the knowledge of all shaft locations in order to be applied effectively and avoid influx of oxygen-containing water [3], and also know the huge water consumption inherent to this technique and the availability of this resource, making this technique unpractical for environment as northern Chile. The improved of this technique is by covering the tailings with a layer of sediment or organic material, to prevent oxygen ingress and some protection against re-suspension of the tailings due to the actions of weather [3,19,20].

Another suggested approach for minimizing AMD production is producing environmentally benign composites which acid consuming materials [21,22]. For example, in order to precipitate iron (III), as ferric phosphate is to add solid-phase phosphates (such as apatite) to pyritic mine waste; thereby reducing it's potential to oxidant of sulfide minerals [12]. But may only be temporary, then the application of soluble phosphate is with hydrogen peroxide, the peroxide oxidizes pyrite, producing ferric iron, which reacts with the phosphate to produce a surface protective coating of ferric phosphate.

The lithotrophic iron and sulfur-oxidizing bacteria the principle agents that perpetuate the generation of AMD are catalysts these reactions. Several laboratory and field tests using biocides have been carried out to inhibit their activities in mineral spoils and tailings, but the effectiveness has been found to be highly variable results and at best, only affords short-term control. Also the biocides are quite toxic and the effectiveness requires repeated applications. Is important further research for obtain a good tool against this kind of bacteria [3,11].

Even when these proposals work well in theory, the required quantities of resources such as water, effective sealing, biocides, etc., makes it unlikely to be applied everywhere. Most of Chilean operations are open pits located in the Atacama Dessert, driest desert on earth with an raining average of 2 mm/year. Therefore it is not practical to flood these sites in order to avoid contact between the minerals, water and oxygen. In fact there is water accumulating in the bottom of the open pit due its proximity to the water table.

This "sealing layer" that covers the spoil is usually constructed with clay, although in areas of the world that experience acute wet and dry seasons, drying and cracking of the cover can render it less effective than in temperate zones [23].

Given the practical difficulties mentioned above in inhibiting the

formation of AMD at the source, there is the alternative of minimizing the polluting impact in receiving streams and rivers and the greater environment by controlling the spread of AMD and polluted water. In order to classify this processes the terms "active" and "passive" are commonly used. Conventional application consisting in the application of chemical products to neutralize and precipitate metals present in acid mine west, and latter to the use of natural and constructed wetland ecosystems. The difference between the process if is the application is continue active) or not (passive), in this case, obviously the active process is more expensive and require more maintenance than the passive process. In Figure 1, we show a brief summary of the remediation technology [24].

Even when mining industries adopt just a few processes to treat all kinds of ores, the diversity of minerals it is possible to find in any given operation has lead to unique adaptations of these accepted processes, which results in unique waste. This type of wastes has forced researchers to adapt treatment processes as well, resulting in an endless list of possible treatments because each is focused on a particular contaminant or pollutant. The tables below show some of the considerations and comments about the main treatment alternatives (Tables 4 and 5) [25].

A summary of the techniques known in the remediation of AMD are displayed:

Abiotic process: The acid produced through mineral weathering can be buffered with alkalinity from treatment or natural sources; typically calcium carbonate, calcium hydroxide, or calcium oxide, are the neutralizing materials [26].

a. Active process: Maintain the stream with a pH over 6,5 and net alkaline conditions is the traditional goal for AMD treatment. This appears to be evident and still valid, even when the greater goal should be take back the stream characteristics to a pre-mining quality from a chemical and biological standpoint. Usually the design of AMD

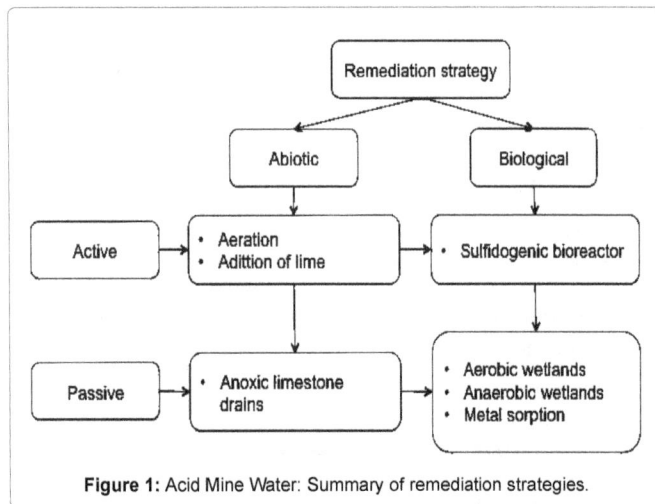

Figure 1: Acid Mine Water: Summary of remediation strategies.

Alternative	Considerations
Biotechnology	Usage of anaerobic sulfate reducing bacteria requires special conditions with constant nutrient feeding, pH, and temperature, is not economic option, which make it unlikely applicable at low costs without an important research work.
Separation using membranes (electrodialisis, reverse osmosis, ultrafiltration) and ionic interchanging resins.	These alternatives don't seem to be feasible because the effluent nature affects the membranes as much as the resins; it's expensive in large volumes. The main problems are high calcium and sulfate concentration that obstruct and destroy the base material of resins and membranes (polymeric or ceramic).
Evaporations and crystallization.	Even when these alternatives are actually possible from a technical point of view, it can be extremely costly given the large volume of effluent that needs to be treated.
Sulfate precipitation techniques (bario salts with/without regeneration and via etringgite formation.	Bario salt precipitation is feasible, but quite expensive. In former studies indicating FAD is efficient isolating $BsSO_4$, when adding sodium oleate. When forming etringgite, there are some patent and pilot tests, but none with an industrial application.

Table 4: Considerations about applying technologies [3,4,25].

Preventive methods	-Remove and/or isolate sulfur.
	-Remove oxygen from water sealing.
	-Remove oxygen from dry sealing.
	-Alkaline additives.
	-Bactericides.
Containing methods	-Avoid water flow.
	-Permeable reactive barriers (PRB) of reactive material located in the path of contaminants, considered as a passive treatment method.
	-Disposal in containment structure.
Remediation and treatment techniques	-Neutralization and precipitation of hydroxides and sulphur.
	-Separation by flocculation and flotation or lamellar settler.
	-Wetlands.
	-Remove sulfate and Mn^{+2} ions through co-precipitation.

Table 5: Prevention, control, treatment, and water usage from AMD [3,4,25].

treatment are based on the adding of enough alkali to buffer the stream, even when a neutral pH and a net alkaline conditions are necessary to achieve biological recovery, it is not always sufficient. This treatment strategy leads allows to recover the water chemistry and alkalinity, but produces lots of precipitated colloids and mineral as sulfates and iron hydroxides, affecting the habitat underwater trough the sediments cover of the streambed. Phenomena like this required to identify the possible depositional areas and settlement mechanisms to achieve a successful stream remediation [8].

b. Passive process: The Anoxic Limestone Drains (ALDs) can also be used to treat acidic waters, however, are not suitable of the treating all AMD, in the case of important concentrations of ferric iron, the half life of the process decrease until a few months. ALDs are limestone filled trenches through which acidic water is directed so the limestone can produce bicarbonate alkalinity via dissolution (Eq.5).

$$CaCO_3 + H^+ \rightarrow Ca^{2+} + HCO_3^- (Eq.5)$$

This kind of process is anoxic to prevent the contact of AMD with the oxygen and subsequent iron oxidation. The ALDs are capped with clay or compacted soil, to maintainer the anoxic conditions. Generally, ALD can be used as one component in the passive treatment, and also combined with another passive process by wetlands [3,7].

Biotic process: The biotic techniques are an option to remove contaminants using natural biological activity with the advantage of lower operating costs and clean technology.

a. Active process: Sulfogenic bioreactor: Sulfide minerals are also oxidized in a similar way, releasing metals and acidic sulfate in solution. Sulfate reducing bioreactors have become an economically viable alternative to conventional chemical processes for the treatment of wastewater that contains acid and metals. Sulfate reducing bacteria (SRB) have an ability to reduce sulfate to hydrogen sulfide, consumes protons while increased the alkalinity which leads it neutralization of the acid and produces stable precipitates of heavy metals, (Eq. 6 and 7) [11,27].

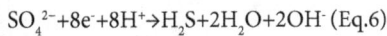

$$SO_4^{2-} + 8e^- + 8H^+ \rightarrow H_2S + 2H_2O + 2OH^- (Eq.6)$$

$$HS^- + M^{2+} \rightarrow MS_{(s)} + H^+ (Eq.7)$$

Different kinds of anaerobic reactor have been employed in the biological removal of sulfate, using microorganism suspended and attached growth. The anaerobic baffled reactor, present an advantage because have a compartmentalized structure, which protect biomass of adverse environmental conditions, such as low pH and high metal concentrations.

b. Passive process: Passive treatment systems for acid drainage are intended to renovate and improve the quality of water that passes through them. In this point we including the use of the wetland by treated the AMD, for increased the pH and reduced iron and metal, present in the liquid as well, the potential use of biomass, for sequestering the metal present in the AMD [7,28].

Aerobic wetlands: This process is a kind of passive treatment system, simplest but is limited in the types of water they can treat effectively. These systems allow aeration to the mine water flowing among the vegetation. Dissolved Fe to oxidize, and to provide residence time where the water is slowed for Fe oxide products to precipitate. The AMD aerobic wetlands are used to treat mildly acidic or net-alkaline water containing elevated Fe concentrations, but have limited capacity to neutralize acidity; even the wastewater leaving the treatment may be pH lower than that entering [7].

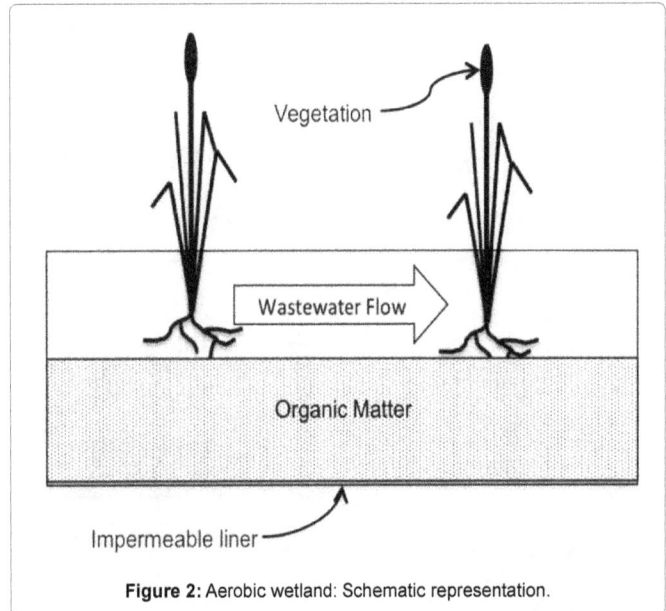

Figure 2: Aerobic wetland: Schematic representation.

A typical aerobic wetland system is a shallow trench planted with cattails (*Typha sp.*) (Figure 2). The depression that holds the wetland may or may not be lined with a synthetic or clay barrier. The iron-oxidizing bacteria oxidize iron at interfaces of aerobic and anaerobic zones. The plantation of macrophytes (e.g., *Phragmites australis, Typha latifolia, Juncus effusus*), the operation of the aerobic wetland is improved. In addition to achieving a regulation of the water flow, the iron oxidation is enhanced by the oxygen flow to the roots and by iron uptake of the macrophytes [7].

Anaerobic wetlands: To improve the treatment of the acid water, some changes have been made such as the addition of a bed of limestone beneath or mixed with an organic substrate, which encourages generation of alkalinity as bicarbonate. The biological sulfate reduction is a process that occurs under anoxic conditions (low O_2), whereby sulfate serves as the electron acceptor, this process also consumes protons while producing hydrogen sulfide (H_2S or HS^-), which leads to neutralization of the acidic pH by producing alkaline, and the precipitation of heavy metals, especially Fe and Al (Eq. 8). These systems commonly require large surface areas and long retention times because their effectiveness is limited by the slow mixing of the alkaline substrate water with acidic waters near the surface [7].

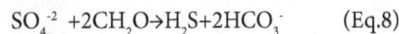

$$SO_4^{-2} + 2CH_2O \rightarrow H_2S + 2HCO_3^- \qquad (Eq.8)$$

The bicarbonate produced in the sulfidogenic oxidation increases the wastewater pH (Eq.8). Hence, metals and sulfate can be concomitantly removed and pH increased from acidic to neutral or alkaline in a single reactor. (Eq.9)

$$M + H_2S \rightarrow MS(s) + H^+ \qquad (Eq.9)$$

Biosorption: Conventional methods for removing metal ions from aqueous solution have been studied in detail, such as ion exchange, chemical precipitation, adsorption on activated carbon, electrochemical treatment, membrane technologies, etc. However, electrochemical treatment and chemical precipitation are ineffective, especially when metal ion concentration in aqueous solutions is as low as 1 to 100 mg L^{-1}, they also produce large amounts of sludge that can be difficult to treat, reaching the 40% of a urban wastewater reatment plan operational costs. Activated carbon adsorption process, ion exchange,

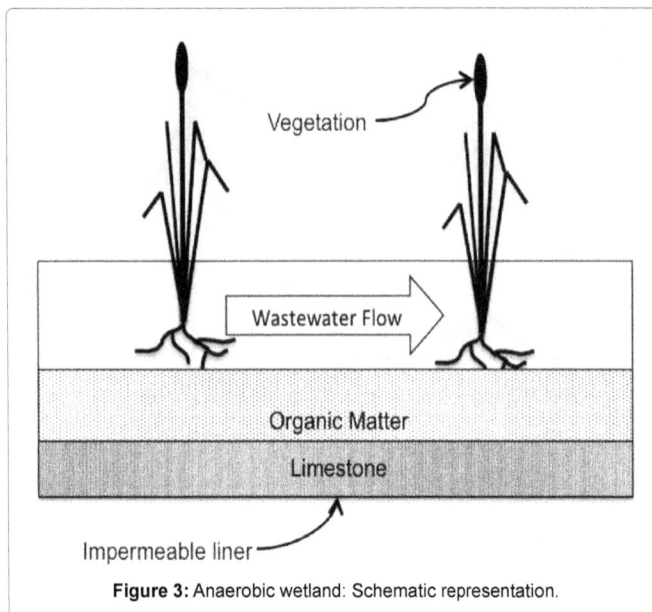

Figure 3: Anaerobic wetland: Schematic representation.

and membrane technologies are expensive, especially when treating a large amount of wastewater and water containing heavy metals at low concentrations [28].

An alternative process is biosorption. The bio-treatment for heavy metals generally includes bioaccumulation by dead or living biomass and the application of living biomass such as bacteria, algae, fungi, and seaweed as a biosorbent; this type of bio-treatment has emerged as a more cost effective process than other chemical processes and also as an eco-friendly process, which minimize byproducts. When dead biomass is immobilized in a polymeric matrix such as a bio-carrier, its presence confers high resistance to chemical environments and also provides additional advantages such as efficient regeneration and easy separation from solution [29] (Figure 3).

The biosorption process including the use of two phases, first one is a solid composed by biological material and a liquid phase (water). The liquid phase containing a species, witches are capable to be sorb, such as metal ion. The use of biomass for wastewater treatment has the advantage to be cheaper than traditional methods because of its availability in large quantities at low price [28].

The use of biosorption for a heavy metal remediation process in groundwater originating from AMD (pH lower than 4) has only been tested in laboratory tests, so further evaluation is required in the field to validate if the technique works effectively to remove heavy metals from AMD.

Phytoremediation: The phytoremediation is the use of green plants and their associated to microbiota, to remove the metals, contains ore renders environmental contaminants harmless [2,30]. Studies of phytoremediation have mainly focused on the removal of metals from contaminated soils, however there are new studies that have used hydroponics to assess the ability of plants to remove metals from aqueous environments (Klein). Indeed the *Thypa sp.* Used in the wetland methodology [25], several studies using *Phragmites australis* for phytoremediation, due is a typical wetland plant that can grow in aquatic system. The hydroponic experiments using synthetic AMD solution, concluding *Phragmites* is a species suitable for phytoremediation of polluted wetlands, because of its large biomass production, can remove

considerable amounts of nitrogen and other macro elements, which accumulate in the leaves [31,32,33].

Copper Mining: Current Treatments

As mentioned previously, acid mine drainage (AMD) in Chile is one of the most important environmental issues facing the mining industry, as more than 95% of the copper mines in the country correspond to porphyry copper deposits, which are characterized by deposits and low grade bulk tonnage. These mines have generated and will generate millions of tons of waste rock dumps and debris, plus numerous large tailings dams. The issue of AMD in Chile is particularly relevant due to the new N°20.551 (modified by 20.819) Act of "mine closure", tending to establish rules about the closure process and responsibilities on the future, forcing the companies to take care of this issue.

Chile is home of El Teniente, the largest underground mine in the world, Chuquicamata, the biggest open pit operation worldwide, and Minera Escondida, the most productive copper mine in the world. The open pit walls and fractures (craters) in underground operations alone are expected to generate massive quantities of AMD. Some studies suggest that there is a production of sterile ore that is diverted directly to sterile ore dumps (<0.2% Cu) at a rate of 3,000,000 (ton/day) [34]. Flotation tails generate 1,000,000 (ton/day) and leaching dumps contribute 0.2-0.4% Cu all being important sources of AMD production.

Based on a study by the National Service of Geology and Mining (known as SERNAGEOMIN: *Servicio Nacional de Geologia y Minería*) within the framework of a collaborative project between Japan and Chile, known as FOCIGAM (Strengthening Institutional Capacity In Environmental Management In Mining) and published in 2007 [35], performed the first "National Registry of abandoned Mining Operations and/or Paralyzed", detecting 213 places where the level of pollution, rising to a total of 520 tasks abandoned by December 2013 [36].

Either as volunteer, forced by law or an early start for the closing, Chile has just a few examples and little experience in closing big mining operations, as listed below:

Mine Closure Plan "El Indio"

In 2002, "El Indio" mine ceased operations by depleting its reserves. After 20 years of operation and a total production of 5.5 million ounces of gold, 24 million ounces of silver, and 500,000 tons of copper, the mine, operated by Barrick, voluntary ceased production and closed the mine. Barrick invested more than US$80 million in the closure. Their plan considered [37].

- Ensure post-closure physical and chemical stability of the facility, in the long term.

- Minimize the impact on the quality and quantity of water from the Bad River, which flowed through the area.

- Ensuring the security of long term operation of the mine.

Mine Closure Plan "Lo Aguirre"

Pudahuel Mining Company Ltd., owner of "Lo Aguirre" mine, presented a voluntary closure plan in 2000. The plan considered the physical and chemical stability of the materials removed in order to protect the integrity and health of the surrounding population, natural resources, and environment. Pudahuel also considered other activities to mitigate the visual impact of the mine, which is near a major highway. This mine closure plan was funded by the remaining solutions containing copper, which is being recovered as a precipitate of 65% copper from scrap iron law [38].

Metal Recovery feasibility at "El Teniente", CODELCO, Chile

The origin of acidic water in the "*El Teniente*" division is due three factors inherent in an underground mining:

1. Abandoned sectors inside the mine (crater), a product of the progress of the operation.

2. The mineralized low grades overload this between the end of the exploitation of block and surface.

3. Rainfall, mainly the melting of snow in the craters formed over time.

Melting snow in the mining site provokes a natural leaching process of the exposed minerals on the surface. The acidic water contains between 1 and 2.5 g per pound with a pH of 2 or 3 and pools at the bottom of the abandoned pit.

The acidic water treatment (El Teniente) is performed in acidic water plant Sx-Ew (Solvent extraction-Electrowinning), which can operate in parallel (flow: 17oLjseg) or in series (flow: 25oLiseg); gg to produce cathodes, Cu g%. Investment Total of this project in 1984 was 17,350 KUSDI [39].

Sulfate Removal Plant "Punta Chungos", Los Pelambres.

The Punta Chungo plant in the mine Los Pelambres has the first plant in Chile to treat effluents using flotation to reuse water for irrigation. This plant removes sulfate ions, suspended solids, and molybdates, among others contaminants. The mine floor has an area of 760 m², the treated effluent concentrate filtration unit Los Pelambres flow in the order of 90 to 120 m³/h. The unit is composed of two stages (two cells 6 × 4 × 1.5 m), one for the removal of sulfate ions and suspended solids in the second stage and the adsorption-ion molybdates co-precipitation of $Fe(OH)_3$ is removed. The ultimate goal is to reduce the molybdenum content of the effluent to 0.01 mg /L, the amount allowed by Chilean Standard Irrigation and Liquid Industrial Wasted (LIW) regulations. The separated sludge during flotation can be mixed with the copper concentrate, copper, and recovering stabilizing iron [4].

Conclusion

Chile has a huge environmental passive related to the AMD sources, due the lack of an standardization and law frame for the 520 mining abandoned facilities, but those are not the only sources as discussed above, active facilities has an important risk of becoming perpetual AMD sources too as sulfured ore copper grade decrease and low grade stock growth.

In the last decade Chile has made great strides in protecting the environment by establishing a series of laws that establish standards and guidelines for companies that impact the environment during their production processes. The Mine Closure Act lays the foundation for a sustainable mining development throughout the life cycle of a mining task.

Even when AMD production risks increase in the last years, due the precipitation increment as well as required copper grade to be processed (increasing production of low grade stocks), actual knowledge about the AMD production mechanism and industrial microbiology shows really interesting possibilities to treat and reuse AMD as well as avoiding its production. Nevertheless, and strong polity to support applied research appears as a requirement to integrate multidisciplinary research teams to considerer academic knowledge with mining experience in order to produce innovative and feasible solutions.

Even when the practical need of take care about AMD productions was always there, in almost every mining operation, the new law framework officially create the need for active mining operations, which is essential to avoid environmental passives increments and start being responsibly and unquestionably developing and sustainable mining.

Acknowledgments

We would like to express our appreciation to Grace Donavan for critical checking and revising the paper's English. This study was supported by FIC-R Antofagasta-Chile: BIP 30412847-0 Project.

References

1. Bahamóndez C (2012) "Importancia de la actividad microbiológica en predicción del drenaje ácido de minas", Thesis works for Biochemistry degree, Universidad of Chile.

2. Costello C (2003) Acid mine Drainage: Innovative Treatment Technologies. US. Environmental Protection Agency, Office of solid waste and emergency Response Technology innovation Office, Washington, D.C.

3. Johnson D K, Hallberg K (2005) "Acid mine drainage remediation options: a review" Science of the total Environment 338: 3-14

4. Santander M, Paiva M, Silva R, Rubio (2011) "Tratamiento de riles del sector minero-metalúrgico y reutilización de las aguas"; Revista de la facultad de Ingeniería de la Universidad de Atacama 25: 10-26.

5. Simate GS, Ndlovu S (2014) Acid mine drainage: Challenges and opportunities Journal of Environmental Chemical Engineering 2: 1785-1803.

6. Sheoran AS, Sheoran V (2006) Heavy metal removal mechanism of acid mine drainage in wetlands: a critical review. Minerals Engineering 19: 105-116.

7. Zipper C, Skousen J, Jage C (2011) Passive Treatment of Acid Mine Drainage. Virginia Cooperative Extension, Virginia Tech. Publication 460-133, Blacksburg 1-14.

8. Kruse NA, DeRose L, Korenowsky R, Bowman JR, Lopez D, et al (2013) The role of remediation, natural alkalinity sources and physical stream parameters in stream recovery. Journal of Environmental Management 128: 1000-1011.

9. Neal C, Whitehead PG, Jeffery H, Neal M (2005) The water quality of the River Carnon, west Cornwall, November 1992 to March 1994: the impacts of Wheal Jane discharges. Science of the Total Environment 338: 23-39.

10. Clarke LB (1995) Coal Mining and Water Quality. London7 IEA Coal Research. pp. 99.

11. Rose P (2013) Long-term sustainability in the management of acid mine drainage wastewaters-development of the Rhodes BioSURE Process. Water SA 39: 583-592.

12. Kontopoulos A (1998) Acid Mine Drainage Control. In: Effluent Treatment in the Mining Industry. Castro, S.H, Vergara, F, Sánchez MA (Eds). University of Concepción. 57-118.

13. Hallberg KB (2010) New perspectives in acid mine drainage microbiology. Hydrometallurgy 104: 448-453.

14. Hedin R, Weaver T, Wolfe N, Weaver K (2010) Passive Treatment of acidic coalmine drainage: The Anna S mine passive treatment complex. Mine Water Environmental 29: 165-175.

15. Van Hille RP, Boshoff GA, Rose PD, Duncam JR (1999) A continuous process for the biological treatment of heavy metal contaminated acid mine water. Resources, Conservation and Recycling 27: 157-167.

16. Bai H, Kang Y, Quan H, Han Y, Sun J, Feng Y (2013) Treatment of acid mine drainage by sulfate reducing bacteria with iron in bech scale runs. Bioresource Technology 128: 818-822.

17. Biblioteca Nacional de Chile.

18. Sernageomin (2011) Environmental Management and Law closure.

19. Chang S, Shin P, Kim B (2000) Biological treatment of acid mine drainage under sulfate reducing conditions with solid waste material as substrate. Water Research 34: 1269-1277.

20. Li MG, Aube BC, St-Arnaud LC (1997) Considerations in the use of shallow water cover for decommissioning reactive tailings. Proceedings of the Fourth International Conference on Acid Rock Drainage, May 30–June 6, Vancouver, BC 1: 115-130.

21. Mehling PE, Day SJ, Sexsmith KS (1997) Blending and layering waste rock to delay mitigate or prevent acid generation: a case review study. Proceedings of the Fourth International Conference on Acid Rock Drainage, May 30–June 6, Vancouver, BC 2: 953-970.

22. Evangelou VP (1998) Pyrite chemistry: the key for abatement of acid mine drainage. In: Geller A, Klapper H, Salomons W, editors. Acidic Mining Lakes: Acid Mine Drainage, Limnology and Reclamation. Berlin7 Springer; p. 197–222.

23. Swanson DA, Barbour SL, Wilson GW (1997) Dry-site versus wet-site cover design. Proceedings of the Fourth International Conference on Acid Rock Drainage, May 30–June 6, Vancouver, BC4: 1595-1610.

24. Klein R, Tischler JS, Mühling M, Schlömann M (2014) Bioremediation of mine water. Advances in Biochemical Engineering/Biotechnology. 141: 109-172.

25. Consejo Minero (2002) Consejo Minero de Chile "Guía Metodológica sobre Drenaje Ácido en la Indutria Minera", Acuerdo de producción limpia sector gran Minería.

26. Younger PL, Banwart SA, Hedin RS (2002) Mine water, Hydrology, pollution, remediation. Kluwer Academic Publishers, Dordrecht.

27. Bekmezci OK, Ucar D, Kaksonen AH, Sahinkaya E (2011) Sulfidogenic biotreatment of synthetic acid mine drainage and sulfide oxidation in anaerobic baffled reactor. Journal of Hazardous Materials. 189: 670-676.

28. Mosbah R, Sahmoune M (2013) Biosorption of heavy metals by Streptomyces species-an overview. Central European Journal of Chemistry 1412-1422.

29. Kim I, Lee M, Wang S (2014) Heavy metal removal in groundwater originating from acid mine drainage using dead Bacillus drentensis sp. immobilized in polysulfone polymer. Journal Environmental Management. 146: 568-574.

30. White S (2003)Wetland Use in Acid Mine Drainage Remediation.

31. Mani D, Kumar C (2014) Biotechnological advances in bioremediation of heavy metals contaminated ecosystems: an overview with special reference to phytoremediation. International Journal of Environmental Science and Technology 11: 843-872.

32. Guo L, Cutright TJ (2014) Effect of citric acid and rhizosphere bacteria on metal plaque formation and metal accumulation in reeds in synthetic acid mine drainage solution. Ecotoxicology and Environmental Safety. 104: 72-78.

33. Baldantoni D, Ligrone R, Alfani A (2009) Macro- and trace element concentrations in leaves and roots of Phragmites australis in a volcanic lake in Southern Italy. Journal of Geochemical Exploration. 101: 166-174.

34. Dueñas C (2010) Generación de Drenajes Ácidos, Fundación Chile.

35. Sernageomin (2007) Catastro de faenas mineras abandonadas o paralizadas y análisis preliminaries.

36. Sernageomin (2011) Management and Law closure.

37. http://barricklatam.com/cierre-el-indio/

38. Sociedad Minera Pudahuel (2011).

39. Dueñas C (2010) Normativas y Guias aplicables al Drenaje Ácido, Fundación Chile.

16

Removal of Trihalo Methanes Using Activated Carbon Prepared from Agricultural Solid Wastes

El-Demerdash FM[1]*, Abdullah AM[2] and Ibrahim DA[3]

[1]*University of Alexandria, Institute of Graduate Studies and Research, Department of Environmental Studies, Alexandria, Egypt*
[2]*Holding Company for water and wastewater, Alexandria, Egypt*
[3]*Alexandria Water Company, Alexandria, Egypt*

Abstract

High chlorine dosages are used in some drinking water plants to overcome the deficiencies in the treatment to at least ensure a supply of microbiologically safe water to the population. This fact and the increment of natural organic matter (NOM) in the aquatic resources due to rainfall increases and anthropogenic activities are becoming a critical concern, due to the formation of chlorination by-products such as trihalomethanes (THM), which are carcinogenic substances. Egypt drinking water system using new treatments is essential to meet the quality guidelines. Trihalomethanes are carcinogenic by-products of disinfection that are present in drinking water. In the present research, adsorption was employed for the removal of THMs found in water supply systems. The effects of pH, contact time, adsorbents and adsorbate concentration on the adsorption system were investigated. The Langmuir and Freundlich adsorption isotherm models were used to analyse the resulting adsorption data. The kinetics of THM removal was found to follow the pseudo-second-order model rather than the Langmuir–Hinshelwood pseudo-first-order model.

Keywords: Trihalomethanes; Activated carbon; Agricultural solid wastes

Introduction

Adsorption is a mass transfer operation in which substances present in a liquid phase are adsorbed or accumulated on a solid phase and thus removed from the liquid. Adsorption processes are used in drinking water treatment for the removal of taste and odour causing compounds, synthetic organic chemicals (SOCs), colour forming organics, and disinfection by-product (DBP) precursors. Inorganic constituents, including some that represent a health hazard, such as perchlorate, arsenic, and some heavy metals, are also removed by adsorption [1,2]. Granular activated carbon (GAC) is used in columns or beds for gas and vapour systems, and also for processing a number of liquids. The carbon must possess sufficient mechanical strength to withstand the abrasion incident to continued use. The development of high adsorptive power is accompanied by loss of mechanical strength and density [1,3]. Therefore the activation stage cannot be too short because the carbon would lack needed adsorptive power; conversely, it cannot be too long for then the carbon would be too soft and bulky. Few materials, in their natural state, can be converted into activated carbon with high density and low attrition. Less dense material, however, can be made dense and yield a hard carbon when mixed with a binder. The binder should be a substance which when carbonized does not liquefy or expand. However, some shrinkage is desirable. The tarry by-products from woods and certain grades of anthracite and bituminous coal have been found to be good binders. To be suitable as a binder, a substance should liquefy or soften during carbonization and swell sufficiently to give a porous structure. Suitable binders include sugars, tar, pitch, and lignin [4].

Powered activated carbon (PAC) used in membrane bioreactor (MBR) employed in the treatment of bleach pulp mill effluents was evaluated. The MBR was operated with hydraulic residence time of 9.5 h and PAC concentration of 10 g/L. The addition of PAC to the MBR reduced the average concentration of chemical oxygen demand (COD) the permeate from 215 mg/L (82% removal efficiency) to 135 mg/L (88% removal efficiency), producing an effluent that can be reused on bleaching stage. Moreover, the addition of PAC to the MBR resulted in the reduction in applied pressure and provided a more stable operation during the monitoring period. This occurrence was probably due to the increase of critical flux after the addition of PAC. The fouling mechanism was investigated and the results showed that controlling the concentration of soluble microbial products (SMP) and extracellular polymeric substance (EPS) by using PAC and keeping the operational flux below critical flux is of major importance for MBR operational sustainability [5].

Low cost adsorbents from agricultural waste like rice husk was developed with various activation methods and tested for the removal of aqueous contaminants. Adsorption of a basic dye, malachite green (MG), from aqueous solution onto nitric acid treated (NRH), and peroxide treated rice husk (PRH) have been investigated. Various experiments were studied using batch adsorption technique under different conditions of pH, adsorbent dosage, initial dye concentration, and temperature. The adsorption capacities of MG by the NRH and PRH were essentially due to electrostatic forces. The NRH and PRH adsorbents had a relatively large adsorption capacity (18.1 and 26.6 mg/g). The adsorbent PRH had a higher surface charge at alkaline pH and enhanced removal of MG was obtained under alkaline conditions. Typical adsorption kinetics indicated the pseudo second-order kinetics behaviour. The adsorption isotherms obeys Langmuir isotherm model.

*Corresponding author: Fatma M. El-Demerdash, University of Alexandria, Institute of Graduate Studies and Research, Department of Environmental Studies, 163 Horreya Avenue, P.O. Box 832, Alexandria 21526, Egypt
E-mail: eldemerdashf@yahoo.com

It was observed that the rate of adsorption improves with increasing temperature and the process is endothermic nature. The negative value of the change in Gibbs free energy ($\Delta G°$) indicates that the adsorption of MG on PRH and NRH is feasible and spontaneous [6].

Currently, problems in drinking water treatment extend beyond the scope of taste and odour control. Much attention is being paid to the regulation and control of numerous organic and inorganic compounds in water. Concerns about the presence of Synthetic Organic Compounds (SOC) arose in 1960s. Beginning in the 1970s it was recognized that disinfection of water with chlorine gas or chlorine-containing compounds led to the generation of organic compounds, collectively termed Disinfection By-Products (DBPs), which were suspected of having adverse effects on health [1,2]. In this regard, Natural Organic Matter (NOM) constitutes the key group of organics acting as precursors for DBP formation. It was also shown that pre-treatment of water with ozone led to inorganic hazardous by-products such as bromates. For many decades, adsorption onto activated carbon has appeared to be one of the most reliable methods of NOM and DBP control. This type of treatment is usually conducted in GAC filters. These are usually placed after sand filtration and before disinfection, but, depending on the characteristics of the water and the object of the treatment, GAC filters may also be positioned at other locations within the treatment train [1,2].

Agricultural and food industries create various waste matters that need to be utilized and convert in value added product. Carbonaceous materials such as coconut shell, palm shell, saw dust and tropical wood are some of the most common agricultural wastes shells used to produce activated carbon. In the present study the activated carbon was prepared from two carbonaceous agriculture wastes by chemical activation using ortho-phosphoric acid (H_3PO_4) at 800 ± 5°C. Agricultural solid waste is one of the rich sources for low cost adsorbents besides industrial by-product or natural material. Due to abundant availability, agricultural waste such as orange peel poses little economic value and moreover, creates serious disposal problems. Many agricultural by-products that are available at little or no cost for example chicken feathers, cassava waste, rice husks have been reported to be capable of removing substantial amounts of metal ion and organic pollutants from aqueous solutions [4,7]. Therefore, the present study was designed to obtain activated carbon from rise husk and wood and to analyse their internal structure through Scanning Electronic Microscopy (SEM). Moreover, it was proposed to measure their DPBs adsorption capacity.

Material and Methods

The activated carbons (commercial and synthetic) used in this study for water purification were four types:

1- The Commercial Activated Carbon (CAC): Provided by Al-Gomhoria Chemical Company; (Alexandria, Egypt).

2- Woody Activated Carbon (WAC): Wood was used as precursor for the preparation of activated carbon. The wood was cut into pieces of 2 to 3 cm in size, dried in sunlight for 2 days. The dried material was then soaked in solution of H_3PO_4 for one hour and kept at room temperature for 24 hours. Thereafter, the wood material was separated, air dried at room temperature and carbonized in muffle furnace at 320°C for 90 min. The carbon was ground to powder then the activated carbon was washed with plenty of distilled water to remove residual acid, dried, sieved into 300 to 850 μm (20-50 ASTM mesh) sizes and stored in a tight lid container for further adsorption studies [8].

3- Rice Straw Activated Carbon (RAC): Rice Straw, after collection

dried at 105 ± 5°C for 6 hours and cooled to ambient temperature in desiccators and was cut into pieces of 2 to 3 cm in size, then soaked in solution of H_3PO_4 for one hour and kept at room temperature for 24 hour. The straw material was separated; air dried at room temperature and was carbonized in muffle furnace at 320°C for 90 min. The carbon was ground to powder then the activated carbon. The prepared AC from rice husk was magnified using electron microscope as shown in Figure 1 [9].

4- Activated Carbon activated by H_2O_2 (CCH): Rice Straw, after collection dried at 105 ± 5°C for 6 hours and cooled to ambient temperature in desiccators and was cut into pieces of 2 to 3 cm in size, then soaked in solution of H_2O_2 for one hour and kept at room temperature for 24 hour. The straw material was separated; air dried at room temperature and was carbonized in muffle furnace at 320°C for 90 min. The carbon was ground to powder then the activated carbon [8]. THM and TOC were determined according to APHA using GC-ECD, *Agilent* 5890 and TOC analyser, respectively (APHA, 2012) [8].

Results and Discussion

Total Organic Carbon (TOC)

Results presented in Tables 1-3 and Figures 2-6 showed that, as the AC dose in mg increased, the removal of TOC increased; at 10 mg dose of AC, the order of removal efficiency of TOC was

Parameter	AC Dose(mg/L)	Commercial carbon (CAC)	Synthesized AC		
			CCH	RAC	WAC
TOC (mg/L) Removal by different type of AC	0	5.7	5.9	5.8	5.2
	2	4.3	4.3	5.1	4.9
	6	4.0	4.2	4.8	4.7
	10	3.6	4.1	4.6	4.3
	14	3.2	3.9	4.2	4.0
	16	2.9	3.7	3.6	3.9
	18	2.4	3.3	2.8	3.7
	20	1.9	3.1	2.6	3.1
	22	1.6	2.9	2.0	2.9
	24	1.4	2.5	1.7	2.7
	26	1.2	2.2	1.4	2.4
Average	15.80	2.93	3.65	3.51	3.80
Minimum	2.00	1.20	2.20	1.40	2.40
Maximum	26.00	5.70	5.90	5.80	5.20

Table 1: Removal of TOC using different low cost adsorpant (AC).

Parameter	AC Dose(mg/L)	Commercial Carbon (CAC)	Synthesized AC		
			CCH	RAC	WAC
TOC Removal % by different type of AC	2	24.3	24.3	9.7	14.4
	6	29.6	25.7	15.3	17.4
	10	37.3	27.8	18.3	23.8
	14	43.7	32.0	26.1	29.9
	16	49.6	34.2	35.7	31.9
	18	58.1	41.9	50.0	35.6
	20	66.7	45.4	53.5	44.9
	22	71.8	49.3	65.1	49.6
	24	76.1	56.9	71.0	52.8
	26	78.3	60.9	76.1	58.8
Average	15.8	53.55	39.84	42.08	35.91
Minimum	6.00	24.30	24.30	9.70	14.40
Maximum	26.00	78.30	60.90	76.10	58.80

Table 2: Removal percentage of (TOC) using different low cost adsorpant(AC).

Max % (Removal)	Min % (Removal)	K	$R^2(1/n)$	AC
78.3	24.3	1.9914	0.955	CAC
60.9	24.3	2.6075	0.9905	CCH
76.1	9.7	1.347	0.9503	RAC
58.8	14.4	2.314	0.9634	WAC

Table 3: Freundlich coefficient for the TOC adsorption isotherms.

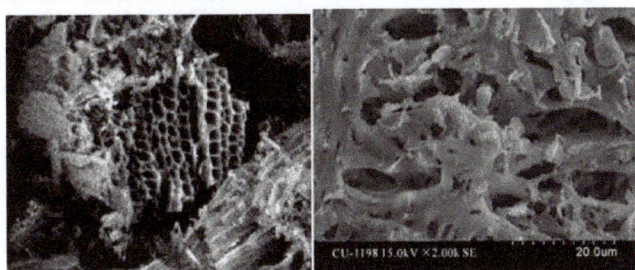

Figure 1: Scanning Electronic Microscopy of AC prepared from rice husk.

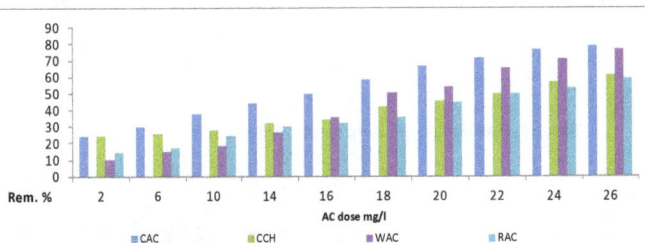

Figure 2: TOC removal % using different low cost of AC.

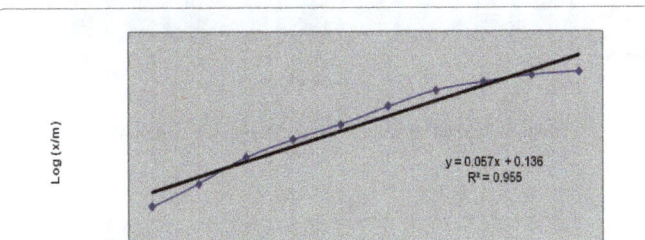

Figure 3: Frendlich isotherms for the TOC adsorption by CAC.

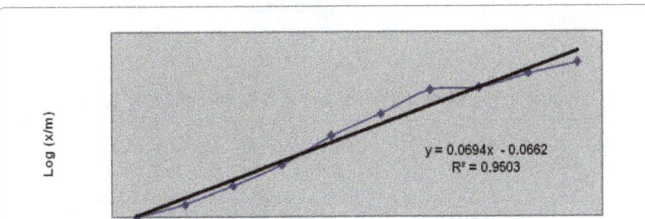

Figure 4: Frendlich isotherms for the TOC adsorption by RAC.

RAC (76.1%), CCH (60.9%), and WAC (58.8%). Activated carbon activated with hydrogen peroxide (CCH) had the highest 1/n value (0.9905) followed by WAC (0.9634), and commercial activated carbon without activation CAC (0.955) and RAC (0.9503). These values of 1/n corresponding to the respective carbons indicate that CCH has the highest rate of adsorption of the solute and RAC the least. These values of 1/n for CCH and RAC also suggest that within the chemical activation, carbon dioxide-activation improves 1/n compared to the other types of carbons. Furthermore, CCH with 1/n≈1.0 (0.9905) indicates that the change observed in adsorbed solute (organic contaminants) concentration is greater than that occurred in the solute concentration. It is evident that, the commercial CAC were exceeded in removal of TOC composite organic compounds than that in case of prepared AC (CCH, RAC and WAC) (Figures 2 a-e) [10].

The removal of organic precursors of disinfection by-products (DBPs), i.e. natural organic matter (NOM), prior to disinfection is considered as the most effective approach to minimize the formation of DBPs. Trihalomethanes (THMs) as the main group of DBPs are categorized and considered to have the potentiality of increasing the rate of liver, kidney and central nervous system adverse effects [11]. NOM is primarily composed of humic substances, such as humic acids (HA) and fulvic acids (FA) that result from decomposition of terrestrial and aquatic biomass, but it can contain a range of organic species and microorganisms and their discharges [10]. The present results are in agreement with the finding of Carrière et al. [12]; who used powdered activated carbon in conjunction with polyaluminum chloride in TOC removal and proved that with adding only 11 mg/l of powdered activated carbon, TOC removal efficiency was improved and taste and odour agents were decreased (Figure 3).

Among the four experimental carbons, the commercial carbon CAC exceeded RAC and WAC in their adsorption performance. The physically activated carbon (steam and carbon dioxide) were superior in their adsorption to chemically activated, this could be due to the inhibition of small organics from penetrating into the micro-pores due to the presence of greater number of polar compounds on the mesopore and macropore surface of the acid activated carbons. These

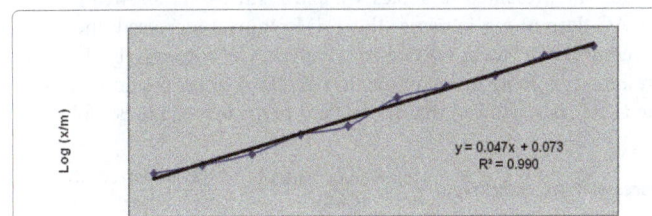

Figure 5: Freundlich isotherms for the TOC adsorption by CCH.

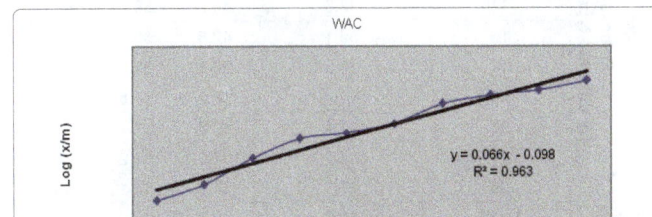

Figure 6: Frendlich isotherms for the TOC adsorption by WAC.

CAC>RAC>CCH>WAC, respectively. Among the four experimental carbons (CCH, CAC, RAC and WAC) The data from Table 3 indicates that in terms of adsorption of TOC organic compounds in the CAC tests, showed that, the highest performance (78.3%) followed by

data infer that the method of activation and precursors selected for the preparation of activated carbons do affect the adsorption of organic compounds and hence are factors to be considered in any adsorption process [13,14]. Also, in agreement with the present results, Abdulla [2] reported that AC removed TOC by 78%.

Pre-chlorination has potential to increase DBP levels during water treatment process. Many treatments really on pre-chlorination to solve operational problems, including turbidity, algae growth control, inorganic oxidation and microbial inactivation as well as taste and odour [1,2]. So, in the present study, AC was used to show the effect of treatment on THM without the necessity to pre-chlorination. Five characteristics of carbon are important in the adsorption of organic molecules from solution. Yenisoy-Karakas et al. [14] reviewed several articles in the literature and suggested that these factors are (1) particle size, (2) surface area, (3) chemisorbed oxygen, (4) hardness and uniformity, and (5) pore structure. These adsorptive characteristics of carbon depend upon the sourcematerial of the carbon and the activation process [14].

Although adding powdered activated carbon enhanced organic matter removal, it caused decreasing coagulant consumption. For the application of poly aluminium chloride, by increasing the amount of powdered activated carbon, TOC was decreased and the least rate of TOC achieved at 100 mg/l powdered activated carbon [14]. Activated carbon adsorption has been widely applied in removing organic matters from wastewater as it has a strong affinity for attaching organic substances even at low concentration. Having large surface area for adsorption, granular activated carbon (GAC) is one of the best adsorbents for removing various organic contaminants [2,14]. The GAC adsorption systems are therefore considered to apply for producing the effluent of high quality from sewage treatment plant which can be reused for various purposes. However, even though it has high adsorption capacity, GAC can only maintain its adsorption for a short time after its available adsorption site becomes exhausted with adsorbed organic pollutants [15].

THM

Results presented in Tables 4-6 and Figures 7-11 showed that, as the AC dose in mg increases the THM removal increased; the order of removal of THM is CAC>RAC>WAC>CCH respectively. The data presented in Table 6 the adsorption of THM organic compounds in the CAC tests, showed that the highest performance (45 %) followed

Parameter	AC Dose(mg/L)	Commercial Carbon (CAC)	Synthesized AC		
			CCH	RAC	WAC
THM Removal % by different type of PAC	2	16.2	7.5	12.0	7.0
	6	26.1	8.9	22.2	10.4
	10	35.8	11.5	32.6	14.6
	14	44.1	17.1	41.2	17.4
	16	49.1	22.0	46.6	19.5
	18	54.3	30.2	52.1	25.9
	20	54.3	31.5	52.1	29.6
	22	54.5	32.1	52.2	34.7
	24	54.8	32.6	52.5	34.4
	26	60.7	32.4	58.8	38.3
Average	15.80	42.37	21.21	39.48	21.71
Minimum	2.00	16.20	7.50	12.00	7.00
Maximum	26.00	60.70	32.60	58.80	38.30

Table 5: Removal percentage of THM using different low cost adsorbent (AC).

Max % (Removal)	Min % (Removal)	K	R²(1/n)	AC
60.7	16.2	0.2268	0.7688	CAC
22.6	7.5	0.2688	0.8968	CCH
58.8	12	0.2265	0.7689	RAC
23.6	7	0.0753	0.9703	WAC

Table 6: Frendlich coefficient for the THM adsorption isotherms.

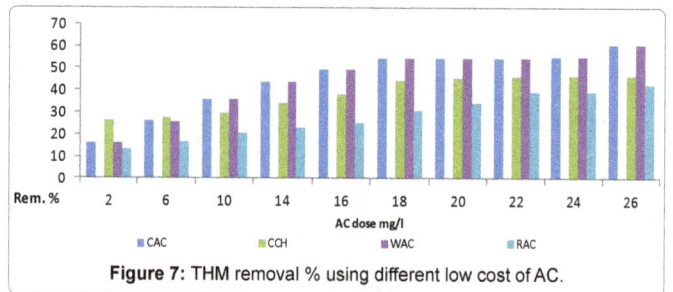

Figure 7: THM removal % using different low cost of AC.

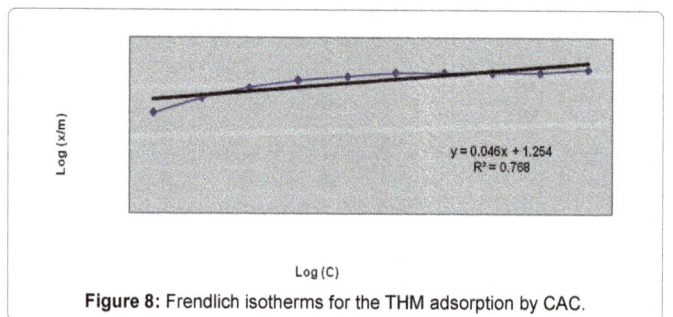

Figure 8: Frendlich isotherms for the THM adsorption by CAC.

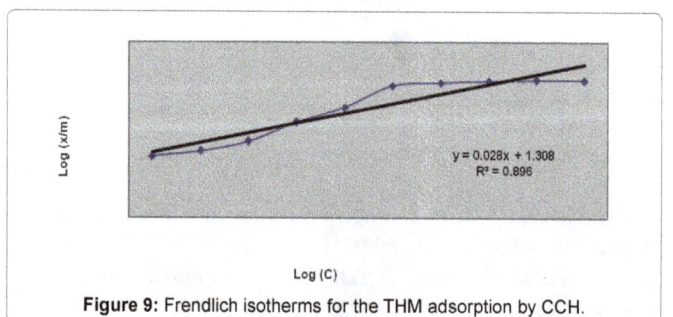

Figure 9: Frendlich isotherms for the THM adsorption by CCH.

Parameter	AC Dose(mg/L)	Commercial Carbon (CAC)	Synthesized AC		
			CCH	RAC	WAC
THM Removal By different type of PAC	0	69.0	54.9	65.7	64.2
	2	57.8	50.8	57.8	59.7
	6	51.0	50.0	51.1	57.5
	10	44.3	48.6	44.3	54.8
	14	38.6	45.5	38.6	53.0
	16	35.1	42.8	35.1	51.7
	18	31.5	38.3	31.5	47.6
	20	31.51	37.6	31.5	45.2
	22	31.4	37.3	31.4	41.9
	24	31.2	37.0	31.2	42.1
	26	27.1	37.1	27.1	39.6
Average	14.36	40.8	43.6	40.48	50.66
Minimum	2.00	27.10	37.00	27.10	39.60
Maximum	26.00	57.80	50.80	57.80	59.70

Table 4: Removal of THM using different low cost adsorpant (AC).

by RAC (42.2%), WAC (23.2%) and CCH (22.6 %). Woody activated carbon (WAC) had the highest 1/n value (0.9703) followed by CCH (0.8968), RAC (0.7689) and CAC (0.7688). These values of 1/n corresponding to the respective carbons indicate that WAC has the highest rate of adsorption of the solute while RAC and CAC have the lowest rate. These values of 1/n for WAC and CCH also suggest that within the chemical activation carbon improves 1/n compared to the other types of carbons. Furthermore, WAC with 1/n≈1.0 (0.0.9703) indicated that the change in adsorbed solute (organic contaminants) concentration is greater than the change in the solute concentration [1]. Also, this is in agreement with Abdullah [2] who reported that AC removed 84% of THM. Moreover, Uyat et al. [16] studied the removing DBPs precursors by enhanced coagulation and powdered activated carbon (PAC) adsorption and proved that supplementing enhanced coagulation with PAC adsorption increased the removal of TOC to 76% and PAC adsorption removed mostly low molecular weight and uncharged natural organic matter (NOM) substances as indicated in Figures 7-11 [12,16].

Factors affecting AC adsorption capacity

Effect of solution pH: The pH of the solution is one of the major factors influencing the adsorption capacity of the compounds that can be ionized. In order to study the effect of the pH, several experiments were carried out with carbon CCH at a temperature of 25°C using different initial solution pH of values (2.0, 7 and 9). The pH of the solutions are maintained by means of ionic buffer solutions composed with NH_4OH, HCl, KH_2PO_4, Na_2HPO_4, NaH_2PO_4 and Citric acid. The effect of pH on the isotherms CCH for TOC is presented in Figure 12. It indicated that at very low pH (pH=2), the amount of TOC adsorbed is lower than those obtained with other pH due to the competition with H^+ on carbonyl sites [4,7]. The effects of pH on the elimination efficiency of TOC using powdered activated carbon showed that the optimum pH range between 5.5 and 6 [2,17].

As a result, at pH 7, the rate of TOC reduction was 76%. Taking into consideration the closeness to the natural pH of water, this showed the high efficiency of powdered activated carbon in the natural pH of water. It has also shown that a considerable reduction in total organic carbon concentration is obtained in pH ranges of 5.5 to 7. This is due to the fact that TOC solubility decreased and organic matter adsorption increased in this ambient pH. The increase of pH favoured the ionization of organic matter and decreasing the pH could redound to neutralize the negative charge on the surface of powdered activated carbon using hydrogen ions as indicated in Figure 4 [2,17].

Effect of solution temperature: It is clear from the present results that the TOC adsorption decreases with increasing temperature as shown in Figure 13. This well-known effect of temperature is expected for physical adsorption that is exothermic in nature in most cases (USEPA and Hoang [11,15]. Due to this temperature dependence of adsorption capacity, the adsorption isotherms have been determined in a thermostatic bath at 25°C. Ramaraju et al. showed that, the rate of adsorption improves with increasing temperature and the process is endothermic nature [6].

Conclusion

Drinking water treatment using activated carbon (AC) is one option for treatment of drinking water problems. AC is an effective method for treating certain organic compounds, unpleasant tastes and odours, and chlorine, as well as its effective for metals, microbial contaminants, algae and other inorganic contaminants. Adsorption

capacity of activated carbon had been found to be dependent on the adsorbent dosage but other variables such as adsorbate concentration, pH variation, activating reagents, temperature have their own effect on adsorption capacity. Hence this prompts the use of different model to carry out analysis of the adsorption capacity (Fruendlich isotherm). Adsorption process can be well understood through isotherms resulting between adsorbate concentration in liquid and amount of adsorbate adsorbed by unit mass of adsorbent at a constant temperature. The present results obtained by the adsorption experiment were analysed by Freundlich, isotherms. TOC and THM could be effectively removed by AC adsorption from drinking water. The fitting result indicated that the adsorption of THM on RAC represents the linear consistency. External mass from TOC from tap water, contact time and mixing

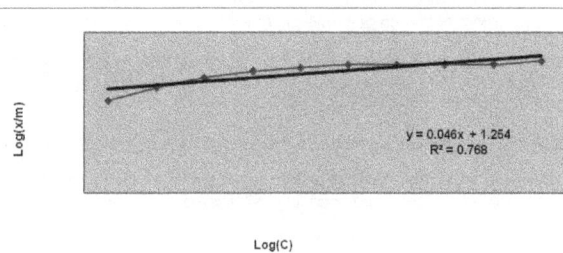

Figure 10: Frendlich isotherms for the THM adsorption by RAC.

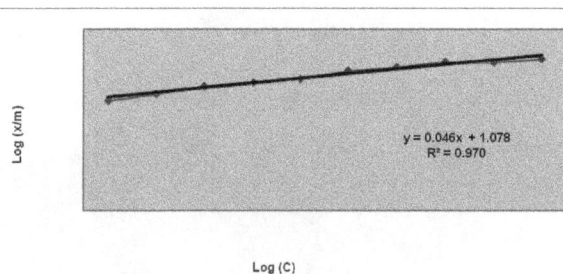

Figure 11: Frendlich isotherms for the THM adsorption by WAC.

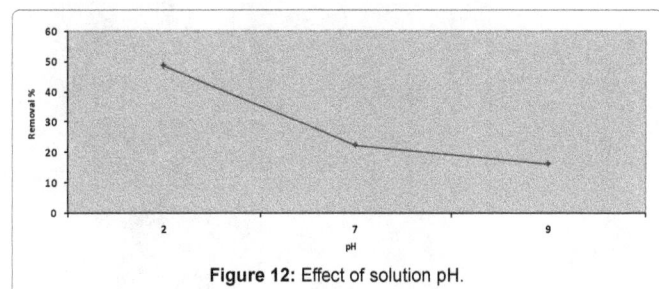

Figure 12: Effect of solution pH.

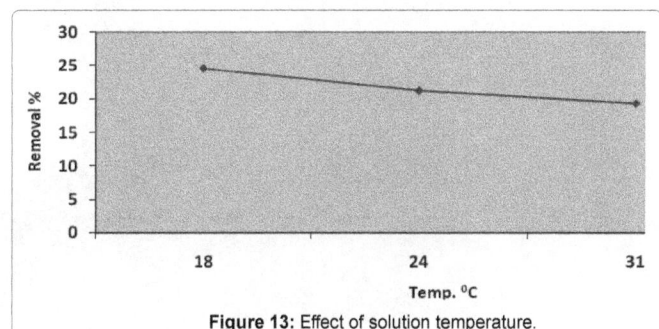

Figure 13: Effect of solution temperature.

were confirmed as controlling step in the sorption process. Under the prevailing conditions, the maximum RAC removal efficiency was found to be 58.8% for 65.7 µg/l THM with contact of 0.026 g of RAC.

References

1. Slavinskaya GV (1991) Chlorination effect on quality of drinking water. Khimiya I Tekhnol. Vody 13: 1013-1022.

2. Abdullah AM (2012) Alexandria University, Alexandria, Egypt.

3. Dibinin MM, Plavnik GM, Zevarina EF (1964) Integrated Study of the Porous Structure of Activated Carbon from Carbonized Sucrose. Carbon 2, 261.

4. Yang T, Lua AC (2003) Characteristics of activated carbons prepared from pistachio-nut shells by physical activation. J Colloid Interface Sci 267: 408-417.

5. Amaral CS, Míriam C, Lange C, Liséte Borges, Cristiano P, et al. (2014) Evaluation of the use of powdered activated carbon in membrane bioreactor for the treatment of bleach pulp mill effluent. Water Environ. Res 86: 788-799.

6. Ramaraju B, Manoj Kumar Reddy P, Subrahmanyam C (2014) Low cost adsorbents from agricultural waste for removal of dyes. Environ Prog Sust Energy 33: 38-46.

7. Bansal RC, Goyal M (2005) Middle East Technical University, Ankara, Turkey.

8. Daud WM, Ali WS (2004) Comparison on pore development of activated carbon produced from palm shell and coconut shell. Bioresour Technol 93: 63-69.

9. Daud WMAW, Houshamnd AH (2010) Textural characteristics, surface chemistry and oxidation of activated carbon. Journal of Natural Gas Chemistry 19: 267-279

10. Alvarez-Uriarte JI, Iriarte-Velasco U, Chimeno-Alanis N,Gonzales-Velasco JR (2010) The effect of mixed oxidants and powdered activated carbon on the removal of natural organic matter. J Hazard Mater 181: 426-431.

11. USEPA, United States Environmental Protection Agency (2001) Controlling disinfection by-products and microbial contaminants in drinking water. Washington: USEPA, Office of research and development.

12. Carrière A, Vachon M, Bélisle JL, Barbeau B (2009) Supplementing coagulation with powdered activated carbon as a control strategy for trihalomethanes: application to an existing utility. J. Water Supply Res 58: 363-371.

13. Kurosaki F, Ishimaru K, Hata T, Bronsveld P, Kobayashi E, et al. (2003) Microstructure of wood charcoal prepared by flash heating. Carbon 41: 3057-3062.

14. Yenisoy-Karakas v, Aygün A, Günes M, Tahtasakal E (2004) Physical and chemical characteristics of polymer-based spherical activated carbon and its ability to adsorb organics. Carbon 42: 477-484.

15. Hoang TTL (2005) University of Technology, Sydney.

16. Uyat V, Yavuz S, Toroaz I, Ozaydin S, Genceli EA, et al. (2001) Disinfection by-products precusors removal by enhanced coagulation and PAC adsorption. Desalination 216: 334-344.

17. Reed BE, Jensen JN, Matsumoto MR (1993) Acid-base characteristics of powdered-activated-carbon surface. J. Environ Eng ASCE 119: 585-590.

17

Assessing the Land Use/Cover Dynamics and its Impact on the Low Flow of Gumara Watershed, Upper Blue Nile Basin, Ethiopia

Gashaw G Chakilu[1]* and Mamaru A Moges[2]

[1]*College of Agriculture, Wolaita Sodo University, Sodo 138, Ethiopia*
[2]*Faculty of Civil and Water Resources Engineering, Bahir Dar Institute of Technology, Bahir Dar University, Bahir Dar, Ethiopia*

Abstract

Land cover and Climate change are very important issues in terms of global context and their responses to environmental and socio-economic drivers. The dynamic of these two factors is currently affecting the environment in unbalanced way including watershed hydrology. In this paper the impact of land use/cover change on stream flow particularly on low flow were evaluated through application of the model Soil and Water Assessment Tool (SWAT) in Gumara watershed, Upper Blue Nile basin Ethiopia. The land use/cover data were obtained from Land Sat image and processed by ERDAS IMAGINE 2010 software. Three land use land cover data; 1973, 1986, and 2013 were prepared and these data were used for base map, model calibration and change study respectively. So, as to evaluate the effect of land use/cover change on low flow of the catchment, the stream flow was simulated by changing 1973 and 2013 LULC but the climate data, which is 1973-1982, was used and it was constant. The low flow of the catchment for these two decades was extracted in simulated flows by Seven Day Sustained (SDS) low flow separation method. The model (SWAT) was calibrated by 1986-1991 climate data and 1986 land use land cover data by using 11 important model parameters selected by sensitivity analysis. The consistency of values of those calibrated parameters was also validated by 1992-1995 climates and with the same land use land cover data. Based on the result, the extreme low flow of Gumara watershed has been decreasing from 0.53 m^3/s to 0.43 m^3/s which showed decreasing by 0.1 m^3/s that is 18.87%. From the overall results of the study, it is possible to conclude that land use land cover change has been influencing the low flow or dry season flow of the catchment. This study has been designed to show how much the land use/cover has been changed and affects the low flow or dry season environmental flow of the catchment. The result has showed some indications that there has to be restoration activities on the land use cover nature of the study area.

Keywords: LULC; Low flow; SWAT; Gumara; Blue Nile; Ethiopia

Introduction

Land cover change and climate change and associated impacts on water resources are being the hot issues in recent years. This is due to the direct or indirect impacts brought by land use and climate change both have contributed to some water problems, such as water shortage, flooding, and water logging to different extent.

The response of hydrology to land use/cover change is an integrated in ecosystem, and may affect the overall healthy functions of a watershed and its ecosystems. Direct and powerful linkages exist among spatially distributed watershed properties and watershed processes [1]. Water shortages and degradation of water supplies threaten the food security and health of people in many parts of the world. This is particularly happened in developing countries that are experiencing rapid population growth and inefficient means to manage water resources [2]. Land use land cover (LULC) changes, particularly those caused by human activities-for example deforestation to clear land for agriculture, are considered to be the most important factor in global environmental change, exerting effects possibly greater than those of other global changes [3].

In Ethiopian, on the head of Blue Nile and in lake Tana basin, over the past few decades there have been a lot of activities that have modified the land use/land cover. Moreover, the hydrological dynamics has been strongly modified by intensive agricultural activities. This has a direct impact on the lake and the flow condition of Abay (upper Blue Nile) as well. Therefore, it is very important to understand the functioning of the lake catchments and their hydrological response under different historical land use and climate change scenario conditions and the water resources development of the basin requires a judicious planning for the protection of the fragile ecosystem. Thus, this study will focus on responses of low flow for the dynamics of land use/cover of Gumera watershed which is part of Lake Tana and upper Blue Nile basin.

Materials and Methods

Study area description

The Gumara River is located to the east direction of Lake Tana; it is found between latitude of 11° 35' and 11° 55' N and longitude of 37° 40' and 38° 10' E. And it has a total drainage area of 127186 ha up to the gauging station (near Woreta), a head of 25 km before it joins the lake. The total main stream length from its origin (near mount Guna) is approximately 132.5 km before the river joins Lake Tana (Figure 1).

General research methods

The effect of land use/cover change on the low river flow of the study area has been simulated by using SWAT hydrological model. The simulated flow generated by the SWAT model was compared with the observed one by using Nash Sutcliff efficiency and relative volume error by the help of excel sheet. To enhance the agreement of simulated

---

*Corresponding author: Gashaw G Chakilu, College of Agriculture, Wolaita Sodo University, Sodo 138, Ethiopia, E-mail: ggismu@yahoo.com

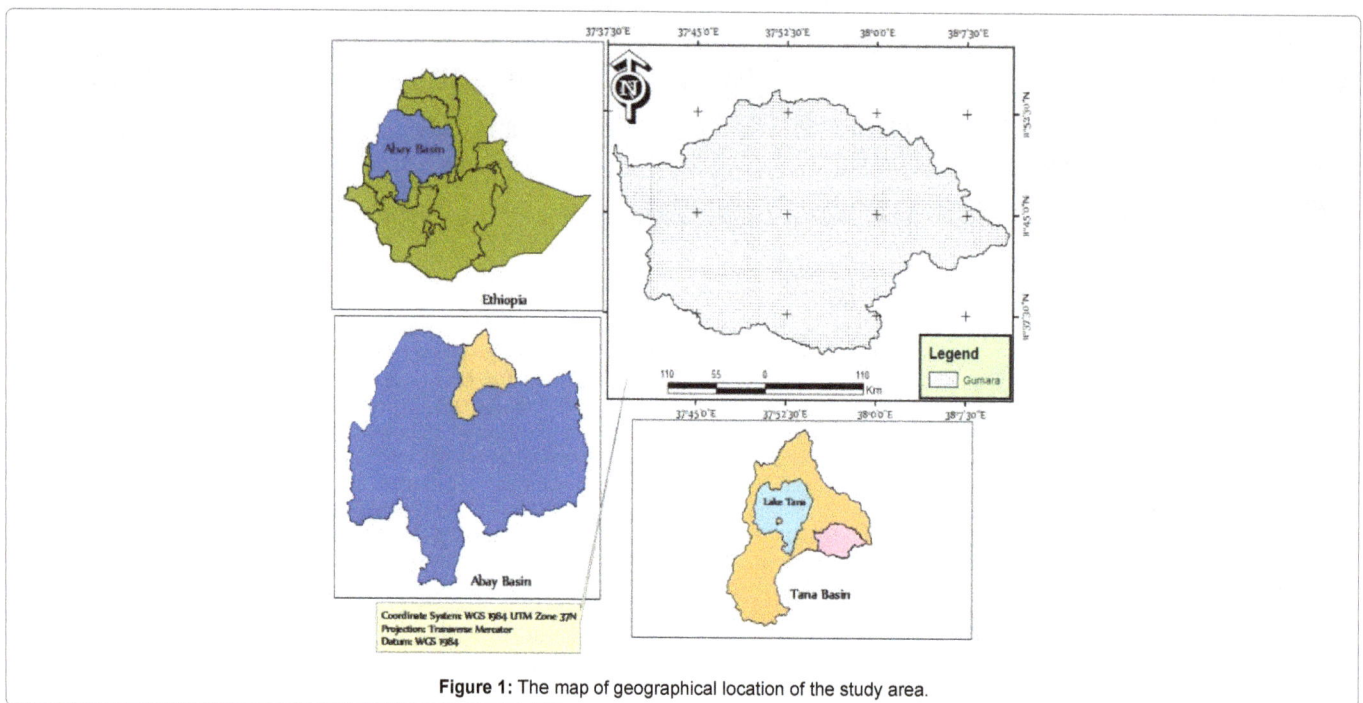

Figure 1: The map of geographical location of the study area.

versus observed flow of the catchment, SWAT CUP model was used for calibration and validation of model parameters selected by sensitivity analysis processes.

SWAT model description

The Soil and Water Assessment Tool (SWAT) is a versatile, physically semi - distributed model with spatial and temporal variability consideration, for simulating runoff and sediment transportation of small and large watersheds. The model is a physical based, semi-distributed and operating on daily time step [4]. As a physical based model, SWAT create Hydrological Response Units (HRUs) to represent spatial heterogeneity based on the specified threshold percentage of the watershed land use, soil types and slope.

The model provides two methods of surface runoff calculation: one is the runoff curve number method developed by the Soil Conservation Service (SCS) of the United States Department of Agriculture (USDA, 1986) and the other is through the Green-Ampt infiltration method. Runoff is generated from given watersheds that are become saturated or during a storm period. Infiltration as initial abstraction measurement and plot studies in the Ethiopian highlands watersheds has shown that infiltration rate on hillsides with dominant sand and gravel cover can be higher than the greatest rainfall intensity with the same magnitude [5].

The hydrologic water balance and cycle simulated by SWAT model is based on the following water balance equation:

$$SW_t = SW_o + (R_{day} - Q_{surf} - E_a - W_{seep} - Q_{gw})$$

Where,

SW_t is the final soil water content (mm)

SW_o is the Initial Soil water content on daily bases (mm)

t is the time in days

R is the amount of rainfall in daily bases (mm)

Q_i is the surface runoff on day (mm)

E_a is the amount of evapotranspiration in daily bases (mm)

W_{seep} is the amount the Vadose zone from soil profile on day (mm)

Q_{gw} is the amount of return flow on daily bases.

Data collection and analysis

Different data types that have been used to conduct this work were obtained from different sources, the study has required basically land use land cover, climate, and stream flow data of the study area to evaluate the relationship of land use change with the low stream flow of Gumara river. But to generate simulated flow and to evaluate the effect land use/cover change from the combined effects of climate and land use/ cover change on low stream flow of the catchment, SWAT model was used. The model uses various input data; these required data were: climate, land use/cover, Soil, DEM, and whether generator [6].

Land use/land cover data

The land use/cover images of the study area have been obtained from USGS earth explorer between path and row of 169 and 52 respectively. And this data has been processed by using ERDAS IMAGINE 2010. Two satellite images were taken in 1973 on 02 March and on the same day and month in 2013. The 1986 land use/cover data was also processed and taken for model calibration.

The three time period land sat satellite images were obtained from Landsat MSS, Landsat TM, and Landsat ETM+ respectively. The preprocessing and processing of theses land use/ cover data were done by using ERDAS IMAGINE 2010 and GIS 10.1 software.

Besides to this, the real field land use/cover data was collected by using GPS device to compare with the extracted 2013 land use/cover data by accuracy assessment tool of ERDAS IMAGINE 2010. Even if it was difficult to be sure, the historical land use/cover information was also collected by informal interviewing the elder people to compare the current and 1973 land use/cover data [7].

Sensor	Date of acquisition	Path and row	Spatial resolution (m)	Source
Landsat 1 MMS	01-02-1973	181/52	57*57	USGS
Landsat 5 TM	28-03-1986	169/52	30*30	USGS
Landsat 8 OLI	06-01-2013	169/52	30*30	USGS

Table 1: Sensor, Acquisition dates, Path/Row, and Resolution of the study area image.

Land use/ land cover classification

Data pre-processing: The land use/cover images of the study area have been obtained from USGS earth explorer between path and row of 169 and 52 respectively. These images were free from cloud cover and they have been geometrically corrected and projected to the World Geodetic System (WGS-84 UTM Zone 37 N). The Landsat images were also resampled using the nearest neighbor algorithm to keep the original brightness and pixels values. Table 1 shows the acquisition dates, sensor, path/row, resolution, and the providers of the images. The acquisition dates of the 1973, 1986 and 2013 images correspond to the dry season of the study area while the images have resolutions of 57 m for 1973 and 30 m for 1986 and for 2013.

This data has been processed by using ERDAS IMAGINE 2010. Among the three land use/cover data, the 1986 were used to calibrate SWAT model parameters, whereas the others 1973 and 2013 land use/cover data were used for the study of change and its effect on low flow of the catchment [8].

The three time period land sat satellite images were obtained from Landsat MSS, Landsat TM, and Landsat OLI respectively. The Landsat multispectral scanner (MSS) bands 1, 2, 3, and 4 cover the spectral range between 0.45-1.10 μm. Both the Landsat thematic mapper (TM) and enhanced thematic mapper (ETM+) bands 1, 2, 3, 4, 5, and 7 cover the spectral range between 0.45-2.5 μm. Observations by bands 1-3 represent visible electromagnetic (EM) radiances at wavelengths 0.45-0.52, 0.52-0.60, and 0.63-0.69 μm, respectively. Band 4 corresponds to the near infrared wavelengths at 0.76–0.90 μm while bands 5 and 7 correspond to the mid-infrared wavelengths at 1.55-1.75 and 2.08-2.35 μm, respectively. The land cover images were created using the band combination of 7, 4, 2 (Landsat TM and OLI) images of 1986 and 2013) and 4, 2, 1 (Landsat MSS image of 1973) to allow visual interpretation of the images in their true color. The preprocessing and processing of theses land use/ cover data were done by using ERDAS IMAGINE 2010 and GIS 10.1 software.

Besides to this, the real field land use/cover data was collected by using GPS device to compare with the extracted 2013 land use/cover data by using accuracy assessment tool of ERDAS IMAGINE 2010. Even if it was difficult to be sure, the historical land use/cover information was also collected by informal interviewing the elder people to compare the current and 1973 land use/cover data [9].

Climate data: The climate data were collected from National Metrological Agency of Ethiopia for five stations including, Bahir dar, Debretabore, Werota, Wanzaye and Amed ber. However, only three metrological stations (Bahir dar, Debretabore and Werota) were used; this was due to long period of record like more than 30 years' data is needed.

The missing data in both rain fall and temperature data were filled by long time daily average value. There *were* also some outliers in the raw data, identified by filtering and drawing the graph by the help of excel sheet and it was corrected by replacing long time average daily

data. Then, the whole years of daily rain fall and temperature data were stacked and prepared with respect to the SWAT using format [10].

Model calibration and validation: The model has been calibrated by (1986-1991) climate and 1986 land use land cover data so as to determine the representative value of parameters on the study area through changing the values of selected parameters until the maximum efficiency of SWAT model was obtained. All objective functions compared the simulated and observed value of the model with different statistical formula.

For many numbers of iterations, the manual calibration is better than automatic one in a way that it takes in to account the consideration of the real physical characteristics of the catchment. Every parameter has their value ranges to be iterated; the model has iterated the calibration process 2000 times of simulation which is the maximum possible number of iteration that SWAT CUP model can do.

After the model, has been calibrated, the consistency of the model parameter values has also been verified by using four-year (1992-1995) climate data and 1986 land use/cover data [11].

SWAT model efficiency: In doing the calibration and validation, the performance of the model was assessed by using coefficient of determination, relative volume error and most importantly Nash-Sutcliffe efficiency. The efficiency E proposed by Nash and Sutcliffe is defined as one minus the sum of the absolute squared differences between the predicted and observed values normalized by the variance of the observed values during the period under study. It is computed as:

The range of NS lies between 1 and -∞; 1 indicates that the best fit of the model or the model generates similar values of flow with the measured values. To accept the model, the NS values should be more than 0.5.

$$NS = 1 - \frac{\sum_{i=1}^{n} \left(Q_{sim(i)} - Q_{obs(i)} \right)^2}{\sum_{i=1}^{n} \left(Q_{obs(i)} - Q^{-}_{obs} \right)^2}$$

RV_E determines the ratios of differences of total simulated and total observed value of flow; it doesn't show us the difference of values with in that corresponding time of measured and simulated discharge values. In other words, even if the value of RV_E of the model is small enough, it is not necessarily mean that the model has good efficiency [12].

$$RV_E = \left(\frac{\sum_{i=1}^{n} Q_{sim(i)} - \sum_{i=1}^{n} Q_{obs(i)}}{\sum_{i=1}^{n} Q_{obs(i)}} \right) 100\%$$

SWAT model structure: To simulate stream flow, SWAT model was constructed starting from watershed delineation. Digital elevation model (DEM) with 30*30 m resolution is the basic and primary ground for watershed delineation and it was used as an input to compute the slope of the catchment. After the watershed, had been delineated, other hydrological response units like, slope, soil and land use characteristics of the model were masked and overlaid with respect to delineated watershed. The model was classified the catchment in to 24 HRUs or sub basins [13].

On the other side, climate data was used as another input in SWAT model for simulation of flow. Daily maximum and minimum temperature, daily rainfall and whether generator was used as in climatic input for the model, the others like solar radiation, relative humidity and wind speed data which are designed to compute the evapotranspiration of the catchment by the model itself were generated on the whether generator data obtained from IWMI.

Then after the process of defining HRUs and importing whether station data, set up of SWAT model was arranged and it had simulated stream flow by default parameters. To fit the simulated flow with observed flow of the catchment, important parameters were selected by the processes of sensitivity analysis and by those parameters the model was calibrated by using automatic calibration (SWAT CUP 2012) until the maximum value of model efficiency was obtained [14,15].

Impact of land use/cover change on low stream flow: Once the model was calibrated by 11 most important parameters and validated the values of these parameters, SWAT model has simulated stream flow two times to evaluate the impacts of land use/cover change on stream flow. The values of those 11 parameters were constant for all simulation processes. First stream flow was simulated by using 1973 land use land cover data and 1973-1982 climate data which was considered as the base line for change study. Secondly, the flow was simulated by 2013 land use land cover data and 1973-1982 climate data [16].

The effects of land use land cover and combined changes were evaluated in such a way that the difference between:

Simulation one and simulation two=effect of land use land cover change.

After the flow has been simulated, the effect of land use/cover change on the normal stream flow of the catchment was determined. The low flow of the base line and change study periods were selected with in the simulated flow generated by using 7 day sustained method.

Results and Discussion

Land use/cover classification accuracy assessment

This assessment is carried out to evaluate the classification efficiency of the ERDAS IMAGINE2010 software so as to represent the current land use /cover map of the study area. The process had determined how effectively pixels were grouped in to the correct feature classes under investigation. It is accomplished the classified image with the ground by the help of field survey or reference map.

To assess the efficiency of classification accuracy, confusion matrix was developed. To conduct this assessment 120 random points were decided to be collected using thump rule; states that 30 points for each land use land cover class and the image was containing four classes.

Reference					
	Class	Grass land	Forest	Cultivation	Bush land	Total
Classified	Grass land	7	0	3	0	10
	Forest	0	5	0	1	6
	Cultivation	6	0	81	3	90
	Bush land	0	1	0	13	14
	Total	13	6	84	17	120

Table 2: LULC confusion matrix.

Class	Producer's accuracy (%)	Omission error (%)	User's accuracy (%)	Commission error (%)
Grass land	7/13=53.85%	46.15	7/10=70.00%	30
Forest	5/6=83.33%	16.66	5/6=83.33%	16.66
Cultivation	81/84=96.43%	3.57	81/90=88.89%	11.11
Bush land	13/17=76.47%	23.53	13/14=92.86%	7.34

Table 3: The LULC classification accuracy assessment.

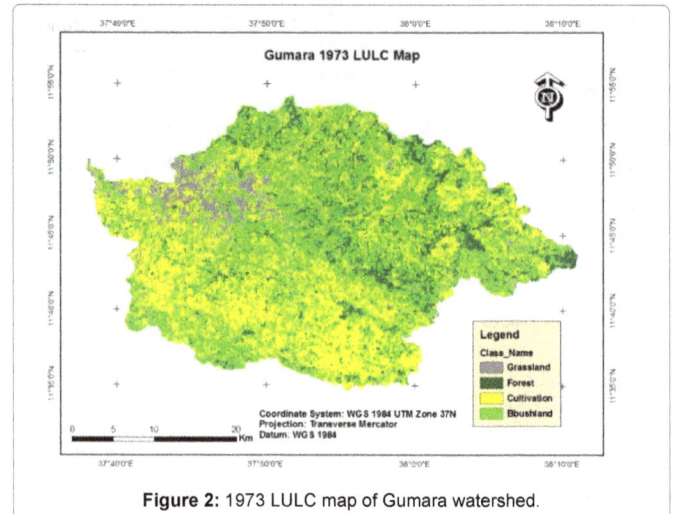

Figure 2: 1973 LULC map of Gumara watershed.

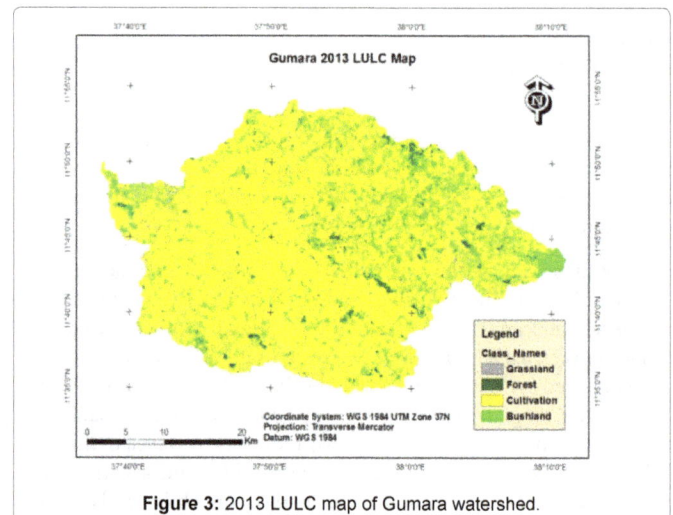

Figure 3: 2013 LULC map of Gumara watershed.

Random points were generated using random point generation tool of Arc GIS software [17].

The numbers of points for each class were determined by their proportional area coverage of the total area of the map. These random points were compared to the real ground land use land cover classes through field survey in areas where it was accessible and reference map like quick bird and Google earth for the areas which were not accessible like mountain area, gorges, and very distant areas of the catchment. On the Table 2, the bold text cells are the numbers of points in each land use land cover classes correctly found on the same classes of the ground. The two totals or the last row and column of Table 2 comprises the numbers of points in each classes of the classified map and the numbers of points of each class on the ground respectively which is proportional out of the total (120) control points.

Land use/Cover classification result

Each land use land cover map has been classified in to four classifications which are forest, bush land, cultivation and grass land (Table 3). The land use land cover map of 1973 is considered as the base line or zero change of the study. The land use land cover classification of 1973 is shown on the map (Figure 2).

The study has tried to show the current land use land cover situation of the study area, 2013 land use land cover map had been prepared to show the extent of change by comparing with the base map (1973). It showed the significant conversion of one land use land cover to the other; agriculture has been expanding and highly dominating other classes. It is because of the increment of population and consequently the society had gone to destruct and expand the area of cultivation land to maintain the need of food security [18].

This land use land cover map was also used for accuracy assessment of the whole land use land cover classification of the catchment. It had been compared with the actual land use land cover nature. The current land use/land cover classes of the study area are shown by the following map (Figure 3).

The accuracy assessment of these land use land cover classification has been done and the result was evaluated based on the values of Kappa statics and the overall accuracy by developing confusion (error) matrix of four classes. It was having 73.33% and 88.33% of Kappa statistic and overall classification accuracy respectively.

Land use /cover change pattern and rate

The pattern and rate of land use/cover change of the study area, has been shown by comparing 1973 and 2013 land use/cover of the study area. The former (1973) land use/cover classification was considered as the base map. Based on that, the extent, and the rate of how much parts of each land use/cover classes has been changed to the other classes had been computed by comparing with 2013 land use/cover map. The extent and rate of land use / cover change between 1973 and 2013 is shown on the Table 4.

Land use/cover change effect on the low flow of the catchment

The annual low flow values were selected from the two decadal periods from already generated flow by land use land cover change of the catchment. Seven day sustained low flow model was used to extract representative low flows for each year in decadal time period [19].

LULC classes	1973 Area (ha)	2013 Area (ha)	Change (ha)	(%)
Grassland	19464.9	11110.4	8354.5	75.19
Forest	22924.5	4638.13	18286.37	394.3
Cultivation	47936.5	96092.9	48156.4	100.5
Bush land	36860.2	15344.6	21515.6	140.2
Total	127186.1	127186.03		

Table 4: Land Use /Cover Change Pattern and Rate.

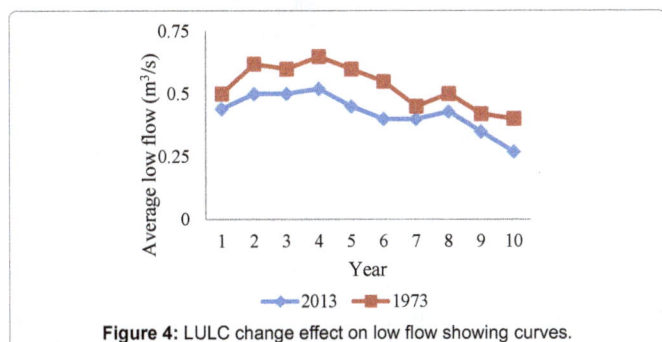

Figure 4: LULC change effect on low flow showing curves.

Year	1	2	3	4	5	6	7	8	9	10	Average
1973 LULC (m³/s)	0.5	0.62	0.6	0.65	0.6	0.55	0.45	0.5	0.42	0.4	0.529
2013 LULC (m³/s)	0.4	0.5	0.5	0.52	0.45	0.4	0.4	0.43	0.35	0.27	0.426

Table 5: The effect of LULC change on the flow of Gumara watershed.

The result had shown that most of the area of the catchment has been covered by cultivation land because of agricultural expansion and consequently the other land use/cover classes had severely been depleted. Especially the forest and bush land classes which are very important physical catchment characteristics to reducing the overland flow and soil erosion of the catchment have been degraded. Due to decreasing of infiltration process, ground water recharge is significantly reduced and consequently stream flow during dry seasons which is generated from ground water has been severely affected. Currently these two-land use/cover classes have covered very small areas i.e. 3.65% and 12.06% respectively of the total area of the catchment. Due to this, the low flow trend of the catchment is decreasing (Figure 4).

The overall gap of these two curves showed the decreasing trend of low flow because of land use land cover change of the catchment. The result indicates that the degradation of vegetation or expansion of agricultural land has been considerably affecting the environmental flow of the catchment. The decreasing of environmental flow has also been affecting the socio-economic situation of the study area. In terms of magnitude, the result has also been evaluated to quantify that exactly how much the land use land cover change has been affecting the low flow trend of the catchment [20].

The mean annual minimum flow for both decades are showed by the last column of the Table 5; which accounts 0.53 m³/s by the base map and 0.43 m³/s by 2013 land use land cover. The mean value difference is 0.1 m³/s; the trend is decreased by 18.87%. Besides to the mean value difference, the minimum of low flow values within ten values of each decade have also been used for evaluation of the change of effect study. The values obtained by the base map and 2013 land use land cover map are 0.4 m³/s and 0.27 m³/s respectively; the difference is 0.13 m³/s which is reduced by 32.5%

The reason is stated by Kuchment [21] the higher infiltration capacity of forest soils increases the opportunity for groundwater recharge, and the flow of rivers tends to be more sustained; which indicates that in Gumara watershed on 1970s there was more or less better vegetation coverage than the current time and therefore the base flow or environmental flow of the catchment has been decreasing.

Conclusions

Landsat images were collected on USGS earth explorer and processed by ERDAS IMAGINE 2010 and ArcGIS 10.1 software. The change of land use land cover of the catchment has been evaluated by comparing the 1973 and 2013 year of land use land cover data. The land use/cover analysis has indicated that most of the previous land use/cover types like forest and bush land have been changed in to agriculture lands. This is due to the expansion of population and competition for the natural resources. This distractive change of land use cover has been significantly affecting the dry season flow or base flow of the catchment.

References

1. Miller S (2002) Integrating Landscape Assessment and Hydrological Modeling for Land Cover Change Analysis. Journal of American Water Ressource Association 38 : 915-929.

2. Luijten JC, Jones JW, Knapp EB (2000) Dynamic modeling of strategic water availability in the Cabuyal River, Columbia : the impact of Land Cover change on the Hydrological Balance. Advances in Environmental Monitoring and Modelling 1: 36-60.

3. Turner BL, Skole D, Snderson S, Fischer G, Fresco L et al. (1995) Land Use and Land Cover Change Science/Research Plan. HDP Reprt Series No 7.

4. Neitsch S, Arnold J, Kiniry J, Williams J (2005) Soil and Water Assessment Tool

Theoretical Documentation-Version 2005. Grassland, Soil & Water Research Laboratory, Agricultural Research Service, and Blackland Agricultural Research Station, Temple, Texas.

5. Ma X, Xu J, Luo Y, Aggarwal SP, Li J (2009) Responses of hydrological processes to land-cover and climate changes in Kejie watershed, south-west China. Hydrological Processes 23: 1179-1191.

6. Yeshaneh E, Wagner W, Exner-Kittridge M, Legesse D , Blöschl G (2013) Identifying Land Use/Cover Dynamics In the Koga Catchment Ethiopia, from Multi-scale Data, and Implications for Environmental Change. ISPRS International Journal of Geo-Information 2: 302-323.

7. FAO-Food and Agriculture Organization of the United Nations: Global Forest Resources Assessment Main report, FAO Forestry Paper 163, Food and Agriculture Organization of the United Nations, Rome, 2010.

8. Dwivedi R, Sreenivas K, Ramana K (2005) Land Use Land Cover Change Analysis in Part of Ethiopia Using Landsat Thematic Mapper Data. International Journal of Remote Sensing 26: 1285-1287.

9. Chong Xu (2005) Hydrology Text Book of Hydrological Models. Uppsala University Department of Earth Sciences Hydrology Earth Sciences Centre.

10. DeFries R, Eshleman KN (2004) Land-use change and hydrologic processes: a major focus for the future. Hydrol Process 18 : 2183-2186.

11. Mati BM, Mutie S, Gadain H, Hame P, Mtalo F (2008) Impacts of land-use/ cover changes on the hydrology of the trans boundary Mara River, Kenya/ Tanzania, Lakes Reserv. Res Manage 13: 169-177.

12. Brandon R, Bottomley BA (1998) Mapping Rural Land Use & Land Cover Change In Carroll County, Arkansas Utilizing Multi-Temporal Landsat Thematic Mapper Satellite Imagery: 1984-1999, University of Arkansas.

13. Moriasi DN, Arnold JG, Van Liew MW, Bingner RL, Harmel RD et al. (2007) Model Evaluation Guidelines for Systematic Quantification of Accuracy in Watershed Simulations. Transactions of the ASABE 50: 885-900.

14. Nash JE, Sutcliffe JV (1970) River flow forecasting through conceptual models, Part I - A discussion of principles. J Hydrol 10 : 282-290.

15. Legesse D, Vallet-colomb C, Gasse F (2003) Hydrological response of a catchment and land use change in tropical Africa: Case study south central Ethiopia. Journal of hydrology 275: 67-85.

16. Hu Q, Willson GD, Chen X, Akyuz A (2005) Effects of climate and land cover change on stream discharge in the Ozark Highlands, USA. Environmental Modeling & Assessment 10: 9-19.

17. Sherbinin D (2002) Land-Use and Land-Cover Change, A CIESIN Thematic Guide. Retrieved 6 15, 2013, International Earth Science Information Network (CIESIN) of Columbia University

18. Smakhtin VU (2001) Low flow hydrology: a review. J Hydrol 240: 147-186.

19. Homdee T, Pongput K, Kanae S (2011) Impacts of land cover changes on hydrologic responses: a case study of chi river basin, Thailand. Journal of Japan Society of Civil Engineers, Ser. B1 (Hydraulic Engineering) 67: I_31-I_36.

20. Zhang YK, Schilling KE (2006) Increasing stream flow and base flow in Mississippi River since the 1940s: Effect of land use change. Journal of Hydrology 324 : 1-4.

21. Kuchment LS (2008) Runoff generation (genesis, models, prediction). Water Problems Institute of RAN, p: 394.

18

Microbial Diversity in Continuous Flow Constructed a Wetland for the Treatment of Swine Waste

Ibekwe AM[1]*, Ma J[1,2], Murinda S[3] and Reddy GB[4]

[1]USDA-ARS, U.S. Salinity Laboratory, 450 W Big Springs Rd, Riverside, CA 92507, USA
[2]College of Environment and Resources, Jilin University, Changchun, Jilin Province, 130021, P. R. China
[3]Department of Animal and Veterinary Sciences, California State Polytechnic University, Pomona, CA 91768, USA
[4]Department of Natural Resources and Environmental Design, North Carolina Agricultural and Technical State University, Greensboro, NC 27411, USA

Abstract

Contaminant removal may largely be a function of many microbial processes in constructed wetlands. However, the role of microbial diversity for the removal of swine waste in constructed wetlands is limited. Here, we used 454/ GS-FLX pyrosequencing to assess archaeal, bacterial, and fungal composition within a surface flow constructed wetland to determine their spatial dynamics and contaminant removal within the wetland. We analyzed our data using UniFrac and principal coordinate analysis (PCoA) to compare community structure and specific functional groups of bacteria, archaea, and fungi in different sections of the wetland. PCoA analysis showed that, bacterial, archaeal, and fungal composition were significantly different ($p=0.001$) for the influent compared to the final effluent. Our results showed that, the wetland system contained relatively higher proportions of bacteria and fungi than archaea. Most of the bacteria and archaea that were associated with nitrogen removal were affiliated with *Nitrosomonas* which are ammonia oxidizing bacteria (AOB), *Candidatus Solibacter*, an anaerobic ammonia oxidizing bacteria (Anammox), as well as *Nitrosopumilus*, ammonia oxidizing archaea (AOA). The detection of AOB, Anammox, and AOA in this wetland shows abundance and diversity of these microorganisms that are responsible for nitrification processes in constructed wetlands.

Keywords: Constructed wetland; Contaminants; Pyrosequencing; Microbial communities; Swine; Wastewater

Introduction

Constructed wetland (CW) is a natural process for the treatment of waste from a variety of sources [1-3]. They are cost-effective, ecologically-friendly, and a simple alternative to conventional technologies for wastewater treatment [3,4]. It has also been shown that, significant benefits to human populations in both [2] developed and developing countries can be achieved through constructed wetlands, and these benefits may include water-quality improvement, water reclamation, conservation of habitat for species, flood control, recreational and educational activities [1,4-7]. North Carolina is one of the largest swine producing states in the United States [2], and waste disposal was traditionally done by flushing into anaerobic lagoons and then later sprayed on agricultural fields. This resulted in swine waste from large swine farms polluting surface and well waters [8]. During hurricane Floyd in 1999 many North Carolina hog waste lagoons overflew [2] and polluted many surface water systems with fecal indicator bacteria from compromised septic and municipal sewage systems, and livestock waste lagoons [8]. After the storm, North Carolina invested resources [2] into new, cost-effective technologies for waste disposal, and one such technology was constructed wetlands for the treatment of swine waste. As a result, a pilot constructed wetland for the treatment of swine waste was evaluated at [2] North Carolina Agricultural and Technical State University (Greensboro NC, USA). The wetland system uses [2] natural plants for the removal of nitrogen (N), phosphorus (P), solids, and Chemical Oxygen Demand (COD) from treated swine wastewater, resulting in a cleaning final effluent [9].

The removal efficiency by constructed wetlands has been shown to be a function of diverse microbial communities [10]. However, there is a general lack of information on the diversity and changes of the microbial communities in long-term constructed wetlands treating swine waste [2]. Work on bacterial diversity in constructed wetlands is well documented, whereas the role played by archaea and fungi is not clear. The main contaminants from swine waste [2] may include nutrients, salts, microbes, including pharmaceutically active compounds, and their removal involves complex physical, chemical, and biological processes [2]. The aim of the current study was to compare the composition of microbial communities in a continuous surface flow constructed wetland used for the treatment of swine waste using pyrosequencing.

Materials and Methods

Experimental site and sampling

The experiment was conducted at a swine research facility at North Carolina Agricultural and Technical State University farm in Greensboro, NC, USA. This was a continuous flow constructed wetland originally built in March 1996 [11] (Figure 1), and planted with *Typha latifolia* L. (broadleaf cattail) and *Schoenoplectus americanus* (Pers-American bulrush). The wetland has six cells and each cell is 40 m long and 11 m wide. In 2003, a modification was made to cells 5 and 6 [11] to achieve a continuous marsh system with a slope of 0.33% (Figure 1). The new sections were planted with giant bulrushes (*Scirpus californicus*) as previously noted [11]. Sixty-five to 115 pigs were managed in the swine house between January 2007 and January 2012. Samples were collected from eight different points from different sections of the wetland in

*Corresponding author: Abasiofiok Mark Ibekwe, USDA-ARS-U. S. Salinity Laboratory, 450 W Big Springs Rd, Riverside, CA 92507
Email: Mark.Ibekwe@ars.usda.gov

duplicate in November 2012. Grab samples were collected sequentially from effluent from the swine house, lagoon 1, lagoon 2, and the 8,000 L storage tank. More samples were collected from continuous wetland cell influent, final effluent samples, storage pond, and the final recycled effluent. The wetland cell received 10 kg/ha/day with a hydraulic load of 3.8 m^3/day. Samples were stored on ice and taken to the laboratory for further analysis. All samples were analyzed for ammonium (NH^+_4-N), nitrate (NO_3-N), total-phosphate (TP) and available-phosphate (PO_4^{3-}) using a flow injection analysis instrument (Lachat-QuikChem 8000, Loveland, CO, USA), as well as carbon (C) and nitrogen (N) concentrations using the Perkin-Elmer 2400, CHNS/O series II Analyzer (Shelton, CT, USA) [4].

DNA extraction and Pyrosequencing

Total bacterial DNA was extracted from effluent samples using Power Water DNA kits (MO BIO, Inc., Solana Beach, CA), according to the manufacturers' protocol. DNA extracts from duplicate samples (200 ng) were pooled. Extracted DNA (2µL) was quantified using the NanoDrop ND-1000 spectrophotometer (NanoDrop Technologies, Wilmington DE), and run on a 1.0% agarose gel before it was used for pyrosequencing. DNA was stored at -20° C prior to pyrosequencing analysis. The DNA samples (15.0 µl) were then submitted to Research and Testing Laboratories (Lubbock, TX) for polymerase chain reaction (PCR) optimization and pyrosequencing analysis. Bacterial tag-encoded FLX amplicon pyrosequencing was conducted on the 16S rRNA gene for amplification of bacteria and archaea sequences, whereas the microbial tag-encoded FLX amplicon was used for amplification of 18S rRNA gene sequences in fungi. Pyrosequencing (TEFAP) procedures described earlier [12] based on 16S rRNA genes for archaea (V3-V5 regions) and bacteria (V1-V3 regions) with primer pairs, 340F90 (GYGCASCAGKCGMGAAW)/806R96 (GGACATCVSGGGTATCTAAT) and 28F

(GAGTTTGATCNTGGCTCAG)/519R (GTNTTACNGCGGCKGCTG), respectively, were employed. The fungal 18S rRNA gene was amplified using the primer pair, SSUF (TGGAGGGCAAGTCTGGTG)/ funSSUR(TCGGCATAGTTTATGGTTAAG). Pyrosequencing data were analyzed using the dist. seqs function in MOTHUR, version 1.9.1 [13]. Raw reads were treated as previous described [14] using the Pyrosequencing Pipeline Initial Process of the Ribosomal Database Project (RDP), to sort the data, trim off the adapters, barcodes and primers using the default parameters, and to remove ambiguous 'N' [15]. Sequence libraries were further resampled to obtain similar numbers of sequences for diversity and richness estimations [16]. All sequence reads with a quality score 20 and a read length 200 bp were removed. Shannon's diversity index values (H'), and Chao estimates were calculated from MOTHUR as well as operational taxonomic units (OTUs, 97% similarity). We used RDP Classifier with a bootstrap cutoff of 80% for taxonomic classification of the bacterial sequences. We used the keep first 200 bp commands in MOTHUR to eliminate sequencing noise resulting in a sequence read fragment covering the 18S region and aligned to the SILVA database for further analysis of data as stated above.

Statistics and analysis of pyrosequencing data

SAS version 9.1 [17] was used to conduct analysis of variance (ANOVA) to determine differences in wetland properties. PCoA and UniFrac analysis were carried out using MOTHUR to group microbial communities of different samples into different taxonomic groups using the RDP Classifier. RDP Complete Linkage Clustering was also used to generate OTUs and weighted UniFrac from all samples [18]. The relaxed neighbor-joining algorithm in Clearcut (version 1.0.9) [19] was used for the construction of phylogenetic trees and the parsimony test (P-values of 0.05) in Treeclimber [20] was used for between-site comparisons.

Results

Nutrient removal

There were significant decreases (P<0.05) in total N, ammonium (NH_4^+), and total suspended solids (TS) in wetlands from the lagoon to the final effluent (Figure 2). The removal rates of N and NH_4^+ in this study were above 70% from the influent to the effluent and this was similar to what had been previously reported [9,21]. The removal rate of total and organic phosphorus was low between the lagoon and the effluent samples. The removal rates of total and suspended solids were significantly higher (P<0.01) between lagoon and the final effluent [18].

Community composition, diversity, and estimated richness

A total of 6354, 81234, and 50719 sequence tags for archaea, bacteria, and fungi, respectively, were generated through 454 pyrosequencing (Tables 1-3). The total numbers of OTUs were 661, 8429, and 1946 for archaea, bacteria, and fungi, respectively (Table S1-S3). For the archaea, the highest number of sequences was found in storage pond while the lowest was in the final effluent (Table 1). For bacterial the highest sequence tag was from lagoon while the lowest was from mid marsh (Table 2). Furthermore, for fungi, the highest sequence tag was found in samples collected from storage pond and the lowest from mid marsh (Table 3). We normalized our data to the smallest sequence tag, for reanalysis to show normal distribution of variances (Table S1-S3). Shannon diversity index (H') showed variations in diversity among the different wetland segments with the highest archaeal and fungal

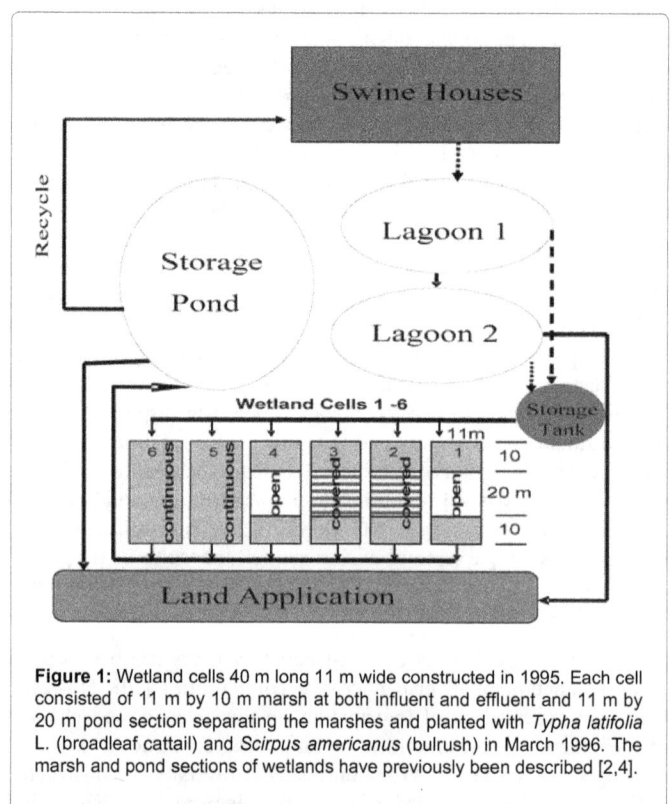

Figure 1: Wetland cells 40 m long 11 m wide constructed in 1995. Each cell consisted of 11 m by 10 m marsh at both influent and effluent and 11 m by 20 m pond section separating the marshes and planted with *Typha latifolia* L. (broadleaf cattail) and *Scirpus americanus* (bulrush) in March 1996. The marsh and pond sections of wetlands have previously been described [2,4].

Group	nseqs	OTUs	chao	invsimpson	npshannon	simpson	coverage
S1	630	51	132.2	5.333953	2.361805	0.187478	0.953968
S2	952	111	354.83333	7.096459	3.043943	0.140915	0.919118
S3	619	92	202.5	16.108388	3.530358	0.062079	0.915994
S4	3318	171	381	3.398484	2.161431	0.294249	0.968354
S5	257	76	130.66667	16.472709	3.830253	0.060706	0.840467
S6	280	89	149.05556	33.759723	4.180865	0.029621	0.832143
S7	163	41	167.5	11.094958	3.192706	0.090131	0.858896
S8	135	30	48.2	7.97619	2.842945	0.125373	0.896296

Table 1: Total sequence tags for archaea generated from pyrosequencing from the wetland land units.

Group	nseqs	OTUs	chao	invsimpson	npshannon	simpson	coverage
S1	9201	751	1440.81	14.230392	4.141051	0.070272	0.953918
S2	11967	1683	3677	22.425723	5.147448	0.044592	0.916688
S3	12385	1233	2765.64	7.080288	4.102311	0.141237	0.940654
S4	10181	1222	2585.45	19.936472	4.683588	0.050159	0.927414
S5	6934	1268	2907.43	20.243401	5.272591	0.049399	0.88852
S6	10800	1758	3380.38	64.356874	6.053123	0.015538	0.914907
S7	10795	822	1809.5	25.159356	4.438501	0.039747	0.955998
S8	8971	1719	3653.54	87.078426	6.120541	0.011484	0.890313

Table 2: Total sequence tags for bacteria generated from pyrosequencing from the wetland land units.

Group	nseqs	OTUs	chao	invsimpson	npshannon	simpson	coverage
S1	10340	135	345.789474	2.212453	1.255139	0.451987	0.991296
S2	8814	102	224.0625	1.167906	0.52713	0.856234	0.992852
S3	6987	556	1186.123457	27.501675	4.283528	0.036361	0.954201
S4	1761	180	390.416667	11.555549	3.552191	0.086539	0.942646
S5	4609	472	947.561644	12.476529	4.024419	0.08015	0.942721
S6	5554	201	439.214286	3.572661	2.243241	0.279903	0.979114
S7	9204	117	289.117647	1.770473	1.058795	0.564821	0.991634
S8	3450	183	414.875	5.192079	2.690632	0.192601	0.969275

Table 3: Total sequence tags for fungi generated from pyrosequencing from the wetland land units.

diversity in the continuous flow effluent, while the highest bacterial diversity was from manure influent to the lagoon 1.

From our pyrosequencing data, the majority of archaeal sequences belonged to the phyla, *Euryarchaeota* (88.7%) and *Crenarchaeota* (5.21%), and at the class level *Methanomicrobia* (42.1%), *Thermoplasmata* (26.25%), and *Methanobacteria* (17.35%) were dominant (Table 4). Bacterial sequences primarily comprised of the phyla, *Proteobacteria* (36.58%), *Bacteroidetes* (18.15%), *Firmicutes* (11.86%) (Table 5). The most abundant fungal phylum was the Basidiomycota (86.3%) (Table 6).

Using the greengenes database, we identified potential sequences at the genus level for specific functional groups of microorganisms that could perform specific functions in constructed wetlands (Table 7). These sequences were mainly from bacteria and archaea. We had a relatively low abundance of *Nitrosomonas*, which are ammonia oxidizing bacteria (AMO) in the midsection of the wetland. However, high sequences of a carbon degrader (CD) *Roseburia* were detected in this section of the wetland. *Flavobacterium, Rhodobacter, Thauera*, and *Methylophilus* are all denitrifying bacteria (DN), and as expected were found in relatively high abundance in this section of the wetland. *Hydrogenophaga* (hydrogen oxidizing bacteria) was also found in relatively high abundance. *Candidatus Solibacter*, an anaerobic ammonia oxidizing bacterium (Anammox), and *Nitrosopumilus*, an

ammonia oxidizing archaea (AOA), were detected in this section of the wetland in relatively low numbers. The detection of both Anammox and AOA was not surprising since both are known to perform major functions in major water columns [22].

Spatial phylogenetic structure of microbial community from different segments of the wetland

We applied PCoA (Figure 3), and the UPGMA hierarchical clustering analysis (data not shown) to determine the distribution of microbial phylogenetic similarities and sorted them into different groups by applying UniFrac distance matrix using the UniFrac web interface in MOTHUR. The pattern for archaeal community dynamics was different from that of bacteria (Figure 3A). An archaeal sample from the swine house and lagoons clustered to the right, whereas, samples from storage tanks, wetland cells effluent, mid marsh effluent, and the final effluent for land application clustered to the left. These segments clustered closer to each other, which was the opposite for bacteria. As stated earlier, these four segments have already gone through the wetland lagoons and wetland cells, where most of the biological interactions with complex communities are occurring for the degradation of major compounds from the swine house waste (Figure 1). Bacterial community structures from wetland samples collected from the swine house and lagoons 1and 2 were significantly different (p<0.0001) based on parsimony tests (Figure

3B) as wetland compounds were strong structural factors influencing bacterial assemblages (R^2=0.63, p=0.0024). The PCoA (Figure 3B) showed samples from swine house and lagoons clustering to the left while samples from storage tanks, wetland cells effluent, mid marsh effluent, and the final effluent for land application clustered to the right. Furthermore, bacterial community structures from these four segments of the wetland were significantly different based on the parsimony test (p<0.001), and this was confirmed by hierarchical clustering analysis (data not shown) with Jackknife supporting values. In general, the pattern of separations found in bacteria was also observed with fungal community structures in the different wetland segments.

Discussion

We employed 454 pyrosequencing techniques to quantify archaea, bacteria, and fungi community structures in a continuous flow constructed wetland. One major advantage of pyrosequencing is high-throughput and a large dataset of sequences, which can identify many microorganisms in a single analysis. As shown in this study, and from our previous study with bacteria [3], 16S rDNA sequences from different segments of the wetland may be used to determine the efficiency of the constructed wetland. This approach can monitor changes of microbial communities as waste flows through the wetland and has high potential for improving water quality based on bacterial concentrations. However, this technique does not quantify the members in the community in terms of cell number in the water samples, nor does it imply viability of the organism, but may demonstrate the presence of potential DNA sequences as shown in our previous study [23]. Here, we showed the diversity of microbial communities in relation to nutrient content in different sections of the constructed wetland fed with anaerobic lagoon swine wastewater. It has also been reported that, the bulk of the water quality improvement in constructed wetlands is due to microbial activities [1,2,24]. Archaeal, bacterial, and fungal OTUs from the influent were higher than from the effluent of the wetland. These results were associated with nutrient content differences in the different segments of the wetland where the concentrations of TN, NH_4^-, TP and PO_4^{3-} decreased from lagoon to the final effluent (Figure 2). The results can be partly explained by the decrease in nitrogen and phosphorous, which are essential nutrients for microbial growth. This is different from the research results reported by Ipsilantis and Sylvia [25], where microbial counts did not consistently increase under elevated nutrient conditions due to carbon limitation. Nitrogen removal in wetlands has been documented to be the result of microbial activities, which play roles in the pathways of Anammox [26] and nitrification-denitrification [27]. In addition, P removal may partially depend on microbial activities via mineralization [26,28] and immobilization.

We used pyrosequencing in this study to investigate the microbial diversity in the wetlands system because changes in environmental variables will likely have some influence on changes in OTUs. Archaea from storage tank samples displayed the highest species richness, diversity and evenness than samples from other segments of the wetland while results from the final effluent showed the lowest number of sequences; OTUs, and other diversity indices (Table 1). However, for bacteria and fungi, the highest diversity came from samples from wetland effluent, which has gone through the continuous marsh section of the wetland. The abundance, richness, and diversity of bacterial and fungal communities were considerably higher than archaea in a majority of the wetland cells (Table 2 and 3). This was similar to wetlands planted with *Vetiveria zizanioides* or *Juncus effusus* L. that showed much higher bacterial abundance but lower archaeal abundance [29]. In our study bacteria and fungi outnumbered archaea in all the wetland cells.

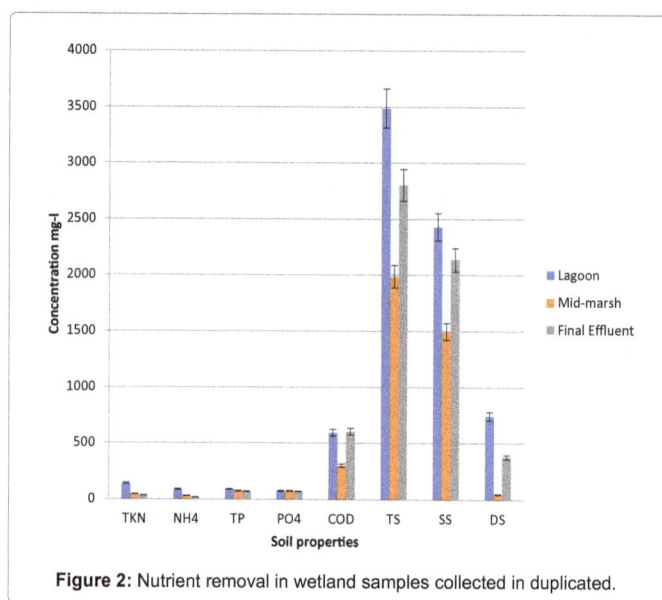

Figure 2: Nutrient removal in wetland samples collected in duplicated.

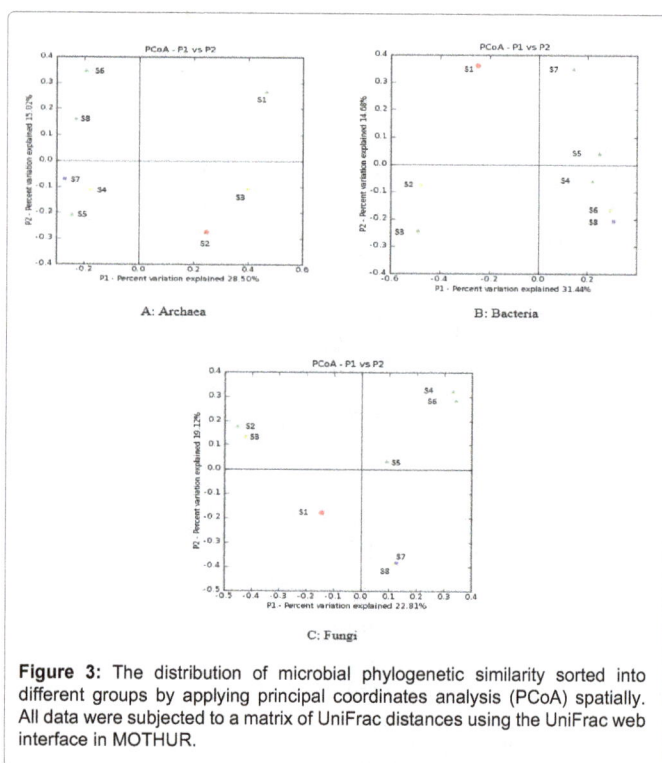

Figure 3: The distribution of microbial phylogenetic similarity sorted into different groups by applying principal coordinates analysis (PCoA) spatially. All data were subjected to a matrix of UniFrac distances using the UniFrac web interface in MOTHUR.

Results from PCoA analysis confirmed that, TN, NH_4^-, TP and PO_4^{3-}, were strongly correlated with the distribution of microbial species on which NH_4^-, TP and PO_4^{3-} concentrations had significant effects. These results suggested that, high levels of nutrient status promoted diversity and distribution of microbial species within the community. It has also been shown that, shifts in the structure of bacterial communities can be associated with changes in a number of soil properties including soil texture and soil nitrogen availability [30,31]. On the other hand, Calheiros et al. [32] reported that, bacterial diversity in constructed wetland may be a significant driver influencing the final effluent quality. Ogier et al. [33] and Dubernet et al. [34] reported that, the difference in the relative abundance of community members may affect the detection of certain species due to competition during PCR. However, there may well be some species whose presence was obscured because they were not detected by pyrosequencing. Our study showed that, when wetland samples were analyzed spatially, community structures from wetland samples associated with the swine house and lagoons 1 and 2 were significantly different (P<0.0001) based on parsimony test of UniFrac data (Figure 3A) from storage tanks, wetland cells effluent, storage pond, and the final effluent. Microbial communities in wetland cells are highly responsive to perturbation, dissolved organic matter concentration, and chemical stress, among others [35-38].

Pyrosequencing data showed high levels of *Bacillus* which is one of the largest Eubacteria found in soil. Some bacilli are phosphate solubilizing and are capable of surviving in extreme conditions due to their ability to form spores. Han and Lee [39] reported that, the inoculation of certain species of *Bacillus* may increase soil P availability. The high concentration of P in swine wastewater will limit the assimilative capacity of wetland soil and the presence of such bacteria in our constructed wetlands for swine wastewater treatment could impair the capabilities of P nutrient removal. A high relative abundance of potential denitrifying bacteria, *Flavobacterium*, was detected in mid-section of the wetland. *Flavobacterium* associated with this wetland had been shown to be closely related to nitrifying bacteria [2]. However, these authors noted that even though these bacteria demonstrated phylogenic similarity, it does not mean that they have denitrifying potential. Therefore, the existence of denitrifying bacteria may be a contributing factor to the decreasing trend of nitrogen concentration from the influent to the effluent end in our wetland.

Another important pathway for the removal of nitrogen from this wetland is through the Anammox process. During this process, ammonium and nitrite are converted to dinitrogen gas [40]. Our research findings suggested the existence of Anammox bacterial sequences, *Candidatus Solibacter*, and this bacterium is a strict anaerobic autotroph. It is also known for its extremely slow growth resulting in limited applications [41]. Partial-nitrification with Anammox in constructed wetlands with higher removal efficiency of total nitrogen than conventional methods has also been reported [42],

thus confirming the role of these group of bacteria in water quality improvements in constructed wetlands. It is possible that Anammox bacteria played a role in nitrogen removal in our wetland. It has been reported that Anammox uses carbon dioxide as its carbon source to produce biomass ($CH_2O_{0.5}N_{0.15}$) and nitrite as electron acceptor for ammonium oxidation, and electron donor for the reduction of carbon dioxide [40]. In our study, we identify six OTUs of *Candidatus Solibacter*, whose sequences were 100% similar to *Candidatus Solibacter* from different sections of our wetland. This wetland has a high concentration of ammonium and nitrite, and as mentioned above [40], the Anammox bacteria require both ammonium and nitrite, which can be found at or near the aerobic–anaerobic interface of sediments and water bodies to function efficiently. Our samples were collected in the continuous flow section of the wetland which provides the most ideal region for the enrichment of Anammox. In constructed wetlands, ammonium diffuses upwards and meets the oxygen that is diffusing downwards, and create aerobe-anaerobe zone for most of the microbial activities associated aerobic bacteria or crenarchaeal ammonium oxidizers. In our study, about 5.1% of archaea were *Crenarchaeota* (Table 4). We also detected by pyrosequencing *Nitrosopumilus* which is a major ammonia oxidizing archaea [22,29,43]. These authors have shown that both bacteria and archaea communities can play important roles in biogeochemical processes in constructed wetland system.

In conclusion, the different sections of the wetland had different nutrient status reflecting different bacterial communities and diversities, permitting different bacteria to play important roles in nutrient removal within the wetland. The removal rates of N and NH_4^+ in this study were above 70% from the influent to the effluent and this signifies the important roles of different microbial groups, played in contaminant removal. Most important, is the improvement of water quality from the final effluent since this water could be spread on the pasture. Therefore, the removal of the main pollutants from the swine wastewater would have a beneficial effect on the ground and surface water and in most of the swine producing regions of the world. This project is an innovative model for waste management for the swine industry and other confined-animal facilities.

Taxon	Total	S1	S2	S3	S4	S5	S6	S7	S8
Archaea	100	100	100	100	100	100	100	100	100
Crenarchaeota	1.2	0.16	0	0.16	1.45	1.97	2.5	1.84	8.15
Euryarchaeota	97.64	99.84	100	99.68	98.01	94.49	92.14	90.18	78.52
pMC2A384	0.08	0	0	0	0.09	0	0	0.61	0.74
unclassified archaea	1.09	0	0	0.16	0.45	3.54	5.36	7.36	12.59
Full sequence percentages of archaea classes									
Taxon	Total	A.MU	A.SW	A.LA	A.MI	A.MM	A.ME	A.SP	A.FE
C2	1.01	0	0	0.16	1.33	1.97	1.79	1.84	4.44
Methanobacteria	15.62	62.38	20.59	22.46	2.53	18.9	18.57	33.13	19.26
Methanomicrobia	59.34	0	3.36	25.04	93.64	65.35	54.29	52.76	52.59
pMC2A209	0.02	0	0	0	0.03	0	0	0	0
Thaumarchaeota	0.16	0.16	0	0	0.09	0	0.71	0	2.96
Thermoplasmata	22.3	37.46	76.05	52.02	1.84	4.72	17.86	4.29	2.96
unclassified archaea	1.09	0	0	0.16	0.45	3.54	5.36	7.36	12.59
unclassified Crenarchaeota	0.02	0	0	0	0	0	0	0	0.74
unclassified euryarchaeota	0.38	0	0	0.16	0	5.51	1.43	0	3.7
unclassified pMC2A384	0.08	0	0	0	0.09	0	0	0.61	0.74

Table 4: Full sequence percentages of archaea at the phylum and class levels.

Taxon	Total	S1	S2	S3	S4	S5	S6	S7	S8
Acidobacteria	0.07	0.08	0	0.06	0.04	0.5	0	0.02	0
Actinobacteria	1.31	2	0.39	3.21	1.05	0.74	0.08	2.43	0.12
Armatimonadetes	0.01	0.01	0	0	0.09	0.03	0	0	0
Bacteroidetes	18.15	10.02	14.08	10.38	11.14	18.23	28.73	17.01	39.15
Chlorobi	0.02	0	0	0.03	0.09	0	0	0	0
Chloroflexi	0.13	0.07	0.08	0.24	0.39	0.23	0	0.01	0
Deferribacteres	0	0	0	0	0	0	0	0	0.01
Fibrobacteres	0.05	0	0.03	0	0	0	0.3	0	0.09
Firmicutes	11.86	0.22	8.44	0.6	1.97	0.42	47.56	0.29	34.89
Fusobacteria	0.01	0	0.04	0	0	0	0	0	0.04
Gemmatimonadetes	0.02	0.02	0.01	0.01	0	0.12	0	0.01	0
Lentisphaerae	0.01	0	0.03	0.01	0.02	0	0	0	0.01
OD1	0	0	0	0.01	0	0	0	0.01	0
Planctomycetes	0.01	0.02	0	0.01	0.02	0.06	0	0.02	0
Proteobacteria	36.58	27.39	58.49	60.1	66.36	29.59	1.19	29.13	7.44
Spirochaetes	0.54	0	0.79	0.05	0.12	0.09	2.17	0	0.94
SR1	0	0	0.01	0	0	0	0	0	0
Tenericutes	0.12	0	0.43	0.15	0.06	0.03	0.02	0	0.18
TM7	0.06	0.11	0	0.06	0.04	0.06	0	0.19	0
unclassified bacteria	30.33	57.98	17.13	24.24	18.38	47.59	19.94	49.95	17.13
Verrucomicrobia	0.73	2.09	0.04	0.86	0.24	2.32	0.01	0.94	0
Total	100	100	100	100	100	100	100	100	100

Table 5: Full sequence percentages of bacteria at the phylum level.

Taxon	Total	S1	S2	S3	S4	S5	S6	S7	S8
Basidiomycota	87.26	97.26	92.58	76.48	81.03	85.98	89.09	95.56	45.36
unclassified fungi	12.74	2.74	7.42	23.52	18.97	14.02	10.91	4.44	54.64

Full taxonomy RDP by order									
	Total	F.FE	F.LA	F.ME	F.MI	F.MM	F.MU	F.SP	F.SW
Agaricomycetes	85.84	96.87	91.89	70.92	80.18	83.9	88.13	95.3	44.09
unclassified basidiomycota	1.43	0.4	0.69	5.57	0.85	2.08	0.95	0.26	1.28
unclassified fungi	12.74	2.74	7.42	23.52	18.97	14.02	10.91	4.44	54.64

Full taxonomy RDP by family									
	Total	F.FE	F.LA	F.ME	F.MI	F.MM	F.MU	F.SP	F.SW
Agaricales	83.74	96.66	90.89	60.46	78.48	81.58	87.95	94.99	42.7
unclassified Agaricomycetes	2.1	0.2	1	10.46	1.7	2.32	0.18	0.3	1.39
unclassified basidiomycota	1.43	0.4	0.69	5.57	0.85	2.08	0.95	0.26	1.28
unclassified fungi	12.74	2.74	7.42	23.52	18.97	14.02	10.91	4.44	54.64

Full taxonomy RDP by genus									
	Total	F.FE	F.LA	F.ME	F.MI	F.MM	F.MU	F.SP	F.SW
Pleurotaceae	81.84	95.66	90.67	51.94	75.41	80.17	87.94	93.67	42.55
unclassified Agaricales	1.9	1.01	0.22	8.52	3.07	1.41	0.02	1.33	0.14
unclassified Agaricomycetes	2.1	0.2	1	10.46	1.7	2.32	0.18	0.3	1.39
unclassified basidiomycota	1.43	0.4	0.69	5.57	0.85	2.08	0.95	0.26	1.28
unclassified fungi	12.74	2.74	7.42	23.52	18.97	14.02	10.91	4.44	54.64

Full taxonomy RDP by species									
	Total	F.FE	F.LA	F.ME	F.MI	F.MM	F.MU	F.SP	F.SW
Hohenbuehelia	81.84	95.66	90.67	51.94	75.41	80.17	87.94	93.67	42.55
unclassified Agaricales	1.9	1.01	0.22	8.52	3.07	1.41	0.02	1.33	0.14
unclassified Agaricomycetes	2.1	0.2	1	10.46	1.7	2.32	0.18	0.3	1.39
unclassified basidiomycota	1.43	0.4	0.69	5.57	0.85	2.08	0.95	0.26	1.28
unclassified fungi	12.74	2.74	7.42	23.52	18.97	14.02	10.91	4.44	54.64

Table 6: Full sequence percentages of some dominant Fungi from phylum to species.

AOB	Nitrosomonas	11
CD	Roseburia	124
DN	Flavobacterium	1626
DN	Rhodobacter	28
DN	Thauera	57
DN	Methylophilus	3
HOB	Hydrogenophaga	7116
anammox	CandidatusSolibacter	6
AOA	Nitrosopumilus	4

AOB- ammonia oxidizing bacteria, CD- carbon degrader, DN- denitrifying bacteria, HOB-hydrogen oxidizing bacteria, anammox- anaerobic ammonia oxidizing bacterium, AOA- ammonia oxidizing archaea.

Table 7: Bacteria and archaea with potential contribution for removal of contaminants from constructed wetlands.

Acknowledgements

This research was supported by the 212 Manure and Byproduct Utilization Project of the USDA-ARS. We thank Damon Baptista for technical assistance. Mention of trade names or commercial products in this publication is solely for the purpose of providing specific information and does not imply recommendation or endorsement by the U.S. Department of Agriculture. The US Department of Agriculture (USDA) prohibits discrimination in all its programs and activities on the basis of race, color, national origin, age, disability, and where applicable, sex, marital status, familial status, parental status, religion, sexual orientation, genetic information, political beliefs, reprisal, or because all or part of an individual's income is derived from any public assistance program.

References

1. Ibekwe AM, Grieve CM, Lyon SR (2003) Characterization of microbial communities and composition in constructed dairy wetland wastewater effluent. Appl Environ Microbiol 69: 5060-5069.

2. Dong X, Reddy GB (2010) Soil bacterial communities in constructed wetlands treated with swine wastewater using PCR-DGGE technique. Bioresour Technol 101: 1175-1182.

3. Oopkaup K, Truu M, Nõlvak H, Ligi T, Preem JK (2016) Dynamics of bacterial community abundance and structure in horizontal subsurface flow wetland mesocosms treating municipal wastewater.Water 8: 457.

4. Ibekwe AM, Ma J, Murinda SE, Reddy GB (2016) Bacterial community dynamics in surface flow constructed wetlands for the treatment of swine waste. Sc Total Environ 544: 68-76.

5. Ibekwe AM, Murinda SE, DebRoy C, Reddy GB (2016) Potential pathogens, antimicrobial patterns, and genotypic diversity of Escherichia coli isolates in constructed wetlands treating swine wastewater. FEMS Microbiol Ecol 92: 1-14.

6. Ibekwe AM, Lyon SR, Leddy M, Jacobson-Meyer M (2007) Impact of plant density and microbial composition on water quality from a free water surface constructed wetlands. J Appl Microbiol 102: 921-936.

7. Kadlec RH, Knight RL (1996) Treatment wetlands. Lewis Publishers, New York 1996.

8. Casteel MJ, Sobsey MD, Mueller JP (2006) Fecal contamination of agricultural soils before and after hurricane-associated flooding in North Carolina. J Environ Sci Health Part A 41: 173-184.

9. Reddy GB, Hunt PG, Phillips R, Stone K, Grubbs A (2001) Treatment of swine wastewater in marsh-pond-marsh constructed wetlands. Water Sci Technol 44: 545.

10. Scholz M, Lee BH (2005) Constructed wetlands: a review. Int J Environ Stud 62: 1256-1261.

11. Forbes DA, Reddy GB, Hunt PG, Poach ME, Ro KS et al. (2010) Comparison of aerated marsh-pond-marsh and continuous marsh constructed wetlands for treating swine wastewater. J Environ Sc Health 45: 803-809.

12. Dowd SE, Callaway TR, Wolcott RD, Sun Y, McKeehan T et al. (2008) Evaluation of the bacterial diversity in the feces of cattle using 16S rDNA bacterial tag-encoded FLX amplicon pyrosequencing (bTEFAP). BMC Microbiol 8: 125.

13. Schloss PD, Westcott SL, Ryabin T, Hall JR, Hartmann M, et al. (2009) Introducing mothur: open-source, platform-independent, community-supported software for describing and comparing microbial communities. Appl Environ Microbiol 75: 7537-7541.

14. Cole JR, Wang Q, Cardenas E, Fish J, Chai B, et al. (2009) The Ribosomal Database Project: improved alignments and new tools for rRNA analysis. Nucleic Acids Res 37: D141–D145.

15. Claesson M, O'Sullivan O, Wang Q, Nikkila J, Marchesi J, et al. (2009) Comparative analysis of pyrosequencing and a phylogenetic microarray for exploring microbial community structures in the human distal intestine. PloS One 4: e6669.

16. Champely S, Chessel D (2002) Measuring biological diversity using Euclidean metrics. Environ Ecol Stat 9: 167-177.

17. SAS Institute (2009). JMP User guide. Release 9.1. SAS Inst Cary, NC.

18. Hamady M, Lozupone C, Knight R (2010) Fast UniFrac: facilitating high-throughput phylogenetic analyses of microbial communities including analysis of pyrosequencing and PhyloChip data. ISMEJ 4: 17-27.

19. Sheneman L, Evans J, Foster JA (2006) Clearcut: a fast implementation of relaxed neighbor joining. Bioinformatics 22: 2823-2824.

20. Schloss PD, Handelsman J (2006) Introducing treeclimber, a test to compare microbial community structures. Appl Environ Microbiol 72: 2379-2384.

21. Poach ME, Hunt PG, Reddy GB, Stone KC, Johnson MH, et al. (2004) Swine wastewater treatment by marsh-pond-marsh constructed wetlands under varying nitrogen loads. Ecol Eng 23: 165-175.

22. Hatzenpichler R (2012) Diversity, physiology, and niche differentiation of ammonia-oxidizing archaea. Appl Environ Microbiol 78: 7501-7510.

23. Ibekwe AM, Leddy M, Murinda SE (2013) Potential human pathogenic bacteria in a mixed urban watershed as revealed by pyrosequencing. PLoS ONE 11: e79490.

24. Gersberg RM, Elkins BV, Lyon SR, Goldman CR (1986) Role of aquatic plants in wastewater treatment by artificial wetlands. Water Res 20: 363-368.

25. Ipsilantis I, Sylvia DM (2007) Abundance of fungi and bacteria in a nutrient-impacted Florida wetland. Appl Soil Ecol 35: 272-280.

26. Oehl F, Frossard E, Fliessbach A, Dubois D, Oberson A (2004) Basal organic phosphorus mineralization in soils under different farming systems. Soil Biol Biochem 36: 667-675.

27. Sundberg C, Tonderski K, Lindgren PE (2007) Potential nitrification and denitrification and the corresponding composition of the bacterial communities in a compact constructed wetland treating landfill leachates. Water Sci Technol 56: 159-166.

28. Truu J, Nurk K, Juhanson J, Mander U (2005) Variation of microbiological parameters within planted soil filter for domestic wastewater treatment. J Environ Sci Health Part A: Toxic/Hazard Subst Environ Eng 40: 1191-200.

29. Long Y, Yi H, Chen S, Zhang Z, Cui K, et al. (2016) Influences of plant type on bacterial and archaeal communities in constructed wetland treating polluted river water. Environ Sci Pollut Res 23: 19570-19579.

30. Frey SD, Knorr M, Parrent JL, Simpson RT (2004) Chronic nitrogen enrichment affects thee structure and function of the soil microbial community in temperate hardwood and pine forests. Forest Ecol Manage 196: 159-171.

31. Lauber CL, Strickland MS, Bradford MA, Fierer N (2008) The influence of soil properties on the structure of bacterial and fungal communities across land-use types. Soil Biol Biochem 40: 2407-2415.

32. Calheiros CSC, Duque AF, Moura A, Henriques IS, Correia A, et al. (2009) Substrate effect on bacterial communities from constructed wetlands planted with Typha latifoliatreating industrial wastewater. Ecol Eng 35: 744-753.

33. Ogier JC, Son O, Gruss A, Tailliez P, Delacroix-Buchet A (2002) Identification of the bacterial microflora in dairy products by temporal temperature gradient gel electrophoresis. Appl Environ Microbiol 68: 3691-3701.

34. Dubernet SH, Desmasuress N, Guéguen M (2004) Culture-dependent and culture-independent methods for molecular analysis of the diversity of lactobacilli in "Camembert de Normandie" cheese. Lait 84: 179-189.

35. Bodtker G, Thorstenson T, Lillebo BLP, Thorbjornsen BE, Ulvoen RH et al. (2008) The effect of long-term nitrate treatment on SRB activity, corrosion rate and bacterial community composition in offshore water injection systems. J Ind Microbiol Biotechnol 35: 1625-1636.

36. Wassel RA, Mills AL (1983) Changes in water and sediment bacterial community structure in a lake receiving acid-mine drainage. Microbial Ecology 9: 155-169.

37. Nelson CE (2009) Phenology of high-elevation pelagic bacteria: the roles of meteorologic variability, catchment inputs and thermal stratification in structuring communities. ISME J 3: 13-30.

38. Hirayama H, Takai K, Inagaki F, Yamato Y, Suzuki M, et al. (2005) Bacterial community shift along a subsurface geothermal water stream in a Japanese gold mine. Extremophiles 9: 169-184.

39. Han HS, Lee KD (2005) Phosphate and potassium solubilizing bacteria effect on mineral uptake, soil availability and growth of Eggplant. Res J Agric Biol Sci 1: 176-180.

40. Kuenen JG (2008) Anammox bacteria: From discovery to application. Nat Rev Microbiol 4: 320-326.

41. Pathak BK, Kazama FF, Tanaka Y, Mori K, Sumino T (2007) Quantification of anammox populations enriched in an immobilized microbial consortium with low levels of ammonium nitrogen and at low temperature. Appl Microbiol Biotechnol 76: 1173-1179.

42. Dong Z, Sun T (2007) A potential new process for improving nitrogen removal in constructed wetlands-promoting coexistence of partial-nitrification and ANAMMOX. Ecol Eng 31: 69-78.

43. Stahl DA, de la Torre JR (2012) Physiology and diversity of ammonia-oxidizing archaea. Annu Rev Microbiol 66: 83-110.

19

Investigation of Sewage and Drinking Water in Major Healthcare Centres for Bacterial and Viral Pathogens

Khan Suliman[1,2], Rabeea Siddique[3], Ghulam Nabi[1], Wasim Sajjad[4], Pathiranage Prajani Mahesha Heenatigala[1,2], Yang Jingjing[1,2], Qingman Li[1,2], Hongwei Hou[1,2]* and Ijaz Ali[5]*

[1]The Key Laboratory of Aquatic Biodiversity and Conservation of Chinese Academy of Sciences, Institute of Hydrobiology, Chinese Academy of Sciences, Wuhan, Hubei,430072, PR China
[2]State Key Laboratory of Freshwater Ecology and Biotechnology, Institute of Hydrobiology, Chinese Academy of Sciences, Wuhan, 430072, PR China
[3]Department of Biomedical Engineering, Huazhong University of Science and Technology, Wuhan 430074, PR China
[4]Key laboratory of petroleum resources, Gansu Province/Key laboratory of petroleum resources research, Institute of Geology and Geophysics, Chinese Academy of sciences, Lanzhou 730000, PR China.
[5]Department of Biosciences, COMSATS, Islamabad, 44000, Pakistan

Abstract

Water is a major source of microbes, including pathogens that can cause critical pathological conditions and outbreak of epidemics. Due to lack of proper medical waste-management system in Peshawar, most of the waste is disposed of near sewage lines which run parallel to drinking water supply increasing the chances of water contamination. This study was undertaken to examine bacterial and viral pathogens in fresh and waste water in major Health care units. Conventional culturing techniques were used to identify bacterial pathogens followed by biochemical analysis, whereas viral pathogens were detected by Polymerase Chain Reaction (PCR). Analysis of sewage and drinking water supply in major health care facilities of Peshawar city indicated that *Klebsiella pneumoniae* and *Staphylococcus aureus* were found in all water samples whereas serious health risk causing bacteria including *Mycobactirium tuberculosis* were also detected in some regions. Two viral pathogens, Hepatitis C virus (HCV) and Hepatitis B virus (HBV) were found in open sewage water of Khyber Teaching Hospital and Dabgari Garden (G). The presence of these pathogens in water is a serious threat to public health and the environment and calls for immediate action to enforce proper medical waste-management to eliminate the risks to human health.

Keywords: Water; Major healthcare centers; Pathogens; Public health

Introduction

Water pollution is one of the most pervasive problem afflicting people throughout the world. Waterborne illness and multiple epidemics are related to the consumption of contaminated or inadequately treated water is a global public health concern. As a developing country, Pakistan has poor water treatment system and ranked 80th among 122 nations in terms of providing good quality drinking water [1]. However, scenario is even worst in many cities of Pakistan where drinking water is unsafe for direct human consumption and severely contaminated with bacterial and viral pathogens. In Pakistan, waterborne diseases and parasitic infections due to contaminated water accounts for nearly 60% and 80% of children's death respectively. Every year approximately 250,000 children die due to water-borne diarrhoea solely and 1.2 million people get affected by waterborne pathogens in Pakistan [2]. Furthermore, scientific data and evidence considering role of waterborne pathogens in the epidemiology of hospital-acquired infections are insufficient [3]. Recent reports on pathological conditions of identified water borne pathogens have provided novel insights into the understanding of pathology and effect of diseases [4-6], which persist in numerous aquatic systems due to the advantage of resistance to various environmental factors [7].

Health facilities, mainly health care centres, hospitals, clinics and laboratories pose higher risk of water contamination since these are more likely to be the sources of viral and bacterial pathogens [8-11]. Although numerous research has been conducted to address the detection and origination of pathogens in both drinking water and wastewater [12,13], however, insufficient studies have done explicitly about tracing the occurrence of pathogens in water sources near healthcare facilities [14,15].

Our current study was conducted to detect the presence of pathogenic microbes both in drinking and waste water samples collected from Khyber Teaching Hospital (KTH), Hayatabad Medical Complex (HMC) and Dabgari Garden (DBG/DG), the major healthcare facilities of Peshawar, capital city of province Khyber Pakhtunkhwa (KP), Pakistan. Everyday these hospitals provide health facilities to thousands of local people, patients coming from far flung areas of KP, and from Afghanistan as well. During personal visits to these hospitals for sample collection improper sewage systems allowing stagnant water retention for several days and mass of untreated disposed materials were observed, which may stir up risk of contamination in drinking water. In addition, the disposal of waste materials from diagnostic laboratories and pharmaceutical centres poses a significant threat to public health.

Inadequate information is available about sewage and drinking water quality near major health care units of Peshawar, KP, Pakistan and no investigation for pathogens has been done particularly considering water sources of the healthcare centers. Additionally, no pre-defined rules and laws presented by WHO are set and applied by healthcare management and higher authorities for such investigation and providing good quality treated water. Therefore, the focus of current study is determination of viral and

---

*Corresponding author: Hongwei Hou, The Key Laboratory of Aquatic Biodiversity and Conservation of Chinese Academy of Sciences, Institute of Hydrobiology, Chinese Academy of Sciences, Wuhan, Hubei, 430072, PR China
E-mail: houhw@ihb.ac.cn

bacterial pathogens in drinking and sewage water of major health care units of Peshawar to highlight critical role contaminated water plays in waterborne diseases. Culture techniques and Polymerase Chain Reaction (PCR) as the most commonly used methods for monitoring and detection of bacterial and viral pathogens [16-19], are applied in this research.

Materials and Methods

Study site description and sampling

Khyber Pakhtunkhwa (KP) province -with population of 26.9 million and area of 74,521 square kilometer- is located in the north-western region of Pakistan and by the size of population is the 3rd biggest province of Pakistan. Peshawar (33° 99' 16 "N, 71° 51 '36 "E) is its provincial capital and largest city and hub of hospitals where patients come not only from all around the KP but also from neighbor country Afghanistan. It is crucial that hospitals providing health facility to thousands of patient's everyday have accessibility to pathogen free drinking water. In this study, total of 252 drinking and sewage water samples were examined over a period of one year from January 2013 to December 2013, samples were collected 3 times a year in 2013, in January, June and October respectively. First session of sampling was completed during last two weeks of January, second in first two weeks of June and third time samplings were done during the last two weeks of October. Totally, 126 drinking and 126 sewage water samples were collected from three major hospitals of Peshawar i.e., KTH, HMC and DG, with the interval of 4 months except the rainy days. All samples were collected in sterile bottles from the premises of these health care sites (42 different sites in total, 6 samples were collected from each site, 3-drinking, 3-sewage water samples).

Filtration and DNA/RNA isolation

Water samples were taken to lab soon after collection in ice containers and filtered through sterile filter membranes, 0.22 μm, (Science laboratory, Islamabad, Pakistan) to concentrate the samples for investigation of bacterial pathogens. The whole water samples were processed through DNA/RNA isolation kit (Norgen Biotek, Canada) for the detection of viral pathogens.

Screening and selective media preparation for incubation

Screening was done for the samples collected (nutrients media) and positive samples were further tested through selective media. Selective media (Merck, Rawalpindi, Pakistan) were prepared (according to the prescription provided by Merck, Rawalpindi, PK) for the culturing of bacterial colonies (selective media shown in Table 1). Sterilized media were poured into Petri plates, followed by spreading of concentrated water samples with the help of a sterilized loop. Petri plates were incubated at 35°C for 24 h (Also 48 hours to get the correct density) and 5 weeks for MTB, followed by sub-culturing of colonies on fresh selective media at 34°C and 36°C [14]. This step of colony sub-culturing was repeated 3 times for confirmation of the resultant colonies. The colonies which showed consistent growth were noted and non-consistent growing bacterial pathogens were neglected to avoid false positive results.

Biochemical analysis

Biochemical tests conducted for identification of bacterial species grown previously on selective media were Catalase, Oxidase, Tube coagulase, Alkaline phosphatase, Motility, Arginine,

Bacterial Pathogens	Catalase test	Oxidase test	Citrate utilization	Lactose utilization	H2S production	Methyl Red test	Voges-Proskauer test	Nitrate test	Indole test	Ureaase test	Motility test	Galactose test	Ornithine test	Maltose test	Mannose test	Inositol test	Trehalose test	Acetate test	Sucrose test	Triple Sugar Iron test	Lysine Iron Agar test	Gelatin hydroly-sis test	Lysine decarboxylase test	Coagulase test	Pyruvate test	Arginine test	Niacine test	Tellurite test	Manitol test	Furazolidone test
E. coli	+	-	+	-	+	+	+	+		+	+		+						+	+	-	-								
E. aerugenes	+	-	-	+	-	-	+	+	-	+	+																			
H. influenzae		+						+	+		+			-								-								
K. pneumoniae	+	-	+	+	-	-	+	+	-	+	-																			
S. marcescens	+	-	+	-	-	-	+	+	-	+	+											+	+							
H. alvei				+						+																				N
P. vulgaris	-	+	+	+	-	+	+	+	+	+	+	+	-	+	-	-													-	
P. mirabilis	+	+	-					-		+	+	-	-	-	+														-	
S. enterica	+	-	+	-	+	+	-	+	+	-	+									-	-									
S. dysentriae	-	-							-	-									-									-	-	
P. stuartii	-	-	+	-	-	+	-	+	+	+	+	+	-	-	+	+	-													
P. aeruginosa	+	+	+	-	-	-	-	+	-	-	+								-				-							
S. typhimurium	+	-			+		-			-	-																			
S. aureus	-	-	+		-					+														+						S
E. faecalis	-	-	+	+	-	+	+	+	-	-																				S
A. hydrophila	+	+	-	+	+	-	+	+	+	-	+		-									-	-			+				
A. sobria	+	+		-		-		+	+	-	+	-	-						+			-	+			+		+		
S. epidermidis	+	+	-	+	-	-	-	+	-	+	-													+						S
M. tuberculosis	-							+																			+	-		

Table 1: Biochemical tests used for identification of the microorganisms at specie level.

Pyruvate, Mannitol, Sucrose and Ornithine, Esculin, fermentation of Sucrose and Lysine decarboxylase (Table 1). Sterile loop was used to pick the bacterial colony from the selective media and tested for biochemical test. The result was noted and process was repeated for three times. Only those bacterial pathogens were noted which gave same result every time.

The instructions provided by Merck science lab Rawalpindi were followed to get the results. For Lactose test color change was noted as positive after broth culture. Indole test was noted as positive by appearance of pink red layer. Red color formation after addition of alpha-naphthol+sodium hydroxide while shaking the tube for 10 minutes was an indication of positive result for Voges-Proskauer test. Green color change to blue confirmed the positive result for Citrate test. For Nitrate test the color changed into dark red within 5-10 minutes. This test was carried with the addition of N,N-dimethyl-1-naphthylamine and sulphuric acid. The Oxidase test gave positive by appearance of purple color after applying 1% tetramethyl-p-phenylenediamine dihydrochloride on filter paper. For Catalase test the oxygen bubbles demonstrated positive result. Black precipitates affirmed positive test for H_2S. Appearance of reddish color during Methyl Red test confirmed positive results for presence of E. coli. The yellow color was commuted to red for urease test indicating positive result. Visualizing under microscope, a hazy zone (irregular movement) formation confirmed the positive result for motile bacteria and a single line of growth formation indicated presence of non-motile bacteria. Regain of purple color from yellow color after 48 hours' incubation confirmed positive result for Ornithine test. Maltose test showed positive result after conversion of red color to yellow color. Here phenol red was used as PH indicator. In case of Mannose test the normal red color (phenol red indicator) commuted to yellow or pink, an indication of positive result. Similarly, Inositol test was noted positive by color transformation from red (phenol red indicator) to yellow or pink. For the Trehalose test the transformation of red color to yellow affirmed positive result. For sucrose test the color change from red to yellow was observed as an indication of positive result. For acetate test the clear zone formation was an indication of the acetic acid producing bacteria so it was considered positive for Acetobacter. For Triple Sugar Iron and Lysine Iron Agar tests the color change, butt and gas production was noted to the slants and compared the information available in the list provided by science lab Rawalpindi Pakistan. Gelatine hydrolysis test was performed and the starch hydrolization by making clear zones in surrounding was noted for positive results after addition of iodine. Lysine decarboxylase test was noted as positive by the color change to purple. A small amount of oil was added to prevent oxygen from moving out. In Coagulase test the clot formation indicated positive result. For Pyruvate test the change of blue green color to yellow was taken as confirmation of positive result. In Arginine test the purple color was changed to yellow, which is acquired as an indication of positive result. Tellurite test was confirmed as positive by the appearance of grey color on the growing colonies. In mannitol test the red color change to yellow confirmed the positive result. Furazolidone test was performed and the resistance or sensitivity was observed and compared to the list of information provided by science lab Rawalpindi Pakistan.

RT-PCR and gel electrophoresis

HCV RNA isolation from concentrated water samples were carried out using the Water RNA/DNA purification kit (NORGEN Biotek, Canada), according to manufacturer's instructions. The extracted RNA was reverse transcribed into cDNA. The amplified cDNA/DNA was subjected to PCR amplification. The PCR product was run on gel electrophoresis followed by observation of the obtained bands through gel documentation.

Results

Identification of bacterial pathogens

Determination of pathogenic bacteria and viruses by conventional culturing and molecular techniques, respectively, is a reliable approach for assessment of water quality. Only those pathogens were included in final results which were found throughout the year in collected samples to make sure the presence of microbes regardless of physiological effect and environmental changes. Therefore, it is concluded that these pathogens made the studied sites their permanent habitat. Out of all the samples analysed, 42% pathogens were identified in drinking water and 58% pathogens in sewage water samples. Considering overall results (drinking and sewage water samples), KTH samples were highly contaminated (40%), followed by DG (31%) and HMC (29%) with little difference in results. Common bacterial pathogens traced in drinking water samples collected from all sites indicated that KTH water being highly contaminated had 10 different pathogenic species, HMC had 6, whereas, 4 different pathogenic species were detected in DG water. However, in case of sewage water, high species diversity was observed in DG samples that was contaminated with 14 different pathogenic species, as compared to 13 and 11 different pathogenic species investigated from HMC and KTH respectively. Besides common bacterial pathogens, some other important but seldom bacterial (Mycobacterium tuberculosis) pathogens were also identified in sewage water samples. Among the identified pathogens, Klebsiella pneumoniae and S. aureus were detected frequently, as compared to Proteus mirabilis, Psudomonas aeruginosa and Enterococcus faecalis, which were least common observed pathogens in all samples. Paradoxically, fresh water samples collected from DG had shown presence of Proteus vulgaris, and M. tuberculosis in sewage water that was present in almost 80% of all the samples collected from different locations of DG. The largest number of pathogenic bacterial species in fresh water systems was found in KTH samples, while the lowest number of pathogenic bacteria species in fresh water sources was found at HMC. However, in sewage water systems the largest numbers of bacterial species were observed at DG and the lowest numbers of bacterial pathogens were detected at HMC. A detailed list of bacterial pathogens identified in each sampling site is given in Table 2.

Identification of viral pathogens

Water samples collected from multiple sites of DG, KTH and HMC was further investigated for the presence of viral pathogens i.e., HCV and HBV. Sewage water samples collected from KTH and DG determined presence of HBV, whereas, HCV was only detected in the sewage water samples collected from KTH. However, no viral pathogens were detected in fresh water samples collected from any studied area (Table 2; Figure 1).

Comparative analysis for pathogens identified in healthcare centres

Based on type of species, comparatively more pathogens were detected in sewage water, that is, total 17 different types of pathogens were ascertained in sewage water and 11 in fresh water systems (Figure 2).

Pathogen	Health hazard caused	Detection Method	Selective media	Colony appearance	KTH F.W	KTH S.W	DG F.W	DG S.W	HMC F.W	HMC S.W
Escherichia coli	Urinary tract infection, neonatal meningitis and food poisoning	Culture/ biochemical	EMB agar	Metallic green sheen	+	+	+	+	-	+
Enterobacteraerogenes	Urinary and respiratory tract infections	Culture/ biochemical	EMB agar	Pink color of bacterial growth	-	-	-	+	-	+
Haemophilusinfluenza	Pneumonia, septicemia, meningitis and skin infection	Culture/ biochemical	Chocolate agar	Offwhite distinct colonies with pungent smell	+	-	-	+	-	+
Klebsiellapneumoniae	Pneumonia, septicemia and ankylosing	Culture/ biochemical	McConkey agar	Pink colored round and long branched colonies	+	+	+	+	+	+
Serratiamarcescens	Urinary tract infection, septicemia, meningitis and endocarditis	Culture/ biochemical	XLD agar	Light pink round colonies	+	-	-	+	+	+
Hafniaalvei	Urinary tract infection and diarrhea	Culture/ biochemical	McConkey agar	Light brown rough colonies	-	+		+		
Proteus vulgaris	Urinary tract infections	Culture/ biochemical	HE & XLD agar	Yellow orange colonies	-		+			
Proteus mirabilis	Urinary tract infections	Culture/ biochemical	XLD agar	Orange colonies with white center	-	+	-	+	-	+
Salmonella enterica	Enteric fever, paratyphoid fever, septicemia and salmonellosis	Culture/ biochemical	MSRV agar & XLD agar	Red colonies with black center, straw colonies with hallow zone around		+		+		
Shigelladysentriae	Shigellosis	Culture/ biochemical	HE agar & XLD agar	Red to pink colonies	+	+		+	+	+
Providenciastuartii	Gastroenteritis and bacteremia	Culture/ biochemical	MU/SC agar	Yellowish orange centred colonies	-	-		+		
Pseudomonas aeruginosa	Urinary tract infection, pulmonary tract infection, wounds and burns infections	Culture/ biochemical	TSA agar, PI agar	Brown colonies, Blue, Green colonies	-	+	-	-	-	+
Salmonella typhi	Typhoid	Culture/ biochemical	HE agar	Blue green colonies with dark center	+	+	-	-	-	+
Staphylococcus aureus	Meningitis, pneumonia, gastroenteritis and wound infections	Culture/ biochemical	TS agar,MSA	White round colonies	+	+	-	-	-	+
Enterococcus faecalis	Bacterimia, urinary tract infection and meningitis	Culture/ biochemical	Blood agar	circular, convex white colonies	-	-	-	+	-	-
Aeromonashydrophila	Aerolysin Cytotoxic Enterotoxin (ACT) production causes tissue damage	Culture/ biochemical	CLED/TCBS agar	Yellow round with full margin colonies	+	-	-	-	+	-
Aeromonassobria	Foodbrone diseases, diarrhea and wound infections	Culture/ biochemical	CLED agar	Round colonies with white margins	+	+	-	-	+	-
Staphylococcus epidermidis	Skin infections, meningitis, osteomyelitis, toxic shock syndrome and pneumonia	Culture/ biochemical	Blood agar, MS agar	White colored full margined colonies	+	+	+	+	+	+
Mycobacterium tuberculosis	Tuberculosis	Culture/ biochemical	LJ medium	Granular light brown colonies with rough surface	-	-	-	-	+	-
Virus	Hepatitis C	PCR	N/A		-	+	-	-	-	-
Virus	Hepatitis B	PCR	N/A		-	+	-	+	-	-

Table 2: Identification of pathogenic microorganisms in fresh water (F.W) and sewage water (S.W) samples collected from different healthcare centers. The table also illustrates the health hazards caused by the detected microorganisms and the detection methods and media used. The +, - sings indicate the presence and absence of microorganism in the specified region respectively.

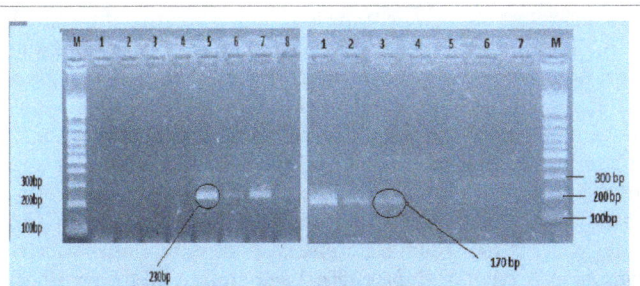

Figure 1: Both HCV and HBV targeted amplicons of size 170 and 230 bp respectively, were isolated from sewage water of major healthcare centers in Peshawar. HCV was detected in water samples collected from KTH, whereas HBV was detected in sewage water samples collected from KTH and DG.

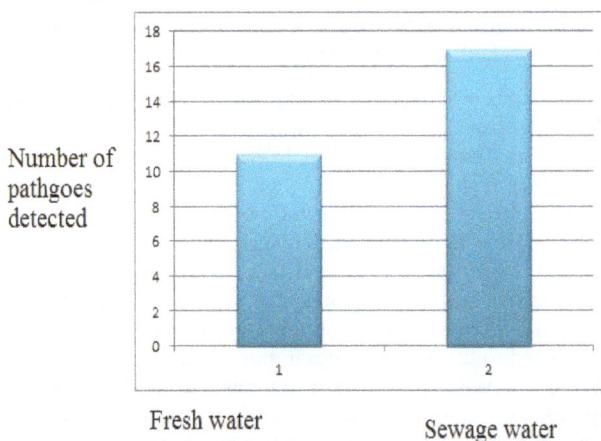

Figure 2: Comparative analysis of fresh water and sewage water samples regarding numbers of identified bacterial pathogens. This figure shows that out of the total identified bacterial species 17 types of species were present in the fresh water whereas 11 types of species were detected in fresh water.

DG sewage water contains the most diverse species of pathogens while its fresh water sources contain the least pathogenic species, as shown in Figure 3.

Most frequently observed pathogens in either fresh water or sewage water samples from all sample collecting sites were *klebsiella* and *Staphylococcus epidermidis,* whilst the least common pathogens were *Proteus vulgaris, Providencia, Enterobacter faecalis* and *Mycobacterium tuberculosis* (Figure 4).

The overall result of both fresh and sewage water sources confirmed that KTH samples were comparatively more contaminated than DG/DBG and HMC (Figure 5). The least number of the bacterial species in DBG makes it safer than others and it might be due the privatized sector is taking better care to dispose the materials. Although there was no cleaning and burning systems but the lower bacterial burdens in water samples indicated the better treatment of wastes comparatively the other two sectors. KTH water indicated the most risk posing among all the investigated healthcare centers. It shows that lesser attention is provided to the treatment.

Considering samples (both fresh and waste water) collected from each sample collecting site/health care units, the highest numbers of pathogens were observed in sewage water i.e., 11.77[a] ± 0.57. Besides in sewage water the maximum numbers of pathogens were found in DG i.e., 13.33 ± 0.66 as compared to HMC i.e., 10.33 ± 0.66 where the

lowest numbers of pathogens were detected. However, in case of fresh water, the maximum numbers of pathogens were found in KTH, whilst the minimum numbers of pathogens were identified in DG (Table 2). From overall result, it is evident that KTH water is highly contaminated and inadequate for consumption having highest number of pathogenic species i.e., 10.66[a] ± 0.61, whilst the lowest number of pathogens identified in HMC i.e., 8.00[b] ± 1.09 is also not safe to use.

Discussion

For all living organisms water is the most vital and important factor of survival. Inadequate access to clean water, inappropriate water treatment and bad sanitation systems is one of the most pervasive issues distressing people throughout the globe, causing waterborne infectious diseases, cause approximately 10 million deaths per year [3,20,21]. Human health is prone to microbial risks caused by enteric viruses and bacteria [22]. Studies have shown that contaminated drinking water has been source of several critical diseases, for instance, diarrhea, nausea, Cholera, typhoid, dysentery, abdominal pain and food poisoning. Situation is even worst at health care centres, where drinking water is source of pathogens transmission showing negligence of managerial authority towards supplying properly treated water. Variant pathogens are observed in ground and surface water, flood and dam water [23-25]. Furthermore, presence of bacterial pathogens is associated with physiochemical characters and location of drinking water sites [26].

To the best of our knowledge, the present study is the first systematic analysis on water sources of healthcare centres of Peshawar, KPK, Pakistan highlighting the presence of multiple substantial bacterial pathogens in hospital's drinking and sewage water [27]. List of variant infectious bacterial and viral pathogens identified in water samples that were present consistently throughout the year at KTH, DB and HMC are given in Table 1. The abundancy of these pathogens in water sources calls out for appropriate initiatives to be taken to curb outbreak of waterborne epidemics associated with contaminated water consumption [28]. In different sites the variation is characterized by physiochemical differences of the water sources [26]. Presence of Hepatitis B and C viruses in open water sources causes death of 60% of the affected people if persists for a longer time and proliferate continuously [29,30], are associated with serious public health issues [29,31]. Most frequently reported pathogenic species considering all water samples are *E. coli, S. auerus, K. pneumonia, S. typhi* and *P. aeruoginosa* [2,26,32], on the basis of current study it is suggested that consumption of such water is threat to public health.

In our analysis the pathogens investigated can cause severe health problems in humans [9,33]. Most of the bacterial pathogens detected have been reported previously to be present in common water sources or home based drinking water sources [23,26] but their presence in the water sources of healthcare centers was not considered to be investigated. Furthermore, DG fresh water sources were contaminated with one third of pathogens number to that of KTH. We suggested that this high number of pathogens might be because of the improper water supply sources where sewage water can get entered into drinking water sources because of leakages in pipelines. Interestingly P. Vulgaris was only found in fresh water of DG but it was not detected in sewage water sources or other fresh water sources. Analysis of sewage water allowed us to detect diverse numbers of pathogens, the highest number in DG. The presence of *K. pneumonia* and *S. auerus* in all

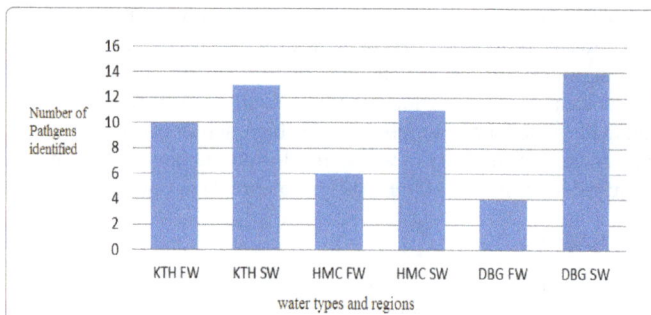

Figure 3: Comparison between different regions regarding specie type based numbers of pathogens identified. This figure illustrates that fresh water of KTH had 10 different bacterial species where its sewage water had 13 different bacterial species. In HMC 6 different types of bacterial species were identified in fresh water sources and 11 bacterial species were found in sewage water sources. The numbers of pathogenic species were 4 and 14 in fresh and sewage water respectively in samples collected from DBG.

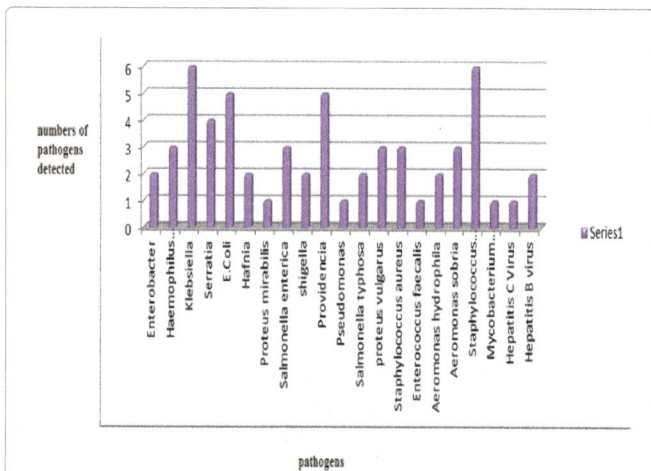

Figure 4: This figure illustrates that *Staphylococcus aureus* and *Klebsiella* were among the abundance bacterial pathogens. These species were found in all of the samples collected from the three healthcare centers. *E. coli* and *Providencia* were the second most abundant bacterial pathogens. In this list, the least abundant bacterial pathogens were *Mycobacterium, Pseudomonas, Hafnia,* HCV and HBV which were found in only one type of the water samples.

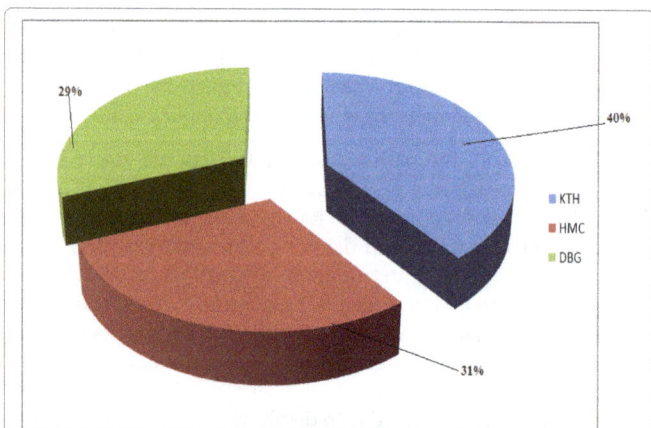

Figure 5: This figure illustrates the pathogens burden in the healthcare centers. The highest bacterial burdens were found KTH which is one of the biggest hospitals of the city. It had 40% of the total bacterial species burden whereas HMC had 31% and was ranked second. The safer among all was DBG where the total bacterial burden was 29%.

sites regardless of the water type is an indication that these are the permanent species as these were also reported in daily used water sources in surrounding regions [26]. *P. mirabilis, P. aeruginosa* and *E. faecalis* were unexpectedly found in the least sites as these are generally found in water sources of diverse locations [23,27]. In addition *P. stuartii* was interestingly found in almost all water sources, which was considered to be present at most in sewage water sources only. *A. sobria* presence was detected in fresh water of KTH too which is an unexpected result. Other bacterial pathogens (except *M. tuberculosis*) were found in diverse sites as they are generally considered to be found in water sources [26,28]. Surprisingly the viral species were detected in KTH and DG sewage water which is a threat to treatment seekers and patients care takers. We investigated that some highly pathogenic bacteria including *M. tuberculosis* were present persistently throughout the study period. Furthermore, the viral pathogens were also detection continuously throughout the year, which indicates that no proper treatment is carried to the water sources.

To our knowledge, in hospitals, fluids from diagnostic tests and laboratories are improperly disposed of allowing pathogenic bacteria and viruses to contaminate water that runs off to the tap water and sewage systems, subsequently contaminating drinking water. The investigated pathogens, however, may be present in fresh water due to the lack of management interest in providing properly treated water wiped off from the pathogenic bacterial species; and treatment of hospital wastes accurately before disposing it. At the same time the laboratories owners are not admonished to throw the wastes in open places. Other sources of bacterial contamination of fresh water are surface runoff through hospitals and urban areas, pastures and agricultural lands, leakage of sewage disposal systems and septic tanks, overloaded sewage treatment plants, disposal systems and raw sewage deep well injection [2,9,34]. Similarly, we propose that contamination of drinking water observed during the present study involves factors like cross-connections, broken or leaking pipes, back-siphonage (backflow of polluted or contaminated water, from a plumbing fixture or cross-connection into a water supply line, due to a lowering of the pressure in the line) and intermittent water supply [9,34,35], and these pathogens have made the studied sites as their permanent habitats.

Our approach offers an unbiased identification of those bacterial and viral pathogens which can lead to serious human health problems. The overall investigated pathogens in hospital's water samples are similar to the investigations of other water sources either drinking water, dam water, flood water or sewage water in KPK [26,27], somehow, our results differ in a way that through our investigation few uncommon bacterial pathogens like *M. tuberculosis,* and exceptional viral species are also identified. The reason might be the investigation site (hospitals) and consideration of only those pathogens in the results which were found in all sample collection sites and present throughout the year.

Conclusions

We came up with a conclusion that if current condition continued water borne illnesses will pose serious threat to public health. Addressing existence of disease causing pathogens in water sources for instance, *E. coli, S. aureus, P. stuartii, K. pneumonia, H. influenzae,* and *P. sobira,* calls out for a tremendous amount of research to be conducted to identify robust new water purifying techniques at lower cost, with minimal use of chemicals. These pathogens can enter into water pipelines through back-siphon age, cross-connections, broken or leaking rusted pipelines, thus intermittent water supply results in contamination of the distribution system. Hospital's waste and patient's fluid should be disposed of properly. It is encouraged to drink boiled water and have drinking utensils autoclaved, since most bacterial and viral pathogens cannot survive in boiled water.

Acknowledgements

This work was supported by Relief International, Pakistan and State Key Laboratory of Freshwater Ecology and Biotechnology (Grant No. Y119011F01) and CAS-TWAS.

Conflict of Interest

The authors declare that there is no conflict of interests regarding the publication of this paper.

References

1. UNESCO (2003).

2. Sardar K, Irfan AS, Said M, Riffat NM, Mohammad TS (2005) Arsenic and Heavy Metal Concentrations in Drinking Water in Pakistan and Risk Assessment: A Case Study. Hum Ecol. Risk. Assess 21: 1020-1031.

3. Akbar A, Sitara U, Khan SA, Muhammad N, Khan MI, et al. (2013) Driniking water quality and risk of waterborne diseases in the rural mountainous area of Azad Kashmir Pakistan. Intl. J. Biosci. 3: 245-251.

4. Jarraud S, Mougel C, Thioulouse J, Lina G, Meugnier H, et al. (2002) Relationships between Staphylococcus aureus genetic background, virulence factors, age groups (alleles) and human disease. Infec. Immunol. 70: 631-641.

5. Herold S, Karch H, Schmidt H (2004) Shiga toxin-encoding bacteriophages-genomes in motion. Intl. J Med Microbiol 294: 115-121.

6. Feigin V, Brainin M, Breteler MMB, Martyn C, Wolfe C, et al. (2004) Teaching of neuroepidemiology in Europe: time for action. Euro J Nerol 11: 795-799.

7. Wang L, He Y, Xia Y, Wang H, Liang S (2014) Investigation of mechanism and molecular epidemiology of linezolid-resistant Enterococcus faecalis in China. Inf Genet Evol 14: 160-169.

8. Rosina G, Ferru MA, Alonso JL, Manzano JR, Calgua B, et al. (2010) Molecular detection of pathogens in water-the pros and cons of molecular techniques. Water Res 44: 4325-4329.

9. Azizullah A, Khattak MN, Richter P, Häder DP (2011) Water pollution in Pakistan and its impact on public health. Environ Intl 37: 479-497.

10. Khan S, Shahnaz M, Jehan N, Rehman S, Shah MT, et al. (2013) Drinking water quality and human health risk in Charsadda district, Pakistan. J Cleaner Prod 60: 93-101.

11. Friedlander LR, Puri N, Schoonen MA, Wali Karzai A (2015) The effect of pyrite on Escherichia coli in water: proof-of-concept for the elimination of waterborne bacteria by reactive minerals. J Water Health 13: 42-53.

12. Gomez-Alvarez V, Humrighouse BW, Revetta RP, Domingo JWS (2015) Bacterial composition in a metropolitan drinking water distribution system utilizing different source waters. J Water Health pp: 57.

13. September SM, Els FA, Venter SN, Brözel VS (2007) Prevalence of bacterial pathogens in biofilms of drinking water distribution systems. J Water Health 5: 219-227.

14. Tassadaq H, Roohi A, Munir S, Ahmed I, Khan J, et al. (2013) Biochemical characterization and identification of bacterial strains isolated from drinking water sources of Kohat, Pakistan. Afr J Microbiol Res 7: 1579-1590.

15. Mody RK, Meyer S, Trees E, White PL, Nguyen T, et al. (2014) Outbreak of Salmonella enterica serotype I 4,5,12 i: infections: the challenges of hypothesis generation and microwave cooking. Epidemiol Infec 142: 1050-1060.

16. Nele W, Frost C, Marre R (2001) Detection of Legionellae in Hospital Water Samples by Quantitative Real-Time Light Cycler PCR. App Environ Microbiol 67: 3985-3993.

17. Thomas S, Kohnen W, Jansen B, Obst U (2002) Detection of antibiotic-resistant bacteria and their resistance genes in wastewater, surface water, and drinking water biofilms. FEMS Microb Ecol 43: 325-335.

18. Audemard C, Reece KS, Burreson EM (2004) Real-Time PCR for detection and quantification of the protistan parasite Perinsusmarinus in environmental waters. App Environ Microbiol 70: 6611-6618.

19. Largus TA, Kelley ST, Amand AS, Pace NR, Hernandez MT (2005) Molecular identification of potential pathogensin water and air of a hospital therapy pool. PNAS 102: 4860-4865.

20. Pindi PK, Yadav PR, Shanker AS (2013) Identification of Opportunistic Pathogenic Bacteria in Drinking Water Samples of Different Rural Health Centers and Their Clinical Impacts on Humans. Biomed Res intl. Article ID 348250.

21. Baudart J, Coallier J, Laurent P, Prevost M (2002) Rapid and Sensitive Enumeration of Viable Diluted Cells of Members of the Family Enterobacteriaceae in fresh water and drinking water. Appl Environ Microbiol 68: 5057-5063.

22. Podschun R, Ullmann U (1989) Klebsiella spp. as nosocomial pathogens: epidemiology, taxonomy, typing methods and pathogenicity factors. Clin Microbiol 11: 589-603.

23. Shar AH, Kazi YF, Kanhar NA, Soomro IH, Zia SM, et al. (2010) Drinking water quality in Rohri City, Sindh, Pakistan. Afr J of Biotech 9: 7102-7107.

24. Ahmed T, Kanwal R, Tahir SS, Rauf N (2004) Bacteriological analysis of water collected from different dams of Rawalpindi/Islamabad region in Pakistan. Pak J Biolo Sci 7: 662-666.

25. Khan FA, Ali J, Ullah R, Ayaz S (2014) Bacteriological quality assessment of drinking water available at the flood affected areas of Peshawar. Toxicol & Environ Chem pp: 1448-1454.

26. Ahmad B, Aquat ML, Ali J, Bashir S, Mohammad S, et al. (2014) Microbiology and Evaluation of Antibiotic Resistant Bacterial Profiles of Drinking Water in Peshawar, Khyber Pakhtunkhwa. World App Sci J 30: 1668-1677.

27. Hussain T, Roohi A, Munir S, Ahmed I, Khan J, et al. (2013) Biochemical characterization and identification of bacterial strains isolated from drinking water sources of Kohat, Pakistan African Journal of Microbiology Research 7: 1579-1590.

28. Fong TL, Shindo M, Feinstone SM, Hoofnagle JH, Bisceglie AM (1991) Detection of replicative intermediates of hepatitis C viral RNA in liver and serum of patients with chronic hepatitis C. J Clin Invest 88: 1058-1060.

29. Liu WC, Liu QY (2014) Molecular mechanisms of gender disparity in hepatitis B virus-associated hepatocellular carcinoma. W J Gastroenterol 20: 6252-6261.

30. Maan R, Veldt BJ, Janssen V (2014) Eltrombopag for Thrombocytopenic Patients with Chronic HCV Infection. Gastroenterol 16: 53-63.

31. Radcliffe RA, Bzixler D, Moorman A, Hogan VA, Greenfield VS, et al. (2013) Hepatitis B virus transmissions associated with a portable dental clinic, West Virginia. J Ameri Dent Asso 144: 1110-1118.

32. PCRWR. Annual Report 2005-2006. Islamabad, Pakistan: Pakistan Council for Research in Water Resources, 2008. Available online at http://www.pcrwr.gov.pk/Annual%20Reports/New%20Annual%20Repot%20200506_2.pdf.

33. Ravasan NM, Oshaghi MA, Sara H, Zahra S, Amir AA, et al. (2014) Aerobic microbial community of insectary population of Phlebotomus papatasi. J Arthr Bor Dis 8: 69-81.

34. Sarah PW, Victor PJG, Katharine GF (2007) Detection of Bacteroidales Fecal Indicators and the Zoonotic Pathogens E. coli O157:H7, Salmonella and Campylobacter in River Water. Environ Sci Technolo 41: 1856-1862.

35. Marta R, Jurek T, Szleszkowski L, Gladysz A (2014) Outbreak of hepatitis C among patients admitted to the Department of Gynecology, Obstetrics, and Oncology. American J Infec Cont 429: 7-10.

20

Long-Term Variations of Aerosols Concentration over Ten Populated Cities in Iran based on Satellite Data

Foroozan Arkian*

Meteorology Department, Marine Science and Technology Faculty, North Tehran Branch, Islamic Azad University, Tehran, Iran

Abstract

In this study, three different sensors of satellites including the Moderate Resolution Imaging Spectroradiometer (MODIS), Multi-angle Imaging Spectroradiometer (MISR), and Total Ozone Mapping Spectrometer (TOMS) were used to study spatial and temporal variations of aerosols over ten populated cities in Iran. Also, the Hybrid Single Particle Lagrangian Integrated Trajectory (HYSPLIT) model was used for analyzing the origins of air masses and their trajectory in the area. An increasing trend in Aerosol concentration was observed in the most studied cities in Iran during 1979-2016. The cities in western part of Iran had the highest annual mean of aerosol concentration. The highest AOD value (0.76 ± 0.51) was recorded in May 2012 over Ahvaz, and lowest value (0.035 ± 0.27) was recorded in December 2013 over Tabriz. After Ahvaz, the highest AOD value was found over Tehran (annual mean: 0.11 ± 0.20). The results show that AOD increases with increasing industrial activities, but the increased frequency of aerosols due to land degradation and desertification is more powerful in Iran. The trajectories analysis by the HYSPLIT model showed that the air masses come from Egypt, Syria, and Lebanon, and passed over the Iraq and then reached to Iran during summer. Aerosol Radiative Forcing (ARF) has been analyzed for Zanjan (AERONET site) during 2010-2013. The ARF at surface and top of the atmosphere were found to be ranging from -79 wm^{-2} to -10 Wm^{-2} (average: -33.45 Wm^{-2}) and -25 wm^{-2} to 6 wm^{-2} (average: -12.80 Wm^{-2}), respectively.

Keywords: Aerosol optical depth; MODIS; MISR; TOMS; AERONET; HYSPLIT; Iran

Introduction

Aerosol Optical Depth (AOD) is an integral measure of the total amount of aerosols contained within a vertical column of air. Atmospheric aerosols are a complex mixture of particulate matter (PM) and liquid/solid particles suspended in the atmosphere. These particles include a combination of organic compounds, inorganic ions, crustal compounds, the biological substance and several trace elements. The diameter of the aerosols varies from nanometres to hundred micrometres, depending on formation mechanisms, the sources, and the meteorological and geographical situations. De Miranda et al. has studied on the relationship between aerosol particles chemical composition and optical properties to identify the biomass burning contribution to fine particles concentration in Brazil. Kosmopoulos et al. [1-4] indicated that aerosols have been a main factor in the global climate change of the last two decades because they play an important role in radiative transfer and the radiative balance of the atmosphere-earth system. Some satellite remote sensing studies have served to monitor the radiative effects of aerosols and the global aerosol budget, as well as their impact on climate [5-8]. Researchers have also used AERONET (Aerosol Robotic Network) remote sensing to show the optical properties of each aerosol type [9-14].

There have been comparatively few studies on aerosols in Iran based solely on ground data [15-17]. Masoumi et al. [15] used ground-based aerosol monitoring network to study optical properties of aerosols such as AOD, AVSD, AE, SSA and refractive index over Zanjan during 2006-2008. Bayat et al. [16] analyzed SSA and the polarized phase function of atmospheric aerosols over Zanjan during 2010-2012. Khoshsima et al. [17] investigated AOD, AE and Angstrom turbidity coefficient during 2009-2010 over Zanjan. Also, a few studies have been conducted based on satellite data [18,19]. Rashki et al. [18] studied on seasonal cycle MISR, MODIS and TOMS AOD over Sistan region in Iran. The results show high aerosol concentration during summer and lower in winter. The Aerosol Index and AOD highlighted the Sistan region as a major

source of dust in southwest Asia. Gharibzadeh et al. [19] studied optical properties and radiative effect of aerosol during two dust events in 2013 over Zanjan.

Aerosols concentration is growing in most of the Middle East due to increasing population and industrialization adjacent to megacities. In this research, we have investigated the spatial and temporal variation of aerosol optical depth over ten populated cities in Iran using the Imaging Spectroradiometer (MODIS), the Multi-angle Imaging Spectroradiometer (MISR), and the Total Ozone Mapping Spectrometer (TOMS). AOD retrievals of MISR and MODIS are validated by ground base AERONET data (Zanjan city site). In addition, The NOAA Hybrid Single Particle Lagrangian Integrated Trajectory (HYSPLIT) model was used to identify the origin and track of dusty air masses to Iran during summer. In addition, we have analyzed Aerosol Radiative Forcing (ARF) at the earth surface and the top of the atmosphere over Zanjan during 2010-2013.

Methodology

Studied area

Iran has a varied diversity of natural resources. The Zagros mountains cover the northwestern and central parts of Iran; the Alborz mountains cover the northern and western parts of Iran, and The Great Salt and Emptiness Deserts (The Dasht-e Lut and the Dasht-e Kavir)

---

*Corresponding author: Arkian F, Meteorology Department, Marine Science and Technology Faculty, North Tehran Branch, Islamic Azad University, Tehran, Iran
E-mail: f.arkian@gmail.com

are located in central and eastern sections. The Caspian Sea is located in the north, and the Persian Gulf and Oman Sea are located in south Iran. The topographic diversity in Iran makes it unique for any study of spatio-temporal patterns. Iran has a continental type of climate, with cold winters and hot summers. Figure 1 shows average monthly temperature and rainfall from 1960-1990 at Tehran. The ten largest and most populated cities have been selected for analysis of the spatial and temporal variations in aerosol concentrations in Iran (Figure 2).

Dataset and analysis

Aerosol properties from three datasets including TOMS, MODIS and MISR sensors were used to understand and analyze the variability of aerosols over different regions of Iran. Table 1 shows satellite and AERONET datasets used in this study.

Aerosol optical depth is an integral measure of the total amount of aerosol contained within a vertical column of air. Thus, it collectively assesses such aerosols as haze, desert dust, sea salt and smoke particles. The sun photometer measure AOD by a voltage (V) that is related to the irradiance (I). The irradiance and sun photometer voltage at the top of the atmosphere are Io and Vo, respectively. The Beer-Lambert-Bouguer law is used to determine total optical depth (τ_{TOT}):

$$V(\lambda)=Vo(\lambda)d^2 exp[-\tau(\lambda)_{TOT}*m] \qquad (1)$$

Where V is voltage (at wavelength λ), τ_{TOT} is the total optical depth and d is the actual average of Earth-Sun distance, and m is the optical air mass [20].

Aerosol Robotic Network (AERONET): This study uses aerosol optical depth (AOD) data from AERONET (Aerosol Robotic NETwork). AERONET is a network of surface monitoring stations, each of which uses the identical automatic Sun-sky scanning spectral radiometer [20]. AERONET provides and archives two kinds of measurements that are useful in assessing aerosols characteristics. One is direct sun radiation extinction across the spectrum and the second is the angular distribution of sky radiance. From this, it is possible to compute in near-real-time aerosol spectral optical depths, aerosol size distributions, etc. There is only one AERONET monitoring station in Iran, in Zanjan city (IASBS), and we have used its AOD, AVSD, AE, SSA, ASY and ARF data in this research.

Total Ozone Mapping Spectrometer (TOMS): We have used TOMS to measure the amount of aerosols suspended in the atmosphere. Three type of TOMS including TOMS Nimbus 7 data (1979-1993), TOMS EP data (1996-2005) and OMTO3d data (2006-2016) were used to identify spatial and temporal variation of the Aerosol Index (AI) during 36 years. For TOMS, AI is defined as

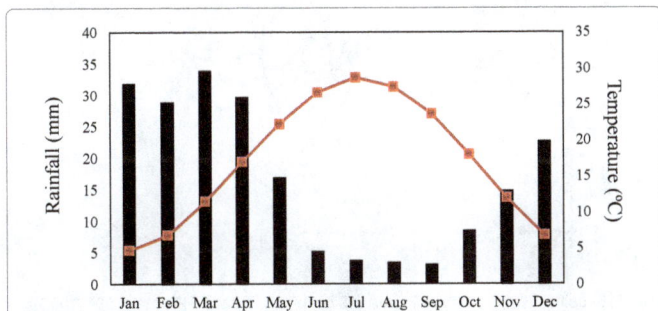

Figure 1: Average monthly temperature and rainfall in Tehran from 1960-1990.

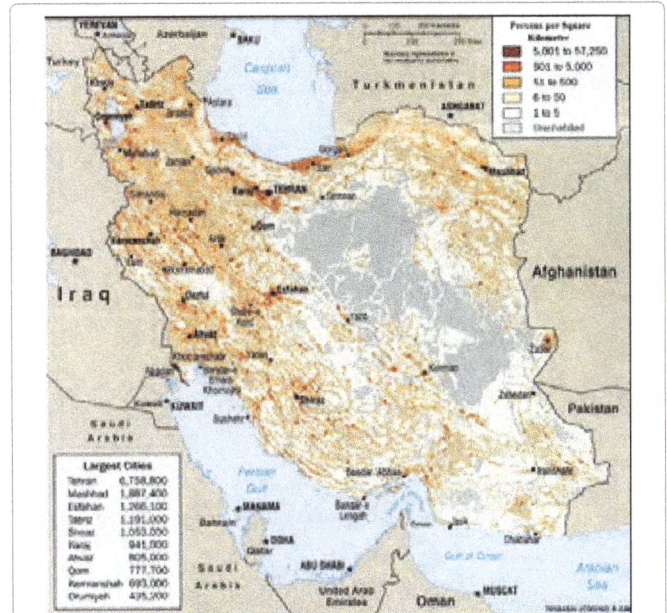

Figure 2: Map of study area. Cities evaluated and their population are indicated in the inset.

$$AI = -100\left\{ log10\left(\frac{I_{331}}{I_{360}}\right)_{measured} - log10\left(\frac{I_{331}}{I_{360}}\right)_{calculated}\right\} \qquad (2)$$

where I_{331} and I_{360} are, TOMS measured and calculated reflectances at 331 nm and 360 nm (using the Lambertian Equivalent Reflectivity derived at 360 nm).

Moderate Resolution Imaging Spectroradiometer (MODIS): The MODIS instrument mounted on the Terra and the Aqua satellites have the spectral bands ranging in wavelength from 250 m to 1.0 km at the nadir and various spatial resolutions. MODIS monitors the AOD with an error of 0.03 ± 0.05 AOD over the ocean and $\pm 0.05 + 0.15$ AOD over land. In this study is based on Deep Blue AOD at 550 nm (MYD08_D3) for MODIS with spectral resolution from 0.415 to 14.235 micron.

Multi-angle Imaging Spectro Radiometer (MISR): The MISR instrument installed on the Terra satellite collects observations at nine different viewing angles. MISR enables to identify different types of atmospheric particles [21], cloud forms, and land surface covers in a sun-synchronous orbit. Kahn et al. [22] mentioned that 70% of MISR AOD data are within 0.05 (or 20% × AOD) of sun-photometer-measured AOD values.

We have used the MISR Level 3 Component Global Aerosol Product (MIL3MAE_v4 AOD) covering a day and month of column aerosol 555 nanometer optical depth. This data product is a global summary of the Level 2 aerosol parameters of interest averaged over a month, with a resolution of 0.5 degree by 0.5 degree. Aerosol Robotic Network (AERONET) data for the Zanjan station is also considered to verify $MISR_{AOD}$ data.

Results and Discussion

Annual trend in TOMS Aerosol Index (AI)

Monitoring of AI trends by TOMS has the advantage of global coverage and applies the same approach to detecting aerosol trends over different sites. An increasing trend in AI can be observed in most selected cities during 1979-2015 (Figure 3). A decreasing trend in

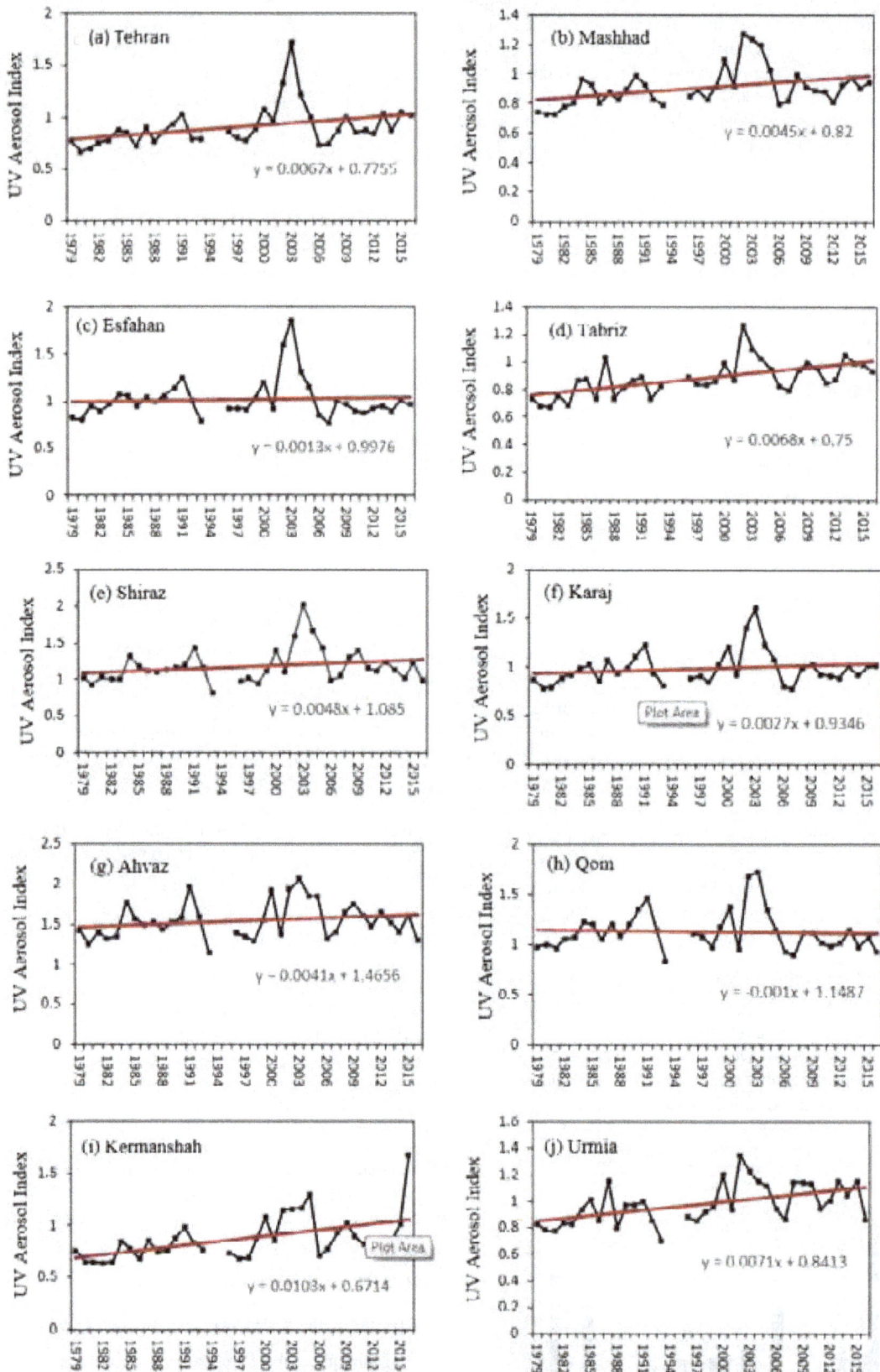

Figure 3: Variation of UV Aerosol Index for the most populated cities in Iran during 1979-2016, using TOMS Nimbus 7 data (1979-1993), TOMS EP data (1996-2005) and OMTO3d data (2006-2016).

Sensor	Data used	Product	Special resolution	Spectral band (nm)
TOMS Nimbus-7	01/1/1979-31/12/1992	Daily level	1 × 1.25°	340-380
TOMS EP	22/07/1996-31/12/2005	Daily level	1 × 1.25°	340-380
TOMS OMI	01/01/2005-07/05/2016	Daily level	1 × 1°	340-380
MISR Terra	2000-2015	Daily level 3	0.5 × 0.5°	555
MODIS Terra	2000-2015	Daily level 2	1 × 1°	550
	2010-2013 (Plots)			
AERONET (Ground base data)	2010	Daily level 2		500

Table 1: Satellite and AERONET datasets used in this study.

AI can be observed just at Qom city. The greatest increases in annual mean AOD is seen in Kermanshah, Urmia and Tabriz, respectively. The increase in anthropogenic aerosol emissions is a result of increasing populations, growing industrialization, and urbanization. Other studies show the high level of anthropogenic aerosol emissions over megacities close to the study area by satellite monitoring [23]. On the other hand, Iran is located in the center of the Northern Hemisphere dust belt. Observed rising trends in most of cities also can be related to the recent increase in the frequency of dust events in Iraq due to land desertification [24]. The Khuzestan Plain in southwest of Iran is the area with the highest frequency of dust events, over which dusty air is almost permanently present in summer, while the coastal plain of the Persian Gulf is the second most affected area [25]. The satellite cannot distinguish differences between natural and anthropogenic aerosol due to the mixing process in atmosphere. But since the long-term changes in natural aerosols are relatively small, the observed increasing and declining trends can be attributed to changes in anthropogenic aerosols. The extreme value in AI is seen during 2003 in most cities of Iran. A lot of dust storms were blowing over Iran in March, May, November, and December during 2003. The MODIS sensor with the maximum spatial resolution of 250 m shows a dust storm over Iran, Saudi Arabia, Iraq, United Arab Emirates, and Kuwait on 28th May 2003 (Figure 4a). The dust is thicker over the Persian Gulf. The time averaged map of daily AI shows that the origin of the dust storm is from the eastern part of Saudi Arabia and United Arab Emirates (Figure 4b).

Annual trend in MISR aerosol optical depth

The MISR views the whole Earth's surface every nine days. Depending on latitude, repeat coverage occurs between 2 and 9 days. MISR has a ~400 km swath (MISR Technical Document). To confirm the conclusions from TOMS data, we have analyzed MISR AOD data. Similar increasing trends in aerosol concentrations over the ten Iranian cities have been detected from MISR data over the period 2000-2015 (Figure 5). Tehran and Karaj very similar variation in AOD because they are geographically close. The annual mean and rising trend of AOD in Ahvaz (capital of Khuzestan province and 7th populated city) are higher than in other cities. The MISR$_{AOD}$ slightly showed high dust amount in 2003 (Like TOMS$_{AOD}$) but there is higher AOD's extreme value over western cities of Iran including Ahvaz, Kermanshah, Urmia, Tabriz, and Shiraz in 2008. Other research showed the same result in increasing dust storm events over the study area in 2008 [25,26]. Iran experienced the worst dust storm of the last decade in July 2008. The storm formed over Iraq and then spread eastward to Iran and covered the western and northern part of Iran and the northern part of Saudi Arabia as well. The satellite image of absorption optical depth shows that the thickness of the dust is enough to completely obscure the underlying surface of the eastern part of Iraq and western part of Iran (Figure 6a). Solid or liquid particles suspended in the air can absorb or reflect sunlight before it reaches the Earth's surface. The Ozone Monitoring Instrument (OMI) observed aerosols over Iran, Iraq, and the Saudi Arabia on July 5, 2008 (Figure 6b).

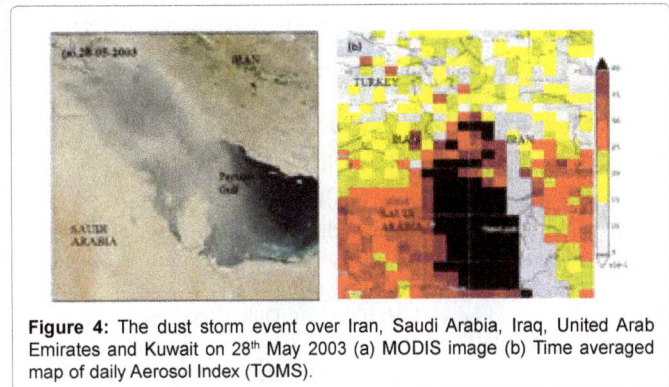

Figure 4: The dust storm event over Iran, Saudi Arabia, Iraq, United Arab Emirates and Kuwait on 28th May 2003 (a) MODIS image (b) Time averaged map of daily Aerosol Index (TOMS).

Annual means and seasonal variations in MODIS AOD

Table 2 shows seasonal and annual values of AOD using MODIS data over Iran. Data were extracted using 550 nm and were analyzed for years 2000-2015. The highest values of AOD were observed in Ahvaz in the southwestern part of Iran. The annual mean AOD for Ahvaz was 0.39 ± 0.25 over this period. The drying of Hor al-Azim, one of the most important wetlands in southwestern Iran, is probably the primary reason for high concentration of aerosols in this city. After the war with Iraq, oil extraction around the wetland decreased the water level and increased the frequency of dust storms in the Khuzestan. The second reason is that some dust sources in the Middle East such as Iraq, Syria and Arabian Peninsula contribute to dust production in the area. The wind direction in the warm/dry months can transport the dust from these regions to the western part of Iran [27]. The number of dust storms striking western Iran increased markedly between 2000 and 2009 by as much as 70 to 175 percent [28]. After Ahvaz, the highest AOD values are at Tehran (Table 2). This result probably is due to high industrial activities and urbanization in this Mega-city. The results show that AOD increases with increasing industrial activities, but the increased frequency of aerosols due to land degradation and desertification is more powerful in Iran. The highest mean AODs were recorded during the spring and the lowest AODs in the autumn for most of the cities (Table 2).

Figure 7 confirms that AOD values are higher over Ahvaz. The highest AOD value (0.76 ± 0.51) was found in May 2012 over Ahvaz and, the lowest AOD value (0.035 ± 0.27) was found in December 2013 over Tabriz. Seasonal AOD variations can be found in the all cities with the highest values in spring and summer; and the lowest values in winter (Figure 7). High wind speeds mobilizing greater quantities of dust in the area during warm months and picks up the dust. In addition, the lowest AOD value might be due to precipitation in the cold season [29]. Kim et al. [30] found that AOD is high in March (0.44±0.25) and low in September (0.24±0.21) in the East Asian region in 2009.

City name	Population	Annual and seasonal mean and standard deviation of AOD at 550 nm				
		Annual mean	Winter	Spring	Summer	Autumn
Tehran	67,58,800	0.11 ± 0.2051	0.11 ± 0.0132	0.15 ± 0.0783	0.11 ± 0.2472	0.09 ± 0.1132
Mashhad	18,87,400	0.08 ± 0.2823	0.07 ± 0.0167	0.08 ± 0.0531	0.08 ± 0.2408	0.08 ± 0.1673
Esfahan	12,66,100	0.10 ± 0.2342	0.12 ± 0.0283	0.16 ± 0.0209	0.07 ± 0.1056	0.07 ± 0.01481
Tabriz	11,91,000	0.07 ± 0.2131	0.09 ± 0.0821	0.09 ± 0.1064	0.06 ± 0.3925	0.06 ± 0.1222
Shiraz	10,53,000	0.08 ± 0.1761	0.09 ± 0.1256	0.10 ± 0.0423	0.07 ± 0.1802	0.07 ± 0.1516
Karaj	9,41,000	0.10 ± 0.1439	0.07 ± 0.0253	0.13 ± 0.0687	0.11 ± 0.0470	0.07 ± 0.1811
Ahvaz	8,05,000	0.39 ± 0.2497	0.31 ± 0.1536	0.48 ± 0.2436	0.50 ± 0.0631	0.29 ± 0.0483
Qom	7,77,700	0.10 ± 0.1015	0.09 ± 0.0349	0.14 ± 0.0205	0.10 ± 0.0202	0.08 ± 0.0704
Kermanshah	6,93,000	0.08 ± 0.2331	0.11 ± 0.1541	0.09 ± 0.0580	0.08 ± 0.2504	0.06 ± 0.1534
Urmia	4,35,200	0.09 ± 0.2514	0.09 ± 0.0361	0.10 ± 0.0821	0.07 ± 0.2236	0.09 ± 0.1758

Table 2: Annual and seasonal means of AOD at 550 nm in different cities of Iran for the period 2000-2015.

Comparison of MODIS$_{AOD}$ and MISR$_{AOD}$ and AERONET$_{AOD}$ data

Intercomparison is necessary to evaluate the accuracy of a long-term database for climatological studies. In this study, the area average AOD correlation between MODIS and MISR over Iran is calculated. The correlation has been examined using the daily average of MISR (MIL3DAE v4) and Level-3 MODIS (MOD08_D3.v6) and data gotten from the GIOVANNI site. The two datasets have the different spatial resolution. The MODIS data are available with $1 \times 1°$ whereas MISR data are available at $0.5 \times 0.5°$. MISR were converted to spatial resolution $1 \times 1°$ using a box averaging algorithm in the Giovanni site.

The correlation between MODIS and MISR AOD data was about 0.7 in winter, spring and autumn seasons of 2010 (Figure 8a, 8b and 8d). The correlation coefficient was calculated to be 0.36 for summer of 2010 (Figure 8c). The same studies were conducted in Southeastern Asia [31] and Pakistan [32,33], and they found high correlation coefficients (≥ 0.7) between MODIS and MISR AOD. The spatial map correlation between MODIS and MISR AOD data showed low correlation over Alborz and Zagros Mountains range in Iran (not shown). This result could be a result of insufficient data points in the region due to snow cover in high elevation of the mountains range. We extracted the AOD data of the sun photometer site (IASBS) for 2010 from the AERONET level 2 data archive and then we used it for assessing the accuracy of MISR$_{AOD}$ and MODIS$_{AOD}$ data. This sun photometer site is located in Zanjan city in Northwest Iran. Since, the MISR observation repeat time is only 3 or 4 visits per month for the IASBS site, we compare AERONETAOD and MISRAOD daily data when MISR swept this site. Validation of MISR$_{AOD}$ and MODIS$_{AOD}$ using AERONET$_{AOD}$ shows a high correlation coefficient of 0.87 and 0.76 at the 0.05 significance level, respectively.

Air mass trajectory analysis

To better understand the origins of the air masses arriving in Iran from dust sources during warm and cold months, we did back trajectory analyses using HYSPLIT model [34]. These back trajectories were analyzed for seven days at four altitudes (3003 m, 2000 m, 1000 m, and 500 m). The paths of air masses that reached Ahvaz and Tehran cities in 10 May and 10 Jun 2011 are shown in Figure 9. The trajectories show that air masses come from Syria, Lebanon, and passed over the Iraq and then reached to Ahvaz during 10 May 2011 (Figure 9a). On 10 Jun 2011, the air masses reached Tehran from Egypt, Lebanon, and then Syria and Iraq (Figure 9b). The air masses in summer reach to Iran from dry deserts in the north of Africa such as the Sahara. In addition, some dust sources in the Middle East including Syria and Iraq increase in the aerosol concentrations over Iran. In winter, the air masses propagate from Europe and Mediterranean Sea regions (Not shown) and cause a lot of rainfall over Iran.

Aerosol radiative forcing

Solar radiation as the main forcing of atmospheric circulation controls the weather and climate system. Atmospheric aerosols can cause a decrease in solar radiation flux by radiative effects. Ramaswamy et al. [35] defined Aerosol Radiative forcing (ARF, in wm^{-2}) as the net imbalance in irradiance (Solar plus long wave) and can be shown as following equation:

$$\Delta F = (F_a \uparrow - F_a \downarrow)_{with} - (F_0 \uparrow - F_0 \downarrow)_{without} \qquad (3)$$

Where ΔF is irradiance, the first and second sentences in the right of the equation are the net irradiance with and without aerosol, respectively. Indirectly, these changes effect on specific ARF changes. Figure 10 shows the variation of ARF at the earth's surface (BOA) and top of atmosphere (TOA) also, Aerosol Optical Depth (AOT) at Zanjan during the 2010-2013 period. Seasonal ARF variations was found over Zanjan with the highest values in warm seasons and the lowest values in cold seasons (Figure 10). The deep cooling effect has been occurred on May 2011 due to mineral dust's effect of reflecting and cooling at the visible wavelengths. However, Dust's warming counters half of its cooling effect due to absorbing and warming at the longer infrared wavelengths [36]. The ARF at surface and TOA were found to be ranging from -79 Wm^{-2} to -10 Wm^{-2} (average: -33.45 Wm^{-2}) and -25 wm^{-2} to 6 wm^{-2} (average: -12.80 Wm^{-2}), respectively. The difference between ARF at surface and TOA is higher in Jun than other months. Radiative forcing at the BOA is much larger (3 times) than that at the TOA. An increase in AOT would lead to the decrease in radiative forcing (more negative value) at the surface and top of the atmosphere. The negative values of ARF indicate solar radiative cooling effect and positive values of ARF indicate radiative warming effect [37]. Strong correlation between ARF and Aerosol Optical Thickness was calculated -0.92 at 0.05 significance level.

Conclusion

Three satellite sensors MODIS, TOMS and MISR have been used to study the seasonal and annual variability of aerosols concentration over the ten populated cities in Iran. Both TOMS and MISR data showed an increase in aerosols concentration in most cities with time. The maximum AOD value (0.76 ± 0.51) was found in May 2012 over Ahvaz city in southwestern Iran. Also, the highest annual AOD means were found over Ahvaz by 0.39 ± 0.24. The cities in the western part of Iran had highest annual means of aerosol concentration during 1979-2016. Since the western part of Iran does not include the biggest and most populated cities, it can be concluded that the highest value of aerosol concentration in this area is not the result of increasing urbanization and industrialization. The increase in the frequency of dust outbreaks in the Hor al-Azim and some dry areas in Iraq due to

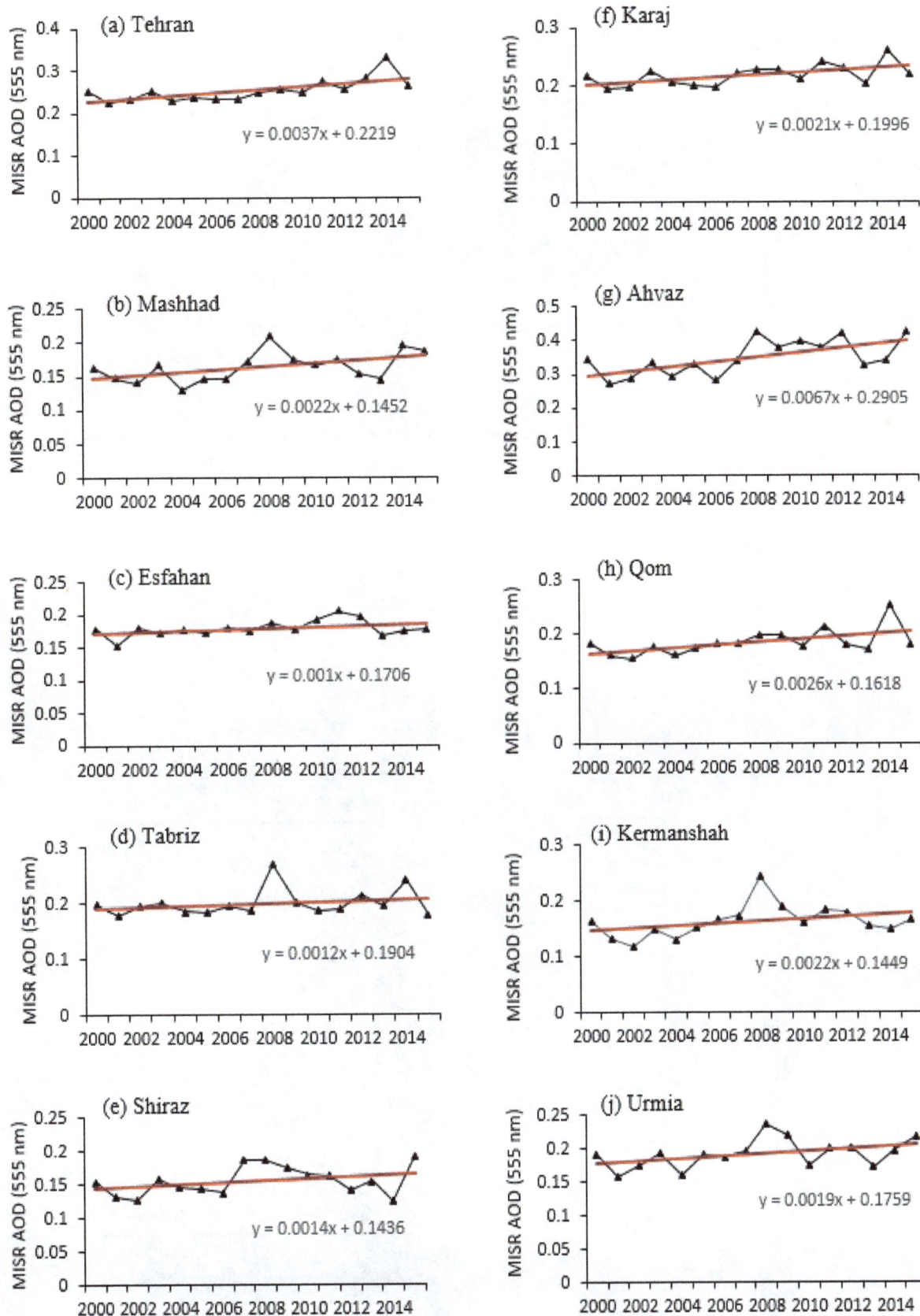

Figure 5: Trends in MISR AOD over selected cities in Iran.

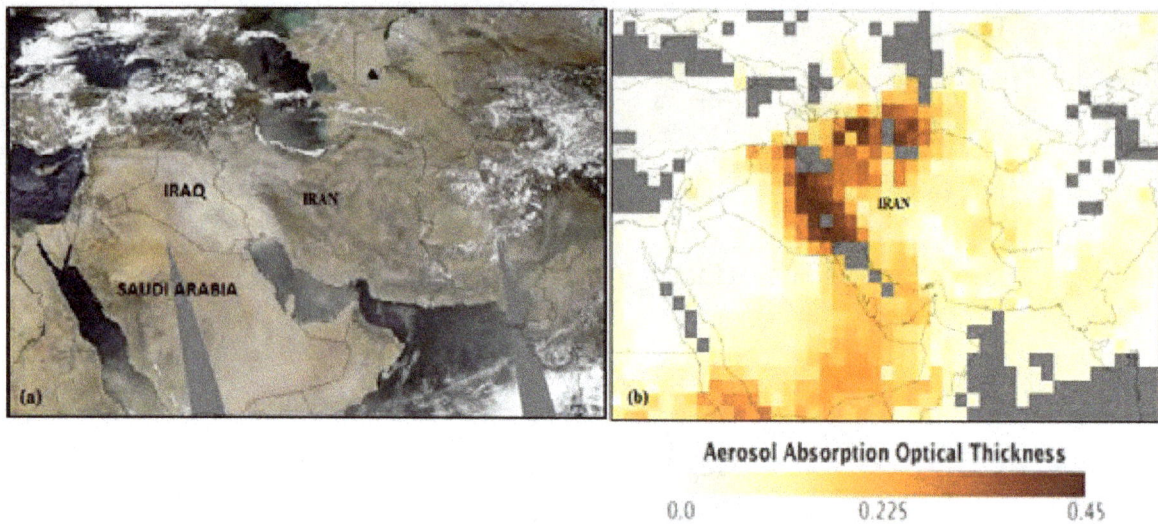

Figure 6: The dust storm event over Iran, 5th July 2008 (a) Aqua MODIS image (b) Aerosol Absorption Optical Thickness by Ozone Monitoring Instrument (OMI) on NASA's Aura satellite.

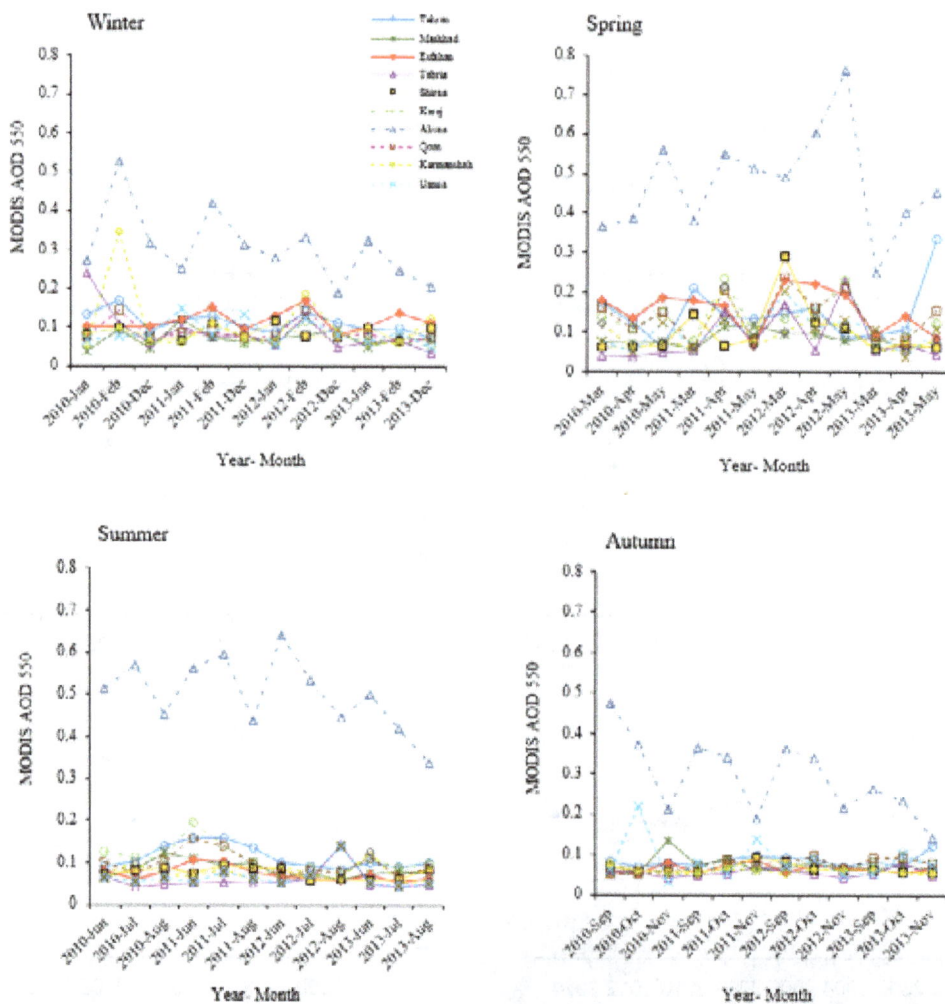

Figure 7: Seasonal variations in MODIS AOD (550 nm) in (a) winter (b) spring (c) summer (d) autumn, for the period covering 2010-2013, over various cities in Iran.

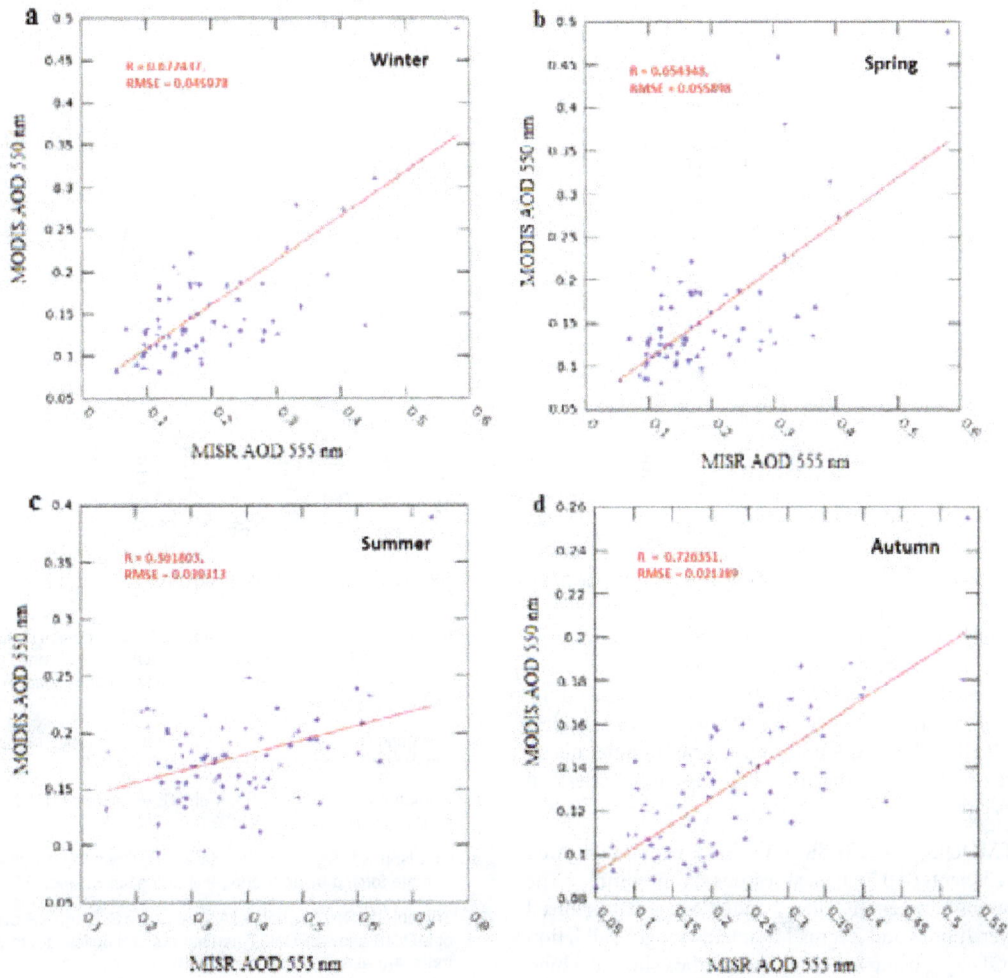

Figure 8: Area average correlations between MODIS and MISR AOD for (a) winter (b) spring (c) summer (d) Autumn over regions of Iran in 2010.

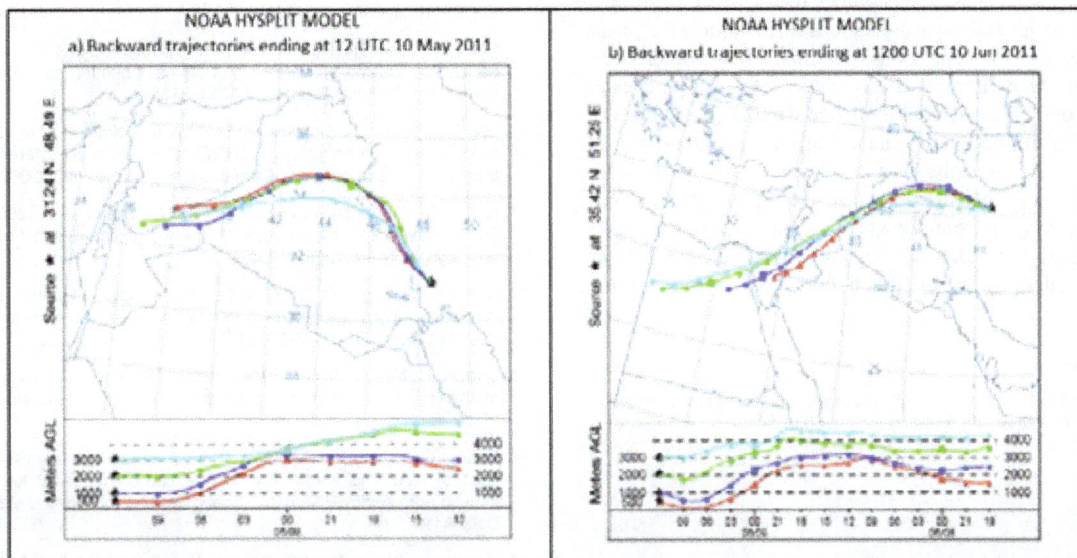

Figure 9: Seven-day back trajectories (a) For Ahvaz city 10th May and (b) For Tehran city 10th June 2011.

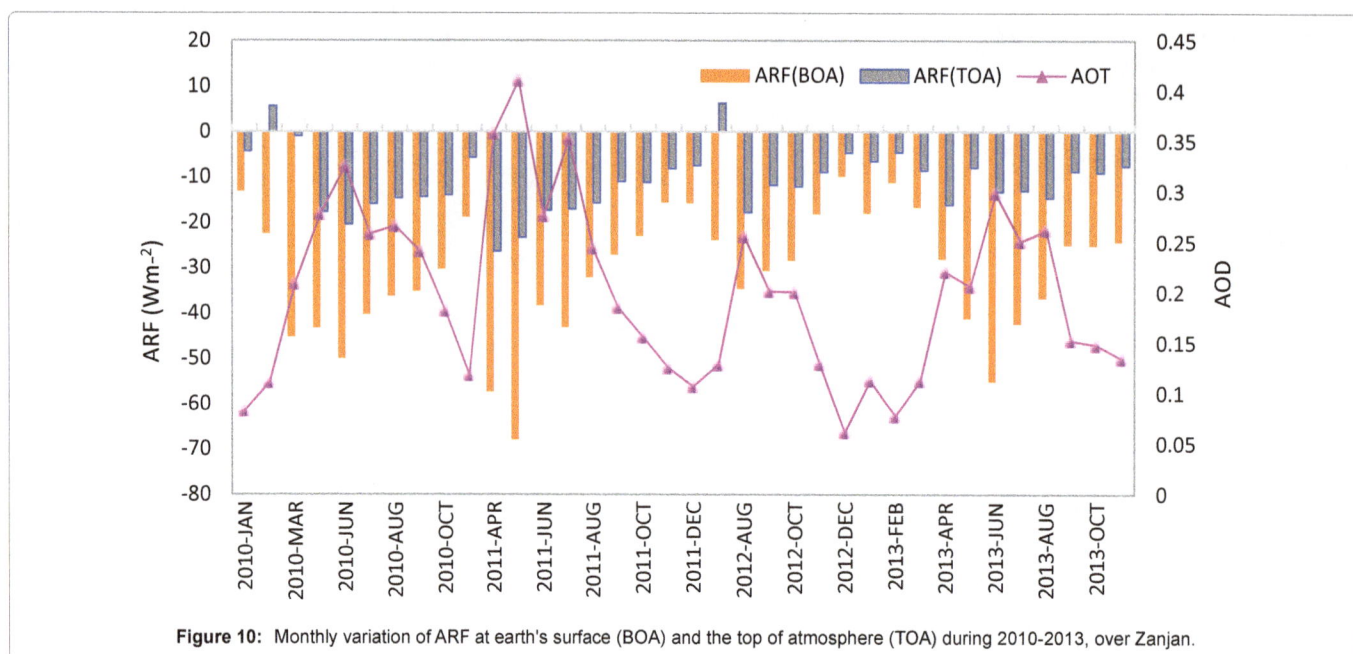

Figure 10: Monthly variation of ARF at earth's surface (BOA) and the top of atmosphere (TOA) during 2010-2013, over Zanjan.

land degradation and desertification are the main factors that affected aerosol concentration in western Iran. According to seasonal analysis, the highest aerosol concentration was found during the spring, and the lowest concentration was found in winter over the selected cities. After Ahvaz, the highest AOD value was found over Tehran (annual mean: 0.11 ± 0.20). This result probably is due to high industrial activities and urbanization in this city.

Comparison of $MODIS_{AOD}$ and $MISR_{AOD}$ retrievals over Iran showed high correlation coefficients (≥ 0.7) in most seasons except summer. The spatial map correlations between $MODIS_{AOD}$ and $MISR_{AOD}$ data showed low correlations over Alborz and Zagros Mountains range. Validation of $MISR_{AOD}$ and $MODIS_{AOD}$ using AERONET AOD data showed a high correlation coefficient of 0.87 and 0.91, respectively.

Back trajectory examination of AOD showed that air masses come from Egypt, Syria, Lebanon, and pass over the Iraq and then reach Iran during summer. The air masses in summer reached to Iran from the Sahara Desert in Africa and some dry regions in the Middle East causes the increase in aerosol concentrations over Iran; but in winter, the air masses propagate from Europe and Mediterranean Sea and causes a lot of rainfall over Iran. Aerosol Radiative Forcing (ARF) has been analyzed over Zanjan (AERONET site in Iran) during 2010-2013. The highest values of ARF were observed in warm seasons and the lowest values in cold seasons. The ARF at surface and top of the atmosphere were found to be ranging from -79 wm^{-2} to -10 Wm^{-2} and -25 wm^{-2} to 6 wm^{-2} over Zanjan, respectively. Strong correlation between ARF and Aerosol Optical Thickness was calculated -0.92 at 0.05 significance level [38,39].

Acknowledgements

We are grateful to Douglas Klotter for his assistance in graphics preparation and data retrieval. Analyses and visualizations used in this study were produced with the Giovanni online data system, developed, and maintained by the NASA GES DISC.

References

1. Kosmopoulos PG, Kaskaoutis DG, Nastos PT, Kambezidis HD (2008) Seasonal variation of columnar aerosol optical properties over Athens, Greece, based on MODIS data. Remote Sensing of Environment 112: 2354-2366.

2. Papadimas CD, Hatzianastassiou N, Mihaloppoulos N, Kanakidou M, Katsoulis BD, et al. (2008) Assessment of the MODIS collections C005 and C004 aerosol optical depth products over the Mediterranean basin. Atmospheric Chemistry and Physics 9: 2987-2999.

3. Lohmann U, Feichter J (2005) Global indirect aerosol effects: a review. Atmospheric Chemistry and Physics 5: 715-737.

4. Ramanathan V, Crutzen PJ, Kiehl JT, Rosenfeld D (2001) Aerosol, climate, and hydrological cycle. Science 294: 2119-2124.

5. Charlson RJ, Schwartz SE, Hales JH, Cess RD, Coakley Jr JA, et al. (1992) Climate forcing by anthropogenic aerosols. Science 255: 423-430.

6. Tripathi SN, Dey S, Chandel A, Srivastava S, Singh RP, et al. (2005) Comparison of MODIS and AERONET derived aerosol optical depth over the Ganga Basin, India. Annales Geophysicae 23: 1093-1101.

7. Kaufman YJ, Tanre D, Dubovik O, Karnieli A, Remer LA (2001) Absorption of sunlight by dust as inferred from satellite and ground-based remote sensing. Geophysical Research Letter 28: 1479-1482.

8. Kaufman YJ, Tanré D, Boucher O (2002) A satellite view of aerosols in climate system. Nature 419: 215-223.

9. Eck TF, Holben BN, Reid JS, Dubovik O, Smirnov A, et al. (1999) Wavelength dependence of the optical depth of biomass burning, urban, and desert dust aerosols. J Geophys Res 104: 31333–31349.

10. Eck TF, Holben BN, Dubovik O, Smirnov A, Goloub P, et al. (2005) Columnar aerosol optical properties at AERONET sites in central eastern Asia and aerosol transport to the tropical mid-Pacific. J Geophys Res 110: D06202.

11. Holben BN, Tanré D, Smirnov A, Eck TF, Slutsker I, et al. (2001) An emerging ground-based aerosol climatology: Aerosol optical depth from AERONET. J Geophys Res 106: 12067-12097.

12. Dubovik O, King MD (2000) A flexible inversion algorithm for the retrieval of aerosol optical properties from Sun and sky radiance measurements. J Geophys Res 105: 20673-20696.

13. Dubovik O, Holben BN, Eck TF, Smirnov A, Kaufman YJ, et al. (2002) Variability of absorption and optical properties of key aerosol types observed in worldwide locations. J Atmos Sci 59: 590-608.

14. Cattrall C, Reagan J, Thome K, Dubovik O (2005) Variability of aerosol and spectral lidar and backscatter and extinction ratios of key aerosol types derived from selected Aerosol Robotic Network locations. J Geophys Res 110: D10SA11.

15. Masoumi A, Khalesifard HR, Bayat A, Moradhaseli R (2013) Retrieval of aerosol optical and physical properties from ground-based measurements for Zanjan, a city in Northwest Iran. Atmos Res 120–121: 343-355.

16. Bayat A, Khalesifard HR, Masoumi A (2013) Retrieval of aerosol single-scattering albedo and polarized phase function from polarized sun-photometer measurements for Zanjan's atmosphere. Atmos Meas Tech 6: 2659-2669.

17. Khoshsima M, Ahmadi-Givi F, Bidokhti AA, Sabetghadam S (2014) Impact of meteorological parameters on relation between aerosol optical indices and air pollution in a sub-urban area. Journal of Aerosol Science 68: 46-57.

18. Rashki A, Kaskaoutis DG, Eriksson PG, Rautenbach CJdeW, Flamant C, et al. (2013) Spatio-temporal variability of dust aerosols over the Sistan region in Iran based on satellite observation. Nat Hazards 71: 563-585.

19. Gharibzadeh M, Alam KH, Bidokhti AA, Abedini Y, Masoumi A (2016) Radiative Effects and Optical Properties of Aerosol during Two Dust Events in 2013 over Zanjan, Iran. Aerosol and Air Quality Research 17: 888-898.

20. Holben BN, Eck TF, Slutsker I, Tanre D, Buis JP, et al. (1998) AERONET- A federated instrument network and data archive for aerosol characterization. Remote Sensing Environ 66: 1-16.

21. Diner DJ, Beckert JC, Reilly TH, Bruegge CJ, Conel JE, et al. (1998) Multi-angle Imaging Spectroradiometer (MISR) instrument description and experiment overview. IEEE Trans Geosci Remote Sens 36: 1072-108.

22. Kahn R, Li WH, Martonchik J, Bruegge C, Diner D, et al. (2005) MISR low-light-level calibration and implications for aerosol retrieval over dark water. J Atmosph Sci 62: 1032-1062.

23. Alam K, Qureshi S, Blaschke T (2011) Monitoring spatio-temporal aerosol patterns over Pakistan based on MODIS, TOMS and MISR satellite data and a HYSPLIT model. Atmospheric Environment 45: 4641-4651.

24. Goudie AS, Middleton NJ (2006) Desert Dust in the Global System.

25. Choobari OA, Ghafarian P, Owlad E (2015) Temporal variations in the frequency and concentration of dust events over Iran based on surface observations. International Journal of Climatology 36: 2050-2062.

26. Gerivani H, Lashkaripour GR, Ghafoor IM, Jalili N (2010) The source of dust storm in Iran: a case study based on geological information and rainfall data. Carpathian J Earth Environ Sci 6: 297-308.

27. Zoljoodi M, Didevarasl A Saadatabadi AR (2013) Dust Events in the Western Parts of Iran and the Relationship with Drought Expansion over the Dust-Source Areas in Iraq and Syria. Atmospheric and Climate Sciences 3: 321-336.

28. Hamidi M, Kavianpour MR, Shao Y (2013) Synoptic analysis of dust storms in the Middle Est, Asia Pac. J Atmos Sci 49: 279-286.

29. Ranjan RR, Joshi HP, Iyer KN (2007) Spectral variation of total column aerosol optical depth over Rajkot: a tropical semi-arid Indian station. Aerosol and Air Quality Research 7: 33-45.

30. Kim HS, Chung YS, Lee SG (2013) Analysis of spatial and seasonal distributions of MODIS aerosol optical properties and ground-based measurements of mass concentrations in the Yellow Sea region in 2009. Environ Monit Assess 185: 369.

31. Xiao N, Shi T, Calder CA, Munroe DK, Berrett C, et al. (2009) Spatial characteristics of the difference between MISR and MODIS aerosol optical depth retrievals over mainland Southeast Asia. Remote Sensing of Environment 113: 1-9.

32. Prasad AK, Singh S, Chauhan SS, Srivastava MK, Singh RP, et al. (2007) Aerosol radiative forcing over the Indo-Gangetic plains during major dust storms. Atmospheric Environment 41: 6289-6301.

33. Alam K, Trautmann T, Blaschke T, Majid H (2012) Aerosol optical and radiative properties during summer and winter seasons over Lahore and Karachi. Atmos Environ 50: 234-245.

34. Draxler RR, Rolph GD (2003) HYSPLIT (HYbrid Single-Particle Lagrangian Integrated Trajectory) Model access via the NOAA ARL READY Website, NOAA Air Resour. Lab, Silver Spring, MD, USA. Available from: http://www.arl.noaa.gov/ready/hysplit4.html

35. Ramaswamy V, Chen CT (1997) Climate forcing-response relationships for greenhouse and shortwave radiative perturbations. Geophys Res Lett 24: 667-670.

36. Hansen J, Sato M, Ruedy R (1997) Radiative forcing and climate response. J Geophys Res 102: 6831-6864.

37. Hatzianastassiou N, Matsoukas C, Drakakis E, Stackhouse Jr PW, Koepke P, et al. (2007) The direct effect of aerosols on solar radiation based on satellite observations, reanalysis datasets, and spectral aerosol optical properties from Global Aerosol Data Set (GADS). Atmos Chem Phys 7: 2585-2599.

38. Chu D, Kaufman Y, Ichoku C, Remer L, Tanr'd D, et al. (2002) Validation of MODIS aerosol optical depth retrieval over land. Geophysical Research Letters 29: 8007.

39. Remer LA, Kaufman YJ, Tanr´e D, Mattoo S, Chu DA, et al (2005) The MODIS aerosol algorithm, products and validation. Journal of Atmospheric Science 62: 947-973.

21

Climate Projection Outlook in Lake Haramaya Watershed, Eastern Ethiopia

Eba Muluneh Sorecha*

School of Natural Resources Management and Environmental Sciences, Haramaya University, Dire Dawa, Ethiopia

Abstract

Smallholder farmers in Ethiopia generally face widespread problems driven by climate change. For this reason, the study of climate change at watershed level might be critical to solve the problem from its root. The study was conducted in Lake Haramaya Watershed, Eastern Ethiopia to project ad characterize the climatic condition of the coming thirty years (2020-2050). Thirty-four years of rainfall, maximum and minimum temperature baseline data were collected from National Meteorological Agency of Ethiopia. Whereas, thirty years (2020-2050) projected rainfall, maximum and minimum temperatures were downscaled from MarkSim web version for IPCC AR5 data (CMIP5) using five climate models namely: BCC-CSM1-1, CSIRO-Mk3-6-0, HadGEM2-ES, MIROC-ESM, MIROC-ESM-CHEM, and MIROC5 under two Representative Concentration Pathways (RCPs): RCP4.5 and RCP8.5. The results of the study revealed that the annual mean rainfall will be increased by 20.70 and 24.14% under RCP4.5 and RCP8.5, respectively compared to baseline average value of 777.51 mm/yr. The annual rainfall under RCP4.5 ranges from 769.6 to 1090 mm/yr having the CV value of about 11%; whereas, under RCP8.5, it will range from 771.9 to 1129 mm/yr with the CV value of 13.17%. Kiremt (JJAS) season rainfall will increase from the baseline of 107.55 mm/yr to 135.79 and 136.27 mm/yr under RCP4.5 and RCP8.5, respectively. Moreover, a study indicated that the annual and seasonal temperature under RCP4.5 and RCP8.5 will be expected to increase during 2020-2050 period. Under high emission scenario of RCP8.5, the annual maximum temperature could rise from 24.73°C baseline to 25.41°C.

Keywords: Climate projection; Climate change; Lake Haramaya; Ethiopia

Introduction

Intergovernmental Panel on Climate Change (IPCC) has reported that by 21st century, the average global temperature would rise between 1.4 and 5.8°C and rainfall would vary up to ± 20% from the 1990 level. Despite uncertainties on the directions and magnitude of climate changes, there is significant scientific evidence that shows an increase in average temperature and overall climate variability in the semi-arid tropics, with subsequent increases in the occurrence of droughts, floods and heat waves that affect people, their crops, and livestock [1,2].

Future climate projections for Eastern Africa countries vary, with high altitude areas of Ethiopia potentially benefiting from warming temperatures. However, without adequate adaptation measures, most of the region is likely to be deleteriously affected by rising temperatures due to increasing rates of evaporation and transpiration [3]. Moreover, spatial and temporal variability of rains, and an increase in the intensity of rainfall events, frequency and duration of droughts will affect the farming communities at large [4]. Some of these changes are already being experienced across the region; others are predicted in the near future [5-7].

Ethiopia's mean annual temperature is showing a significant warming trend leading to increasing rates of evapotranspiration and crop water requirements, further adding to the already frequent water stress of crops [8]. Future projections show that the mean maximum temperature will increase by 2-2.3 °C until 2030 and by 2.2-2.7 °C until 2050, while the mean minimum temperature will rise by 0.8-0.9 °C until 2030 and 1.4-1.7 °C until 2050, all in conjunction with a surge of hot days and nights and a decrease of cold days and nights [9].

Smallholder farmers in Ethiopia generally face widespread problems driven by climate conditions: water scarcity for drinking, industry, agricultural crop production, lack of pasture and livestock feed. This vicious cycle of poverty, food insecurity and natural resources degradation is basically caused because of higher population growth but is being exacerbated by increasing weather variability and climate change. Specifically, the impacts of climate change on water bodies in Africa have become a prominent and hot agenda. Of Africa's total population, about 60% live in rural areas and almost more than 80% rely on water from seasonal rainfall for domestic and other needs [10].

In line with this, except few studies [4,8,9] conducted in Central and northern parts of Ethiopia, nothing has been done in eastern Hararghe on climate projection and its implications on the livelihoods of the farming communities. Therefore, this study aims to investigate climate projection outlook in Eastern Hararghe at watershed level, so as to enhance the resilience of the farming community against the impacts of climate change. This gives also an opportunity to plan appropriate adaptation measures that must be taken ahead of time. Moreover, it will give enough room to consider possible future risks in all phases of climate related projects.

Materials and Methods

Description of the study area

Lake Haramaya Watershed is located in Haramaya and partly in

***Corresponding author:** Eba Muluneh Sorecha, School of Natural Resources Management and Environmental Sciences, Haramaya University, PO Box: 138, Dire Dawa, Ethiopia, E-mail: ebamule1@gmail.com*

Kombolcha districts, Eastern Hararghe Zone, Oromia National Region State, and East Ethiopia. The Watershed lies between 9°23′12.27′′-9°31′9.85′′ N and 41°58′28.02′′-42°8′h10.26′′ E (UTM Zone 38) (Figure 1) and covers an area of 15,329.96 ha. The elevation ranges from 1800 to 2345 meters above sea level. Information obtained from Ethiopian National Meteorological Agency indicates that the mean annual rainfall and mean maximum and minimum temperatures of Haramaya watershed are 800.9 mm, 24.18°C, and 9.9°C, respectively. Mixed farming system: cash crop production such as coffee (Coffee arabica) and chat (Catha edulis), vegetable crop production like: potatoes, carrots, onion, and green pepper are widely produced in the area. Animal fattening is observed with a number of households.

Research approach

Long term meteorological data, thirty-four years of rainfall, maximum and minimum temperature baseline data were collected from National Meteorological Agency of Ethiopia. Whereas, thirty years 2020 to 2050) projected climate data: rainfall maximum and minimum temperatures were downscaled from MarkSim web version for IPCC AR5 data (CMIP5). In the processes of downscaling, five climate models namely: BCC-CSM1-1; CSIRO-Mk3-6-0; HadGEM2-ES; MIROC-ESM; MIROC-ESM-CHEM; MIROC5 under two Representative Concentration Pathways (RCPs): RCP4.5 and RCP8.5. The models selected are highly applicable for African climate studies [3,11]. Thus, changes in climate over specific area, Lake Haramaya watershed over the determined period of 2020 to 2050 were noticed as monthly temperature changes in (°C) and monthly precipitation changes in (%) from the base period of 1980-2013.

Results and Discussion

Climate projection

Annual and seasonal rainfall projection: The results of the analysis of the projected data over Lake Haramaya revealed that the annual mean rainfall will be about 938.5 and 965.2 mm/yr under RCP4.5 and RCP8.5, respectively compared with the baseline average value of 777.51 mm/yr (Table 1). The annual rainfall under RCP4.5 ranges from 769.6 to 1090 mm/yr having the CV value of about 11%. Whereas, under high representative concentration pathway (RCP8.5), the rainfall in the upcoming thirty years will range from 771.9 to 1129 mm/yr with the CV value of 13.17% (Table 1). This implies that the annual rainfall will increase certainly under both climate scenarios considered for this particular study.

Moreover, it has been pinpointed through this study that the seasonal rainfall over the specific watershed will rises up. For instance, the Kiremt (JJAS) season rainfall will increase from the baseline of 107.55 mm/yr to 135.79 and 136.27 mm/yr, respectively. However, comparing the CV values for both scenarios, rainfall under RCP8.5 could be highly variable, CV value of 56% (Table 2). Thus, Kiremt season rainfall under this scenario will not be promising to work with particularly, for those farming communities. The same trend could be observed for Belg (FMAM) season the watershed, where the highest CV value of 103% (Table 2).

The result of study also showed that there will be increments in the seasonal rainfall of Bega (ONDJ) season compared to the baseline. It will rise up from 22 mm/yr to 135 mm/yr and 18 mm/yr under RCP4.5 and RCP8.5 scenarios, respectively (Table 2). This indicates that there is probability that Bega season or dry period in the watershed could

	Baseline	RCP4.5	RCP8.5
Mean	777.51	938.45	965.2
Minimum	309.4	769.6	771.9
Maximum	1667	1090	1129
SD	321.83	104.31	127.12
CV (%)	41.39	11.12	13.17

Table 1: Descriptive statistics of annual rainfall totals in the upcoming thirty years (2020-2050) at Lake Haramaya watershed, Eastern Ethiopia.

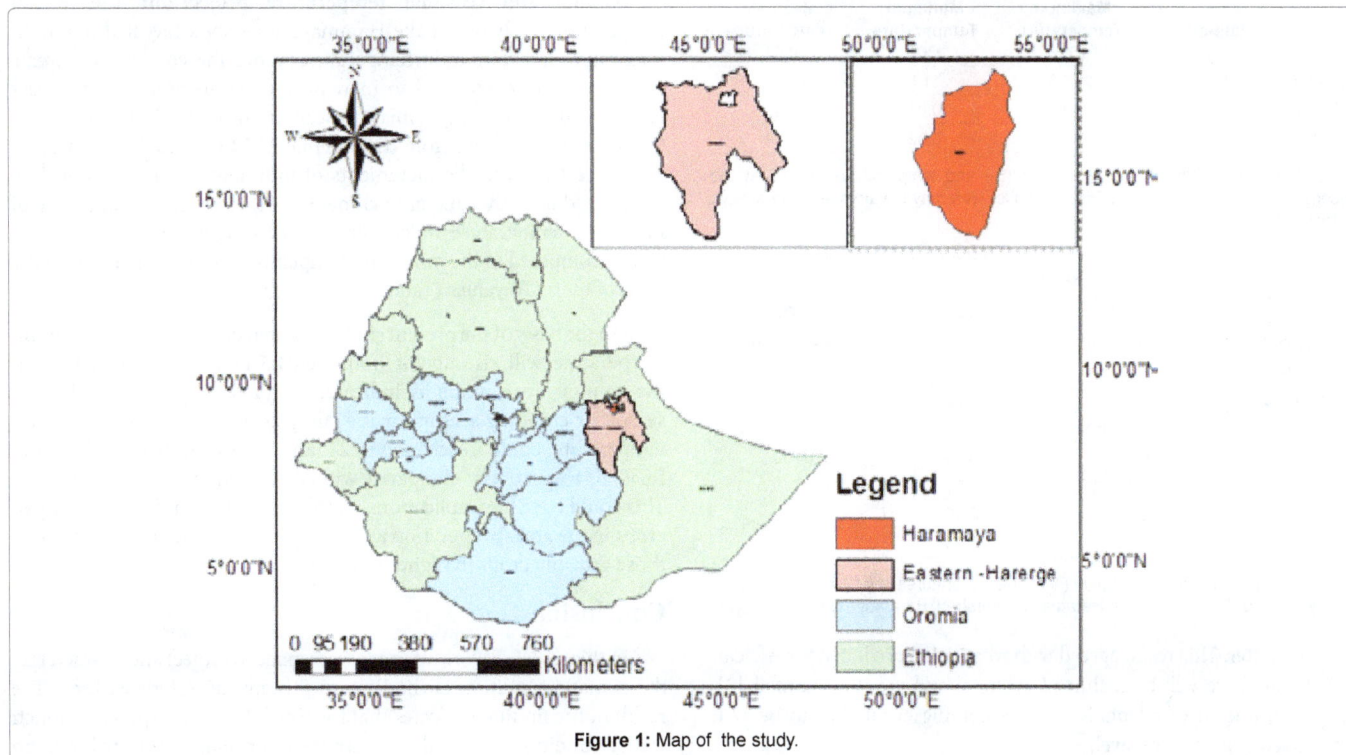

Figure 1: Map of the study.

JJAS RF (mm)	Baseline	RCP4.5	RCP8.5
Mean	107.55	135.79	136.27
SD	16.3	2.7	76.68
CV (%)	15.15	1.98	56.27
FMAM (mm)	Baseline	RCP4.5	RCP8.5
MEAN	82	120.03	86.33
SD	60.77	0.69	89.24
CV (%)	74.11	0.58	103.37
ONDJ (mm)	Baseline	RCP4.5	RCP8.5
Mean	22.41	135.21	18.7
SD	21.85	1.01	7.02
CV (%)	97.49	0.75	37.53

Table 2: Descriptive statistics of seasonal rainfall in the upcoming thirty years (2020-2050) at Lake Haramaya watershed, Haramaya district, Ethiopia.

Annual Temp (°C)	Tmax-baseline	Tmin-baseline	Tmax-RCP4.5	Tmin-RCP4.5	Tmax-RCP8.5	Tmin-RCP8.5
Mean	24.73	11.5	24.9	12.01	25.41	11.8
SD	1.26	1.85	1.73	2.26	1.83	2.38
CV (%)	5.09	16.05	7.43	20.35	7.84	21.11

Table 3: Descriptive statistics of seasonal rainfall in the upcoming thirty years (2020-2050) at Lake Haramaya watershed, Haramaya district, Ethiopia.

JJAS Temp (°C)	Tmax-baseline	Tmin-baseline	Tmax-RCP4.5	Tmin-RCP4.5	Tmax-RCP8.5	Tmin-RCP8.5
Mean	23.89	12.66	24.06	12.91	25.9	13.1
SD	1.74	0.14	0.82	0.99	0.66	0.99
CV (%)	7.27	1.11	3.56	8.32	2.84	8.2
FMAM Temp (°C)	Tmax-baseline	Tmin-baseline	Tmax-RCP4.5	Tmin-RCP4.5	Tmax-RCP8.5	Tmin-RCP8.5
Mean	25.76	12.51	25.94	12.6	25.97	12.89
SD	0.5	1.36	1.53	2.31	1.67	2.53
CV (%)	1.93	10.9	6.15	19.1	6.69	20.76
ONDJ Temp (°C)	Tmax-baseline	Tmin- baseline	Tmax-RCP4.5	Tmin-RCP4.5	Tmax-RCP8.5	Tmin-RCP8.5
Mean	24.53	9.32	25.99	9.5	24.7	9.51
SD	0.38	1.07	1.44	2.37	1.81	2.68
CV (%)	1.57	11.51	6.53	25.65	8.18	28.2

Table 4: Descriptive statistics of seasonal temperatures in the upcoming thirty years (2020-2050) at Lake Haramaya watershed, Haramaya district, Ethiopia.

	Rainfall (%)	Maximum Temperature (°C)	Minimum Temperature (°C)	Mean air Temperature (°C)
Baseline	777.51(mm)	24.73	11.5	18.11
RCP4.5	20.7	0.06	0.04	0.05
RCP8.5	24.14	0.05	0.02	0.04

Table 5: Annual rainfall totals change in (%) and temperatures in (°C) in the upcoming thirty years (2020-2050) at Lake Haramaya watershed, Haramaya district, Ethiopia.

Rainfall (%)			
	JJAS	FMAM	ONDJ
Baseline	107.55 (mm)	82 (mm)	22.41 (mm)
RCP4.5	26.27	46.38	503
RCP8.5	26.7	5.28	-14.73
Air temperature (°C)			
	JJAS	FMAM	ONDJ
Baseline	18.27 (°C)	19.14 (°C)	16.93 (°C)
RCP4.5	0.04	0.03	0.08
RCP8.5	0.04	0.03	0.07

Table 6: Seasonal rainfall change in (%) and temperatures in (°C) in the upcoming thirty years (2020-2050) at Lake Haramaya watershed, Haramaya district, Ethiopia.

become wetter. This result agreed with what IPCC predicted for African climate, as there will be a shift of seasonal and monthly rainfall [2]. Therefore, a kind of adaptation to the coming events has to be well recognized at grass root level.

Annual and seasonal temperature projection: The annual temperature study over Lake Haramaya watershed revealed that both maximum and minimum temperature under the considered climate scenarios will be expected to increase for the upcoming thirty years (2020-2050). Under high emission scenario of RCP8.5, the maximum temperature could rise from the baseline 24.73°C to 25.41°C (Table 3). The same trends in the increments of minimum temperature will be expected due to the changing climate. Furthermore, the variability of maximum temperature under both scenarios (RCP4.5 and RCP8.5) will be less compared to the minimum temperature condition; this is on the basis of their CV values (Table 3).

On the basis of the present study, seasonal maximum and minimum temperature will rise up for both RCP4.5 and RCP8.5 scenarios in the coming thirty years as indicated in Table 4. During the Kiremt season or cropping season for the study area in particular, maximum temperature could reach about 24 and 25.9°C under RCP4.5 and RCP8.5, respectively, compared with the baseline 23.89°C (Table 4). This could have big implications on the production and productivity of crops in the area. Evaporation and transpiration could increase, leading those growing crops in higher water demand.

Conclusion

In this research work attempts were made to project and characterize the climatic condition of the Eastern Ethiopia at watershed level. The result of the finding indicates that the watershed will experience much wetter condition than today. In contrast, air temperature will rise up relative to the baseline temperature condition. Tables 5 and 6 shows the annual and seasonal extent of changes in rainfall in % and temperature in °C in the upcoming thirty years relative to the baseline rainfall and temperature.

Generally, rainfall and temperatures will increase under both climate scenarios considered in the coming periods 2020-2050. Therefore, a typical adaptation mechanism to the event have to be well understood and undertaken at all levels; to be resilient against the consequences of climate change in Lake Haramaya watershed.

Acknowledgements

I am thankful to the National Meteorological Agency of Ethiopia for free of charge data required for this study. My regards also go to the staff of School of Natural Resources Management and Environmental Sciences for moral support and editing the manuscript.

References

1. IPCC (Intergovernmental Panel on Climate Change) (2001) Climate Change 2001: The Scientific Basis. Contribution of Working Group I to the Third Assessment Report of the Intergovernmental Panel on Climate. Cambridge University Press, Cambridge, UK, p: 881.

2. IPCC (Intergovernmental Panel on Climate Change) (2007) Summary for Policy Makers. Chapter 11 of the 4th IPCC Report on Regional Climate Projections. Chapter 9 on Africa.

3. Thornton PK, Jones PG, Alagarswamy A, Andresen J (2009) Spatial variation of crop yield responses to climate change in East Africa. Global Environmental Change 19: 54-65.

4. Yemenu F, Chemeda D (2010) Climate resources analysis for use of planning in crop production and rainfall water management in the Central Highlands of Ethiopia, the case of Bishoftu District, Oromia Region. Journal of Hydrology and Earth System Science 4: 10-45.

5. FAO (Food and Agriculture Organization) (2008) The State of Food Insecurity in the World 2008. High food prices and food security-threats and opportunities. FAO, Rome.

6. Oguge O (2012) Environment, Conservation and Livelihoods in Eastern Africa: challenges and opportunities. Oxfam Workshop on Livelihoods and Humanities in Eastern and Central Africa, 2nd April 2012, Naivasha, Kenya.

7. World Bank (2013) Annual Report on End extreme poverty and shared Prosperity by 2030. Washington DC, USA.

8. Kassie BT, Rotter RP, Hengsdijk H, Asseng S, Van Ittersum MK, et al. (2014) Climate variability and change in the Central Rift Valley of Ethiopia: challenges for rainfed crop production. Journal of Agricultural Science 152: 58-74.

9. Hadgu G, Tesfaye K, Mamo G (2015) Analysis of climate change in Northern Ethiopia: implications for agricultural production. Theoretical and Applied Climatology 121: 733-747.

10. JMP (2008) Global water supply and sanitation 2008 report. Joint Monitoring Programme WHO/UNICEF. Geneva: World Health Organization.

11. Yang W, Seager RA, Cane M (2014) The East African Long Rains in Observations and Models. Journal of Climate 27: 345-436.

22

The Generation of Mega Glacial Meltwater Floods and Their Geologic Impact

Paul A LaViolette*

The Starburst Foundation, 1176 Hedgewood Lane, Niskayuna, New York 12309, United States

Abstract

A mechanism is presented explaining how mega meltwater avalanches could be generated on the surface of a continental ice sheet. It is shown that during periods of excessive climatic warmth when the continental ice sheet surface was melting at an accelerated rate, self-amplifying, translating waves of glacial meltwater emerge as a distinct mechanism of meltwater transport. It is shown that such glacier waves would have been capable of attaining kinetic energies per kilometer of wave front equivalent to 12 million tons of TNT, to have achieved heights of 100 to 300 meters, and forward velocities as great as 900 km/hr. Glacier waves would not have been restricted to a particular locale, but could have been produced wherever continental ice sheets were present. Catastrophic floods produced by waves of such size and kinetic energy would be able to account for the character of the permafrost deposits found in Alaska and Siberia, flood features and numerous drumlin field formations seen in North America, and many of the lignite deposits found in Europe, Siberia, and North America. They also could account for how continental debris was transported thousands of kilometers into the mid North Atlantic to form Heinrich layers. It is proposed that such layers' form at times when a North Atlantic sea ice shelf borders the ice sheet and when climate warms abruptly producing accelerated meltwater discharge.

Keywords: Glacial meltwater floods; Glacier bursts; Permafrost deposits; Drumlins; Lignite deposits; Heinrich event; Supraglacial lakes

Introduction

Over a century ago Dana [1] in his Manual of Geology proposed that during periods of rapid glacial melting, the ice sheets may have released tremendous meltwater floods that reshaped the continental landscape. He stated (p. 553) that a flood "vast beyond conception" was the final event in the history of the North American ice sheet as indicated in the peculiar stratification of the flood deposits and in the dispersal of the stratified drift southward along the Mississippi Valley to the Gulf. He said that only under the rapid contribution of immense amounts of sand and gravel, and of water from such extensive a source could such deposits have been accumulated.

Some years later in 1893, Sir Howorth [2] voiced skepticism about Dana's idea. He wrote that if the climate was so cold as to accumulate enormous sheets of ice, he could not understand how the summer melting of this ice could be sufficiently rapid to produce such floods. He said that if floods were due to the melting of the ice, at the close of the glacial age, the change of climate involved must have been very sudden, or very rapid, much more sudden, and rapid than is consistent with any uniformitarian theory.

Howorth [2] makes a good point that meltwater would have been produced at an insufficient rate to result in catastrophic flooding. Even if the ice sheet surface were to have melted at a rate sufficient to raise sea level by ~4 cm per day, the resulting subglacial meltwater outflow from the glacier's edge would not have been of very great depth. Take as an example the Laurentide Ice Sheet, which measured about 3,200 km from centre to edge along a north to south transect. During MWP 1A, it is estimated to have discharged 14,000 km^3 of meltwater per year, or about 0.002 km^3 per day from each kilometer of its perimeter. If meltwater were flowing continuously from its periphery at a speed of 10 m/s, this discharge would only have been about two meters deep. Greater flow speeds would result in even lesser depths. Certainly, a continuous meltwater discharge mechanism is not sufficient to account for the morphology of the flood deposits seen in regions bordering the ice sheet.

As one solution, it has been customary to suggest that meltwater had become impounded in proglacial lakes retained behind an ice dam and that once breached the resulting flood would have been forceful enough to sculpt the surrounding land features. Such a mechanism, for example, was first proposed by Bretz in a series of papers he wrote between 1923 and 1932 to explain the formation of the channelled topographic features in the Scabland region of eastern Washington. This flood, variously named the Spokane Flood or Missoula Flood, was attributed to the sudden emptying of proglacial Lake Missoula that once covered western Montana; Baker VR [3] for a review. Also, Kehew A et al. [4] have suggested a domino effect mechanism in which proglacial Lake Regina, which once resided in southern Saskatchewan, may have suddenly emptied, and created a catastrophic flood that moved south-eastward to trigger the sequential emptying of a series of interconnected proglacial lakes in Manitoba, North Dakota, and Minnesota (Lake Souris, Lake Hind, and Lake Agassiz).

But the above mechanisms, deal with relatively local, "small-scale" catastrophic floods, and have difficulty in explaining the great extent of the flood deposits found all over the world in regions lying more than a thousand kilometers from the ice sheet margin. They also have difficulty accounting for the presence of high altitude flood deposits, for example, the Alaskan upland silts which are found on ridges situated 250 to 650 meters above the valley floor [5]. Due to the fact that proglacial lakes lie at relatively low elevation, near the base of the

***Corresponding author:** Paul A. LaViolette, The Starburst Foundation, 1176 Hedgewood Lane, Niskayuna, New York 12309, United States
E-mail: plaviolette@starburstfound.org

ice sheet, floods from such lakes would have been unable to mount to such heights.

Shaw et al. [6,7] have proposed another mechanism in which water stored in subglacial reservoirs could have been catastrophically released from beneath the edge of the ice sheet to form sheet-like floods that were able to form the extensive drumlin fields observed in North America and Europe. Drumlins are oval or chevron-shaped hills seen to cover large areas of land that once lay at the border of the North American ice sheet. They measure about 60 to 100 feet in height, 1200 to 1800 feet in width and can range in length from less than half a mile to several miles. One such drumlin field in central-western New York is estimated to contain as many as 10,000 drumlins. Earlier theorists had suggested that drumlins have been created by glacial surges in which advancing glaciers smeared a mixture of clay, gravel, and boulders into streamlined mounds. Shaw, however, points out that drumlins were most likely produced by forceful discharges of glacial meltwater hundreds of feet deep and estimates that meltwater shaping the Livinstone Lake drumlin field in northern Saskatchewan was formed by at least 84,000 km³ of meltwater discharging at rates as high as 60 million m³/s. He points out that a meltwater flood origin is more likely as drumlins have a shape similar to erosional forms produced by turbulent flows and resemble landforms normally ascribed to glaciofluvial action. He notes that the floodwaters forming these drumlin fields must have been at least as wide as the fields themselves, 20 to 150 km, and must have been sufficiently deep to submerge the drumlin tops, in some cases a minimum of hundreds of feet deep.

The present paper investigates a novel mechanism by which mega floods, comparable to large scale tsunamis, could have been produced during the last ice age. It is proposed that during periods of major climatic warming, glacial meltwater was able to accumulate on an ice sheet's surface forming a multiplicity of supraglacial lakes. A dam failure of one such perched lake near the ice sheet summit then could trigger a consecutive domino-effect discharge of other perched lakes to creates a self-amplifying traveling meltwater wave, or water avalanche, of enormous height, mass, velocity, and kinetic energy, and consequently having great destructive force. This has been termed a "continental glacial meltwater wave," or "glacier wave" for short (Figure 1) [8]. It is proposed that this phenomenon occurred during the ice age partly because of the vast extent of the ice sheets, which existed at that time and partly because supraglacial lakes would have been present in

great numbers during the dramatic climatic warmings that periodically occurred at that time.

Supraglacial Lakes Now and Then

It is known that glaciers have the ability to store a considerable amount of meltwater on their surface in the form of perched ponds or lakes called supraglacial lakes. Also, meltwater can be stored on their surface in the form of runoff that moves laterally over its surface. Satellite observations show that a substantial number of supraglacial lakes are present on the Greenland Ice Sheet particularly during the summer melting season. Fitzpatrick et al. [9] studied several sectors of the Russel Glacier on the southwestern edge of the Greenland Ice Sheet extending from the ice sheet margin to ~2000 m elevation. An area totalling 6500 km² was studied over 11 melt seasons from 2002 to 2012 and found to contain an average of 200 supraglacial lakes. The lakes had an average area of 0.68 km², but were observed to grow as large as 8 km². Depending on the particular year of observation, total meltwater volume for these supraglacial lakes was found to range from 0.09 km³ to 0.25 km³ with the largest volume having occurred during the unusually warm summer of July 2012. When averaged across the 6500 km² study area, the 2012 meltwater volume would amount to an average depth of 3.8 cm.

Fitzpatrick et al. [9] found that supraglacial lakes last an average of 10 days before rapidly draining due to the formation of hydraulic fractures, which form moulins that channel the water to the ice sheet bed. However, Koenig et al. [10] report that some lakes survive through the winter season as buried lakes located on average about 2 meters below the surface. They found that these could re-emerge as supraglacial lakes during the following melt season or in other cases can remain buried for multiple seasons before re-emerging.

A large amount of meltwater can also exist on the surface of the ice sheet in the form of supraglacial streams and rivers. Smith et al. [11] studied a 5,328 km² area on the Greenland Ice Sheet and logged 5,928 km of large streams forming 523 supraglacial stream networks. They found that these rivers always terminated in moulins where the water was rapidly conveyed to the bottom of the ice sheet. In some cases, supraglacial streams were observed to pass through supraglacial lakes. They found that during the July 2012 summer, the total supraglacial storage, including supraglacial lakes, streams, and rivers, amounted to about 0.19 km³, equivalent to an average depth of 3.6 ± 0.9 cm, comparable to the findings of Fitzpatrick et al. [9].

Fitzpatrick et al. [9] found that in warmer years' supraglacial lakes were present at higher ice sheet elevations. In particular, Tedesco et al. [12] report that over the period of July 11-13, 2012 about 97% of the Greenland Ice Sheet underwent surface melting and that the melt extent exceeded 60% over a 10-day period. In previous summer seasons, glacial lakes had covered about 40% of the ice sheet. Surface temperature during this 2012 melt period was observed to be -5.5° C, the warmest in four decades. This may be compared with the period 2002-2009 when surface temperature averaged -7.5° C. Hence current observations show that there is a direct correlation between temperature at the ice sheet surface and the prevalence of supraglacial lakes and surface melting.

Studies of the ice record shows that, during the last ice age, periods of cold climate were abruptly interrupted by numerous warming events termed Dansgaard/ Oeschger (D/O) events. Also, the Earth's coral reef records show that during some of these warmings the ice sheets were melting far more rapidly than rates seen today in Greenland and

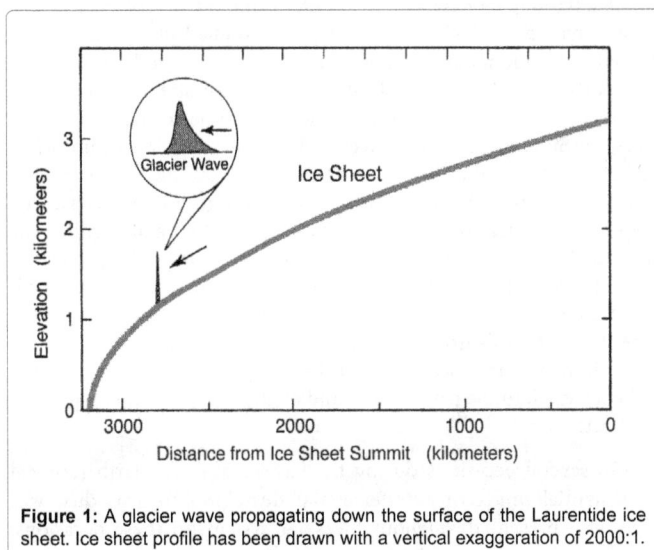

Figure 1: A glacier wave propagating down the surface of the Laurentide ice sheet. Ice sheet profile has been drawn with a vertical exaggeration of 2000:1.

Antarctica. One significant period was during the Alleröd/Bölling interstadial when the formerly glacial climate had warmed up to interglacial temperatures. Another period of interest is the abrupt warming event, which occurred at the end of the Younger Dryas (YD) that terminated the ice age. Borings of the Barbados coral reef, which have been dated by AMS ^{14}C and U-Th reveal that during both of these climatic warmings sea level was rising at exceedingly high rates due to the influx of glacial meltwater [13,14]. These two intervals of rapid meltwater influx were termed meltwater pulse 1a (MWP 1A) and MWP 1B. Liu et al. [14] date these as spanning 14.2 to 13.8 k cal yrs BP and 11.5-11.1 k cal yrs BP respectively. During MWP 1A sea level is estimated to have risen at a rate of at least 4 cm per year [14,15]. Also, the data of Clark et al. [16] from coral reefs of Barbados and the Sunanda Shelf show an increase of 18 m over the period from 14.5-14.2 k cal yrs BP, implying a rate of sea level rise of 6 cm per year, although their dates for MWP 1A fall about 300 years earlier. These indicate that sea level was rising about 12 to 18 times faster than the current average rate of 3.4 mm per year as reported by Nerem et al. [17]. Actually, the rates estimated for MWP 1A represent averages for these periods, the data being also consistent with sea level having risen even faster at specific times within these intervals. For MWP 1B event, Emiliani et al. [18] has reported a sea level rise of several meters per decade occurring around 11.5 k cal yr BP and has noted that this date coincides with Plato's flood date. Emiliani's [18] findings imply a sea level rise of about 20 cm per year, or about 60 times that of the current average. Although, Abdul et al. [19] report a lower rate of about 4 cm per year for this period, 11.45-11.1 k cal yr BP. Also, other meltwater pulses have been found which predate MWP 1A. Yokoyama et al. [20] report finding an abrupt rise in sea level occurring around 19 k cal yrs BP in the Bonaparte Gulf of Australia.

Given that the North American ice sheet at certain times during the last ice age was undergoing a rate of melting far greater than is seen today in Greenland, it is reasonable to conclude that supraglacial lakes were at those times far more prevalent than what was observed in Greenland during the 2012 melt season. Since ice melting would have primarily occurred at the ice sheet surface, it is reasonable to expect that there was a 12 to 60-fold greater prevalence of supraglacial lakes during these periods of rapid sea level rise, as well as a comparable increase in the prevalence of surface meltwater runoff. Extrapolating from the average meltwater depths of 3.6-3.8 cm estimated for the Greenland Ice Sheet in 2012, this implies that the North American ice sheet during those times could have had an average equivalent distributed meltwater depth ranging from 0.5 to 2 meters. With this excessive output of meltwater, moulins that normally channel surface water to the ice sheet bedrock may have been filled to capacity and unable to convey meltwater to the bedrock surface as readily as those studied today. This would have resulted in an increased amount of water being impounded on the ice sheet surface. Also, one should consider that the summit of the North American ice sheet was then located around where the Hudson Bay is currently situated (lat. 53° N). Hence, it would have lain 14° further south from the latitude of the Russel Glacier region of Greenland (67° N). Even though a large portion of the ice sheet surface would have been above 3000 meters' elevation, the North American ice sheet would have been receiving a greater amount of insolation facilitating its melting during interstadial periods.

With this supraglacial environment prevailing during episodes of extreme climatic warming, it is likely that on occasion one of the supraglacial lakes whose water was retained behind an ice jam could have melted sufficiently to cause a dam breach resulting in the sudden release of its water. Upon descending the ice sheet, such an outburst could trigger the successive emptying of supraglacial lakes resulting in

a progressive growth of the flood wave according to the mechanism outlined in the next section. Meltwater outbursts from perched reservoirs are occasionally observed in modern times and have been variously termed glacier bursts or glacier floods [21,22]. Grimsvötn, Iceland is one such location that is prone to recurrent glacier bursts. Its floods issue from a large subglacial lake of area 35 to 40 km[2] and depth of about 200 meters perched at a height of ~1500 meters above sea level. The lake is about three-fourths of the way to the summit of the Vatnajokull glacier at latitude 64.4° N. Geothermal heat from an underlying active volcano causes ice to melt over a 300 km[2] region and feed this perched lake. When the stored meltwater capacity becomes excessive, progressive failure of the lake's ice walls causes the catastrophic release of meltwater, during which about 7-8 km[3] of glacial meltwater become discharged in the space of just a few days. Ward [22] notes that the exponential increase in discharge rate may be explained by a feedback process in which increased flood flow rate through the water transmitting channels progressively enlarges the channels which in turn progressively increases the flow rate.

Another example of a proglacial lake that periodically produces glacier bursts is the ice dammed Lake Merzbacher located on the Inylchek glacier in the Tien Shan Mountains of east Kyrgyzstan (42° N). The lake does not drain through most of the year, but when a hole finally melts through its dam, the lake drains within three days producing a discharge of up to 1000 m[3]/s that creates major flooding downstream [23,24].

Although various mechanisms have been suggested for the cause of the sudden climatic warmings that occurred during the ice age, there is considerable evidence that these warmings had a solar cause. Several studies find there to be a link between solar variability and major variations of terrestrial ice age climate as in the production of D/O events and other climatic oscillations [25-33].

Furthermore, there is evidence that the Sun was far more active during the last ice age than it is today. A study of solar flare tracks etched in lunar rock micrometeorite craters indicates that around 16 k ago the average solar cosmic ray intensity was 50 times higher than at present, declining to 15 times higher by 12 kyrs BP, and eventually reaching its present activity level early in the Holocene [34]. Also, according to Zook [35] elevated radiocarbon levels found in the surfaces of lunar rocks indicate that between 17-12 k yrs BP the Moon was being exposed to a solar cosmic ray flux that averaged 30 times higher than its present flux [34]. He concluded that this unusually high solar activity, unlike anything we have seen in modern times, played a critical role in terminating the last ice age. Gold [36] studied glassy patches on the bottoms of lunar craters and found their surface to be covered by dust particles that had been heated to a molten state by an episode of intense heat. He concluded that, at some time in the past 30,000 years, the Sun must have produced a very large solar flare or nova-like outburst that caused its luminosity to increase 100-fold for a period of 10 to 100 seconds. Another theory is that this glazing was caused by one or more super-sized coronal mass ejections that had engulfed the Earth and Moon [8,37]. Of course, the Earth has far more protection from charged particles and UV radiation than does the Moon. But if the ozone layer were to be partially destroyed or the geomagnetic field were to significantly drop in intensity due to the occurrence of a main phase decrease during a super solar flare, this energetic radiation could make its way into the troposphere and contribute to the melting of the ice sheets.

On several occasions during the last ice age, the Earth received a substantial influx of extraterrestrial dust [38-40]. This dust was likely also present throughout the solar system on these occasions.

Consequently, it is likely that substantial quantities were accreted by the Sun and to have aggravated the Sun into an extremely active state of flaring. Related to this, Hoyle et al. [41] demonstrated that the Sun's passage through an interstellar dust cloud could have contributed material at a sufficient rate to have increased its luminosity several percent. Dust shrouded stars having a mass similar to the Sun called T Tauri stars are observed to exhibit extremely high flaring activity, excessive luminosity, and infrared excesses. So, it is possible that, during the last ice age, the Sun could have been aggravated into a Tauri-like state which could explain the lunar rock findings of Zook et al. [35,36].

The Sun's luminosity is known to rise during periods of active flaring with the extra output being primarily in the UV and X-ray spectral regions. Bruns et al. [42] have noted that solar UV radiation increased by 25-30% during solar flares occurring in 1967. Since UV wavelengths normally constitute about 10% of the Sun's luminosity, this implies a bolometric solar luminosity increase of 2.5-3% during flare events. Destruction of the ozone layer by the high levels of solar cosmic ray radiation occurring during the time of an extreme solar storm would have allowed this radiation to penetrate the stratosphere and exacerbate the heating of the troposphere particularly in high latitude regions. An active Sun, then, could explain episodes of abrupt climatic warmings and accompanying rapid ice sheet melting that occurred during the last ice age.

There is evidence that MWP 1B was produced by extreme solar events. For example, the GISP2 ice record registers an acidity spike at the end of the YD at a depth of 1678.3 meters (11,571 cal yrs BP), the largest such spike to occur during the entire YD [43]. It coincided with a one-year long cooling, registered as a -4 per mille change in oxygen isotope ratio, and was immediately followed by a 14-year long warming. This was followed one Hale solar cycle later (11,549 cal yrs BP) by an almost equally large acidity spike, which coincided with an abrupt climatic cooling followed by a decade-long warming that ended the YD and heralded the beginning of the Holocene. This terminal event also appears to coincide with a small ^{10}Be deposition rate peak. Also, Mörner [44] has observed that around 11,550 years BP, the Earth's magnetic field intensity increased by more than fivefold. Further, he reports that between 14,500 and 11,550 cal yrs BP the intensity and declination of the Earth's magnetic field underwent major cyclic variations in step with the eleven-year sunspot cycle having an amplitude hundreds of times larger than modern geomagnetic solar cycles. So, the geomagnetic record supports the suggestion that the Sun was at that time perhaps more than an order of magnitude more active than it is today.

MWP 1A (14.2-13.8 k cal yrs BP) may also have been associated with a period of elevated solar activity. For example, studies of the varved record from the Cariaco Basin shows that a large spurt in ^{14}C production occurred around 14,050 cal yrs BP. Also, Ising [45] has reported a large westerly magnetic declination swing occurring between 14.2 and 13.8 k cal yrs BP in southern Sweden.

The Generation of a Glacier Meltwater Wave

Consider now a period during the ice age when climate abruptly warmed to such an extent that the upper surface of the ice sheet was melting at a very fast rate causing supraglacial lakes, rivers, and streams to form in great abundance, over an order of magnitude greater than the prevalence seen today in Greenland. Suppose then that the ice sheet surface was covered with meltwater to such an extent that if dispersed evenly over the ice sheet surface it would have had an average depth of half a meter. This would have produced conditions favoring the generation of a glacier wave.

A glacier wave would initially have been seeded by a glacier burst similar to those we see taking place today at Grimsvötn and at other glacier locations. Suppose that an ice dam retaining a perched meltwater lake located near the center of the Laurentide Ice Sheet had failed resulting in the catastrophic release of 1 km³ of meltwater. Suppose that this produced a flood wave having a frontal length of 1 kilometers, a height of 10 meters, and a forward velocity of 10 m/s. A one-meter length of this wave would have a volume per meter of $v=10^6$ m³, a mass of 10^9 kg, and a kinetic energy of 5×10^{10} joules.

Whether this glacier wave would continue to grow in size as it travelled down the ice sheet slope, or whether it would become dissipated by the pools of water it encountered, would depend on whether its continued forward motion resulted in either a net energy gain or net energy loss for the wave. The kinetic energy required to accelerate the ambient meltwater medium that the wave encounters to its own forward velocity must necessarily come from the wave itself. Moreover, the wave, in turn, acquires its kinetic energy from its store of potential energy, which is gradually released as it travels forward down the glacier slope. If the energy demand on the wave to accelerate the encountered medium were to become so great that this rate of energy loss exceeded the wave's rate of energy gain from the kinetic energy acquired in the course of its descent down the glacier surface, the wave's velocity relative to the ambient medium would tend to diminish, and eventually the wave would dissipate. On the other hand, if the rate of kinetic energy acquisition were to exceed the rate of kinetic energy consumption, the net kinetic energy of the wave would increase and the wave would tend to accelerate.

Consider a 1 meter wide increment of a 10^9 kg glacier burst wave such as that proposed above, having a height of 10 meters and forward velocity of v=10 m/s. As it travels forward over an ice sheet having water ponded at an average depth of 0.5 meters, it would encounter meltwater at the rate of

$$\frac{dm}{dt} = 1\,m \times 0.5\,m \times 10\,m/s \times 1000\,kg/m^3 = 5,000\,kg/s \quad (1)$$

To bring this encountered water up to its own forward velocity, the wave would need to be supplying kinetic energy at the rate of

$$\frac{dE_{out}}{dt} = \frac{1}{2}\frac{dm}{dt}v^2 = \frac{1}{2}(5,000\,kg/s) \times (10\,m/s)^2 = 2.5 \times 10^5\,joules/s \quad (2)$$

Let us assume that the ice sheet has a parabolic profile having a maximum elevation of 3200 meters and north-south distance extent of 3200 kilometers, as shown in Figure 1. Suppose that its elevation (y) varies as a function of its north-south distance (x) as:

$$y = \frac{(3200-x)^{-0.5}}{17.5} \quad (3)$$

Hence the gradient κ of its surface contour on average would vary as: $\kappa = \dfrac{dy}{dx} = \dfrac{1}{35(3200-x)^{-0.5}}$ (4)

Consequently, as this one-meter wave increment advanced down the glacier slope, it would drop in altitude at the rate of:

$$\frac{dh}{dt} = \kappa v = \frac{0.29}{(3200-x)^{-0.5}}\,m/s \quad (5)$$

and would be converting potential energy into kinetic energy at the rate of:

$$\frac{dE_{in}}{dt} = \varepsilon\,m\,g\,\frac{dh}{dt} = 0.9 \times 10^9\,kg \times 9.8\,m/s^2 \times \frac{0.29}{(3200-x)^{-0.5}}\,m/s = \frac{2.52 \times 10^9}{(3200-x)^{-0.5}}\,joules/s \quad (6)$$

where ε is the efficiency with which the advancing wave transfers energy to the ambient ponded meltwater to accelerate it to the wave's forward velocity. Here we take ε to be 90%, which supposes that the majority of the consumed kinetic energy goes into accelerating this water in the forward direction, the remainder being, converted into turbulence and frictional losses. Parameter g is the gravitational constant 9.81 m/s^2.

In this particular case for x=100, we find that $\frac{dE_{in}}{dt} >> \frac{dE_{out}}{dt}$. Hence, the wave will accelerate and its store of forward kinetic energy will increase. The wave is predicted to accelerate even if the conversion efficiency is assumed to be almost an order of magnitude smaller.

As the glacier wave descends the slope of the ice sheet, it will grow in size and mass, and hence be able to release potential energy at a greater rate as it advances. If the wave were to sweep up all of the meltwater encountered in its path in the course of its 3200-km journey, and if this meltwater were formerly stored on the surface of the ice sheet at an average depth of 0.5 meters, then by the time the glacier wave had reached the edge of the ice sheet, every meter along the wavefront would hold a volume of 1.6×10^{-3} km^3; i.e., 1 m × 0.5 m × 3.2 × 10^6 m. This might express in the form of a wave having a height of ~ 100 meters and overall length of ~ 16 kilometers in its travel direction. A one meter wide increment of this wave along its front would have accumulated a mass of M=1.6 × 10^9 kg by the time it reached the bottom of the ice sheet.

If we ignore drag due to air friction and air turbulence, a glacier wave originating from the ice sheet summit (h=3200 m) would theoretically attain a velocity of v ~250 m/s (~900 km/hr) by the time it reached the edge of the ice sheet, where

$$v = (2gh)^{1/2}. \tag{7}$$

Every meter along the front of the wave then would be accelerating forward with a force of 4.5×10^8 Newtons; i.e., $F = mg\kappa$.

A wave traveling at 250 m/s (Mach 0.73), and having a height D=100 meters, will have a Reynold's number Re=vD/v~2 × 10^9 indicating turbulent flow, where v=1.17 × 10^{-5} m^2/s is the air kinematic viscosity at 0°C. The equation for the air drag on the wave is given as:

$$F_d = 1/2 \rho \, v^2 \, C_D A \tag{8}$$

where ρ=1.29 kg/m^3 is the density of air at 0°C, v is the wave velocity, C_D is the drag coefficient, and A is the area of the wave front for an incremental section of 1 meter along the front of the wave transverse to the direction of its motion. We assume that the wave has a drag coefficient of 0.016, similar to that for wind stress acting on the ocean [46]. A 100-meter-high wave that rises to its maximum height over 8 kilometers will present a surface area of 8,000 m^2 per meter of wave front. Hence, Equation (8) predicts an air drag force of 5.2 million Newtons, which is about 1% of the gravitational force acting to accelerate the wave forward. So, air drag will not substantially reduce the wave's velocity. Earlier in its journey when the wave's velocity is lower, this retarding force would be even lower.

Upon reaching the edge of the ice sheet, traveling at the speed of 250 m/s, the wave would be transporting meltwater at the rate of 400 km^3/s per kilometer of wave front. For a wave of mass m=1.6 × 10^{12} kg per kilometer frontage and v=250 m/s, each kilometer length of the wave front would be delivering a sum total kinetic energy of about 5 × 10^{16} joules, equivalent to about 12 million tons of TNT, or about 800 Hiroshima A-bombs. Compared with the initial glacier burst that seeded the glacier wave near the crest of the ice sheet, upon reaching the edge of the ice sheet, this glacier wave, now greatly grown in size, would have a kinetic energy per meter of wave front that was 1,000-fold greater.

The example discussed here, of course, is only a rough estimate made for the purpose of illustration. However, a more precise analysis would still render essentially the same results. Namely, in a nonequilibrium, low-friction regime such as would exist on the sloping surface of a rapidly melting continental glacier, if a meltwater seed wave of sufficient size were generated somewhere on the ice sheet surface as a result of an ice dam breech, this wave would spontaneously grow in size through a "domino effect." Beyond a critical threshold of wave size, a growing glacier wave would become the dominant mechanism of meltwater transport from the ice sheet since a glacier wave would be capable of dissipating potential energy at a greater rate as compared with homogeneous meltwater flow. Therefore, thermodynamic considerations support the theory that such waves would have existed during the ice age period.

Glacier wave growth would take place not only along the direction of travel, but along the length of the wave front as well. A dam failure beginning as a localized event near the summit of the ice sheet could fan out to form an extended wave front, measuring hundreds, perhaps even a thousand kilometers in length, by the time it had reached the ice sheet border. Upon encountering proglacial lakes at the foot of the ice sheet, a glacier wave of the magnitude estimated above would have enough force to jettison the entire lake over its retaining moraines or ice dam. For example, Proglacial Lake Missoula in northern Montana is estimated to have contained about 2000 km^3 of meltwater [47]. By comparison, a 100 km long section of a glacier wave of the size proposed above coming from the Cordilleran Ice Sheet would have contained about 160 km^3 of meltwater (along with chunks of unmelted ice). If a wave of this size were to accelerate the entire lake to its own bulk velocity, the wave would suffer a 3.7-fold drop in velocity to about 70 meters per second. Thus, the flood features of the Channeled Scabland could just as well be attributed to the action of a major glacier wave or a series of such waves causing rapid discharge of Lake Missoula. Alternatively, a glacier wave by itself may have been sufficient to cause such features.

Upon reaching the edge of the ice sheet, a glacier wave consisting of an avalanche of water, unmelted ice chunks, and other debris traveling forward at a speed of several hundred meters per second, would have reduced its forward velocity, shortened its length, and grown in height much like a tsunami does when it reaches land. Perhaps reaching a height of 300 meters and having a frontal length of a thousand kilometers, such a wave could have traversed thousands of kilometers over a continental land mass before becoming dissipated. Upon entering the ocean, such a glacier wave would have caused a considerable volume displacement of the water mass, resulting in a mega tsunami that could have crossed thousands of kilometers of ocean and still had sufficient energy to produce considerable erosion on remote shores.

To demonstrate that avalanches could produce waves of the proposed height, one may cite as an example the Lituya Bay, Alaska mega tsunami [48,49]. This occurred on July 9, 1958 as a result of a rockslide triggered by a 7.7 magnitude earthquake. Approximately 40 million m^3 of rock debris fell from an elevation of 914 meters into the head of the bay displacing its water. The mega tsunami so produced ran up to a height of 524 meters on the opposite side of the bay stripping its trees and eroding the soil down to its bedrock. Some estimate that the wave achieved an average velocity of 240 miles per hour and a height of 30 to 60 meters as it traveled out into the bay. If we assume that the avalanche fell from an average height of 700 meters, equation (6) given above would predict that, by the time it reached sea level, it would have been moving forward at 260 miles per hour which is close to the value given above for the wave velocity in the bay. By this example from Alaska, the suggestion that a wave of far greater kinetic energy launched from a fourfold greater height on the ice sheet surface would have achieved a height of 300 meters and forward velocity of 550 miles per hour does not sound that farfetched. The proposed glacier wave could have had a run-up of almost 1500 meters, able to surmount mountains of moderate height.

Looking at the Barbados coral record and considering other periods during the deglaciation period, other than the MWP 1A and MWP 1B spurts, it is apparent that the ice sheets were still melting far faster than today. For example, between 13-14 k cal yrs BP sea level was rising at the rate of 1 cm per year, or over 3 times faster than the current sea level rise. If we assume that during this period the ice sheet surface had an average meltwater depth over three times more than was observed in 2012 for the Greenland Ice Sheet, this would then amount to 10 cm instead of the 50 cm that we assumed in the calculations above. At this reduced average meltwater depth, a glacier wave would still have been able to grow in size as it progressed down the ice sheet. Only, its kinetic energy would have been about 5 times less than what was estimated above for an average meltwater depth of 50 cm.

Glacier waves could also have formed during the cold Younger Dryas period. The "quarter bag" $\delta^{18}O$ profile obtained from the GRIP ice core record (Figure 2) shows that as many as 20 brief warming episodes occurred during the Younger Dryas in which temperatures approached Holocene temperatures. Meltwater pulses produced during these warm periods each lasting from several years to a few decades would have passed unnoticed in the coral reef sea level record due to not having sufficient resolution. Once the warming event terminated, snow would have accumulated causing sea level to fall once again. During one such climatic warming which occurred in the early Younger Dryas (see arrow in Figure 2), temperatures rose to their maximum warmth within the span of a few years. This event occurred immediately after a dramatic rise in atmospheric ^{14}C concentration that dates around 12,837 years BP in the Cariaco Basin radiocarbon record. It was associated with spikes in ^{10}Be concentration, nitrate ion concentration, and ice acidity as seen in the GISP 2 Greenland ice core record. All of these suggest that the Earth at that time had been impacted by a super solar proton event (SPE) [43].

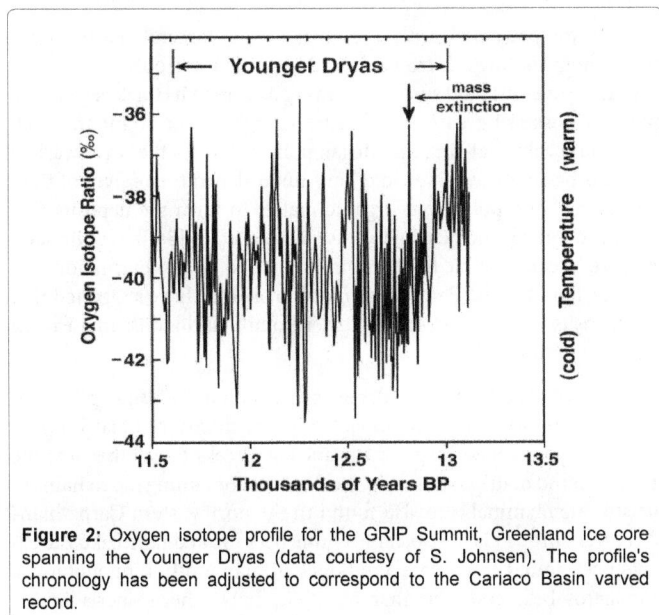

Figure 2: Oxygen isotope profile for the GRIP Summit, Greenland ice core spanning the Younger Dryas (data courtesy of S. Johnsen). The profile's chronology has been adjusted to correspond to the Cariaco Basin varved record.

In summary, it is suggested here that many of the loam-like drift deposits observed in many areas of the world that were once periglacial regions during the Wisconsin Ice Age are flood deposits produced by a series of such glacier waves issuing from the continental ice sheets. If a major climatic amelioration simultaneously affected all of the Earth's ice sheets, these glacier waves from different ice sheet locations would likely have been released all at about the same time.

Explaining the Permafrost Silt Deposits of the Arctic Regions

Almost all areas of Alaska that lie below an altitude of 300-450 meters are covered with a loamy blanket of perennially frozen silt that ranges in thickness from a few millimeters to over 60 meters [50]. For at least 100 years there has been a controversy as to the origin of this material, referred to locally as "muck." Explanations in the geological literature have mostly included gradually acting processes, such as fluvial, lacustrine, marine, and eolian deposition mechanisms and weathering.

Several Alaskan silt formations have been distinguished, dating as far back as the Illinoian glaciation. Of these formations, the Goldstream Formation is of particular interest. This is one of the most widespread formations in central Alaska and is believed to date from the latter portion of the Wisconsin Ice Age [51,52]. It is a valley-bottom accumulation which is present in almost all creek and small river valleys in central Alaska. The Goldstream Formation is regarded as the greatest depository of Pleistocene vertebrate remains in Alaska if not in North America [51]. Most of these fossils are found in valley bottoms and the greatest concentrations occur where small tributaries join large creeks [50].

Péwé [5] reports that in the Fairbanks area (as one example) the silt ranges from 0.5 to 25 meters in thickness on the tops of low hills located 15-45 meters above the valley floor and thins out to a thickness of about a meter on ridges situated 250-600 meters above the valley floor. Clearly, normal fluvial action, such as that proposed by some theoreticians to explain the valley bottom deposits, has considerable difficulty accounting for deposits perched at such altitudes. Again, many theories have been put forth to account for the origin of the upland silt, ranging from an estuarian inundation of Central Alaska to high velocity winds as the transporting agent. Péwé [5] reviews a number of these theories and concludes that an eolian mechanism best accounts for the upland silt. However, if wind were the transporting agent, then one might ask why should such "wind-blown" material often be found to overlie auriferous gravel deposits. For example, Péwé [5] reports that on top of Gold Hill at an altitude of about 220 meters (80 meters above the flood plain) a bore hole revealed the presence of 50 meters of dry upland silt overlying auriferous gravels. The eolian hypothesis must attribute this to chance superposition, the deposition of wind-blown loess over a pre-existing auriferous gravel substrate. The glacier wave flood hypothesis, though, is able to not only account for the origin of the silt and the gravel, but also for their stratigraphic placement relative to one another.

Besides the unique stratigraphic placement of the placer gold, these upland silt deposits also bear a resemblance to the valley bottom deposits in that the silt is often found to overlie the remains of extinct megafauna; e.g., see Taber S [53] or Péwé TL [5]. Péwé [5,50,51] attempts to account for the upland silts in terms of eolian transport and for the valley bottom deposits (e.g., the Goldstream Formation) in terms of fluvial action. But such completely different transport mechanisms fail to explain why these high and low elevation deposits are so strikingly similar. A glacier meltwater wave, on the other hand, would be able to account for the formation of both the valley bottom deposits and the upland silts, and explain their similarity.

Of the various other proposals accounting for the origin of the Alaskan upland silts it is interesting to consider the lacustrine hypothesis as put forth by Eakin [54,55]. Eakin [55] states that the "mechanical analysis of the silt seems to indicate more strongly an aqueous rather than an eolian origin." He cites several geomorphic features to support his argument. For example, he notes that the presence of plains standing at different elevations connected by steep scarps running for several miles in length could have been produced only by aqueous erosion.

He notes that some agency must have eliminated the old base-level of erosion and established a new base-level at a much greater elevation, nearly 300 meters above the present level of the Yukon at Ruby. He proposed that this base-level change involved extensive inundation of the old land surface, a fact indicated by the character, structure, distribution, and topographic expression of the unconsolidated deposits and of the bedrock beneath them. He suggested that a 150,000-250,000 km² region in central Alaska had temporarily impounded glacial meltwater to a height of 380 meters, drainage to the sea being blocked by an ice dam.

Péwé [5] offers several convincing criticisms of Eakin's hypothesis, such as the lack of evidence for the required drainage barrier, the lack of lacustrine varves, clay, or aquatic life in the silt deposits, and the finding that the silts do not have a definite upper limit in their altitude. However, while Péwé's criticisms correctly rule out the lacustrine scenario, they leave untouched Eakin's conclusions that the silts are of aquatic origin. In fact, many of Eakin's suggestions could be rescued from these criticisms if the glacial meltwater inundation of central Alaska was short-lived, involving a translating wave of water. It is interesting to note that the divide between the Laurentide and Cordilleran ice sheets created a natural water course which emptied northward onto the central Alaskan plains. This geographic feature could have served to channel glacial meltwater to this region whether the meltwater originated from the ice sheet surface itself in the form of a glacier wave, or whether it involved (more conventionally) the sudden release of water impounded behind an ice jam such as may have occurred in the ice sheet divide.

The same depositional agent forming the Alaskan muck appears to also have been at work in Northern Siberia forming the so called "mammoth horizon" or "tundra horizon." Péwé et al. [52] call attention to the striking resemblance between the perennially frozen silt deposits in Central Alaska and those found in central Yakutia. They note that both areas consist of unglaciated rolling lowlands and river terraces surrounded by high mountains that were extensively glaciated in Pleistocene times. Thus, they suggest that a similar depositional history for these geographically widely separated regions would not be surprising. It is worth noting that if glacier waves were the depositional agents in these two subarctic regions and were similarly generated from bordering glaciers, one would expect to find a similarity in the deposits found in these two regions, in agreement with the observation.

It is here proposed instead that these flooding episodes manifested as a sequence of large scale, high kinetic energy meltwater waves created during intervals of major climatic warmings. Most regions of Northern Siberia and Alaska were within 2000 km of the edge of the Laurentide Ice Sheet (Siberia by a polar route). In addition, Northern Siberia also had alpine-type glaciers. Thus, in the event of a sudden

climatic amelioration, it is easy to see how flooding could take place on a widespread scale throughout these arctic regions. It could also explain flood deposits found in North America in regions that lay south of the southern edge of the ice sheet as well as features found in regions that once lay south of the European ice sheet.

High kinetic energy, glacier waves could also simultaneously account for many of the puzzling features of the megafaunal deposits found in these regions. The meltwater not only would have drowned, fragmented, and buried the animals without warning, but it also would have tended to preserve their remains from decay. The icy contents of a glacier wave would have been close to 0°C and would probably have been composed of a slushy mixture of cold meltwater, ice chunks, and detritus which together would have helped to refrigerate the carcasses. With the onset of the Holocene, climate changed from boreal to arctic as winter snow was deposited over these near freezing deposits. The change of albedo would have locked this region into a permanently frozen state.

The Formation of lignite deposits

Glacier waves could also account for the accumulation of large quantities of carbonized peat found buried in valleys in various parts of Europe, Canada, and northern United States. As one example, many such lignite deposits are found in both central Peloponnesus and northern Greece. The deposit found in the Peloponnesian town of Megalopolis is believed to be of Pleistocene age. It is situated in a basin that is elongated in a north-south direction and encircled by mountains that rise to elevations of 1250 to 1500 meters. The basin floor itself lies about 400 meters above sea level and at its centre holds an extensive deposit of carbonized peat measuring some 30 square kilometers and in places reaches a depth of as much as 70 meters. Upon close examination, this material is found to consist of a conglomeration of wood pulp, tiny bits of organic matter, and wood chips ranging in size from a few millimetres to several centimetres, all turned dark brown in colour. The condition of the material gives the impression that the wood and vegetable matter had been violently shredded by flood waters and interred while still in a fresh state. Moreover, the stratified appearance of the deposit, with its interleaved horizontal zones of clay and gravel, indicates that this material had been deposited catastrophically, as if having precipitated from the transient presence of flood-waters.

We are presented with the image of a forceful glacier wave proceeding southward from the edge of the European ice sheet advancing over the land surface, scouring it clean of its thick vegetation, and still possessing sufficient kinetic energy to surmount the high mountain peaks that ring the Megalopolis valley. As the wave washed over this basin, its waters would have unburdened themselves of their load of clay and pulverized organic matter to form the deposits that today are mined for their fuel value. Begossi et al. [56] have similarly proposed a catastrophic flood origin for the Rio Bonito Formation coal deposits found in the Paraná Basin of Brazil. They have suggested that the deposits were laid down by a glacial outburst flood issuing from a melting mountain glacier.

At Megalopolis, lignite deposits are found in some places to overlie the bones of mammoths [57,58]. The discovery of such bones in antiquity may have inspired the ancient Greeks to call this area the "Battleground of the Giants." Lignite deposits containing the remains of Pleistocene mammals are also found in the northeastern Carpathians of central Europe [59]. Also, large lignite beds located near Lake Zurich have been found to contain the bones of mammoth, hippopotamus, rhinoceros, bear, and other mammals [60]. In Northern Siberia lignite, has been found embedded together with mammoth bones [61].

In the case of the Megalopolis lignite deposit, common assumption is that these deposits were formed gradually over a period of about a million years and that prevailing bog conditions preserved fallen trees from decay. The brown color of the lignite is typically attributed to sulphurous fumes that had acted on the deposits while they were interred. However, this gradualistic theory fails to account for the character of the wood which has been shredded presumably by the action of violent forces. Also, it doesn't explain the stratified nature of the deposits which present the hallmarks of catastrophic deposition by the action of flood waters. The proposal that glacier waves are often generated during periods of high solar activity could explain the dark color of the lignite. A super solar proton event intense enough to eliminate the protection of the geomagnetic field would have allowed solar cosmic ray particles and their shower of secondary electrons to pass to the Earth's surface where they could have carbonized trees while still standing. The glacier wave arriving shortly afterward would have shredded and transported this material leaving it to settle out in valley areas.

The origin of Heinrich events

Heinrich [62] reported discovering that North Atlantic sediments contain layers composed primarily of rock grains of continental origin with unusually low concentrations of foraminifera plankton. Studies of these so called "Heinrich layers," or "Heinrich events" indicate that continental bedrock debris was somehow transported for distances of up to 3000 kilometers before being deposited at these mid ocean locations. These layers appear only in ice age sediments and recur at intervals of 5000 to 10,000 years. They occur at the endings of ice ages as well as within glacial periods where they herald the onset of D/O events. These layers are found to be most prominent at the mouths of the Gulf of St. Lawrence, the Hudson Bay, and the Denmark Strait as well as a few other locations, which are generally recognized as evidence that this material was somehow being transported by outflow from the continental ice sheet. A 90,000-year record of Heinrich event debris layers found in core V23-81 is shown in Figure 3. Also, Figure 4 shows the foraminifera climate profile for core V23-81 published by Broecker et al. [63] with the positions of Heinrich events H1 and H0 in this core superimposed for comparison.

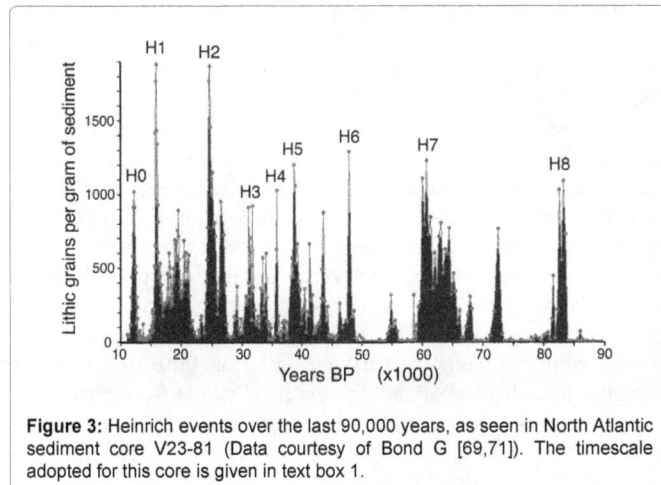

Figure 4: The positions of Heinrich events compared to climatic boundaries. Red curve plots the lithic abundance data provided courtesy of G. Bond showing Heinrich events H0 and H1. Black curve plots the foraminifera ratio climate profile for core V23-81 (from Broecker et al. [63]). Radiocarbon dates are indicated by arrows.

Box 1
Timescale Adopted for Core V23-81

A timescale for dating the Heinrich events in core V23-81 is developed by correlating the climatic boundaries of this profile with corresponding climatic boundaries dated using the Cariaco chronology (Table 1). Dates for intervening depths have been interpolated. For depths, greater than 2.1 meters, the chronology is based on radiocarbon dates for core V23-81 published by Vogelsang et al. [64] and Sarnthein, which have been converted to calendar dates by using the radiocarbon conversions from INTCAL 09. The adopted time-depth model, displayed in Figure 5, gives a date of around 15.8 kyrs BP for the peak of event H1. Other researchers (e.g., Broecker et al. [65,66]) have dated the H1 event as peaking 16.8 to 17 kyrs BP, hence over one thousand years earlier. The radiocarbon dates of Vogelsang et al. [64] in the depth range 2.1-2.4 m, which lie off the trend line, are here judged to be approximately 1200 years too old due to the influx of old carbon during the H1 meltwater discharge event. Similarly, radiocarbon dates in the depth range of 3.8-4.0 m are judged to be 1000-2000 years too old due to old carbon influx associated with the H3 meltwater discharge event.

Heinrich [62] proposed that the bedrock material making up these layers had been conveyed by armadas of icebergs launched from the edge of the ice sheet. He suggested that, during their journey out to sea,

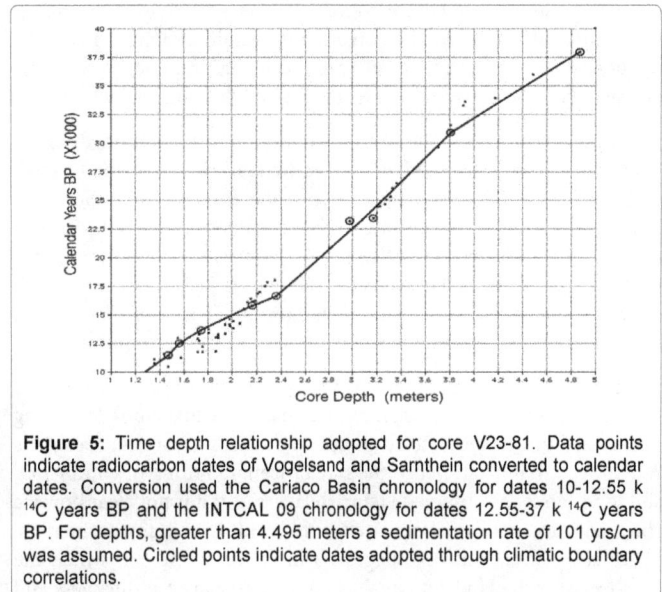

Figure 3: Heinrich events over the last 90,000 years, as seen in North Atlantic sediment core V23-81 (Data courtesy of Bond G [69,71]). The timescale adopted for this core is given in text box 1.

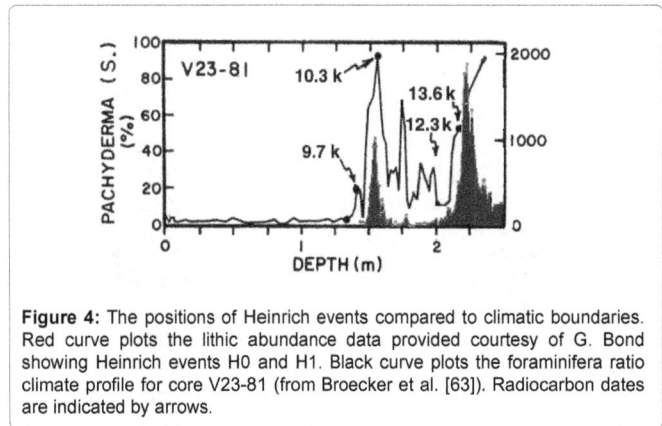

Figure 5: Time depth relationship adopted for core V23-81. Data points indicate radiocarbon dates of Vogelsand and Sarnthein converted to calendar dates. Conversion used the Cariaco Basin chronology for dates 10-12.55 k ^{14}C years BP and the INTCAL 09 chronology for dates 12.55-37 k ^{14}C years BP. For depths, greater than 4.495 meters a sedimentation rate of 101 yrs/cm was assumed. Circled points indicate dates adopted through climatic boundary correlations.

Depth (cm)	Years BP	Climatic zone
148	11,550	Holocene begins
164	12,950	Younger Dryas begins
175	13,250	Intra-Alleröd cold peak max
200	14,130	Older Dryas begins
210	14,700	Bölling begins

Table 1: Dating Core V23-81 (150-210 cm) using Climatic Boundaries

the icebergs gradually melted and progressively released their trapped sediment. One problem with this theory, however, is that Heinrich layers begin with a sharp boundary suggesting that they were deposited far more suddenly and catastrophically than can be explained by rapid increases in iceberg population. X-ray studies carried out by Manighetti et al. [67] show that the bottom-most layer he examined had been deposited so abruptly that it had compressed the ocean bottom fluff layer, the thready surface zone produced by burrowing animals. Normally, this surface fluff layer migrates upward as sediment gradually accumulates. But in this case deposition was so sudden that the layer had no time to migrate. Instead, it became packed and preserved beneath the Heinrich layer as would have occurred if large amounts of sediment were dumped in a matter of hours. Strata consisting primarily of lithic grains and nearly devoid of foraminifera are found within Heinrich layers. The almost total absence of foraminifera in these strata indicates that they were necessarily deposited very rapidly.

It is proposed here that Heinrich layers were instead formed by the action of glacier waves. A sudden climatic warming, possibly triggered by the occurrence of one or more extreme solar events that caused meltwater to pond in great quantities on the ice sheet surface, would have been able to generate one or more glacier waves. These mega avalanches advancing at speeds of hundreds of kilometers per hour and passing in a matter of minutes would have been able to affect the rapid deposition indicated in these layers.

Johnson and Lauritzen [68] have proposed a similar mechanism for Heinrich layers in which repetitive jökulhlaups would have issued from a Hudson Bay lake upon repeated failure of an ice dam at the mouth of the Hudson Straight producing major fresh water pulses into the North Atlantic. The principal difference is that the Johnson-Lauritzen mechanism proposes meltwater floods that issue from reservoirs situated either beneath or adjacent to the ice sheet and involving a single dam failure event. Whereas, according to the glacier wave concept proposed here, meltwater would originate from the upper surface of the ice sheet and produce a high volume, high kinetic energy meltwater wave that grows in size as it descends the ice sheet.

Ocean core studies show that foraminifera population dropped markedly during the period when Heinrich layers were being formed, with arctic species being the only one's present [69]. Rather than indicating a period of cold atmospheric climate, this could instead indicate that cold glacial meltwater was entering the oceans at the time of deposition. This is supported by isotopic studies, which show conspicuous decreases in planktonic $\delta^{18}O$ ratio in these strata, indicating colder water temperatures and lower surface salinity.

Thinking that Heinrich events took place during periods of colder climate, researchers were originally led to assume that climatic cooling had caused the ice sheets to advance and that this in turn increased the rate of ice calving and hence the rate of iceberg rafting of continental soil to produce the observed Heinrich layers. But it is not clear how ice sheet advances or cooler climate would result in a dramatic increase in ice calving. One theory suggests a mechanical effect in which the ice sheet over the Hudson Bay, having reached a critical thickness and

weight, sloughed down the Hudson Strait into the ocean. However, Heinrich layers were being produced by sediment transported in a synchronized fashion via geographically separated outflow channels. For example, there is evidence that some detritus in Heinrich layers is of Icelandic origin hence involving more easterly portions of the ice sheet. Consequently, to explain the sediment transported from the other locations similar ice surges would be needed for these other ice sheet locations. The ice sheet advance theory, however, has difficulty explaining such synchronization since advancing ice sheets would not be expected to necessarily calve ice at the same time. The climatically triggered glacier wave mechanism, on the other hand, accounts for why material was entering from various locations in a coordinated fashion. Some groups advocating the ice berg theory, such as Hulbe CL et al. [70], have instead considered climatic warming as being the trigger of these events. Their mechanism, like that of Heinrich, involves the release of ice berg armadas, although produced during a period of warming and ice melting rather than during a period of cold climate and ice sheet expansion.

Core V23-81, whose Heinrich layer record is displayed in Figures 3 and 4 was taken from a location that would have been about 700 kilometers west of the edge of the European ice sheet and about 1500 kilometers south of the edge of the Greenland Ice Sheet. Core DSDP 609, which also registers Heinrich layers, was taken from an even more remote location that formerly was situated about 1500 kilometers west of the edge of the European ice sheet and about 2000 kilometers south of the Greenland Ice Sheet. Heinrich layers have even been found off the coast of Portugal, at a distance of 3000 km. Glacier waves would have been able to transport continental soil debris such great distances into the ocean since a sea ice shelf is believed to have bordered the ice sheet at those times. The divide between the Labrador Ice Sheet and the Keewatin Ice Sheet would have funneled glacier waves southward allowing them to pick up detritus from this valley and transport it southward over the bordering ice shelf. If comparable to the Ross Sea ice shelf, the North Atlantic ice shelf could have had a thickness of several hundred meters which would have made it strong enough to convey the brief passage of a glacier wave. If its southern margin were to consist of closely packed icebergs, this too could have had sufficient integrity to support a transient glacier wave. The reason why mid ocean Heinrich layers are not observed to occur in the midst of a prolonged interstadial or warm interval is that the ice shelf by that time would have melted away causing glacier waves to drop their debris much closer to shore.

The continental detritus data plotted in Figure 6 is taken from Bond et al. [71] and as with Figure 3, it uses the time-depth chronology presented in Table 1. The data are compared with the GISP2 Summit, Greenland ice core climate profile. Some Heinrich layers are found to correlate with periods of transition from cold to warm climate, as seen for Heinrich events H0, H1, and H2. On the assumption that the detritus was deposited suddenly, the sedimentation rate would have been abnormally high during a glacier wave passage. This implies that some layers would actually be more compressed in time than is evident in the plot shown in Figure 3. Nevertheless, it is also possible that a consecutive series of glacier waves could have blended to form a time-extended Heinrich layer. In fact, there is evidence of stratification being present within the Heinrich layers, suggesting that the material was not deposited by a single event, but by clusters of events. To explain these sharp boundaries and successive stratified layers a sudden depositional mechanism is needed, one capable of recurring at spaced intervals. Recurring glacier waves fit this requirement. The rapid deposition of

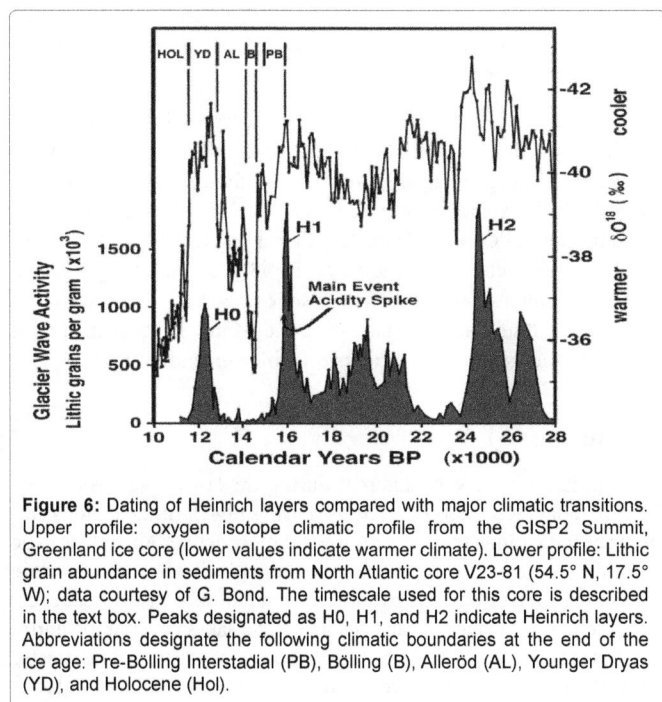

Figure 6: Dating of Heinrich layers compared with major climatic transitions. Upper profile: oxygen isotope climatic profile from the GISP2 Summit, Greenland ice core (lower values indicate warmer climate). Lower profile: Lithic grain abundance in sediments from North Atlantic core V23-81 (54.5° N, 17.5° W); data courtesy of G. Bond. The timescale used for this core is described in the text box. Peaks designated as H0, H1, and H2 indicate Heinrich layers. Abbreviations designate the following climatic boundaries at the end of the ice age: Pre-Bölling Interstadial (PB), Bölling (B), Alleröd (AL), Younger Dryas (YD), and Holocene (Hol).

the detritus during any given event would explain the low prevalence of foraminifera within them.

As the ice shelf melted and broke up into an armada of ice bergs, any debris deposited on their surfaces by the glacier wave would have eventually been deposited on the ocean floor as the ice bergs melted. So, ice bergs could also have played a role in accounting for the delayed deposition of some of the debris.

Correlations of Heinrich events to oxygen isotope climate profile will tend to unintentionally show that a cold climate prevailed at the onset of a Heinrich event since isotope profiles have an inherent time lag in registering climatic changes causing their isotopic ratio to respond slowly to any abrupt warming. Heinrich layer H1 is of particular interest since the core chronology adopted here shows that it peaks near the beginning of the Pre-Bölling interstadial (15.8 kyrs BP), the warming trend that began the deglaciation, which ended the ice age. This warming was global in extent since in Chile, mountain glaciers were rapidly retreating at this time [72]. Also, the Alerce, Chile climatic profile and Byrd Station, Antarctica ice core show that a major deglacial warming trend began at this time synchronous with the northern hemisphere warming [73] (Figures 1 and 5). Naafs et al. [74] have studied Heinrich layers in the North Atlantic between 40° and 50° N and have found evidence that they occur in those locations in conjunction with surface water warming.

The onset in H1 lithic grain deposition shown in Figure 6 also correlates with the time of the Main Event acidity peak found in Antarctic ice [72] and which has been interpreted to indicate a period of increased cosmic dust influx into the solar system [38,75]. As mentioned earlier, a dusty solar environment could have been the cause of the elevated solar activity that appears to have prevailed at that time. A smaller Heinrich event labeled as H0 is seen to have an apparent culmination around 12.1 kyrs BP. It correlates with a Younger Dryas interstadial warming evident in the Byrd Station, Antarctica and Summit, Greenland ice core climate profiles. A moderate rise in lithic

grain deposition occurs at the beginning of this event around 12.7 kyrs BP. This coincides with a major D/O warming seen in the Greenland ice core record (Figure 2). An increase in the radiocarbon excess seen around this date in the Floating Late Glacial Pine dendrochronology record [76] may indicate the occurrence of a super SPE.

The observation that Heinrich events coincide with the onset of climatic warming is even more dramatically illustrated in Figure 7, which correlates the timing of Heinrich events in core V23-81 with shifts in the GISP2 ice core ^{15}N/^{14}N isotope ratio [77] higher ^{15}N/^{14}N values indicating warmer temperatures. As is seen, the correspondence to the Heinrich event peaks is quite striking. The slight deviation for Heinrich event H1 may be attributed to inaccuracies in estimating the core's sedimentation rate for this period. In particular, considering that core sedimentation rate increases during Heinrich events due to the deposition of wave transported debris, peak H1 would span a shorter period of time than is indicated here and its onset would likely move to the left with an onset of ~16 kyrs BP to bring it into coincidence with the warm interval indicated in the Greenland ice core isotope ratio

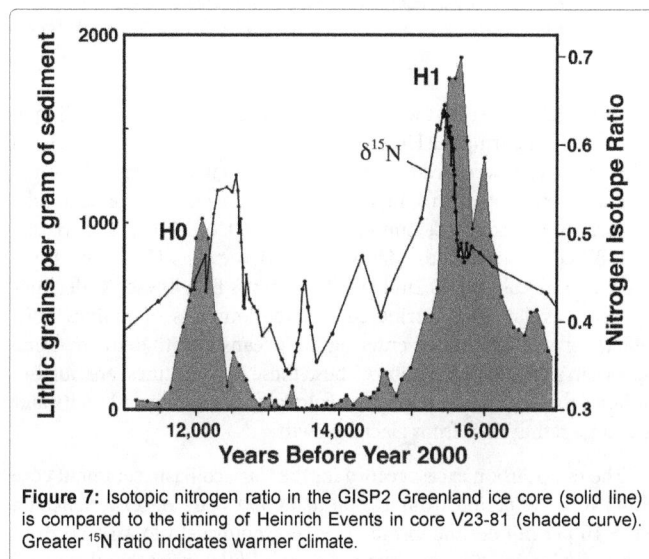

Figure 7: Isotopic nitrogen ratio in the GISP2 Greenland ice core (solid line) is compared to the timing of Heinrich Events in core V23-81 (shaded curve). Greater ^{15}N ratio indicates warmer climate.

record.

Flower et al. [78] have measured Mg/Ca ratios and $\delta^{18}O$ ratios in foraminifera in Orca Basin core EN32-PC6 and find that seawater salinity declined during the interval 16.1 to 15.5 kyrs BP, which preceded the more prominent salinity decline dating between 15.2-13.0 kyrs BP; see Figure 8. They interpret both as an indication of an increased influx of glacial meltwater into the Gulf during this period. They note that the earlier meltwater influx event occurred during Heinrich event H1 which they have marked in their plot with a shaded band. The finding that H1 was associated with glacial melting and meltwater influx into the Gulf, in turn, supports the contention made earlier that Heinrich events are essentially powerful meltwater avalanches that discharge from the ice sheet during the onset of climatic warmings. The meltwater spike evident in the Younger Dryas portion of the Mg/Ca profile, which dates at around 12.5 kyrs BP, may correlate with the detritus spike evident in the early part of Heinrich event H0 discussed earlier.

Over eight major Heinrich events have occurred over the course of the past 90,000 years and the number reaches almost 20 when more moderate events are included (Figure 3). In particular, the moderate detritus peak evident at 19 k cal yrs BP in Figure 3 correlates with the

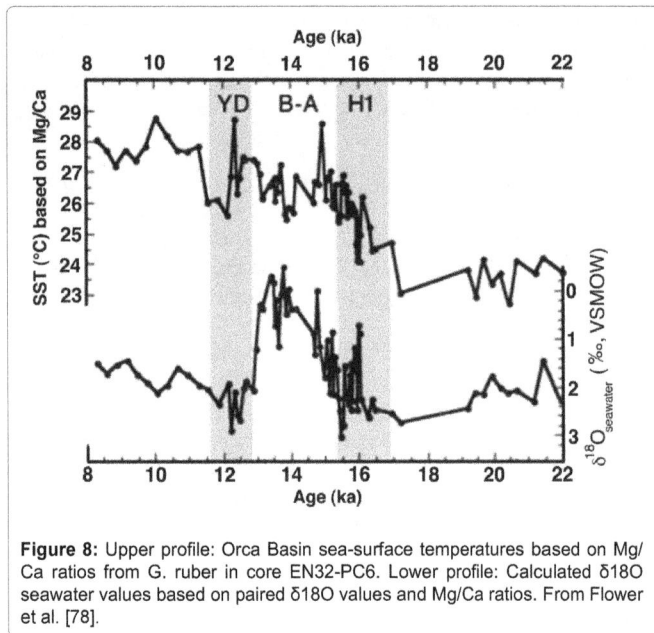

Figure 8: Upper profile: Orca Basin sea-surface temperatures based on Mg/Ca ratios from G. ruber in core EN32-PC6. Lower profile: Calculated δ18O seawater values based on paired δ18O values and Mg/Ca ratios. From Flower et al. [78].

abrupt sea level rise event which Yokoyama et al. [20] report finding in Australia. Comparing the Heinrich event record with the Cariaco Basin radiocarbon excess record it is seen that times of major Heinrich events correlate with downturns in the radiocarbon excess. For example, declines in ^{14}C excess beginning around 39,000, 32,000, and 16,000 years BP correlate with the dates of Heinrich events H5 (39 k cal yrs BP), H3 (32 k cal yrs BP), and H1 (16 k cal yrs BP). These ^{14}C declines are easily explained if attributed to large amounts of radiocarbon-depleted glacial meltwater entering the oceans and thereby lowering the ocean's ^{14}C concentration. Thus, these ^{14}C declines are further evidence that Heinrich events arise from glacial melting and meltwater discharge rather than from glacial growth.

The radiocarbon excess record for the Cariaco Basin sediment core shows that the period from 16,500 to 14,500 years BP experienced a 190 ± 10 per mil decline in the ^{14}C excess, equivalent to a 13.6% drop in the ^{14}C/C ratio [79]. Broecker and Barker [80] suggest that this could be explained if radiocarbon-depleted meltwater had entered the ocean and mixed with surface water, but they could not locate a reasonable source. They consider the possibility that the water may have come from an isolated high-salinity ocean bottom reservoir that subsequently mixed with surface waters, but conclude that the size of the abyssal reservoir is insufficient to explain the decline. They acknowledge that alpine glaciers were rapidly retreating during this time, suggestive of a major climatic warming, but did not consider the possibility that the ^{14}C decline was due to radiocarbon-depleted meltwater entering the oceans from these receding glaciers and ice sheets.

Broecker and Barker [80] say that this period is referred to as the "Mystery Interval" since during this climatic warming unusually cold temperatures prevailed in the North Atlantic and western Mediterranean. But there is no mystery if one acknowledges that glacial meltwater entering the oceans at near 0° centigrade would have kept sea surface temperatures low even during a climatic warming period. They mention that this ^{14}C decline falls close to the time of Heinrich event H1, but make no connection between the two events. One reason is that they date H1 as occurring about 17,500 years BP which would give it an end date prior to the beginning of the Mystery Interval.

However, when core V28-81 is dated as per the chronology adopted here, Heinrich event H1 is found to peak at 15.8 kyrs BP spanning the bulk of the Mystery Interval ^{14}C decline.

There is also evidence that Heinrich events H5 and H3 both correlate with times of elevated solar activity. H5 is found to follow: a) the Laschamp geomagnetic intensity minimum and geomagnetic reversal which extended from 40.8-42.3 k cal yrs BP [81], b) a very dramatic rise in ^{14}C excess from 42-40 k cal yrs BP [79], and c) a very large [10]Be spike centred at 41 k cal yrs BP [82], all indicating a period of extremely high solar and galactic cosmic ray activity. Also, Heinrich event H3 is found to correlate with the Mono Lake geomagnetic minimum and geomagnetic excursion which extended from 35-32 k cal yrs BP [83].

Conclusion

Conditions were very different during the last ice age from what they are today. Ice sheets covered the Earth's high latitude regions and there are indications that the Sun was far more active. Abrupt climatic warmings occurred at recurrent intervals which were much more intense than anything seen in more recent times. As a consequence, there were times when supraglacial meltwater lakes covered the entire surfaces of the North American and European ice sheets to a far greater extent than is seen today in Greenland. Under such conditions, a glacier burst occurring somewhere on the ice sheet surface could trigger the successive emptying of other supraglacial lakes to form a self-amplifying high kinetic energy meltwater wave. Such glacier waves would have rapidly descended the ice sheet and, upon reaching its margin, would have been able to continue their journey creating catastrophic flooding of the surrounding land area. Calculations show that glacier waves reaching hundreds of miles per hour were a reasonable mechanism of meltwater transport at that time. Such waves could account for the upland permafrost silts in Alaska, for similar flood deposits in Siberia, for the many drumlin fields seen in North America, and for many of the lignite valley deposits seen in parts of Europe and North America. During times when an ice shelf existed off the coast of the Laurentide Ice Sheet, these waves could have traveled thousands of kilometers from the ice sheet margin releasing their debris to form the observed Heinrich layers in mid ocean sediments.

Upon crossing land areas that bordered the ice sheets, these immense flood waves would have posed a lethal hazard to land animals and to ice age man. These events could be the source of the many flood myths preserved by cultures all over the world [37].

Acknowledgements

I would like to thank the Starburst Foundation (starburstfound.org) for their help in funding the publication of this paper.

References

1. Dana JD (1880) Manual of Geology. American Book Co., New York.

2. Howorth HH (1887) The Mammoth and the Flood. Sampson Low, Marston, Searle, & Rivington, London.

3. Baker VR (2009) Megafloods and global paleoenvironmental change on Mars and Earth. GSA Special 453: 25-36.

4. Kehew A, Clayton L (1983) Late Wisconsin floods and development of the Souris-Pembina spillway system in Saskatchewan, North Dakota, and Manitoba. In Glacial Lake Agassiz pp: 187-210.

5. Péwé TL (1955) Origin of the upland silt near Fairbanks, Alaska. Bull. Geol Soc Am 67: 699-724.

6. Shaw J (1989) Drumlins, subglacial meltwater floods, and ocean responses.

Geology 17: 853-856.

7. Shaw J, Kvill D, Rains B (1989) Drumlins and catastrophic subglacial floods. Sedimentary Geology 62: 177-202.

8. LaViolette PA (1983) Galactic Explosions, Cosmic Dust Invasions, and Climatic Change. Ph.D. dissertation, Portland State University, Portland, Oregon, p: 763.

9. Fitzpatrick AAW, Hubbard AL, Box JE, Quincey DJ, Van AD, et al. (2014) A decade (2002-2012) of supraglacial lake volume estimates across Russell Glacier, West Greenland. The Cryosphere 8: 107-121.

10. Koenig LS, Lampkin DJ, Montgomery LN, Hamilton SL, Turrin JB, et al. (2015) Wintertime storage of water in buried supraglacial lakes across the Greenland ice sheet. The Cryosphere 9: 1333-1342.

11. Smith LC, Chua VW, Yanga K, Gleasona CJ, Pitchera LH, et al. (2015) Efficient meltwater drainage through supraglacial streams and rivers on the southwest Greenland ice sheet. PNAS 112: 1001-1006.

12. Tedesco M, Fettweis X, Mote T, Wahr J, Alexander P, et al. (2013) Evidence and analysis of 2012 Greenland records from spaceborne observations, a regional climate model and reanalysis data. Cryosphere 7: 615-630.

13. Fairbanks RG (1989) A 17,000 year glacio-eustatic sea level record: Influence of glacial melting rates on the Younger Dryas event and deep ocean circulation. Nature 342: 637-642.

14. Bard E, Hamelin B, Fairbanks RG, Zindler A (1990) Calibration of the ^{14}C timescale over the past 30,000 years using mass spectrometric U-Th ages from Barbados corals. Nature 345: 405-410.

15. Liu JP, Milliman JD (2004) Reconsidering melt-water pulses 1A and 1B: global impacts of rapid sea-level rise. J Ocean University China 3: 183-190.

16. Clark PU, Mitrovica JX, Milne GA, Tamisiea ME (2002) Sea-level fingerprinting as a direct test for the source of global meltwater pulse 1A. Science 295: 2438-2441.

17. Nerem RS, Chambers D, Choe C, Mitchum GT (2010) Estimating mean sea level change from the 32 TOPEX and Jason altimeter missions. Marine Geodesy 33: 435.

18. Emiliani C, Gartner S, Lidz B, Eldridge K, Elvey DE, et al. (1975) Paleoclimatological analysis of late quaternary cores from the northeastern Gulf of Mexico. Science 189: 1083-1088.

19. Abdul NA, Mortlock RA, Wright JD, Fairbanks RG (2016) Younger Dryas sea level and meltwater pulse 1B recorded in Barbados reef crest coral Acropora palmata. Paleoceanography 31: 330-344.

20. Yokoyama Y, Lambeck K, DeDeckker P, Johnston P, Fifield LK (2000) Timing of the Last Glacial Maximum from observed sea-level minima. Nature 406: 713-716.

21. Thorarinsson S (1953) Some new aspects of the Grimsvotn problem. J Glaciol 2: 267.

22. Ward RC (1978) Floods: A Geographical Perspective. New York: John Wiley & Sons.

23. Mayer C, Lambrecht A, Hagg W, Helm A, Scharrer K (2008) Post-drainage ice dam response at Lake Merzbacher, Inylchek glacier, Kyrgyzstan. Geografiska Annaler a Physical Geography 90: 87-96.

24. Qiao L, Mayer C, Liu S (2015) Distribution and interannual variability of supraglacial lakes on debris-covered glaciers in the Khan Tengri-Tumor Mountains, Central Asia. Environ Res Lett 10: 014014.

25. Wagner G, Beer J, Masarik J, Muscheler R, Kubik PW, et al. (2001) Presence of the solar de Vries cycle (~205 years) during the last ice age. Geophys Res Let 28: 303-306.

26. Clemens SC (2005) Millennial-band climate spectrum resolved and linked to centennial-scale solar cycles. Quat Sci Rev 24: 521-531.

27. Raspopov V, Dergachev V, Kozyreva O, Kolstrom T (2005) Climate response to de Vries solar cycles: evidence of Juniperus turkestanica tree rings in Central Asia. Mem SA It 76: 760-765.

28. Versteegh GJM (2005) Solar forcing of climate, 2: Evidence from the past. Space Sci Rev 120: 243-286.

29. Braun H, Christl M, Rahmstorf S, Ganopolski A, Mangini A, et al. (2005) Possible solar origin of the 1470-year glacial climate cycle demonstrated in a coupled model. Nature 438: 208-211.

30. Velasco VM, Mendoza B (2008) Assessing the relationship between solar activity and some large 33 scale climatic phenomena. Adv Space Res 42: 866-878.

31. Braun H (2009) Strong indications for nonlinear dynamics during Dansgaard-Oeschger events. Clim Past Discuss 5: 1751-1762.

32. Braun H, Kurths J (2010) Were Dansgaard-Oeschger events forced by the Sun? Eur Phy J Special Topics 191: 117-129.

33. Braun H, Ditlevsen P, Kurths J, Mudelsee M (2011) A two-parameter stochastic process for Dansgaard Oeschger events. Paleoceanography 26: PA3214.

34. Zook HA, Hartung JB, Storzer D (1977) Solar flare activity: Evidence for large-scale changes in the past. Icarus 32: 106-126.

35. Zook HA (1980) On lunar evidence for a possible large increase in solar flare activity approximately 2×10^4 years ago, In: The ancient sun: Fossil record in the earth, moon and meteorites. Proceed. Conf., Boulder, CO, October 16-19, 1979. New York and Oxford, Pergamon Press, 245-266.

36. Gold T (1969) Apollo II observations of a remarkable glazing phenomenon on the lunar surface. Science 165: 1345-49.

37. LaViolette PA (1997) Earth Under Fire. Rochester, VT: Bear & Co.

38. LaViolette PA (2005) Solar cycle variations in ice acidity at the end of the last ice age. Planetary and Space Science 53: 385-393.

39. Narcisi B, Petit JR, Engrand C (2007) First discovery of meteoritic events in deep Antarctic (EPICA-Dome C) ice cores. Geophys Res Let 34: L15502.

40. LaViolette PA (2015) The episodic influx of tin-rich cosmic dust particles during the last ice age. Adv Space Res 56: 2402-2427.

41. Hoyle F, Lyttleton RA (1950) Variations in solar radiation and the cause of ice ages. J Glaciol 1: 453.

42. Bruns AV, Prokofiev VK, Severny AB (1970) On the contribution of solar activity to the ultraviolet spectrum of the Sun. In: Ultraviolet Stellar Spectra and Ground-based Observations, edited by Houziaux and Butler, pp: 256-259.

43. LaViolette PA (2011) Evidence for a solar flare cause of the Pleistocene mass extinction. Radiocarbon 53: 303-323.

44. Mörner NA (1978) Annual and Inter-Annual Magnetic Variations in Varved Clay. J Interdisci Cycle Res 9: 229-241.

45. Ising J (2001) Pollen analysis, chronology and palaeomagnetism of three Late Weichselian sites in southern Sweden. Thesis, Department of Quaternary Geology, Lund University.

46. Trenberth KE, Large WG, Olson JG (1989) The effective drag coefficient for evaluating wind stress over the oceans. J Climate 2: 1507-1516.

47. Baker VR (2009) Megafloods and global paleoenvironmental change on Mars and Earth. GSA 453: 25-36.

48. Weiss R, Fritz HM, Wünnemann K (2009) Hybrid modeling of the mega-tsunami runup in Lituya Bay after half a century. Geophy Res Lett 36: L09602.

49. Paik J, Shin, C (2015) Multiphase flow modeling of landslide induced impulse wave by VOF method. AGU Fall meeting abstract: NG23A-1764.

50. Péwé TL (1975) Quaternary Geology of Alaska. Geological Survey Professional Paper 835.

51. Péwé TL (1975) Quaternary Stratigraphic Nomenclature in Unglaciated Central Alaska. Geological Survey Professional Paper 862.

52. Péwé TL, Journaux A, Stuckenrath R (1977) Radiocarbon dates and late-Quaternary stratigraphy from Mamontova Gora, unglaciated central Yakutia, Siberia, U.S.S.R. Quaternary Research 8: 51-63.

53. Taber S (1943) Perennially frozen ground in Alaska: Its origin and history. Bulletin of the Geological Society of America 54: 1433-1548.

54. Eakin HM (1916) The Yukon-Koyukuk region, Alaska. U.S. Geological Survey 631.

55. Eakin HM (1918) The Cosna-Nowitna region, Alaska. U. S. Geological Survey 667.

56. Begossi R, Fávera JCD (2002) Catastrophic floods as a possible cause of organic accumulation giving rise to coal, Paraná Basin, Brazil. Intl J Coal Geol 52: 83-89.

57. Melentis JK (1963) Studies on the fossil vertebrates of Greece: 3. The osteology of the Pleistocene Proboscidians of the basin of Megalopolis in the Peloponnese (Greece). Annales 35 Géologiques des Pays Helléniques 14: 1-107.

58. Melentis JK (1965) First find of Palaeoloxodon antiquus germanicus in the Young Pleistocene deposits of the basin of Megalopolis in the (Peloponnese). Internships Akadimias Athinon 40: 197-207.

59. Jack RL (1877) Glacial drift in the northeastern Carpathians, Quarterly Journal Geological Society of London 33: 673-681.

60. Lubbock J (1896) The scenery of Switzerland and the causes to which it is due. Macmillan, New York.

61. Lyell C (1850) Principles of geology. D. Appleton, New York.

62. Heinrich H (1988) Origin and consequences of cyclic ice rafting in the northeast Atlantic Ocean during the past 130,000 years. Quaternary Research 29: 142-152.

63. Broecker WS, Andree M, Wolfli W, Oeschger H, Bonani G, et al. (1988) The chronology of the last deglaciation: Implications to the cause of the Younger Dryas event. Paleoceanography 3: 1-19.

64. Vogelsang E, Sarnthein M (2001) Age control of sediment core V23-81. Pangaea.

65. Broecker W (1994) Massive iceberg discharges as triggers for global climate change. Nature 372: 421-424.

66. Adams J, Maslin M, Thomas E (1999) Sudden climatic transitions during the Quaternary. Progress in Physical Geography 23: 1-36.

67. Manighetti B, McCave IN, Maslin M, Shackleton NJ (1995) Chronology for climate change: Developing age models for the biogeochemical ocean flux study cores. Paleoceanography 10: 513-525.

68. Johnson RG, Lauritzen SE (1995) Hudson Bay-Hudson Strait jökulhlaups and Heinrich events: A hypothesis. Palaeogeogr. Palaeoclimatol. Palaeoecol 117: 123-137.

69. Bond G, Heinrich H, Broecker W, Labeyrie L, Mcmanus J, et al. (1992) Evidence for massive discharges of icebergs into the North Atlantic Ocean during the last glacial period. Nature 360: 245-249.

70. Hulbe CL, MacAyeal DR, Denton GH, Kleman J, Lowell TV (2004) Catastrophic ice shelf breakup as the source of Heinrich event icebergs. Paleoceanography 19: PA1004.

71. Bond GC, Lotti R (1995) Iceberg discharges into the North Atlantic on millennial time scales during the last glaciation. Science 267: 1005-1010.

72. Lowell TV, Heusser CJ, Andersen BG, Moreno PI, Hauser A, et al. (1995) Interhemispheric correlation of late Pleistocene glacial events. Science 269: 541-1549.

73. LaViolette PA (2005) Evidence of global warming at the Termination I boundary and its possible cosmic dust cause.

74. Naafs BDA, Hefter J, Stein R, Haug GH (2011) Evidence for surface water warming in the IRDbelt during Heinrich events. Interim Colloquium in Salamanca, Salamanca, Spain.

75. Hammer CU, Clausen HB, Langway CC Jr. (1997) 50,000 years of recorded global volcanism. Climatic Change 35: 1-15.

76. Hua Q, Barbetti M, Fink D, Kaiser KF, Friedrich M, et al. (2009) Atmospheric ^{14}C variations derived from tree rings during the early Younger Dryas. Quat Sci Rev 28: 2982-2990.

77. Severinghaus JP, Brook EJ (1999) Abrupt climate change at the end of the last glacial period inferred from trapped air in polar ice. Science 286: 930-934.

78. Flower BP, Hastings DW, Hill HW, Quinn TM (2004) Phasing of deglacial warming and Laurentide ice sheet meltwater in the Gulf of Mexico. Geology 32: 597-600.

79. Hughen K, Lehman S, Southon J, Overpeck J, Marchal O et al. (2004) ^{14}C activity and global carbon cycle changes over the past 50,000 years. Science 303: 202-207.

80. Broecker W, Barker S (2007) A 190‰ drop in atmosphere's Δ^{14}C during the "Mystery Interval (17.5 to 14.5 kyr). Earth and Planetary Science Letters 256: 90-99.

81. Bourne MD, Mac Niocaill C, Thomas AL, Henderson GM (2013) High-resolution record of the Laschamp geomagnetic excursion at the Blake-Bahama outer ridge. Geophys J 195: 1519-1533.

82. Raisbeck G, Yiou F, Bourles D, Lorius C, Jouzel J, et al. (1987) Evidence for two intervals of enhanced ^{10}Be deposition in Antarctic ice during the last glacial period. Nature 326: 273-277.

83. Negrini RM, McCuan DT, Horton RA, Lopez JD, Cassata WS, et al. (2014) Nongeocentric axial dipole field behavior during the Mono Lake excursion. J Geophys Res 119: 2567-2581.

23

Stochastic Nature of Salt Mass Transport in Porous Media Under Unstable Conditions

Kamal Mamoua*, Ashok Pandit and Howell Heck

Department of Civil Engineering, Florida Institute of Technology, Melbourne, FL 32901, USA

Abstract

The two main transport mechanisms that occur simultaneously under unstable flow conditions are transport of saltwater from an overlying salt source to the porous media, and transport of salt through the porous media. These mechanisms were simultaneously studied through two fixed mass experiments conducted over 15 days. The transport through the porous media was also studied via three continuous injection experiments lasting between 5 to 29 days. There was no hydraulic gradient across the porous media in any of the experiments. Experiments were conducted in a 1 cm thick plexiglass rectangular sand column (1.70 m × 0.61 m × 0.61 m). The saline source concentration was 36 g/l, and the source heights were 4.5 cm. The sand porosity and hydraulic conductivity were 32% and 9.0 m/d, respectively. The rate of mass transport from the source to the porous media was observed by measuring the salt concentration within the source, while the salt transport through the porous media was documented by measuring breakthrough curves at five locations within the sand column. Fixed mass experiment results, using mass analysis, showed that the salt transport from the source to the porous media was deterministic since both experiments produced identical rates of mass transport from the source to the porous media, the salt transport through the porous media was stochastic since the observed breakthrough curves at the five locations were considerably different. The breakthrough curves measured in three identical continuous injection experiments were also very different supporting the results of the fixed mass experiments. The implications of these findings are that, under unstable conditions, one can predict the salt mass that would enter from a salt source into the underlying porous media with certainty, one cannot predict the rate or pattern of salt transport through the porous media itself.

Keywords: Physical model; Salt transport; Porous media; Breakthrough curves; Stochastic; Fingering; Free convection

Introduction

Forced, free and mixed convective flow

There are many real-world situations where a mass of relatively heavier fluid overlies a porous media containing a lighter fluid. For example, this is fairly common at landfills and waste dump sites [1-3], saline disposal basins [4], seawater inundation along coastal aquifers [5], and in estuaries [6-11]. Typically, when saltwater and freshwater are in contact with each other, the groundwater flow is due to natural hydraulic gradients, and the movement of the fluid is termed "forced" convection. Under forced convective conditions, in the absence of any ambient groundwater velocity, the flux of salt can be regarded as the sum of molecular diffusion and mechanical dispersion [12]. However, when a heavier fluid such as saltwater overlies relatively fresher, and lighter groundwater, the heavier fluid may establish additional hydraulic gradients, which lead to "convective" dispersion; a term used by Bachmat et al. [12]. Subsequently, the term "free" convection has also been used instead of convective dispersion [13]. When both forced convection and free convection operate together in a groundwater system, the resulting flow is termed mixed convective flow [13]. It has been noted by Wooding [14] and others that the presence of a heavier fluid on top of a lighter fluid is potentially unstable, but does not always result in free convective transport.

Past experimental and numerical studies

There are two main transport mechanisms that occur under unstable flow conditions a) the transport of saltwater from the overlying salt source, which is termed as "Source" from here on, to the underlying porous media, and b) the transport of the saltwater through the porous media. These transport processes have been investigated by several researchers via laboratory and/or numerical experiments e.g., [12,14-

23]. In many of these physical model studies, heavier fluids, containing salt, were placed on top of a column or tank filled with saturated porous media containing freshwater. Some of these experiments were continuous injection experiments in which the salt concentration of the heavier fluids was kept constant [7,16,17,23], while others were fixed mass experiments in which the overlying, heavier fluid has a fixed mass of solute which was transported into the porous media over time [12,14,18,20].

Some of the key findings regarding the transport of salt through the porous media are: 1) under free convection conditions, saltwater moves through the porous media in the form of lobe shaped fingers, and convective dispersion, as opposed to molecular and mechanical dispersion, is the key transport mechanism [12,15,20], 2) the number and configuration of these fingers were not reproducible in practical experiments [12,15], 3) fingering causes rapid and erratic redistribution of solutes [15], and the salt travels faster and farther when fingers are formed as opposed to when the transport is due to mechanical dispersion [15,17], 4) salt plumes move faster and farther with increasing source concentrations [4,15], 5) fingers tend to coalesce as they move greater distances [14,16,17], and 6) not only does the salt move faster and farther once it enters the porous media, the total mass of the salt transported

---

*Corresponding author: Kamal Mamoua, Department of Civil Engineering, Florida Institute of Technology, Melbourne, FL 32901, USA
E-mail: kmamoua2012@my.fit.edu

during free convection is typically far greater than transported by diffusion [23]. Fingering was observed in most of these experiments either visually through photographs [12,14,15,17,18], or by some sort of digital processing [20,22]. Mulqueen et al. [24] stated that the non-linear dynamics of the unstable system made it impossible to predict where a finger would grow or how it might develop. This phenomenon was also observed by Schincariol et al. [15] who photographed plume movement under unstable conditions with a continuous NaCl source and noted that position of the instabilities was not the same when an experiment with a source concentration of 2000 mg/l was repeated.

The observations by Mulqueen et al. [24] and Schincariol et al. [15] indicate that when the salt transport through the porous media is due to free convection, it should be expected that when experiments are repeated, the pattern of salt transport through the porous media would be vastly different since even minor differences in the packing of the porous media can create vastly different fingering patterns leading to completely different patterns of transport. In other words, if two experiments are conducted with exactly identical conditions in terms of the type of mass source, and porous media characteristics such as hydraulic conductivity, porosity etc., the transport through the porous media may be different indicating that the salt transport through the porous media is stochastic in nature and cannot be predicted.

To our knowledge, in none of the above-mentioned experiments, except for one study [18], were experiments duplicated to quantitatively assess if the salt transport through the porous media was, or was not, stochastic in nature. In other words, no attempt has been made to repeat an experiment to determine if the mass transport is similar or different as in the previous experiment. Wood et al. [18] did repeat a fixed mass experiment on a 15-cm diameter cylindrical column with a length of 0.9 m, using Calcium Chloride as the source fluid, and determined the breakthrough curves at the outlet of the column in response to the initial concentration in the source fluid which was Calcium Chloride. Wood et al. [18] found that the breakthrough curves from the two experiments were quite similar. However, their experiments did not determine the salt transport inside the column but rather at the outlet of the column. It is likely, that the salt coalesced as it travelled through the column, and the result of the coalescing process may have yielded similar breakthrough curves for both experiments at the column outlet, even though the salt transport within the column may have been vastly different. Furthermore, the experimental system of Wood et al. [18] was a mixed convective system as they applied a hydraulic gradient across the porous medium. It is possible that the presence of the hydraulic gradient may have significantly affected the transport mechanism through the porous media resulting in similar breakthrough curves for both experiments. The present study extends the work of Wood et al. [18] by measuring and comparing the breakthrough curves at multiple ports (five) within the column, instead of just at the outlet, for both fixed and continuous injection experiments. Moreover, no hydraulic gradient was applied in the experiments conducted in this study, so this study is the first one that examines the nature of salt transport within the porous media under strictly free convective flow.

Objectives

The main objective of this paper was to conduct duplicate fixed mass and continuous injection experiments to determine if the salt transport through the porous media is stochastic or deterministic under free convection.

Methods

Description of the physical model

Experiments were conducted in a 1-cm thick plexiglass column which has a height of 1.83 m and a base of 0.61 m by 0.61 m as shown in Figure 1. The column is supported by a metal stand and consists of a source area at the top where the saline solutions were placed, and a drain valve at the bottom. For convenience, the saline solutions placed at the top of the column will be referred to as source solutions in this paper while the column containing the porous media will be referred to as the sand column. The height of the source solutions can be varied to a maximum height of 13 cm using an adjustable overflow pipe. The column has of five sampling ports, P_1 to P_5, located at fairly uniform distances below the source as shown in Figure 1. All ports are connected to a sampling probe that can traverse a distance of 30 cm, i.e., to the center of the column.

The sand used in this study was 40F from Standard Sand and Silica Company, with an effective size (d_{10}) of 0.17 mm, and a uniformity coefficient of 1.8. The sand porosity was obtained by filling a known volume of sand in a graduated cylinder and adding water to the sand in increments until the sand column was fully saturated. The porosity was calculated as the ratio of the volume of water added to the sand volume. The experiment was repeated 5 times and the estimated porosity values ranged from 31.7% to 32.2% with an average porosity of approximately 32% with a standard deviation of 0.21. The hydraulic conductivity of the sand was determined by conducting constant head tests in the previously described sand column. The experiment was repeated five times and the range of hydraulic conductivity values measured were from 8.6 m/d to 9.2 m/d with an average of 9.0 m/d and a standard deviation of 0.20.

Experimental Procedure and Description of Experiments

Sand was placed inside the sand column in three steps. First, sand was filled up to a depth of 1.70 m in increments of 25 cm. Each increment was fully saturated with water before adding the next increment of sand. Saturation was assured by allowing water to stand on top of the sand for a period of one day and observing the water level. The sand was considered saturated if the water level did not drop. This procedure was repeated after adding every 25 cm of sand increment and the entire process took approximately one week. Second, water was passed through the sand column until the inflow was equal to outflow. Water was allowed to flush through the sand column for a period of six hours after inflow became equal to outflow to ensure that there were no structural changes in the sand column as a result of the fluid passing through the porous medium, and to remove any entrapped air bubbles. Third, the drainage valve was closed, and the source area was filled with fresh water to the desired level. The water was allowed to sit in the sand column for a period of one week to ensure that the water loss was only due to evaporation and that the sand column was fully saturated and devoid of any air pockets. The source area was covered by a thin plastic sheet to minimize evaporation losses and no evaporation.

Porous media properties

Losses were observed over the entire duration of the experiments. Salinity and temperature measurements were taken using a YSI salinity meter which measured conductivity and converted it into salinity. The fixed mass experiments were started by adding the saline water solution (Sodium Chloride) to the source area to the desired depth and this procedure took less than one minute. The Continuous Injection Experiments were started by adding saline water at a concentration of

36 g/l into the source area from a storage tank to a depth of 4.5 cm; this procedure took less than one minute. The salinity concentration in the source area was maintained by continuously adding saline water to the source area by a variable-speed Peristatic Minipump. It should be noted that the freshwater salinity in the porous media was 0.3 g/l.

Salinity and temperature measurements at each port were taken by extracting samples from the port locations using the needles attached to each port. Sample sizes were 35 ml. Extracted samples were injected back at the port location after taking salinity and temperature measurements which took less than a minute. Source salinity measurements were taken by inserting the salinity meter into the source and no samples had to be extracted. The sampling schedule for both the fixed mass and continuous injection experiments is shown in Table 1. Salinity measurements at the ports for both experiments 1 and 2 were taken every six hours for the first two days, every 12 hours from the 2nd day to the 10th day, and every 24 hours after that. Salinity samples at the source were taken after 1, 5, 15, 20, 30, 45, 60 minutes, and then at 2, 4, 6, 12, 18, 24 hours. After 24 hours, salinity was measured at intervals of 6 hours on the second day, and at intervals of 12 hours from the 2nd day to the 10th day, and every 24 hours for the rest as shown in the Table 1.

One fixed mass and one continuous injection experiment, having durations (t_d) between 15 to 29 days, were conducted with the source height, h_0, of 4.5 cm, an initial source concentration, C_{so}, of 36 g/l, as shown in Table 2. Experiment 1 was a fixed mass experiment and was repeated once, and the two experiments are termed Experiments 1a and 1b in Table 2, while Experiment 2, a continuous injection experiment, was repeated twice, and the three experiments are termed as Experiments 2a, 2b and 2c in Table 2. Experiment 1 was stopped after 15 days when it was observed that the daily mass exiting from the source into the porous media was less than 0.3%. Experiments 2a and 2b were terminated after 28 and 29 days because there was negligible change in the salinity at each of the five ports and a quasi-steady-state had been reached. Experiment 2c was conducted to ensure that the breakthrough curves were oscillating and unpredictable in the beginning, hence it was stopped after five days.

Analyses and Results

Comparison of breakthrough curves and mass analysis for Experiment 1

The five breakthrough curves measured at each port for Experiments

Duration (day)	At Ports	At Source
0-0.25	Every 6 h	1, 5, 15, 30, 60, 120, 240 and 360
0.25-2	Every 6 h	every 6.0 h
2-10	Every 12 h	every 12 h
10-End	Every 24 h	every 24 h

Table 1: Sampling Schedule at Source and Ports for Experiments 1 and 2.

Experiment No	h_o cm	C_{so} g/l	t_d day	Source condition
1a	4.5	36	15	F
1b	4.5	36	15	F
2a	4.5	36	29	C
2b	4.5	36	28	C
2c	4.5	36	5	C

h_o=source depth; C_{so}=initial source concentration; t_d=duration of experiment; F=fixed source mass; C=continuous source mass

Table 2: Physical Lab Experiment Descriptions.

1a and 1b are shown in Figure 2a and 2b, respectively. The breakthrough curves measured at Ports 1 through 5 during Experiments 1a and 1b are individually compared in Figures 3-7, respectively. It should be noted that since the ports are located in the center of the column, these breakthrough curves reflect the salt transport through the center of the column. The breakthrough curve comparisons indicate that:

- The breakthrough curves for the two experiments are vastly different at Ports 1 to 4 (Figures 3-6) but become fairly similar at Port 5 (Figure 7). The similarity of the two breakthrough curves at Port 5 is comparable to the similarities that were also observed by Wood et al. [18] at their column outlet, since Port 5 is the lowest port and closest to the outlet of the column. These comparisons indicate that the salt distribution in the porous media can be vastly different both temporarily and spatially if the experiments are repeated, but may show similarities near the outlet of the column. The similarity in the breakthrough curves of Experiments 1a and 1b, at Port 5, could be due to the coalescing of the fingers that occurs as the salt moves downward in the column, resulting in the salt moving more as a "front", as in stable systems, instead of via fingers as observed in unstable systems.

- A comparison of the breakthrough curves for Ports 1 and 5 in both Experiment 1a (Figure 2a) and Experiment 1b (Figure 2b) indicates that the concentration at the lower port (Port 5) became higher than that of an upper port (Port 1) towards the end of the experiment. These phenomena indicate that the saline water in the source was "sinking" like a dense non-aqueous phase liquid (DNAPL), and collecting at the bottom of the column, which is very similar to the characteristics exhibited by a DNAPL as it moves down in the porous media. The difference between the DNAPL and the salt in this case is that while the DNAPL moves through the macro-pores in the porous media, the salt moves in the form of fingers.

- The higher concentrations in the lower ports is indicative that the main salt transport mechanism was not molecular diffusion, because in that case the salt concentration at the lower ports would always be smaller than those at the upper ports.

- Another key difference in the salt transport between Experiments 1a and 1b can be demonstrated by the "arrival times" noted at each port as shown in Table 3. Arrival time at a port is defined as the time when the salt concentration at the port was first noticed to be greater than the salt concentration in the fresh water which was 0.3 g/l.

- A comparison of the arrival times for the Experiments 1a and 1b indicates that they are vastly different; for example, the salt reached Port 4 after nine days and four days, respectively. Furthermore, the salt arrived at Port 5 earlier than Port 4 in Experiment 1a whereas that was not the case in Experiment 1b.

A mass analysis was also performed to examine the salt distribution within the porous media in Experiments 1a and 1b, and the results of the mass analysis are shown in Figure 8. Figure 8 shows the initial

Experiment No.	Port 1 Day (hr)	Port 2 Day (hr)	Port 3 Day (hr)	Port 4 Day (hr)	Port 5 Day (hr)
1a	0.25 (6)	1.375 (33)	3.5 (84)	9 (216)	6 (144)
1b	0.25 (6)	1.0 (24)	2 (48)	4 (96)	5 (120)

Table 3: Arrival time at each port for Experiments 1a and 1b.

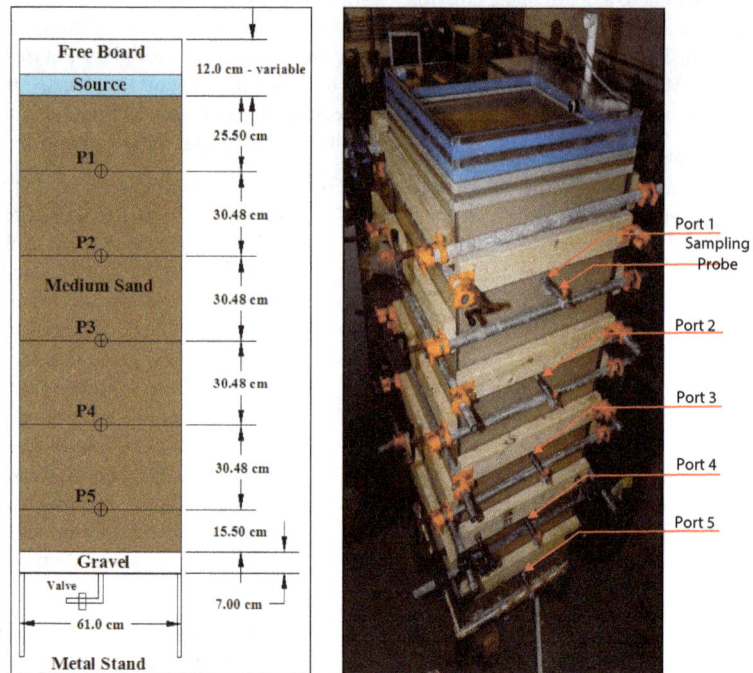

Figure 1: A schematic diagram and photograph of the physical model used for study.

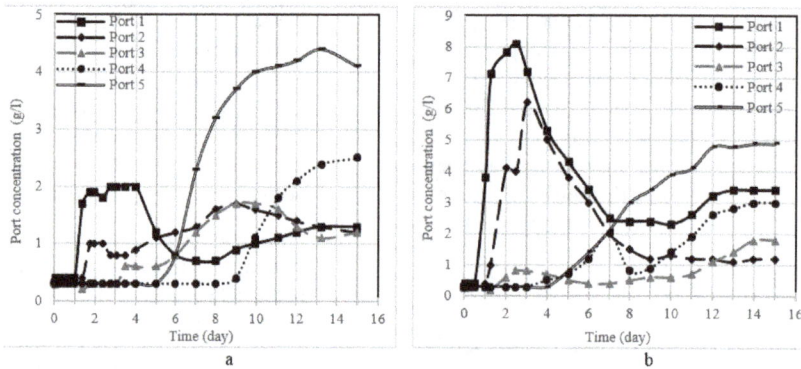

Figure 2: Measured breakthrough curves at five ports for a) Experiment 1a, b) Experiment 1b.

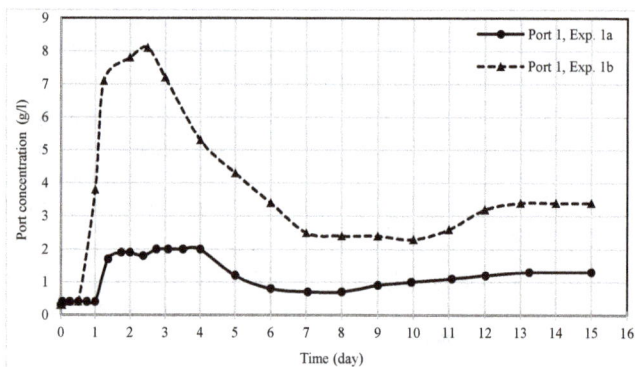

Figure 3: Comparison of breakthrough curves at Port 1 for Experiments 1a, and 1b.

(t=0 days) mass distribution within the column and the source, and the gradual changes in the mass distribution after 1, 5, 10 and 15 days. The mass values shown in these tables were determined by assigning

a volume to each port. For example, the volume assigned to Port 1 extended from the source to halfway between Port 1 and 2. Similarly, the volume assigned to Port 2 extended from the halfway point of Ports

1 and 2 to the halfway points of Ports 2 and 3. The volume assigned to Port 5 extended from the halfway point of Ports 4 and 5 to the bottom of the column. The mass attributed to each port was determined by multiplying the volume of water associated with each port by the salinity concentration measured at the port at the end of the day. The total salt mass for each experiment is sum of the mass of salt added to the source (602 g) and the initial salt amount in the sand column (63 g) due to the fact that the freshwater salt concentration was 0.3 g/l.

The following observations can be made from Figure 8: a) the total initial mass in both columns was 665 g with 602 g residing in the source area; b) the mass of salt recorded within the source in the two experiments was nearly identical at all times; c) the mass distribution within the porous media, after one day, was quite similar for Experiments 1a and 1b (402 g and 409 g); d) the salt mass accounted by the center of the column in Experiments 1a and 1b after one day (402 g and 409 g) were much less than the total salt mass of the system (665 g), and this showed that approximately one third of the salt was moving through other parts of the column and not through the center; e) the salt mass travelling through the center of the column in Experiments 1a and 1b (270 g and 286 g) was much less after five days, compared to what was travelling through the center after one day, and this indicated that the salt was also moving laterally or radially in the porous media which is unlikely to occur if the main transport mechanism was molecular diffusion; f) the total mass recorded in the source and the center of the column in Experiments 1a and 1b after 10 days were 478 g and 233 g, respectively, which shows a remarkable difference in the salt distribution patterns within the porous media during the two experiments; g) a lot more salt mass (197 g) had sunk to Port 5 in Experiment 1a as compared to Experiment 1b (39 g); h) the salt mass distribution in the two experiments was again very similar after 15 days. These observations leads to three important conclusions: 1) Under unstable conditions with free convection, the rate of transport of salt from the source to the porous media is deterministic or predictable as it was nearly equal in the two experiments at all times, 2) the mass transport through the porous media is stochastic or unpredictable, as the salt may follow very different patterns through the porous media even under two identical scenarios, and 3) the stochastic transport can be attributed to the fact that the main transport mechanism of the salt through porous media, under unstable conditions, is fingering and not molecular diffusion, and because the distribution and size of the fingers that may develop in the porous media, could be quite different even under identical hydrologic conditions. In other words, while one may be able to predict the rate of salt mass that is transported from the source to the porous media, it is not possible to predict how this salt mass will be distributed in the porous media after a given time much like it is not possible to accurately predict the movement of a DNAPL through porous media. In the case of the DNAPL, the stochastic nature of transport is due to the random location and sizes of the macro-pores in the porous media, whereas in the case of unstable salt transport, it is because of the random locations and sizes of the fingers.

Comparison of breakthrough curves for Experiment 2

The breakthrough curves measured at different ports for Experiments 2a, 2b, and 2c are shown in Figures 9-11, respectively, and compared for each port in Figures 12-16. These comparisons indicate that:

- The breakthrough curves of Experiments 2a, 2b and 2c are quite different from each other for Ports 1 through 4; only the breakthrough curves at Port 5 are somewhat similar. This

phenomenon is similar to what was observed in the fixed mass experiments in that the breakthrough curves at the lowest port, Port 5, were similar while they were different at the other four ports.

- The breakthrough curves at all ports were not smooth and exhibited that the salt concentration was randomly fluctuating during the first five to ten days at Ports 1 through 4. These fluctuations were not quite as prominent at Port 5.

- All breakthrough curves became relatively smooth after a period of approximately 12 days. This could be because with time the fingers coalesce and the solution moves down more in the form of a traditional plume [25]. In other words, the plume begins to "stabilize" after a certain time.

- As in Experiment 1, the salinity at the two lower ports (Ports 4 and 5) became higher than the salinity at the upper ports (Ports 1 and 2) after a period of 5 days (Figures 9 and 10), and stayed higher till the end of the experiment. Schincariol et al. [15] also observed that the lower part of the column led higher concentration in their plume experiments.

Conclusions and Discussion

Two identical fixed mass experiments, with a salt source on top of a column of porous media, were conducted under unstable conditions to simultaneously study, for the first time: a) the salt transport that occurs from a salt source to the porous media, and b) the vertical salt transport within the porous media itself. The rate of mass transport from the source to the porous media was observed by measuring the salt concentration within the source at various times, while the salt transport through the porous media was documented measuring breakthrough curves at five locations. Results, using mass analysis, showed that while the salt transport from the source to the porous media was deterministic in the sense that both experiments produced identical rates of mass transport from the source to the porous media, the salt transport through the porous media was stochastic since the observed breakthrough curves at the five locations were considerably different. The salt transport through the porous media was also observed by measuring breakthrough curves at five locations in three identical continuous injection experiments. The measured breakthrough curves in the continuous injection experiments were also very different supporting the results of the fixed mass experiments that the mass transport through the porous media, under unstable conditions, is stochastic. The implications of these findings are that while one can predict the amount of salt that would enter from a salt source into the underlying porous media with a great deal of certainty under unstable conditions, one cannot predict the rate or pattern of salt transport through the porous media itself. Other key findings were:

a) The differences in the breakthrough curves were highest in the upper regions of the porous media and decreased as the salt traversed downward. This phenomenon can be attributed to the fact that fingers tend to coalesce as they move downwards resulting in the movement of the salt mass not being as random at the bottom as at the top. The coalescing of fingers, under unstable transport conditions, has also been observed by several other researchers including Bachmat et al. [12] who saw finger coalescing in their dye experiments, and Simmons et al. [26] and Kneafsey et al. [21] who observed the coalescence of fingers in their Hele-Shaw cell experiments.

b) The salt was not just moving vertically but also laterally. This

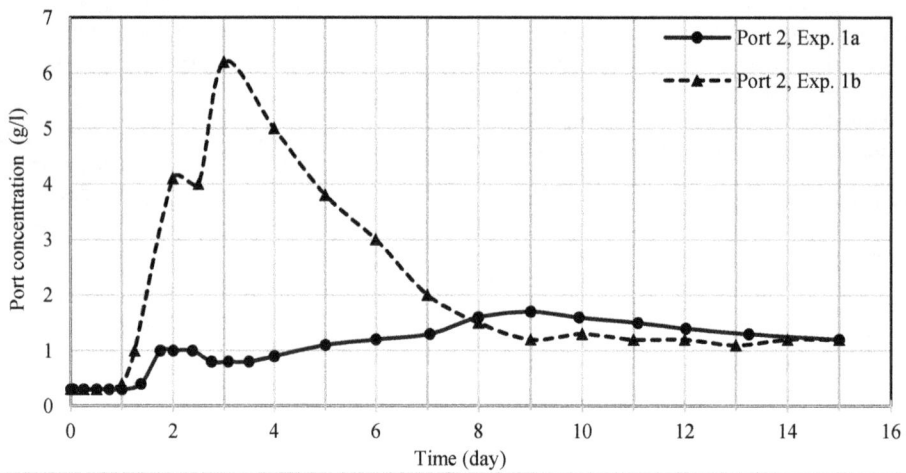

Figure 4: Comparison of breakthrough curves at Port 2 for Experiments 1a, and 1b.

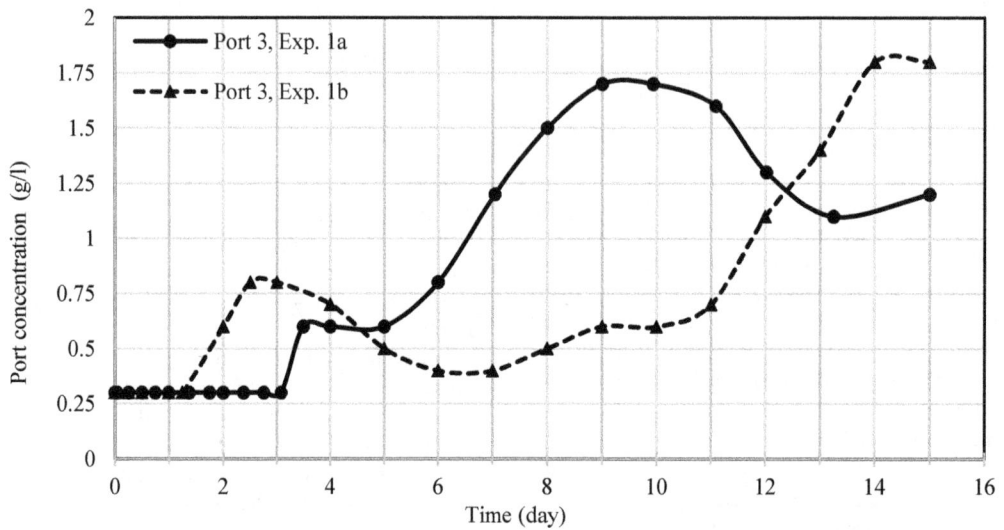

Figure 5: Comparison of breakthrough curves at Port 3 for Experiments 1a, and 1b.

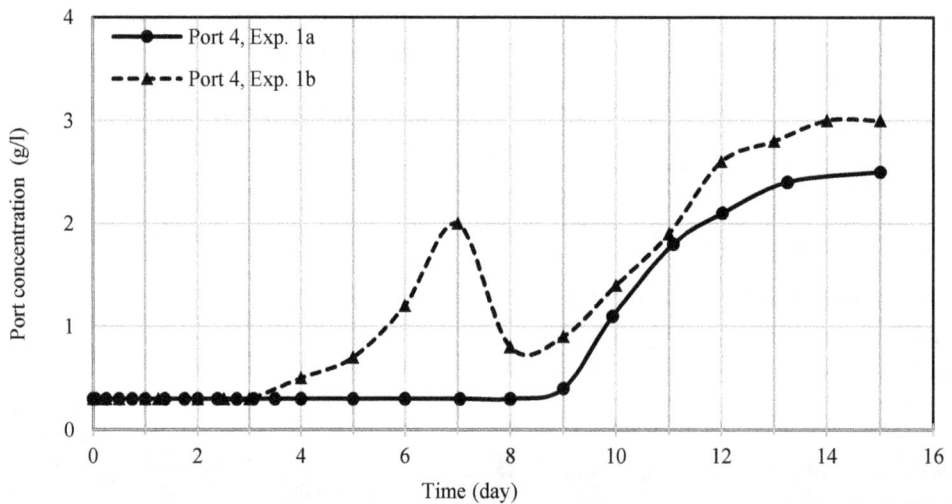

Figure 6: Comparison of breakthrough curves at Port 4 for Experiments 1a, and 1b.

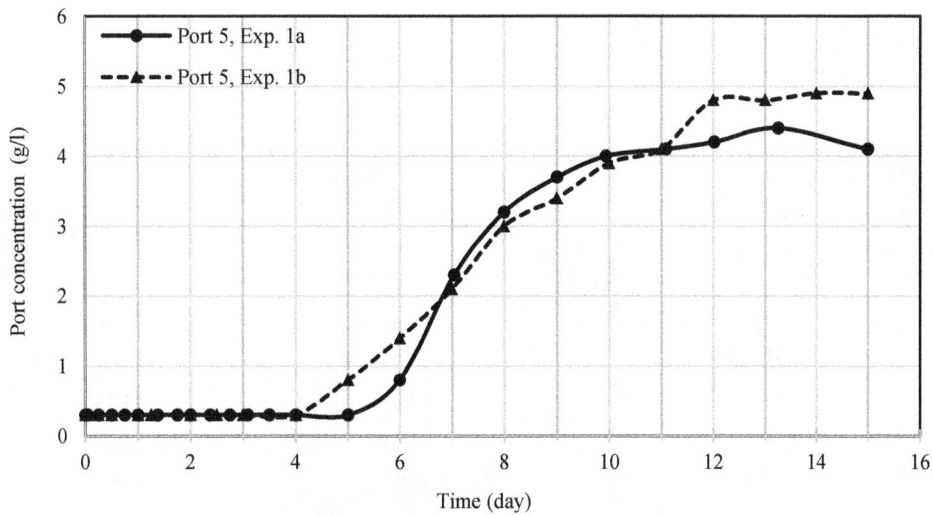

Figure 7: Comparison of breakthrough curves at Port 5 for Experiments 1a, and 1b.

Time (day)	0		1		5		10		15	
	602	602	335	341	124	122	72	72	49	52
	15	15	20	20	59	98	49	39	64	49
	11	11	11	11	40	29	58	43	43	58
	11	11	11	11	22	11	62	29	43	62
	11	11	11	11	11	11	40	11	91	40
	15	15	15	15	15	15	197	39	202	197
Total (g)	665	665	402	409	270	286	478	233	492	458
Exp. No	1a	1b	1a	1b	1a	1b	1a	1b	1a	1b

Figure 8: Mass distribution in the middle sand column for Experiment 1a and 1b in different days; the mass in the top cell represents the salt mass within the source area.

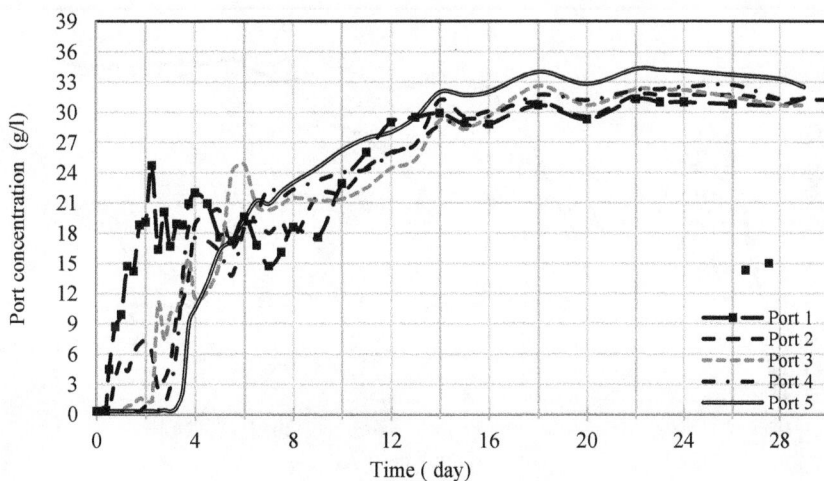

Figure 9: Measured breakthrough curves at five ports for Experiment 2a.

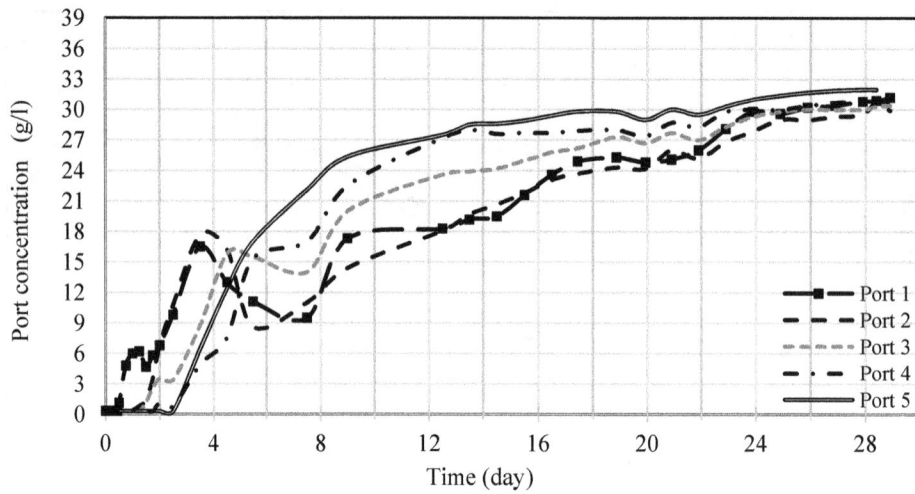

Figure 10: Measured breakthrough curves at five ports for Experiment 2b.

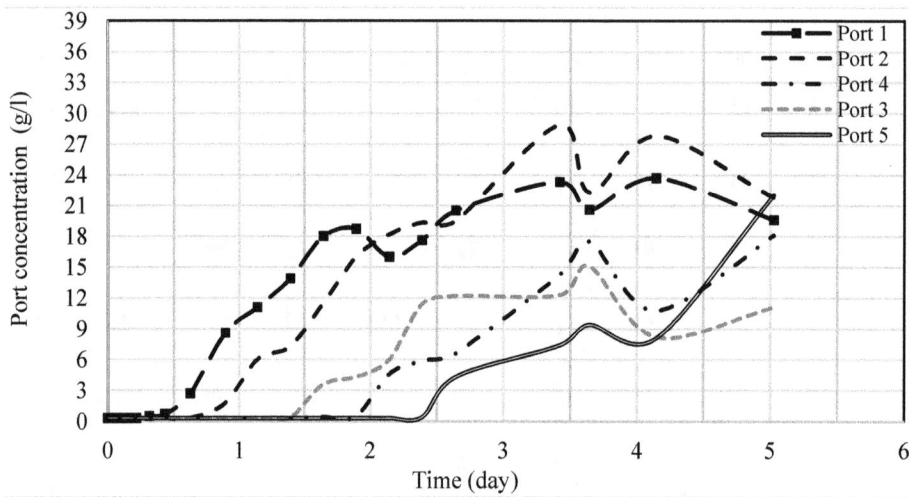

Figure 11: Measured breakthrough curves at five ports for Experiment 2c.

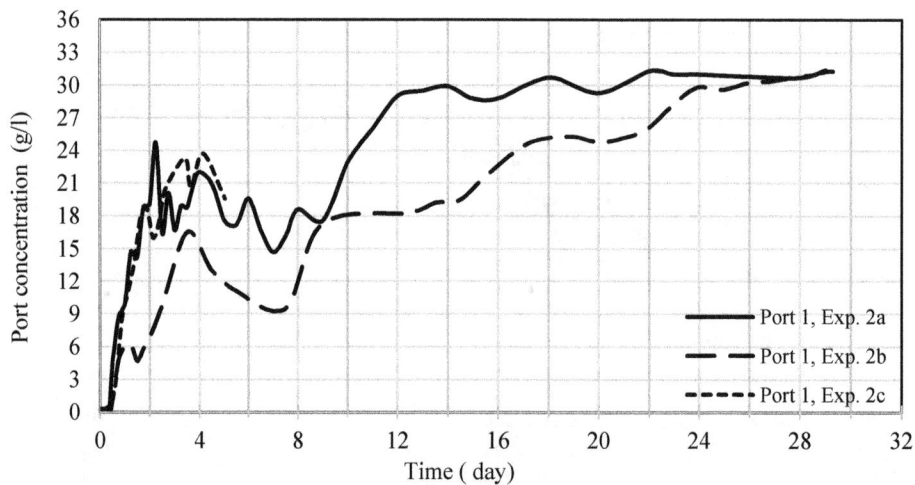

Figure 12: Comparisons of measured breakthrough curves at Port 1 for Experiments 2a, 2b, and 2c.

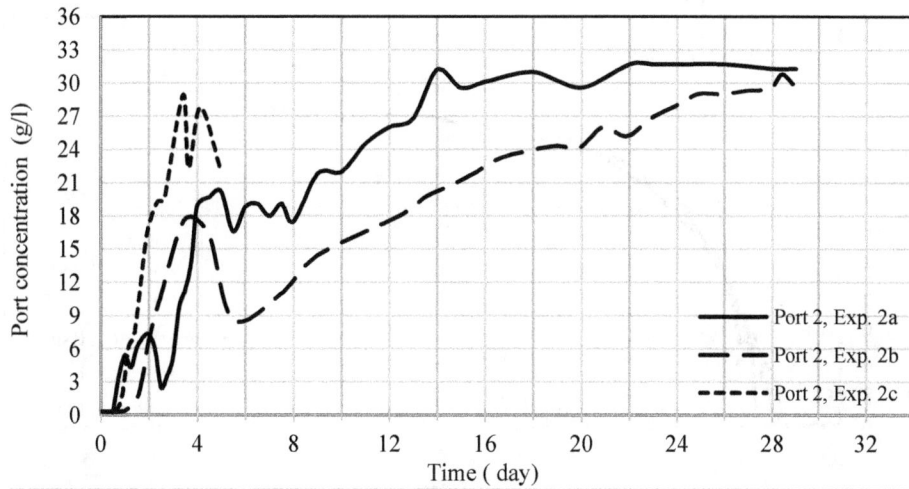

Figure 13: Comparisons of measured breakthrough curves at Port 2 for Experiments 2a, 2b, and 2c.

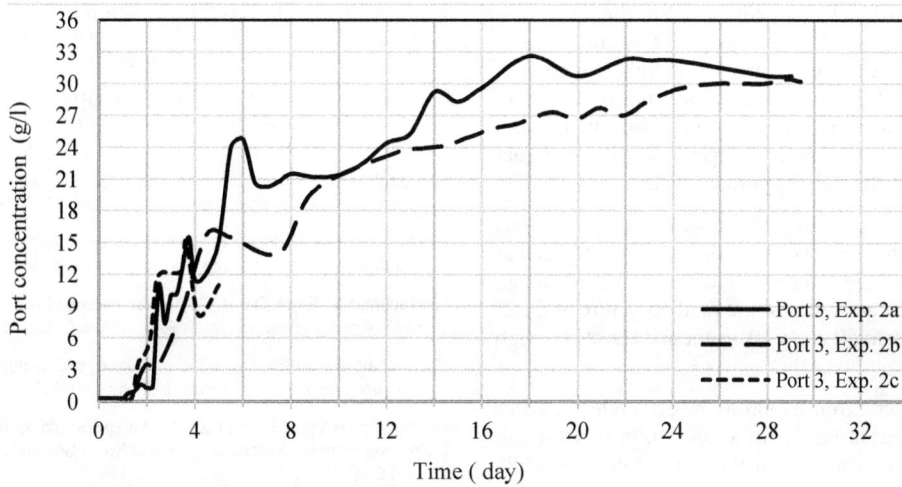

Figure 14: Comparisons of measured breakthrough curves at Port 3 for Experiments 2a, 2b, and 2c.

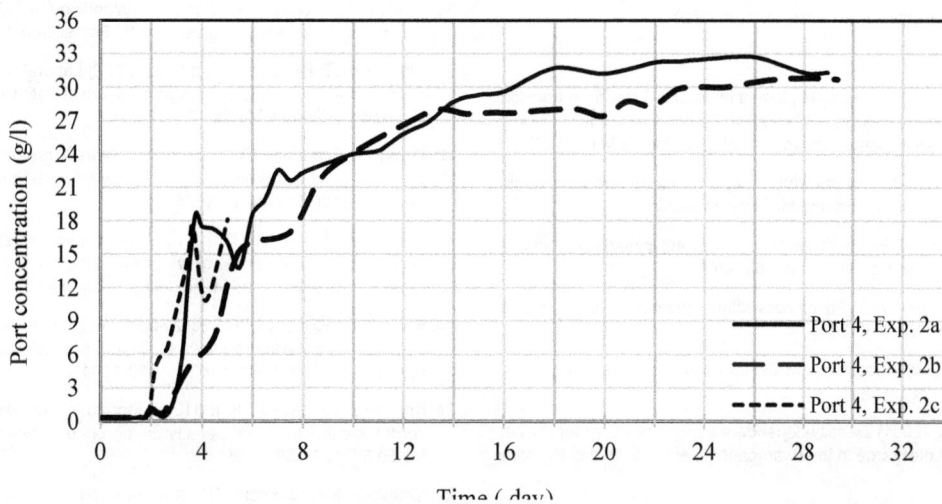

Figure 15: Comparisons of measured breakthrough curves at Port 4 for Experiments 2a, 2b, and 2c.

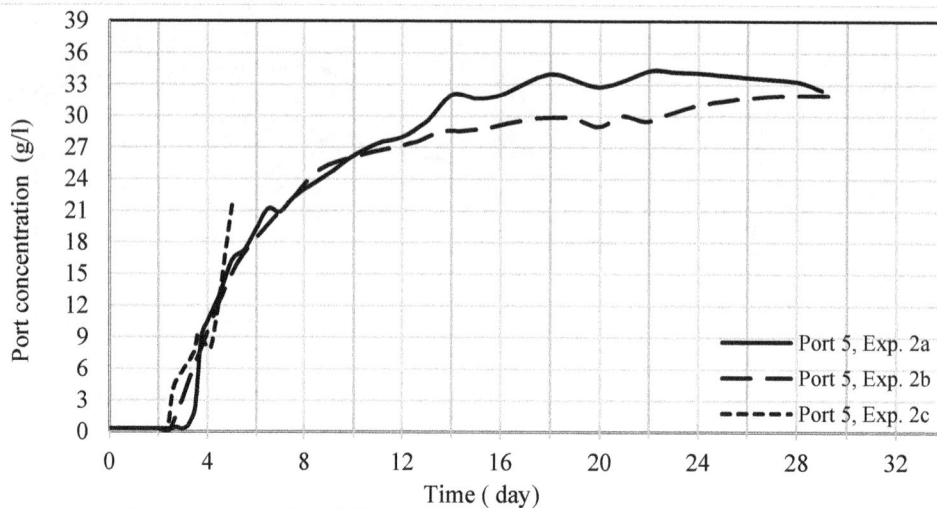

Figure 16: Comparisons of measured breakthrough curves at Port 5 for Experiments 2a, 2b, and 2c.

was evident from mass analyses conducted in the fixed mass experiments which showed that the total mass in the center of the column decreased by more than 100 g. This reduction of mass in the center of the column could only occur because of lateral transport. The lateral transport of the salt can again be attributed to the random nature of finger coalescing that goes on as the salt travels through the porous media.

c) The breakthrough curves in the continuous injection experiments were oscillating in the first five to ten days and smoothened as the experiment continued. The oscillation in the breakthrough curves could also be attributed to the random nature of finger forming and coalescing that occurs as the salt travels through the porous media.

d) The transport of salt through porous media under unstable conditions is somewhat similar to a DNAPL in that the salt takes random pathways in the form of fingers, while the DNAPL takes random pathways due to the presence of macro-pores. The random movement of the salt through the porous media as well as the fact that it sinks to the bottom, similar to a DNAPL, results in an uneven salt concentration within the porous media which cannot be predicted with any certainty.

References

1. Kimmel GE, Braids OC (1980) Leachate plumes in ground water from babylon and islip landfills, long island, new york. Geological Survey (US), Suffolk County (NY), Department of Environmental Control Washington: US Govt Print Off.

2. Kooper WF (1983) Density flow in the water supply package below the dump site in Noordwijk. [S.l.]: National Institute for Drinking Water.

3. Hubao Z, Scwartz FW (1995) Multispecies contaminant plumes in variable density flow systems. Water Resour Res 31: 837-847.

4. Simmons CT, Narayan KA (1997) Mixed convection processes below a saline disposal basin. J Hydrol 194: 263-285.

5. Kooi HJ, Leijnse GA (2000) Modes of seawater intrusion during transgressions. Water Resour Res 36: 3581-3589.

6. Anthony SJ, Turner JV (2001) Density-dependent surface water-groundwater interaction and nutrient discharge in the swan-canning estuary. Hydrol Process 15: 2595-2616.

7. Katsuyuki F, Iba T, Fujihara Y, Watanabe T (2009) Modeling interaction of fluid and salt in an aquifer/lagoon system. Ground Water 47: 35-48.

8. Viezzoli A, Tosi L, Teatini P, Silvestri S (2010) Surface water-groundwater exchange in transitional coastal environments by airborne electromagnetics: The venice lagoon example. Geophys Res Lett p: 37.

9. Casper K, Sonnenborg TO, Auken E, Jorgensen F (2011) Salinity distribution in heterogeneous coastal aquifers mapped by airborne electromagnetics. Vadose Zone J 10: 125-135.

10. Pandit A, Ali N, and Heck H (2011) Spatial calibration of vertical hydraulic conductivity below an estuary. J Hydrol Eng 16: 763-771.

11. Pandit A, Ali N, Heck H, and Mamoua K (2016) Estimation of submarine groundwater discharge into the Indian River lagoon. Austin J Irrigation 2: 1-8.

12. Bachmat Y, Elrick DE (1970) Hydrodynamic instability of miscible fluids in a vertical porous column. Water Resour Res 6: 156-171.

13. Gebhart B (1988) Transient response and disturbance growth in vertical buoyancy-driven flows. J Heat Transfer 110: 1166-1174.

14. Wooding RA (1959) The stability of a viscous liquid in a vertical tube containing porous material. Proceedings of the Royal Society. A, Mathematical, Physical, and Engineering Sciences 252: 120-134.

15. Schincariol RA, Schwartz FW (1990) An experimental investigation of variable density flow and mixing in homogeneous and heterogeneous media. Water Resour Res 26: 2317- 2329.

16. Webster IT, Norquay SJ, Ross FC, Wooding RA (1996) Solute exchange by convection within estuarine sediments. Estuar Coast Shelf Sci 42: 171-83.

17. Simmons CT, Pierini ML, Hutson JL (2002) Laboratory investigation of variable-density flow and solute transport in unsaturated-saturated porous media. Transport in Porous Media 47: 215-244.

18. Wood M, Simmons CT, Hutson JL (2004) A breakthrough curve analysis of unstable density-driven flow and transport in homogeneous porous media. Water Resour Res 40: W035051-9.

19. Post VEA, Simmons CT (2010) Free convective controls on sequestration of salts into low-permeability strata: Insights from sand tank laboratory experiments and numerical modeling. Hydrogeol J 18: 39-54.

20. Klaus J, Oswald S, Held R, Kinzelbach W (2006) Numerical simulation of three-dimensional saltwater-freshwater fingering instabilities observed in a porous medium. Adv Water Resour 29: 1690-1704.

21. Kneafsey TJ, Pruess K (2010) Laboratory flow experiments for visualizing carbon dioxide-induced, density-driven brine convection. Transport in Porous Media 82: 123-139.

22. Goswami RR, Clement TP, Hayworth JH (2012) Comparison of numerical techniques used for simulating variable-density flow and transport experiments. J Hydrol Eng 17: 272-282.

23. Kamal M, Pandit A, Howell H (2016) Physical and Numerical Modeling of Unstable Flow Due to Heavier Saltwater Overlying Freshwater. EWRI Congress West Palm Beach Florida.

24. Mulqueen J, Kirkham D (1972) Leaching of a surface layer of sodium chloride into tile drains in a sand-tank model. Soil Sci Soc Am 36: 3-9.

25. Simmons CT, Fenstemaker TR, Sharp JM (2001) Variable-density groundwater flow and solute transport in heterogeneous porous media: Approaches, resolutions and future challenges. J Contaminant Hydrol 52: 245-275.

26. Simmons CT, Narayan KA, Wooding RA (1999) On a test case for density-dependent groundwater flow and solute transport models: The Salt Lake problem. Water Resour Res 35: 3607-3620.

Permissions

List of Contributors

Mary Ann Pandan
Department of Chemical Engineering, University of St. La Salle, La Salle Avenue, Bacolod City 6100, Negros Occidental, Philippines

Florencio Ballesteros
Department of Chemical Engineering, University of the Philippines, Diliman, Quezon City, 1100, Metro Manila, Philippines

Amba Shetty and Lakshman Nandagiri
Department of Applied Mechanics and Hydraulics, National Institute of Technology Karnataka, Surathkal, Mangalore - 575 025, India

Gebremedhin Kiros
Department of Applied Mechanics and Hydraulics, National Institute of Technology Karnataka, Surathkal, Mangalore - 575 025, India
Department of Soil Resources and Watershed Management, Aksum University, Shire campus: Shire Endaselassie P. O. Box -314, Ethiopia

Mohamed Elhag
Department of Hydrology and Water Resources Management, Faculty of Meteorology, Environment and Arid Land Agriculture, King Abdulaziz University, Jeddah, 21589, Kingdom of Saudi Arabia

Mostafa Said Barseem, Talaat Ali Abd El Lateef, Hosny Mahomud Ezz El Deen and Abd Allah Al Abaseiry Abdel Rahman
Geophysical exploration department, Desert Research Center, 1 Matahaf El Matariya, Cairo, Egypt

Mahasa Pululu S
Department of Geography, Faculty of Natural and Agricultural Sciences, Qwaqwa Campus, University of the Free State, Phuthaditjhaba, South Africa

Palamuleni Lobina G and Ruhiiga Tabukeli M
Department of Geography & Environmental Sciences, School of Environmental and Health Sciences, Mafikeng Campus, North West University, Mmabatho, South Africa

Jian Zhao, Guo Fu and Kun Lei
Chinese Research Academy of Environmental Sciences, Beijing 100012, China

Abhishek Lodh
Indian Institute of Technology Delhi, Centre for Atmospheric Sciences, Hauz Khas, New Delhi 110016, India
National Centre for Medium Range Weather Forecasting (NCMRWF) Earth System Science Organisation, Ministry of Earth Sciences, A-50, Sector-62, NOIDA- 201 309, India

Yitea Seneshaw Getahun
Department of Natural Resources Management, Debre Berhan University, Ethiopia

Van Lanen HAJ
Hydrology and Quantitative Water Management Group, Centre for Water and Climate, Wageningen University, Wageningen, The Netherlands

Camille Risi and Sandrine Bony
LMD/IPSL, CNRS, UPMC, Paris, France

Jerome Ogée and Lisa Wingate
INRA Ephyse, Villenave d'Ornon, France

Thierry Bariac
UMR 7618 Bioemco, CNRS-UPMC-AgroParisTech-ENS Ulm-INRA-IRD-PXII Campus AgroParisTech, Bâtiment EGER, Thiverval-Grignon, 78850 France

Naama Raz-Yaseef
Earth Sciences Division, Lawrence Berkeley National Laboratory, Berkeley, USA Department of Environmental Sciences and Energy Research, Weizmann Institute of Science, PO Box 26, Rehovot 76100, Israel

Jeffrey Welker
Biology Department and Environment and Natural Resources Institute, University of Alaska, Anchorage, AK 99510, USA

Alexander Knohl
Bioclimatology, Faculty of Forest Sciences and Forest Ecology, Georg-August University of Göttingen, 37077 Göttingen, Germany

Cathy Kurz-Besson
Instituto Dom Luiz, Centro de Geofísica IDL-FCUL, Lisboa, Portugal

Monique Leclerc and Gengsheng Zhang
University of Georgia, Griffin, GA 30223, USA

Nina Buchmann
Institute of Agricultural Sciences, ETH Zurich, Zurich, Switzerland

Jiri Santrucek and Marie Hronkova
Biology Centre ASCR, Branisovska 31, Ceske Budejovice, Czech Republic
University of South Bohemia, Faculty of Science, Branisovska 31, Ceske Budejovice, Czech Republic

Teresa David
Instituto Nacional de Investigação Agrária e Veterinária, Quinta do Marquês, Portugal

Philippe Peylin and Francesca Guglielmo
LSCE/IPSL, CNRS, UVSQ, Orme des Merisiers, Gif-sur-Yvette, France

Konchok Dolma, Madhuri S Rishi and Herojeet R
Department of Environment Studies, Punjab University, Chandigarh-160014, India

Eyad Abushandi
Civil Engineering Department, Faculty of Engineering, University of Tabuk, Saudi Arabia

Saleh Alatawi
Vice Rector Office for Graduate Studies and Scientific Research, University of Tabuk, Kingdom of Saudi Arabia

Yasin Goa, Demelash Bassa and Genene Gezahagn
Areka Agricultural Research Center, SARI, Ethiopia

Mekasha Chichaybelew
Debrezeit Agricultural Research Center, EIAR, Ethiopia

Manickum T, John W, and Rachi Rajagopaul
Scientific Services Laboratories: Chemical Sciences, Engineering and Scientific Services Division, Umgeni Water, 310 Burger Street, Pietermaritzburg 3201, KwaZulu- Natal, South Africa

Toolsee N
Wiggins Process Evaluation Facility, Wiggins Water Works, 251 Wiggins Road, Mayville, Engineering & Scientific Services Division, Umgeni Water, KwaZulu-Natal, South Africa

Yeh TY
Department of Civil and Environmental Engineering, National University of Kaohsiung, Taiwan

Johanna Obreque-Contreras, Danilo Pérez-Flores, Pamela Gutiérrez and Pamela Chávez-Crooker
Aguamarina SA. Centro Tecnológico SR-97, Las Colonias 580, Antofagasta, Chile

El-Demerdash FM
University of Alexandria, Institute of Graduate Studies and Research, Department of Environmental Studies, Alexandria, Egypt

Abdullah AM
Holding Company for water and wastewater, Alexandria, Egypt

Ibrahim DA
Alexandria Water Company, Alexandria, Egypt

Gashaw G Chakilu
College of Agriculture, Wolaita Sodo University, Sodo 138, Ethiopia

Mamaru A Moges
Faculty of Civil and Water Resources Engineering, Bahir Dar Institute of Technology, Bahir Dar University, Bahir Dar, Ethiopia

Ibekwe AM
USDA-ARS, U.S. Salinity Laboratory, 450 W Big Springs Rd, Riverside, CA 92507, USA

Ma J
USDA-ARS, U.S. Salinity Laboratory, 450 W Big Springs Rd, Riverside, CA 92507, USA
College of Environment and Resources, Jilin University, Changchun, Jilin Province, 130021, P. R. China

Murinda S
Department of Animal and Veterinary Sciences, California State Polytechnic University, Pomona, CA 91768, USA

Reddy GB
Department of Natural Resources and Environmental Design, North Carolina Agricultural and Technical State University, Greensboro, NC 27411, USA

Ghulam Nabi
The Key Laboratory of Aquatic Biodiversity and Conservation of Chinese Academy of Sciences, Institute of Hydrobiology, Chinese Academy of Sciences, Wuhan, Hubei,430072, PR China

Khan Suliman, Pathiranage Prajani Mahesha Heenatigala, Yang Jingjing, Qingman Li and Hongwei Hou
The Key Laboratory of Aquatic Biodiversity and Conservation of Chinese Academy of Sciences, Institute of Hydrobiology, Chinese Academy of Sciences, Wuhan, Hubei,430072, PR China
State Key Laboratory of Freshwater Ecology and Biotechnology, Institute of Hydrobiology, Chinese Academy of Sciences, Wuhan, 430072, PR China

Rabeea Siddique
Department of Biomedical Engineering, Huazhong University of Science and Technology, Wuhan 430074, PR China

Wasim Sajjad
Key laboratory of petroleum resources, Gansu Province/Key laboratory of petroleum resources research, Institute of Geology and Geophysics, Chinese Academy of sciences, Lanzhou 730000, PR China

Ijaz Ali
Department of Biosciences, COMSATS, Islamabad, 44000, Pakistan

Foroozan Arkian
Meteorology Department, Marine Science and Technology Faculty, North Tehran Branch, Islamic Azad University, Tehran, Iran

Eba Muluneh Sorecha
School of Natural Resources Management and Environmental Sciences, Haramaya University, Dire Dawa, Ethiopia

Paul A LaViolette
The Starburst Foundation, 1176 Hedgewood Lane, Niskayuna, New York 12309, United States

Kamal Mamoua, Ashok Pandit and Howell Heck
Department of Civil Engineering, Florida Institute of Technology, Melbourne, FL 32901, USA

Index